Modern Problems of Mathematical Physics and Their Applications

Modern Problems of Mathematical Physics and Their Applications

Editors

Davron Aslonqulovich Juraev
Samad Noeiaghdam

MDPI • Basel • Beijing • Wuhan • Barcelona • Belgrade • Manchester • Tokyo • Cluj • Tianjin

Editors
Davron Aslonqulovich Juraev
Department of Natural Science Disciplines
Higher Military Aviation School of the Republic of Uzbekistan
Karshi City
Uzbekistan

Samad Noeiaghdam
Industrial Mathematics Laboratory, Baikal School of BRICS
Irkutsk National Research Technical University
Irkutsk
Russia

Editorial Office
MDPI
St. Alban-Anlage 66
4052 Basel, Switzerland

This is a reprint of articles from the Special Issue published online in the open access journal *Axioms* (ISSN 2075-1680) (available at: www.mdpi.com/journal/axioms/special_issues/mathematical_physics_application).

For citation purposes, cite each article independently as indicated on the article page online and as indicated below:

LastName, A.A.; LastName, B.B.; LastName, C.C. Article Title. *Journal Name* **Year**, *Volume Number*, Page Range.

ISBN 978-3-0365-3496-1 (Hbk)
ISBN 978-3-0365-3495-4 (PDF)

© 2022 by the authors. Articles in this book are Open Access and distributed under the Creative Commons Attribution (CC BY) license, which allows users to download, copy and build upon published articles, as long as the author and publisher are properly credited, which ensures maximum dissemination and a wider impact of our publications.

The book as a whole is distributed by MDPI under the terms and conditions of the Creative Commons license CC BY-NC-ND.

Contents

About the Editors vii

Preface to "Modern Problems of Mathematical Physics and Their Applications" ix

Davron Aslonqulovich Juraev and Samad Noeiaghdam
Modern Problems of Mathematical Physics and Their Applications
Reprinted from: *Axioms* 2022, 11, 45, doi:10.3390/axioms11020045 1

Belgees Qaraad, Osama Moaaz, Shyam Sundar Santra, Samad Noeiaghdam, Denis Sidorov and Elmetwally M. Elabbasy
Oscillatory Behavior of Third-Order Quasi-Linear Neutral Differential Equations
Reprinted from: *Axioms* 2021, 10, 346, doi:10.3390/axioms10040346 7

Alessandra Aimi and Chiara Guardasoni
Multi-Asset Barrier Options Pricing by Collocation BEM (with Matlab® Code)
Reprinted from: *Axioms* 2021, 10, 301, doi:10.3390/axioms10040301 25

Yuli D. Chashechkin
Foundations of Engineering Mathematics Applied for Fluid Flows
Reprinted from: *Axioms* 2021, 10, 286, doi:10.3390/axioms10040286 45

Tinggui Chen, Xiaohua Yin, Jianjun Yang, Guodong Cong and Guoping Li
Modeling Multi-Dimensional Public Opinion Process Based on Complex Network Dynamics Model in the Context of Derived Topics
Reprinted from: *Axioms* 2021, 10, 270, doi:10.3390/axioms10040270 79

Amadou Diop and Wei-Shih Du
Existence of Mild Solutions for Multi-Term Time-Fractional Random Integro-Differential Equations with Random Carathéodory Conditions
Reprinted from: *Axioms* 2021, 10, 252, doi:10.3390/axioms10040252 111

Vladimir Vasilyev and Nikolai Eberlein
On Solvability Conditions for a Certain Conjugation Problem
Reprinted from: *Axioms* 2021, 10, 234, doi:10.3390/axioms10030234 129

Efthimios Providas, Stefanos Zaoutsos and Ioannis Faraslis
Closed-Form Solutions of Linear Ordinary Differential Equations with General Boundary Conditions
Reprinted from: *Axioms* 2021, 10, 226, doi:10.3390/axioms10030226 141

Hoang Pham
A Dynamic Model of Multiple Time-Delay Interactions between the Virus-Infected Cells and Body's Immune System with Autoimmune Diseases
Reprinted from: *Axioms* 2021, 10, 216, doi:10.3390/axioms10030216 153

Sittisak Injan, Angkool Wangwongchai, Usa Humphries, Amir Khan and Abdullahi Yusuf
Reinitializing Sea Surface Temperature in the Ensemble Intermediate Coupled Model for Improved Forecasts
Reprinted from: *Axioms* 2021, 10, 189, doi:10.3390/axioms10030189 177

Avram Sidi
Application of a Generalized Secant Method to Nonlinear Equations with Complex Roots
Reprinted from: *Axioms* **2021**, *10*, 169, doi:10.3390/axioms10030169 . . . . **195**

Simon Gluzman
Critical Indices and Self-Similar Power Transform
Reprinted from: *Axioms* **2021**, *10*, 162, doi:10.3390/axioms10030162 . . . . **207**

Mir Hameeda, Angelo Plastino, Mario Carlos Rocca and Javier Zamora
Classical Partition Function for Non-Relativistic Gravity
Reprinted from: *Axioms* **2021**, *10*, 121, doi:10.3390/axioms10020121 . . . . **227**

Pundikala Veeresha, Haci Mehmet Baskonus and Wei Gao
Strong Interacting Internal Waves in Rotating Ocean: Novel Fractional Approach
Reprinted from: *Axioms* **2021**, *10*, 123, doi:10.3390/axioms10020123 . . . . **237**

Savin Treanţă
On a Class of Isoperimetric Constrained Controlled Optimization Problems
Reprinted from: *Axioms* **2021**, *10*, 112, doi:10.3390/axioms10020112 . . . . **253**

Tinggui Chen, Xiaohua Yin, Lijuan Peng, Jingtao Rong, Jianjun Yang and Guodong Cong
Monitoring and Recognizing Enterprise Public Opinion from High-Risk Users Based on User Portrait and Random Forest Algorithm
Reprinted from: *Axioms* **2021**, *10*, 106, doi:10.3390/axioms10020106 . . . . **263**

Alexander G. Ramm
Comments on the Navier–Stokes Problem
Reprinted from: *Axioms* **2021**, *10*, 95, doi:10.3390/axioms10020095 . . . . **279**

Dimitris M. Christodoulou, Eric Kehoe and Qutaibeh D. Katatbeh
Degenerate Canonical Forms of Ordinary Second-Order Linear Homogeneous Differential Equations
Reprinted from: *Axioms* **2021**, *10*, 94, doi:10.3390/axioms10020094 . . . . **289**

Sabahudin Vrtagić, Edis Softić, Marko Subotić, Željko Stević, Milan Dordevic and Mirza Ponjavic
Ranking Road Sections Based on MCDM Model: New Improved Fuzzy SWARA (IMF SWARA)
Reprinted from: *Axioms* **2021**, *10*, 92, doi:10.3390/axioms10020092 . . . . **303**

Davron Aslonqulovich Juraev and Samad Noeiaghdam
Regularization of the Ill-Posed Cauchy Problem for Matrix Factorizations of the Helmholtz Equation on the Plane
Reprinted from: *Axioms* **2021**, *10*, 82, doi:10.3390/axioms10020082 . . . . **327**

About the Editors

Davron Aslonqulovich Juraev

Davron Aslonqulovich Juraev acquired his PhD degree in specialty Differential Equations and Mathematical Physics at the National University of the Republic of Uzbekistan in Tashkent (2018). In 2021, he received the scientific title of Associate Professor in specialty Differential Equations and Mathematical Physics.

He is Associate Professor and Head of the Chair of Natural Science Disciplines Higher Military Aviation School of the Republic of Uzbekistan, Karshi city, Uzbekistan. Through the years, he has published many papers in the field of solving ill-posed problems, Cauchy problem for matrix factorization of the Helmholtz equation, and boundary value problems for elliptic systems.

He is a member of the editorial and reviewer boards of several Scopus-WOS journals. He has also presented numerous communications in international conferences. He is currently an online Postdoctoral Fellow at Anand International College of Engineering under the supervision of Prof. Praveen Agarwal, India.

Samad Noeiaghdam

Samad Noeiaghdam acquired his Ph.D. degree in Applied Mathematics at Central Tehran Branch of Islamic Azad University in Iran. He is Associate Professor of Irkutsk National Research Technical University and senior researcher of South Ural State University. Through the years, he has published many papers in the field of numerical analysis, solving integral equations, ordinary differential equations, partial differential equations, solving ill-posed problems, solving fuzzy problems and bio-mathematical models.

Additionally, he has published some books and chapters in the field of applied mathematics and mathematical modelling. He is a member of editorial and reviewer boards of several Scopus-WOS journals, and he was selected as a winner of outstanding reviewer 2020 by Mathematics (MDPI). He has also presented numerous communications in international conferences.

Preface to "Modern Problems of Mathematical Physics and Their Applications"

There are many applications of mathematical physics in several fields of basic science and engineering. Thus, we have tried to provide the Special Issue "Modern Problems of Mathematical Physics and Their Applications" to cover the new advances of mathematical physics and its applications. In this Special Issue, we have focused on some important and challenging topics, such as integral equations, ill-posed problems, ordinary differential equations, partial differential equations, system of equations, fractional problems, linear and nonlinear problems, fuzzy problems, numerical methods, analytical methods, semi-analytical methods, convergence analysis, error analysis and mathematical models. In response to our invitation, we received 31 papers from more than 17 countries (Russia, Uzbekistan, China, USA, Kuwait, Bosnia and Herzegovina, Thailand, Pakistan, Turkey, Nigeria, Jordan, Romania, India, Iran, Argentina, Israel, Canada, etc.), of which 19 were published and 12 rejected.

Davron Aslonqulovich Juraev, Samad Noeiaghdam
Editors

Editorial

Modern Problems of Mathematical Physics and Their Applications

Davron Aslonqulovich Juraev [1,*,†] **and Samad Noeiaghdam** [2,3,*,†]

1. Higher Military Aviation School of the Republic of Uzbekistan, Karshi City 180100, Uzbekistan
2. Industrial Mathematics Laboratory, Baikal School of BRICS, Irkutsk National Research Technical University, 664074 Irkutsk, Russia
3. Department of Applied Mathematics and Programming, South Ural State University, Lenin Prospect 76, 454080 Chelyabinsk, Russia
* Correspondence: juraev_davron@list.ru (D.A.J.); snoei@istu.edu (S.N.)
† These authors contributed equally to this work.

1. Introduction

There are many applications of mathematical physics in several fields of basic science and engineering. Thus, we have tried to provide the Special Issue "Modern Problems of Mathematical Physics and Their Applications" to cover the new advances of mathematical physics and its applications. In this Special Issue, we have focused on some important and challenging topics, such as integral equations, ill-posed problems, ordinary differential equations, partial differential equations, system of equations, fractional problems, linear and nonlinear problems, fuzzy problems, numerical methods, analytical methods, semi-analytical methods, convergence analysis, error analysis and mathematical models. In response to our invitation, we received 31 papers from more than 17 countries (Russia, Uzbekistan, China, USA, Kuwait, Bosnia and Herzegovina, Thailand, Pakistan, Turkey, Nigeria, Jordan, Romania, India, Iran, Argentina, Israel, Canada, etc.), of which 19 were published and 12 rejected.

2. Brief Overview of the Contributions

Qaraad et al., in [1], have considered a class of quasilinear third-order differential equations with a delay argument. They have established some conditions of a certain third-order quasi-linear neutral differential equation as oscillatory or almost oscillatory. They have solved some examples to demonstrate the importance of the results.

Extending the SABO technique (Semi-Analytical method for Barrier Options), based on the collocation Boundary Element Method (BEM), to the pricing of Barrier Options with a payoff dependent on more than one asset has been discussed by Aimi and Guardasoni in [2]. The numerical results have been presented to show the efficiency and accuracy of the method in the case of a single asset.

With the rapid development of the Internet, the speed with which information can be updated and propagated has accelerated, resulting in wide variations in public opinion. Usually, after the occurrence of some newsworthy event, discussion topics are generated in networks that influence the formation of initial public opinion. After a period of propagation, some of these topics are further derived into new subtopics, which intertwine with the initial public opinion to form a multidimensional public opinion. In [3], Chen et al. were concerned with the formation process of multi-dimensional public opinion in the context of derived topics. Firstly, the initial public opinion variation mechanism was introduced to reveal the formation process of derived subtopics, then Brownian motion was used to determine the subtopic propagation parameters, and their propagation was studied based on complex network dynamics according to the principle of evolution. The formula of the basic reproductive number has been introduced to determine whether derived subtopics

can form derived public opinion, thereby revealing the whole process of multi-dimensional public opinion formation. Secondly, through simulation experiments, the influences of various factors, such as the degree of information alienation, environmental forces, topic correlation coefficients, the amount of information contained in subtopics, and network topology on the formation of multi-dimensional public opinion have been studied by the authors. The simulation results showed that: (1) Environmental forces and the amount of information contained in subtopics are key factors affecting the formation of multidimensional public opinion. Among them, environmental forces have a greater impact on the number of subtopics, and the amount of information contained in subtopics determines whether the subtopic can be the key factor that forms the derived public opinion. (2) Only when the degree of information alienation reaches a certain level will derived subtopics emerge. At the same time, the degree of information alienation has a greater impact on the number of derived subtopics, but it has a small impact on the dimensions of the final public opinion. (3) The network topology does not have much impact on the number of derived subtopics but has a greater impact on the number of individuals participating in the discussion of subtopics. The multidimensional public opinion dimension formed by the network topology with a high aggregation coefficient and small average path length is higher.

In [4], the existence of mild solutions to a multi-term fractional integro-differential equation with random effects has been investigated by Diop and Wu. The results relied on stochastic analysis, Mönch's fixed point theorem combined with a random fixed point theorem with stochastic domain, and the measure of noncompactness and resolvent family theory. The authors have established the existence of random mild solutions under the condition that the nonlinear term is of Carathéodory type and satisfies some weakly compactness condition. By presenting a nontrivial example, they showed the obtained results.

Vasilyev and Eberlein [5] have studied a certain conjugation problem for a pair of elliptic pseudo-differential equations with homogeneous symbols inside and outside of a plane sector. They found that the solution is sought in corresponding Sobolev–Slobodetskii spaces. Using the wave factorization concept for elliptic symbols, they derived a general solution of the conjugation problem. Adding some complementary conditions, a system of linear integral equations has been obtained. For homogeneous symbols, they applied the Mellin transform to such a system to reduce it to a system of linear algebraic equations with respect to unknown functions.

In [6], Providas et al. attempted to find the solution of boundary value problems for ordinary differential equations with general boundary conditions. They have obtained the closed-form solutions in a symbolic form with the general n-th order differential operator, as well as the composition of linear operators. Furthermore, their method is based on the theory of the extensions of linear operators in Banach spaces.

Pham in [7] has presented a mathematical modeling of the virus-infected development in the body's immune system considering the multiple time-delay interactions between the immune cells and virus-infected cells with autoimmune disease. In the proposed model, he tried to determine the dynamic progression of virus-infected cell growth in the immune system. The patterns of how the virus-infected cells spread and the development of the body's immune cells with respect to time delays have been derived in the form of a system of delay partial differential equations. The model can be used to determine whether the virus-infected free state can be reached or not as time progresses. It has been used to predict the number of the body's immune cells at any given time. Several numerical examples have been discussed to illustrate the proposed model. The model provided a real understanding of the transmission dynamics and other significant factors of the virus-infected disease and the body's immune system subject to the time delay, including approaches to reduce the growth rate of virus-infected cells and the autoimmune disease and also enhance the immune effector cells.

The Ensemble Intermediate Coupled Model (EICM) is a model used for studying the El Nino-Southern Oscillation (ENSO) phenomenon in the Pacific Ocean, where anomalies in the Sea Surface Temperature (SST) are observed. In [8], Injan et al. aimed to implement Cressman to improve SST forecasts. The simulation considers two cases in this work: the control case and the Cressman initialized case. These cases are simulations using different inputs where the two inputs differ in terms of their resolution and data source. The Cressman method has been used to initialize the model with an analysis product based on satellite data and in situ data, such as ships, buoys, and Argo floats, with a resolution of 0.25×0.25 degrees. The results of this inclusion are the Cressman Initialized Ensemble Intermediate Coupled Model (CIEICM). Forecasting of the sea surface temperature anomalies was conducted using both the EICM and the CIEICM. The results showed that the calculation of the SST field from the CIEICM was more accurate than that of the EICM. The forecast using the CIEICM initialization with the higher-resolution satellite-based analysis at a 6-month lead time improved the root mean square deviation to 0.794 from 0.808 and the correlation coefficient to 0.630 from 0.611 compared the control model that was directly initialized with the low-resolution in situ analysis.

In [9], Sidi has discussed the secant method, which is a very effective numerical procedure for solving nonlinear equations $f(x) = 0$. In their recent work (A. Sidi, Generalization of the secant method for nonlinear equations. Appl. Math. E-Notes, 8:115–123, 2008), he presented a generalization of the secant method that used only one evaluation of $f(x)$ per iteration, and he provided a local convergence theory for it that concerned real roots. For each integer k, this method has generated a sequence $\{x_n\}$ of approximations to a real root of $f(x)$, where, for $n \geq k$, $x_{n+1} = x_n - f(x_n)/p'_{n,k}(x_n)$, $p_{n,k}(x)$ being the polynomial of degree k that interpolates $f(x)$ at $x_n, x_{n-1}, \ldots, x_{n-k}$, the order s_k of this method satisfies $1 < s_k < 2$. Clearly, when $k = 1$, this method reduces to the secant method with $s_1 = (1 + \sqrt{5})/2$. In addition, $s_1 < s_2 < s_3 < \ldots$, such that $\lim_{k \to \infty} s_k = 2$. The author has studied the application of this method to simple complex roots of a function $f(z)$. He showed that the local convergence theory developed for real roots can be extended almost as is to complex roots, provided suitable assumptions and justifications are made. He has illustrated the theory with two numerical examples.

"Odd" factor approximants of the special form are based on the idea that the critical index by itself should be optimized through the parameters of the power transform to be calculated from the minimal sensitivity (derivative) optimization condition. The critical index is a product of the algebraic self-similar renormalization that contributes to the expressions of control parameters typical to the algebraic self-similar renormalization and of the power transform that corrects them even further. The parameter of power transformation is, in a nutshell, the multiplier connecting the critical exponent and the correction-to-scaling exponent. In [10], Gluzman has studied the minimal model of critical phenomena based on expansions with only two coefficients and critical points. The optimization appears to bring quite accurate, uniquely defined results given by simple formulas. Many important cases of critical phenomena have been covered by the simple formula. For the longer series, the optimization condition possesses multiple solutions, and additional constraints have been applied. In particular, the author constrained the sought solution by requiring it to be the best in the prediction of the coefficients not employed in its construction. In principle, the error and measure of such a prediction have been optimized by itself, with respect to the parameter of the power transform. Methods of calculation based on optimized power-transformed factors have been applied, and results are presented for critical indices of several key models of conductivity and viscosity of random media, swelling of polymers, permeability in two-dimensional channels. The author has discussed several quantum mechanical problems as well.

In [11], Veeresha et al. have tried to analyze the nature and capture the corresponding consequences of the solution obtained for the Gardner–Ostrovsky equation with the help of the q-homotopy analysis transform technique. The fractional operator was used to illustrate its importance in generalizing the models associated with kernel singular. The authors have considered the fixed-point theorem and the Banach space to present the existence and uniqueness within the frame of the Caputo–Fabrizio fractional operator. Furthermore, for different fractional orders, the nature has been captured in plots. The realized consequences confirmed that the considered procedure is reliable and highly methodical for investigating the consequences related to the nonlinear models of both integer and fractional order.

The canonical gravitational partition function Z associated to the classical Boltzmann–Gibbs (BG) distribution $\frac{e^{-\beta H}}{Z}$ has been considered in [12] by Hameeda et al. It is popularly thought that it cannot be built up because the integral involved in constructing Z diverges at the origin. On the contrary, it was shown in (Physica A 497 (2018) 310), by appeal to sophisticated mathematics developed in the second half of the last century, that this is not so. Z has been computed by recourse to (A) the analytical extension treatments of Gradshteyn and Rizhik and Guelfand and Shilov, which permit tackling some divergent integrals, and (B) the dimensional regularization approach. Only one special instance was discussed in the above reference. In [12], the authors obtained the classical partition function for Newton's gravity in the four cases that immediately come to mind.

The Lagrange dynamics generated by a class of isoperimetric constrained controlled optimization problems involving second-order partial derivatives and boundary conditions have been investigated by Treanţă in [13]. The author has derived necessary optimality conditions for the considered class of variational control problems governed by path-independent curvilinear integral functionals. Moreover, the theoretical results are accompanied by an illustrative example. Furthermore, an algorithm has been proposed to emphasize the steps to be followed to solve a control problem.

With the rapid development of "We media" technology, netizens can freely express their opinions regarding enterprise products on a network platform. Consequently, online public opinion about enterprises has become a prominent issue. Negative comments posted by some netizens may trigger negative public opinion, which can have a significant impact on an enterprise's image. In [14], Chen et al. applied the perspective of helping enterprises deal with negative public opinion by combining the user portrait technology and a random forest algorithm to help enterprises identify high-risk users who have posted negative comments and thus may trigger negative public opinion. In this way, enterprises can monitor the public opinion of high-risk users to prevent negative public opinion events. Firstly, they crawled the information of users participating in discussions of product experience, and they constructed a portrait of enterprise public opinion users. Then, the characteristics of the portraits were quantified into indicators, such as the user's activity, the user's influence, and the user's emotional tendency, and then the indicators were sorted. According to the order of the indicators, the users were divided into high-risk, moderate-risk, and low-risk categories. Next, a supervised high-risk user identification model for this classification was established based on a random forest algorithm. In turn, the trained random forest identifier has been used to predict whether the authors of newly published public opinion information are high-risk users. Finally, a back propagation neural network algorithm was used to identify users and compared with the results of model recognition in this paper. The results showed that the average recognition accuracy of the back propagation neural network is only 72.33%, while the average recognition accuracy of the model constructed in this paper is as high as 98.49%, which verifies the feasibility and accuracy of the proposed random forest recognition method.

The aim of [15] was to explain the result concerning the Navier–Stokes problem (NSP) in $\mathbb{R}^3$ without boundaries to a broad audience. Ramm proved that the NSP is contradictory in the following sense: if one assumes that the initial data $v(x,0) \not\equiv 0$, $\nabla \cdot v(x,0) = 0$ and the solution to the NSP exists for all $t \geq 0$, then the author proved that the solution $v(x,t)$ to the NSP has the property $v(x,0) = 0$. This study showed that the NSP is not a correct description of the fluid mechanics problem and the NSP does not have a solution. In the exceptional case, when the data are equal to zero, the solution $v(x,t)$ to the NSP exists for all $t \geq 0$ and is equal to zero, $v(x,t) \equiv 0$. Thus, one of the millennium problems has been solved by the author.

Christodoulou et al. in [16] have derived a family of associated differential equations that share the same "degenerate" canonical form. These equations can be solved easily if the original equation is known to possess analytic solutions; otherwise, their properties and the properties of their solutions are de facto known as they are comparable to those already deduced for the fundamental equation. The authors analyzed several particular cases of new families related to some of the famous differential equations applied to physical problems, and the degenerate eigenstates of the radial Schrödinger equation for the hydrogen atom in N dimensions.

Traffic management is a significantly difficult and demanding task. It is necessary to know the main parameters of road networks in order to adequately meet traffic management requirements. In [17], Vrtagić et al. focused on an integrated fuzzy model for ranking road sections based on four inputs and four outputs. The goal was to determine the safety degree of the observed road sections by the methodology developed. The greatest contribution of the paper was reflected in the development of the improved fuzzy step-wise weight assessment ratio analysis (IMF SWARA) method and integration with the fuzzy measurement alternatives and ranking according to the compromise solution (fuzzy MARCOS) method. First, the data envelopment analysis (DEA) model was applied, showing that three road sections have a high traffic risk. After that, IMF SWARA was used to determine the values of the weight coefficients of the criteria, and the fuzzy MARCOS method was used for the final ranking of the sections. The obtained results were verified through a three-phase sensitivity analysis with an emphasis on forming 40 new scenarios in which input values were simulated. The stability of the model was proven in all phases of sensitivity analysis.

An explicit formula for the approximate solution of the Cauchy problem for the matrix factorizations of the Helmholtz equation in a bounded domain on the plane has been presented by Juraev and Noeiaghdam in [18]. The formula for an approximate solution has included the construction of a family of fundamental solutions for the Helmholtz operator on the plane. This family was parameterized by function $K(w)$, which depends on the space dimension. The authors have improved the results using the function $K(w)$.

In [19], Chashechkin has tried to formulate a list of principles that substantiates the choice of axioms and methods for studying nature. He showed that the axiomatics of fluid flows are based on conservation laws in the frames of engineering mathematics and technical physics. In the theory of fluid flows within the continuous medium model, a key role for the total energy has been distinguished. To describe a fluid flow, a system of fundamental equations has been selected and supplemented by the equations of the state for the Gibbs potential and the medium density. The system has been supplemented by the physically based initial and boundary conditions and analyzed, taking into account the compatibility condition. The complete solutions showed both the structure and dynamics of non-stationary flows. The results of compatible theoretical and experimental studies have been compared for the cases of potential and actual homogeneous and stratified fluid flow past an arbitrarily oriented plate.

Funding: This research received no external funding.

Institutional Review Board Statement: Not applicable.

Informed Consent Statement: Not applicable.

Data Availability Statement: Not applicable.

Acknowledgments: We are thankful the editors and reviewers of the journal *Axioms* for their help and support from our issue.

Conflicts of Interest: The authors declare no conflict of interest.

References

1. Qaraad, B.; Moaaz, O.; Santra, S.S.; Noeiaghdam, S.; Sidorov, D.; Elabbasy, E.M. Oscillatory Behavior of Third-Order Quasi-Linear Neutral Differential Equations. *Axioms* **2021**, *10*, 346. [CrossRef]
2. Aimi, A.; Guardasoni, C. Multi-Asset Barrier Options Pricing by Collocation BEM (with Matlab® Code). *Axioms* **2021**, *10*, 301. [CrossRef]

3. Chen, T.; Yin, X.; Yang, J.; Cong, G.; Li, G. Modeling Multi-Dimensional Public Opinion Process Based on Complex Network Dynamics Model in the Context of Derived Topics. *Axioms* **2021**, *10*, 270. [CrossRef]
4. Diop, A.; Du, W.-S. Existence of Mild Solutions for Multi-Term Time-Fractional Random Integro-Differential Equations with Random Carathéodory Conditions. *Axioms* **2021**, *10*, 252. [CrossRef]
5. Vasilyev, V.; Eberlein, N. On Solvability Conditions for a Certain Conjugation Problem. *Axioms* **2021**, *10*, 234. [CrossRef]
6. Providas, E.; Zaoutsos, S.; Faraslis, I. Closed-Form Solutions of Linear Ordinary Differential Equations with General Boundary Conditions. *Axioms* **2021**, *10*, 226. [CrossRef]
7. Pham, H. A Dynamic Model of Multiple Time-Delay Interactions between the Virus-Infected Cells and Body's Immune System with Autoimmune Diseases. *Axioms* **2021**, *10*, 216. [CrossRef]
8. Injan, S.; Wangwongchai, A.; Humphries, U.; Khan, A.; Yusuf, A. Reinitializing Sea Surface Temperature in the Ensemble Intermediate Coupled Model for Improved Forecasts. *Axioms* **2021**, *10*, 189. [CrossRef]
9. Sidi, A. Application of a Generalized Secant Method to Nonlinear Equations with Complex Roots. *Axioms* **2021**, *10*, 169. [CrossRef]
10. Gluzman, S. Critical Indices and Self-Similar Power Transform. *Axioms* **2021**, *10*, 162. [CrossRef]
11. Veeresha, P.; Baskonus, H.M.; Gao, W. Strong Interacting Internal Waves in Rotating Ocean: Novel Fractional Approach. *Axioms* **2021**, *10*, 123. [CrossRef]
12. Hameeda, M.; Plastino, A.; Rocca, M.C.; Zamora, J. Classical Partition Function for Non-Relativistic Gravity. *Axioms* **2021**, *10*, 121. [CrossRef]
13. Treanţă, S. On a Class of Isoperimetric Constrained Controlled Optimization Problems. *Axioms* **2021**, *10*, 112. [CrossRef]
14. Chen, T.; Yin, X.; Peng, L.; Rong, J.; Yang, J.; Cong, G. Monitoring and Recognizing Enterprise Public Opinion from High-Risk Users Based on User Portrait and Random Forest Algorithm. *Axioms* **2021**, *10*, 106. [CrossRef]
15. Ramm, A.G. Comments on the Navier–Stokes Problem. *Axioms* **2021**, *10*, 95. [CrossRef]
16. Christodoulou, D.M.; Kehoe, E.; Katatbeh, Q.D. Degenerate Canonical Forms of Ordinary Second-Order Linear Homogeneous Differential Equations. *Axioms* **2021**, *10*, 94. [CrossRef]
17. Vrtagić, S.; Softić, E.; Subotić, M.; Stević, Ž.; Dordevic, M.; Ponjavic, M. Ranking Road Sections Based on MCDM Model: New Improved Fuzzy SWARA (IMF SWARA). *Axioms* **2021**, *10*, 92. [CrossRef]
18. Juraev, D.A.; Noeiaghdam, S. Regularization of the Ill-Posed Cauchy Problem for Matrix Factorizations of the Helmholtz Equation on the Plane. *Axioms* **2021**, *10*, 82. [CrossRef]
19. Chashechkin, Y.D. Foundations of Engineering Mathematics Applied for Fluid Flows. *Axioms* **2021**, *10*, 286. [CrossRef]

Article

Oscillatory Behavior of Third-Order Quasi-Linear Neutral Differential Equations

Belgees Qaraad [1,2], Osama Moaaz [1,3,*], Shyam Sundar Santra [4,*], Samad Noeiaghdam [5,6], Denis Sidorov [7,*] and Elmetwally M. Elabbasy [1]

[1] Department of Mathematics, Faculty of Science, Mansoura University, Mansoura 35516, Egypt; belgeesmath2016@students.mans.edu.eg (B.Q.); emelabbasy@mans.edu.eg (E.M.E.)
[2] Department of Mathematics, Faculty of Science, Amran University, Amran 999101, Yemen
[3] Section of Mathematics, International Telematic University Uninettuno, CorsoVittorio Emanuele II, 39, 00186 Roma, Italy
[4] Department of Mathematics, JIS College of Engineering, Kalyani 741235, India
[5] Industrial Mathematics Laboratory, Baikal School of BRICS, Irkutsk National Research Technical University, Irkutsk, 664074, Russia; snoei@istu.edu or noiagdams@susu.ru
[6] Department of Applied Mathematics and Programming, South Ural State University, Lenin Prospect 76, 454080 Chelyabinsk, Russia
[7] Melentiev Energy Systems Institute of Siberian Branch of Russian Academy of Science, 664033 Irkutsk, Russia;
* Correspondence: o_moaaz@mans.edu.eg (O.M.); shyam01.math@gmail.com or shyamsundar.santra@jiscollege.ac.in (S.S.S.); sidorovdn@istu.edu or dsidorov@isem.irk.ru (D.S.)

Abstract: In this paper, we consider a class of quasilinear third-order differential equations with a delay argument. We establish some conditions of such certain third-order quasi-linear neutral differential equation as oscillatory or almost oscillatory. Those criteria improve, complement and simplify a number of existing results in the literature. Some examples are given to illustrate the importance of our results.

Keywords: thrid-order differential equations; delay; oscillation criteria

Citation: Qaraad, B.; Moaaz, O.; Santra S.S.; Noeiaghdam, S.; Sidorov, D.; Elabbasy, E.M.; Oscillatory Behavior of Third-Order Quasi-Linear Neutral Differential Equations. *Axioms* **2021**, *10*, 346. https://doi.org/10.3390/axioms10040346

Academic Editor: Chris Goodrich

Received: 1 September 2021
Accepted: 10 November 2021
Published: 17 December 2021

Publisher's Note: MDPI stays neutral with regard to jurisdictional claims in published maps and institutional affiliations.

Copyright: © 2021 by the authors. Licensee MDPI, Basel, Switzerland. This article is an open access article distributed under the terms and conditions of the Creative Commons Attribution (CC BY) license (https://creativecommons.org/licenses/by/4.0/).

1. Introduction

Consider the third-order neutral delay differential equation of the form

$$\left(r(t)\left(y''(t)\right)^\alpha\right)' + q(t)f(x(\sigma(t))) = 0, \qquad (1)$$

where $y(t) = x(t) + p(t)x(\tau(t))$ and we assume that the following hypotheses are satisfied:

(I_1) $r \in C([t_0, \infty), \mathbb{R})$ is positive and $\pi(t) < \infty$, where

$$\pi(t) = \int_t^\infty r^{-1/\alpha}(s)\mathrm{d}s;$$

(I_2) $p, q \in C([t_0, \infty), \mathbb{R})$, $p \leq p_0 < \infty$, q is non-negative and does not eventually vanish (i.e., $q(t)$ is not eventually zero on any half line $[t_*, \infty)$ for $t_* \geq t_0$);

(I_3) $\sigma, \tau \in C^1([t_0, \infty), \mathbb{R})$, $\sigma(t) < t$, $\tau(t) < t$, $\tau'(t) \geq \tau_0 > 0$, $\sigma \circ \tau = \tau \circ \sigma$ and $\lim_{t\to\infty} \sigma(t) = \lim_{t\to\infty} \tau(t) = \infty$;

(I_4) $f \in C(\mathbb{R}, \mathbb{R})$ and satisfies

$$f(x) > kx^\alpha \text{ for all } k > 0.$$

where α is the quotient of odd positive integers.

By a solution of (1), we mean a nontrivial function $x \in C([T_x, \infty), \mathbb{R})$ with $T_x \geq t_0$, which satisfies the property $r(y'')^\alpha \in C^1([T_x, \infty), \mathbb{R})$, moreover, satisfies (1) on $[T_x, \infty)$. We only consider those solutions of (1) satisfying, on some half-line, $[T_x, \infty)$ and satisfying the condition $\sup\{|x(t)| : T \leq t < \infty\} > 0$ for any $T \geq T_x$. A solution of (1) is oscillatory if it has arbitrarily large zeros on $[T_x, \infty)$; otherwise, it is said to be nonoscillatory. The equation itself is termed oscillatory if all its solutions oscillate, and is said to be almost oscillatory if all its solutions are oscillatory or asymptotically convergent to zero.

The neutral differential equations have numerous applications in electrical engineering, chemical reactions analysis, and economics.

Such equations are essential tools to model and study the dynamics and stability properties of electrical power systems, as in the works of Milano et al. [1,2]. The asymptotic behavior of solutions of associated delay differential equations have been used to describe the behavior of solutions to third-order partial differential equations. Additionally, they are employed for the study of distributed networks containing lossless transmission lines; see [3,4] for more details.

Recently, there has been much research activity concerning the oscillation of second-order differential equations with delay. See, for example, [5,6] and the references cited therein. Compared to the development of the oscillation for the second-order equations, the oscillation for third-order equations has received considerably less attention from researchers; see [7–24].

Baculikova and Dzurina [25,26] and Grace et al. [27] considered the third-order nonlinear delay differential equation

$$\left(r(t)\left(x''(t)\right)^\alpha\right)' + q(t)f(x(\sigma(t))) = 0,$$

in the case where $\pi(t) = \infty$ or $\pi(t) < \infty$.

Saker and Dzurina [28] studied the third-order nonlinear delay differential equation

$$\left(r(t)\left(x''(t)\right)^\alpha\right)' + q(t)x^\beta(\sigma(t)) = 0, \qquad (2)$$

and obtained several oscillation criteria, which guarantee that all non-oscillatory solutions of such Equation (2) tend towards zero.

Ravi et al. [29] investigated a third-order delay differential equation

$$\left(r_2(t)\left(r_1(t)\left(x'(t)\right)\right)'\right)' + q(t)x^\beta(\sigma(t)) = 0,$$

and established some oscillation results that supplemented and improved the results in [27]. Sidorov and Trufanov [30] considered nonlinear operator equations with a functional perturbation of the argument of neutral type and reduced the problem to quasilinear operator equations with a functional perturbation of the argument.

Moaaz, in [11], studied a third-order nonlinear delay differential (2) under the condition $\pi(t) = \infty$; he developed some results of previous references and established several sufficient conditions, which ensure that all solutions of (2) are oscillatory.

In previous papers, the authors used an integral averaging technique and a Riccati transformation to establish some sufficient conditions which ensure that any solution of (1) oscillates or converges to zero. The purpose of this paper is to improve and generalize these results and present some new sufficient conditions, which ensure that any solution of (1) oscillates or, for all its nonoscillatory solutions, tend towards zero as $t \to \infty$.

2. Auxiliary Results

In this section, we state and prove the following lemmas, which will be useful in the proofs of the main results.

Lemma 1 ([29]). *Assume $x(t)$ is nonoscillatory solution of (1). Then, $y(t) > 0$ and there are three possible cases of $y(t)$:*

$$\begin{aligned}
\mathbf{N}_1 \quad & y'(t) > 0, \ y''(t) > 0, \\
\mathbf{N}_2 \quad & y'(t) > 0, \ y''(t) < 0, \\
\mathbf{N}_3 \quad & y'(t) < 0, \ y''(t) > 0.
\end{aligned}$$

Lemma 2 ([31]). *Let $h(u) = Au - B(u - C)^{(\alpha+1)/\alpha}$, where $B > 0$, A and C are constants, α be a ratio of two odd positive numbers. Then, h attains its maximum value on $\mathbb{R}$ at $u^* = C + \left(\frac{\alpha A}{(\alpha+1)B}\right)^\alpha$ such that*

$$\max_{u \in \mathbb{R}} h(u) = h(u^*) = AC + \frac{\alpha^\alpha}{(\alpha+1)^{\alpha+1}} \frac{A^{\alpha+1}}{B^\alpha}.$$

Lemma 3 ([32]). *Assume that $c_1, c_2 \in [0, \infty)$ and $\gamma > 0$. Then*

$$(c_1 + c_1)^\gamma \leq \mu(c_1^\gamma + c_2^\gamma),$$

where

$$\mu := \begin{cases} 1 & \text{if } \gamma \leq 1 \\ 2^{\gamma-1} & \text{if } \gamma > 1 \end{cases}$$

Lemma 4. *Let x be a positive solution of (1), $y'(t) > 0$ and $p(t) \in (0,1)$. Then,*

$$\left(r(t)(y''(t))^\alpha\right)' + kQ(t)y^\alpha(\sigma(t)) \leq 0, \tag{3}$$

where

$$Q(t) = q(t)(1 - p(\sigma(t)))^\alpha.$$

Proof. Assume that x is a positive solution of (1). From hypothesis ($\mathbf{I}_4$), (1) becomes

$$\left(r(t)(y''(t))^\alpha\right)' + kq(t)x^\alpha(\sigma(t)) \leq 0. \tag{4}$$

Since $y'(t) > 0$, we find

$$\begin{aligned}
x(t) &= y(t) - p(t)x(\tau(t)) \geq y(t) - p(t)y(\tau(t)) \\
&\geq y(t) - p(t)y(t) = y(t)(1 - p(t)).
\end{aligned}$$

That is

$$x^\alpha(\sigma(t)) \geq y^\alpha(\sigma(t))(1 - p(\sigma(t)))^\alpha. \tag{5}$$

Combining (4) and (5), we have

$$\left(r(t)(y''(t))^\alpha\right)' + kQ(t)y^\alpha(\sigma(t)) \leq 0.$$

This completes the proof. □

Lemma 5. *Assume that $x(t)$ is a positive solution of (1). Then,*

$$\left(r(t)(y''(t))^\alpha + \frac{p_0^\alpha}{\tau_0}r(\tau(t))(y''(\tau(t)))^\alpha\right)' \leq -\frac{k}{\mu}\hat{Q}(t)y^\alpha(\sigma(t)), \tag{6}$$

where $\hat{Q}(t) := \min\{q(t), q(\tau(t))\}$.

Proof. Let $x(t)$ be a positive solution of (1). Then, there exists $t_1 \geq t_0$ such that $x(\sigma(t)) > 0$ and $x(\tau(t)) > 0$ for all $t \geq t_1$. From Lemma 3 and $\sigma \circ \tau = \tau \circ \sigma$, we obtain

$$y^\alpha(\sigma(t)) = \mu(x^\alpha(\sigma(t)) + p_0^\alpha x^\alpha(\sigma(\tau(t)))). \tag{7}$$

In view of (**I**$_3$), (4) implies

$$0 \geq \frac{p_0^\alpha}{\tau'(t)} \left(r(\tau(t))(y''(\tau(t)))^\alpha \right)' + p_0^\alpha k q(\tau(t)) x^\alpha(\sigma(\tau(t)))$$

$$\geq \frac{p_0^\alpha}{\tau_0} \left(r(\tau(t))(y''(\tau(t)))^\alpha \right)' + p_0^\alpha k q(\tau(t)) x^\alpha(\tau(\sigma(t))).$$

Using (4) with the above inequality, and taking into account (7), we have

$$\left(r(t)(y''(t))^\alpha \right)' + \frac{p_0^\alpha}{\tau_0} \left(r(\tau(t))(y''(\tau(t)))^\alpha \right)' + k\hat{O}(t)(x^\alpha(\sigma(t)) + p_0^\alpha x^\alpha(\tau(\sigma(t)))) \leq 0$$

$$\left(r(t)(y''(t))^\alpha \right)' + \frac{p_0^\alpha}{\tau_0} \left(r(\tau(t))(y''(\tau(t)))^\alpha \right)' + \frac{k}{\mu} \hat{O}(t) y^\alpha(\sigma(t)) \leq 0.$$

Thus,

$$\left(r(t)(y''(t))^\alpha + \frac{p_0^\alpha}{\tau_0} r(\tau(t))(y''(\tau(t)))^\alpha \right)' + \frac{k}{\mu} \hat{O}(t) y^\alpha(\sigma(t)) \leq 0.$$

This completes the proof. □

Lemma 6. *Assume that $x(t)$ is a positive solution of (1) and $y'(t) > 0$. Then*

$$y(\sigma(t)) \geq \acute{c}\sigma(t), \text{ where } \acute{c} := c\varsigma. \tag{8}$$

Proof. Since y' is nondecreasing, this implies that

$$y'(t) \geq y'(t_1) =: c \text{ on } [t_1, \infty).$$

Integrating from $\sigma(t)$ to t_1, we get

$$y(\sigma(t)) \geq c(\sigma(t) - t_1).$$

Hence, for any $\varsigma \in (0, 1)$ and $t \geq t_2$, we see that

$$y(\sigma(t)) \geq \acute{c}\sigma(t).$$

The proof is complete. □

Lemma 7. *Let $x(t)$ be a positive solution of (1). If*

$$\int_{t_0}^\infty \hat{O}(t)\sigma^\alpha(t)dt = \infty, \tag{9}$$

then case $\mathbf{N}_1$ is impossible.

Proof. Assume that $x(t)$ is a positive solution of (1) on $[t_0, \infty)$. Then, there exists $t_1 \geq t_0$ such that $x(\tau(t)) > 0$ and $x(\sigma(t)) > 0$ for all $t \geq t_1$. On the contrary, assume that $y(t)$ satisfies case $\mathbf{N}_1$. Integrating (6) from t_2 to t and using (8), we get

$$r(t)(y''(t))^\alpha + \frac{p_0^\alpha}{\tau_0} r(\tau(t))(y''(\tau(t)))^\alpha$$
$$\leq r(t_2)(y''(t_2))^\alpha + \frac{p_0^\alpha}{\tau_0} r(\tau(t_2))(y''(\tau(t_2)))^\alpha - (\acute{c})^\alpha \frac{k}{\mu} \int_{t_2}^t \hat{O}(t)\sigma^\alpha(t)dt, \quad (10)$$

which is a contradiction. □

Lemma 8. *Let $y(t)$ be a positive increasing solution of (1). If*

$$\int_{t_0}^\infty \frac{1}{r^{1/\alpha}(t)} \left(\int_{t_0}^t \hat{O}(t)\sigma^\alpha(s)ds \right)^{1/\alpha} dt = \infty, \quad (11)$$

then y satisfies case $\mathbf{N}_2$ for $t \geq t_1$ and
(a) $y(t) \geq ty'(t)$ and $y(t)/t$ is decreasing, and $\lim_{t \to \infty} y(t)/t = y' = 0$,
(b) $y'(t) \geq -\pi(t) r^{\frac{1}{\alpha}}(t) y''(t)$ and $y'(t)/\pi(t)$ is increasing.

Proof. Assume that $x(t)$ is a positive solution of (1) on $[t_0, \infty)$. Then, there exists $t_1 \geq t_0$ such that $x(\tau(t)) > 0$ and $x(\sigma(t)) > 0$ for all $t \geq t_1$. Since y is increasing, y satisfies either case $\mathbf{N}_1$ or $\mathbf{N}_2$. In view of $\pi(t) < \infty$ and (11), we see that (9) hold. By Lemma 7, $y(t)$ satisfies case $\mathbf{N}_2$.

On the other hand, it follows from $y'(t)$ is decreasing, such that there exists a constant $\lambda \geq 0$ such that $\lim_{t \to \infty} y'(t) = \lambda \geq 0$. We claim that $\lambda = 0$. As the proof of Lemma 7 we have (10). Take into account $(r(t)(y''(t))^\alpha)' \leq 0$ and $y''(t) < 0$, we have

$$r(t)(y''(t))^\alpha \left(1 + \frac{p_0^\alpha}{\tau_0}\right) \leq -(\acute{c})^\alpha \frac{k}{\mu} \int_{t_1}^t \hat{O}(s)\sigma^\alpha(s)ds.$$

It follows that

$$y''(t) \leq -\acute{c} \left(\frac{1}{r(t)} \left(\frac{k\tau_0}{\mu(\tau_0 + p_0^\alpha)} \right) \right)^{1/\alpha} \left(\int_{t_1}^t \hat{O}(s)\sigma^\alpha(s)ds \right)^{1/\alpha}.$$

Integrating from t_2 to t, we obtain

$$y'(t) \leq y'(t_2) - \acute{c} \left(\frac{k\tau_0}{\mu(\tau_0 + p_0^\alpha)} \right)^{\frac{1}{\alpha}} \int_{t_2}^t \frac{\left(\int_{t_2}^u \hat{O}(s)\sigma^\alpha(s)ds \right)^{\frac{1}{\alpha}}}{r^{\frac{1}{\alpha}}(u)} du.$$

In view of (11), this contradicts the positivity of $y'(t)$. Thus $\lambda = 0$. By "Hospital's rule", we see that

$$\lim_{t \to \infty} \frac{y(t)}{t} = 0 \text{ and } \lim_{t \to \infty} y'(t) = 0.$$

Thus,
$$y(t_1) - y'(t)t_1 > 0. \quad (12)$$

Therefore,
$$y(t) > y'(t)t_1,$$

for $t \geq t_2$. Hence, by the monotonicity of $y'(t)$, one can obtain that

$$y(t) = y(t_1) + \int_{t_1}^t y'(s)ds \geq y(t_1) + y'(t)(t - t_1).$$

By (12), we have
$$\left(\frac{y(t)}{t}\right)' = \frac{ty' - y}{t^2} < 0.$$

Now, it is easy to see that
$$y'(t) \geq -\int_t^\infty \frac{1}{r^{1/\alpha}(s)} r^{1/\alpha}(s) y''(s) ds \geq -r^{1/\alpha}(t) y''(t) \pi(t).$$

Thus,
$$\left(\frac{y'(t)}{\pi(t)}\right)' = \frac{r^{1/\alpha}(t) y''(t) \pi(t) + y'(t)}{r^{1/\alpha}(t) \pi^2(t)} \geq 0.$$

The proof is complete. □

3. Main Results

Theorem 1. *If*
$$\int_{t_0}^\infty \frac{1}{r^{1/\alpha}(t)} \left(\int_{t_0}^t \hat{O}(s) ds\right)^{1/\alpha} dt = \infty, \tag{13}$$
then possible positive solution of (1) satisfies case $\mathbf{N}_3$.

Proof. Assume that $x(t)$ is a positive solution of (1) on $[t_0, \infty)$. Then, there exists $t_1 \geq t_0$ such that $x(\tau(t)) > 0$ and $x(\sigma(t)) > 0$ for all $t \geq t_1$. Suppose that $y(t)$ satisfies case $\mathbf{N}_1$ or $\mathbf{N}_2$. Since y is increasing, then it follows that
$$y(t) \geq y(t_1) = \varrho \text{ for } t \geq t_2. \tag{14}$$

Set
$$\omega(t) = r(t)(y''(t))^\alpha + \frac{p_0^\alpha}{\tau_0} r(\tau(t))(y''(\tau(t)))^\alpha. \tag{15}$$

In (6), we obtain
$$\omega'(t) \leq -\frac{k}{\mu} \hat{O}(t) y^\alpha(\sigma(t)). \tag{16}$$

Since $\omega'(t) \leq 0$, by (15), we have
$$\omega(t) \geq r(t)(y''(t))^\alpha \left(1 + \frac{p_0^\alpha}{\tau_0}\right). \tag{17}$$

Integrating (16) from t_2 to t and using (14), we obtain
$$\omega(t) \leq \omega(t_1) - \frac{\varrho^\alpha k}{\mu} \int_{t_2}^t \hat{O}(s) ds. \tag{18}$$

First, let $y(t)$ satisfies case $\mathbf{N}_1$. We note that $\omega(t) > 0$. Using the fact $\pi(t) < \infty$ together with (13) yields that $\int_{t_0}^t \hat{O}(s) ds$ contradicts the positivity of $\omega(t)$.

If $y(t)$ satisfies case $\mathbf{N}_2$, using (17) in (18) becomes
$$r(t)(y''(t))^\alpha \left(1 + \frac{p_0^\alpha}{\tau_0}\right) \leq -\frac{\varrho^\alpha k}{\mu} \int_{t_2}^t \hat{O}(s) ds,$$

that is
$$y''(t) \leq -\varrho \left(\frac{k\tau_0}{\mu(\tau_0 + p_0^\alpha)}\right)^{\frac{1}{\alpha}} \left(\frac{1}{r(t)} \int_{t_2}^t \hat{O}(s) ds\right)^{\frac{1}{\alpha}}.$$

Integrating from t_2 to t, we have

$$y'(t) \leq y'(t_2) - \varrho \left(\frac{k\tau_0}{\mu(\tau_0 + p_0^\alpha)} \right)^{\frac{1}{\alpha}} \int_{t_2}^{t} \left(\frac{1}{r(u)} \int_{t_2}^{u} \hat{O}(s) ds \right)^{\frac{1}{\alpha}} du.$$

we obtain a cotradiction with the positivity of $y'(t)$. The proof of the theorem is complete. $\square$

Theorem 2. *If*

$$\liminf_{t \to \infty} \int_{\sigma(t)}^{t} \left(\frac{1}{r(u)} \int_{t_2}^{u} \hat{O}(s) \sigma^\alpha(s) ds \right)^{\frac{1}{\alpha}} du > \left(\frac{\mu(\tau_0 + p_0^\alpha)}{k\tau_0 e^\alpha} \right)^{\frac{1}{\alpha}}, \quad (19)$$

then, a possible positive solution to (1) satisfies case $\mathbf{N}_3$.

Proof. Assume that $x(t)$ is a positive solution of (1) on $[t_0, \infty)$. Then, there exists $t_1 \geq t_0$ such that $x(\tau(t)) > 0$ and $x(\sigma(t)) > 0$ for all $t \geq t_1$. Suppose that y satisfies case $\mathbf{N}_1$ or $\mathbf{N}_2$. In view of (19), (11) holds. Hence, by Lemma 8, $y(t)$ satisfies case $\mathbf{N}_2$ and properties (**a**) and (**b**) in Lemma 8. This implies that

$$y(\sigma(t)) \geq \sigma(t) y'(\sigma(t)).$$

Combining the above inequality along with (6), we get

$$\left(r(y''(t))^\alpha + \frac{p_0^\alpha}{\tau_0} r(y''(\tau(t)))^\alpha \right)' \leq -\frac{k}{\mu} \hat{O}(t) \sigma^\alpha(t) (y'(\sigma(t)))^\alpha.$$

Integrating from t_2 to t and using (17), we have

$$-r(t)(y''(t))^\alpha \left(1 + \frac{p_0^\alpha}{\tau_0} \right) \geq \frac{k}{\mu} \int_{t_2}^{t} \hat{O}(s) \sigma^\alpha(s) (y'(\sigma(t)))^\alpha ds. \quad (20)$$

Using the fact that $y''(t) < 0$, we see that

$$-y''(t) \geq \left(\frac{k\tau_0}{\mu(\tau_0 + p_0^\alpha)} \right)^{\frac{1}{\alpha}} y'(\sigma(t)) \left(\frac{1}{r(t)} \int_{t_2}^{t} \hat{O}(s) \sigma^\alpha(s) ds \right)^{\frac{1}{\alpha}}.$$

Now, set $\chi(t) = y'(t)$; we obtain

$$\chi'(t) + \left(\frac{k\tau_0}{\mu(\tau_0 + p_0^\alpha)} \right)^{\frac{1}{\alpha}} \left(\frac{1}{r(t)} \int_{t_2}^{t} \hat{O}(s) \sigma^\alpha(s) ds \right)^{\frac{1}{\alpha}} \chi(\sigma(t)) \leq 0. \quad (21)$$

In view of ([13], Theorem 1), however, the associated delay Equation (21) has a positive solution, which is a contradiction. The proof is complete. $\square$

Remark 1. *Theorem 2 does not require the existence of auxiliary functions such as ([27], Theorem 3), which uses the same principles (compared with first-order delay equations).*

Theorem 3. *Assume that (11) hold. If*

$$\limsup_{t \to \infty} \pi^\alpha(t) \int_{t_0}^{t} \hat{O}(s) \sigma^\alpha(s) ds > \frac{\mu(\tau_0 + p_0^\alpha)}{k\tau_0}, \quad (22)$$

then, the possible positive solution to (1) satisfies case $\mathbf{N}_3$.

Proof. Suppose that y satisfies case N_1 or N_2. We see that (9) holds due to $\pi(t) < \infty$ (this mean that $\lim_{t\to\infty} \pi(t) = 0$) and condition (22). Hence, by Lemma 8, $y(t)$ satisfies case N_2 in addition to properties (a) and (b) in Lemma 8. As in the proof of Theorem 2 with the fact $r(t)(y''(t))^\alpha$ is nonincreasing, and from (20), we obtain

$$-r(t)(y''(t))^\alpha \left(1 + \frac{p_0^\alpha}{\tau_0}\right) \geq -\frac{k}{\mu}\pi^\alpha(t)r(t)(y''(t))^\alpha \int_{t_2}^t \hat{O}(s)\sigma^\alpha(s)ds.$$

That is,

$$\frac{k\tau_0}{\mu(\tau_0 + p_0^\alpha)}\pi^\alpha(t) \int_{t_2}^t \hat{O}(s)\sigma^\alpha(s)ds \leq 1.$$

This contradicts (22). The proof is complete. □

Theorem 4. *Assume that (11) holds. If $\sigma'(t) > 0$ and there exists a nondecreasing function $\rho \in C^1([t_0, \infty), (0, \infty))$, such that*

$$\limsup_{t\to\infty} \int_T^t \left(\left(\frac{k\tau_0}{\mu(\tau_0 + p_0^\alpha)}\right)^{\frac{1}{\alpha}} \frac{\rho(u)}{\sigma(u)r^{\frac{1}{\alpha}}(u)} \int_{t_0}^t \hat{O}(s)\sigma^\alpha(s)ds - \frac{\rho'^2(u)}{4\rho(u)\sigma'(u)}\right)du = \infty, \quad (23)$$

for any $T \in [t_0, \infty)$, then a possible positive solution to (1) satisfies case N_3.

Proof. Assume that $x(t)$ is a positive solution of (1) on $[t_0, \infty)$. Then, there exists $t_1 \geq t_0$ such that $x(\tau(t)) > 0$ and $x(\sigma(t)) > 0$ for all $t \geq t_1$. Suppose that y satisfies case N_1 or N_2. By Lemma 8, $y(t)$ satisfies case N_2 and the properties (a) and (b). Define the function $w(t)$ by

$$w(t) := \rho(t)\frac{y'(t)}{y(\sigma(t))}. \quad (24)$$

Then $w(t) > 0$, and

$$w'(t) = \frac{\rho'(t)}{\rho(t)}w(t) + \frac{\rho(t)y''(t)}{y(\sigma(t))} - \frac{\rho(t)y'(t)y'(\sigma(t))\sigma'(t)}{y^2(\sigma(t))}.$$

Using the fact that $y'(t)$ is decreasing, we have

$$\begin{aligned}
w'(t) &\leq \frac{\rho'(t)w(t)}{\rho(t)} + \rho(t)\frac{y''(t)}{y(\sigma(t))} - \rho(t)\frac{(y'(t))^2\sigma'(t)}{y^2(\sigma(t))} \\
&= \frac{\rho'(t)w(t)}{\rho(t)} + \rho(t)\frac{y''(t)}{y(\sigma(t))} - \frac{\sigma'(t)}{\rho(t)}\left(\rho(t)\frac{y'(t)}{y(\sigma(t))}\right)^2.
\end{aligned}$$

By (24), we obtain

$$w'(t) \leq \frac{\rho'(t)w(t)}{\rho(t)} + \rho(t)\frac{y''(t)}{y(\sigma(t))} - \frac{\sigma'(t)}{\rho(t)}w^2(t). \quad (25)$$

Integrating (6) from t_2 to t and $(y(\sigma(t))/\sigma(t))' < 0$, we have

$$-\left(r(t)(y''(t))^\alpha + \frac{p_0^\alpha}{\tau_0}r(\tau(t))(y''(\tau(t)))^\alpha\right) \geq -\left(r(t_2)(y''(t_2))^\alpha + \frac{p_0^\alpha}{\tau_0}r(t_2)(y''(\tau(t_2)))^\alpha\right)$$
$$+\frac{k}{\mu}\int_{t_2}^t \hat{O}(s)y^\alpha(\sigma(s))ds \qquad (26)$$
$$\geq -\left(r(t_2)(y''(t_2))^\alpha + \frac{p_0^\alpha}{\tau_0}r(t_2)(y''(\tau(t_2)))^\alpha\right)$$
$$+\frac{k}{\mu}\left(\frac{y(\sigma(t))}{\sigma(t)}\right)^\alpha \int_{t_2}^t \hat{O}(s)\sigma^\alpha(s)ds.$$

Since $\lim_{t\to\infty} y(t)/t = 0$, there exists $t_3 > t_2$ such that

$$-\left(r(t_2)(y''(t_2))^\alpha + \frac{p_0^\alpha}{\tau_0}r(t_2)(y''(\tau(t_2)))^\alpha\right) - \frac{k}{\mu}\left(\frac{y(\sigma(t))}{\sigma(t)}\right)^\alpha \int_{t_0}^{t_2} \hat{O}(s)\sigma^\alpha(s)ds > 0.$$

Combining the above inequality in (27) implies

$$-\left(r(t)(y''(t))^\alpha + \frac{p_0^\alpha}{\tau_0}r(\tau(t))(y''(\tau(t)))^\alpha\right) \geq -\left(r(y''(t_2))^\alpha + \frac{p_0^\alpha}{\tau_0}r(y''(\tau(t_2)))\right)^\alpha$$
$$+\frac{k}{\mu}\left(\frac{y(\sigma(t))}{\sigma(t)}\right)^\alpha \int_{t_0}^t \hat{O}(s)\sigma^\alpha(s)ds$$
$$-\frac{k}{\mu}\left(\frac{y(\sigma(t))}{\sigma(t)}\right)^\alpha \int_{t_0}^{t_2} \hat{O}(s)\sigma^\alpha(s)ds$$
$$\geq \frac{k}{\mu}\left(\frac{y(\sigma(t))}{\sigma(t)}\right)^\alpha \int_{t_0}^t \hat{O}(s)\sigma^\alpha(s)ds.$$

Using (17), we have

$$-r(t)(y''(t))^\alpha \left(1 + \frac{p_0^\alpha}{\tau_0}\right) \geq \frac{k}{\mu}\left(\frac{y(\sigma(t))}{\sigma(t)}\right)^\alpha \int_{t_0}^t \hat{O}(s)\sigma^\alpha(s)ds,$$

that is,

$$\frac{y''(t)}{y(\sigma(t))} \leq -\left(\frac{k\tau_0}{\mu(\tau_0 + p_0^\alpha)}\right)^{\frac{1}{\alpha}} \frac{1}{\sigma(t)r^{\frac{1}{\alpha}}(t)} \int_{t_0}^t \hat{O}(s)\sigma^\alpha(s)ds. \qquad (27)$$

Substituting (27) in (25), yields

$$w'(t) \leq \frac{\rho'(t)}{\rho(t)}w(t) - \frac{\rho(t)}{\sigma(s)}\left(\frac{k\tau_0}{\mu(\tau_0 + p_0^\alpha)}\right)^{\frac{1}{\alpha}} \frac{1}{r^{\frac{1}{\alpha}}(t)} \int_{t_0}^t \hat{O}(s)\sigma^\alpha(s)ds - \frac{\sigma'(t)}{\rho(t)}w^2(t)$$
$$= -\frac{\sigma'(t)}{\rho(t)}\left(w(t) - \frac{\rho'(t)}{2\sigma'(t)}\right)^2 + \frac{\rho'^2(t)}{4\rho(t)\sigma'(t)}$$
$$-\frac{\rho(t)}{\sigma(s)}\left(\frac{k\tau_0}{\mu(\tau_0 + p_0^\alpha)}\right)^{\frac{1}{\alpha}} \frac{1}{r^{\frac{1}{\alpha}}(t)} \int_{t_0}^t \hat{O}(s)\sigma^\alpha(s)ds.$$

Hence,

$$w'(t) \leq -\frac{\rho(t)}{\sigma(s)}\left(\frac{k\tau_0}{\mu(\tau_0 + p_0^\alpha)}\right)^{\frac{1}{\alpha}} \frac{1}{r^{\frac{1}{\alpha}}(t)} \int_{t_0}^t \hat{O}(s)\sigma^\alpha(s)ds + \frac{\rho'^2(t)}{4\rho(t)\sigma'(t)}.$$

Integrating from t_3 to t, we have

$$w(t) \leq w(t_3) - \int_{t_3}^{t} \left(\left(\frac{k\tau_0}{\mu(\tau_0 + p_0^\alpha)} \right)^{\frac{1}{\alpha}} \frac{\rho(u)}{\sigma(u) r^{\frac{1}{\alpha}}(u)} \int_{t_0}^{t} \hat{O}(s) \sigma^\alpha(s) ds - \frac{\rho'^2(u)}{4p(u)\sigma'(u)} \right) du,$$

which is a contradiction. The proof is complete. □

By choosing $\rho(t) = \frac{1}{\pi}$, we conclude the following corollary

Corollary 1. *Assume that (11) holds. If there is a nondecreasing function $\rho \in C^1([t_0, \infty), (0, \infty))$ and $\sigma'(t) > 0$, such that*

$$\limsup_{t \to \infty} \int_{T}^{t} \left(\frac{\left(\frac{k\tau_0}{\mu(\tau_0 + p_0^\alpha)} \right)^{\frac{1}{\alpha}}}{\pi(u)\sigma(u) r^{\frac{1}{\alpha}}(u)} \int_{t_0}^{t} \hat{O}(s) \sigma^\alpha(s) ds - \frac{r^{-2/\alpha}(u)}{4\sigma'(u)\pi^3(u)} \right) du = \infty, \qquad (28)$$

for any $T \in [t_0, \infty)$, then possible positive solution of (1) satisfies case $\mathbf{N_3}$.

Theorem 5. *Assume that (11) holds. If there is a nondecreasing function $\delta \in C^1([t_0, \infty), (0, \infty))$, such that*

$$\limsup_{t \to \infty} \int_{t_2}^{t} \left(\delta(s) k \frac{\hat{O}(s)}{\mu} - \frac{(\delta'(s))^{\alpha+1}}{(\alpha+1)^{\alpha+1} (\sigma'(s))^\alpha \pi^\alpha(s) \delta^\alpha(s)} \right) ds > \frac{\delta(t)}{\pi^\alpha(t)\sigma^\alpha(t)}, \qquad (29)$$

then, the possible positive solution to (1) satisfies case $\mathbf{N_3}$.

Proof. Assume that $x(t)$ is a positive solution of (1) on $[t_0, \infty)$; then, there exists $t_1 \geq t_0$ such that $x(\tau(t)) > 0$ and $x(\sigma(t)) > 0$ for all $t \geq t_1$. Suppose that y satisfies case $\mathbf{N_1}$ or $\mathbf{N_2}$. By Lemma 8, $y(t)$ satisfies case $\mathbf{N_2}$ and the properties (**a**) and (**b**).

Define the function $w(t)$ by

$$w(t) := \delta(t) \left(\frac{r(y''(t))^\alpha}{y^\alpha(\sigma(t))} + \frac{1}{\pi^\alpha \sigma^\alpha(t)} \right). \qquad (30)$$

From Lemma 8, it is easy to see that

$$y(\sigma(t)) \geq \sigma(t) y'(\sigma(t)) \geq \sigma(t) y'(t) \geq -\sigma(t) \pi(t) r^{\frac{1}{\alpha}}(t) y''(t). \qquad (31)$$

That is, $w(t) > 0$ and

$$-\frac{\delta(t) r(y''(t))^\alpha}{y^\alpha(\sigma(t))} \leq \frac{\delta(t)}{\pi^\alpha \sigma^\alpha(t)}. \qquad (32)$$

Using (16) and the fact $y'(t)$ is decreasing, we have.

$$w'(t) = \frac{\delta'(t)}{\delta(t)}w(t) + \frac{\delta(t)\left(r(t)(y''(t))^\alpha\right)'}{y^\alpha(\sigma(t))} - \frac{\alpha\delta(t)r(t)(y''(t))^\alpha y'(\sigma(t))\sigma'(t)}{y^{\alpha+1}(\sigma(t))}$$
$$+ \frac{\alpha\delta(t)}{(\pi(t)\sigma(t))^{\alpha+1}}\left(\frac{\sigma(t)}{r^{\frac{1}{\alpha}}(t)} - \sigma'(t)\pi(t)\right)$$
$$\leq \frac{\delta'(t)}{\delta(t)}w(t) - \delta(t)\frac{k}{\mu}\hat{O}(t)$$
$$- \frac{\alpha\sigma'(t)y'(t)}{\delta^{\frac{1}{\alpha}}(t)r^{\frac{1}{\alpha}}(t)y''(t)}\left(w(t) - \frac{\delta(t)}{\pi(t)\sigma(t)}\right)^{\frac{1}{\alpha}+1} + \frac{\alpha\delta(t)}{(\pi(t)\sigma(t))^{\alpha+1}}\left(\frac{\sigma(t)}{r^{\frac{1}{\alpha}}(t)} - \sigma'(t)\pi(t)\right).$$

In view of **(b)** in Lemma 8, we find

$$w'(t) \leq \frac{\delta'(t)}{\delta(t)}w(t) - \delta(t)k\frac{\hat{O}(t)}{\mu}$$
$$- \frac{\alpha\sigma'(t)\pi(t)}{\delta^{\frac{1}{\alpha}}(t)}\left(w(t) - \frac{\delta(t)}{\pi^\alpha(t)\sigma^\alpha(t)}\right)^{\frac{1}{\alpha}+1} + \frac{\alpha\delta(t)}{(\pi(t)\sigma(t))^{\alpha+1}}\left(\frac{\sigma(t)}{r^{\frac{1}{\alpha}}(t)} - \sigma'(t)\pi(t)\right).$$

Set

$$A := \frac{\delta'(t)}{\delta(t)}, \quad B := \frac{\alpha\sigma'(t)\pi(t)}{\delta^{\frac{1}{\alpha}}(t)}, \quad C := \frac{\delta(t)}{\pi^\alpha(t)\sigma^\alpha(t)}.$$

Using Lemma 2, we obtain

$$w'(t) = -\delta(t)k\frac{\hat{O}(t)}{\mu} + \frac{\delta'(t)}{\pi^\alpha(t)\sigma^\alpha(t)} + \frac{1}{(\alpha+1)^{\alpha+1}}\frac{(\delta'(t))^{\alpha+1}}{(\sigma'(t))^\alpha \pi^\alpha(t)\delta^\alpha(t)}$$
$$+ \frac{\alpha\delta(t)}{(\pi(t)\sigma(t))^{\alpha+1}}\left(\frac{\sigma(t)}{r^{\frac{1}{\alpha}}(t)} - \sigma'(t)\pi(t)\right). \tag{33}$$

It is clear that

$$\left(\frac{\delta(t)}{\pi^\alpha(t)\sigma^\alpha(t)}\right)' = \frac{\delta'(t)}{\pi^\alpha(t)\sigma^\alpha(t)} + \frac{\alpha\delta(t)}{(\pi(t)\sigma(t))^{\alpha+1}}\left(\frac{\sigma(t)}{r^{\frac{1}{\alpha}}(t)} - \sigma'(t)\pi(t)\right).$$

In (33), we obtain

$$w'(t) = -\delta(t)k\frac{\hat{O}(t)}{\mu} + \left(\frac{\delta(t)}{\pi^\alpha(t)\sigma^\alpha(t)}\right)' + \frac{1}{(\alpha+1)^{\alpha+1}}\frac{(\delta'(t))^{\alpha+1}}{(\sigma'(t))^\alpha \pi^\alpha(t)\delta^\alpha(t)}.$$

Integrating the above inequality from t_2 to t yields

$$\int_{t_2}^{t}\left(\delta(s)k\frac{\hat{O}(s)}{\mu} - \frac{(\delta'(s))^{\alpha+1}}{(\alpha+1)^{\alpha+1}(\sigma'(s))^\alpha \pi^\alpha(s)\delta^\alpha(s)}\right)ds + \frac{\delta(t)}{\pi^\alpha(t)\sigma^\alpha(t)} - \frac{\delta(t_2)}{\pi^\alpha(t_2)\sigma^\alpha(t_2)}$$
$$\leq w(t_2) - w(t).$$

From (30), we are led to

$$\int_{t_2}^{t} \left(\delta(s) k \frac{\hat{O}(s)}{\mu} - \frac{(\delta'(s))^{\alpha+1}}{(\alpha+1)^{\alpha+1} (\sigma'(s))^\alpha \pi^\alpha(s) \delta^\alpha(s)} \right) ds + \frac{\delta(t)}{\pi^\alpha(t) \sigma^\alpha(t)}$$
$$- \frac{\delta(t_2)}{\pi^\alpha(t_2) \sigma^\alpha(t_2)} \tag{34}$$
$$\leq \delta(t_2) \left(\frac{r(y''(t_2))^\alpha}{y^\alpha(\sigma(t_2))} \right) - \delta(t) \left(\frac{r(y''(t))^\alpha}{y^\alpha(\sigma(t))} \right).$$

By (32), (35) becomes

$$\int_{t_2}^{t} \left(\delta(s) k \frac{\hat{O}(s)}{\mu} - \frac{(\delta'(s))^{\alpha+1}}{(\alpha+1)^{\alpha+1} (\sigma'(s))^\alpha \pi^\alpha(s) \delta^\alpha(s)} \right) ds \leq \frac{\delta(t)}{\pi^\alpha \sigma^\alpha(t)}.$$

The proof is complete. □

Lemma 9. *Let $x(t)$ be a positive solution to (1) and $y(t)$, satisfying case N_3. If*

$$\int_{t_0}^{\infty} \hat{O}(s) ds = \infty \tag{35}$$

or

$$\int_{t_0}^{\infty} \frac{1}{r(s)} \left(\int_{t}^{\infty} \hat{O}(u) du \right)^{\frac{1}{\alpha}} ds = \infty, \tag{36}$$

then $\lim_{t \to \infty} y(t) = 0$.

Proof. Assume that $x(t)$ is a positive solution of (1) on $[t_0, \infty)$, there exists $t_1 \geq t_0$ such that $x(\tau(t)) > 0$ and $x(\sigma(t)) > 0$ for all $t \geq t_1$. Since $y(t) > 0$ and $y'(t) < 0$, there is $\lambda \geq 0$, such that $\lim_{t \to \infty} y(t) = \lambda$. Assume that $\lambda > 0$. Integrating (6) from t_2 to t, we have

$$r(t)(y''(t))^\alpha + \frac{p_0^\alpha}{\tau_0} r(y''(\tau(t)))^\alpha \leq r(t_1)(y''(t_1))^\alpha + \frac{p_0^\alpha}{\tau_0} r(\tau(t_1))(y''(\tau(t_1)))^\alpha$$
$$- \frac{k}{\mu} \int_{t_2}^{t} \hat{O}(s) y^\alpha(\sigma(s)) ds$$
$$\leq r(t_1)(y''(t_1))^\alpha + \frac{p_0^\alpha}{\tau_0} r(\tau(t_1))(y''(\tau(t_1)))^\alpha$$
$$- \frac{k}{\mu} \lambda^\alpha \int_{t_2}^{t} \hat{O}(s) ds.$$

This contradicts (35). Hence $\lambda = 0$. The proof is complete. □

Theorem 6. *Let $x(t)$ be a positive solution of (1). If*

$$\limsup_{t \to \infty} \int_{\sigma(t)}^{t} \hat{O}(s) R(\sigma(t), \sigma(s)) ds > \frac{\tau_0}{\tau_0 + p_0^\alpha}, \tag{37}$$

then case N_3 is impossible, where $R(v,u) = \int_u^v \int_\zeta^v \frac{1}{r^{\frac{1}{\alpha}}(\zeta)} d\zeta$.

Proof. Since $r(y''(t))^\alpha$ is nonincreasing, pick $t_1 \in [t_0, \infty)$ for $t \geq t_1$, we see that

$$-y'(u) \geq \int_u^v \frac{1}{r^{\frac{1}{\alpha}}(s)} r^{\frac{1}{\alpha}}(s) y''(s) ds \geq r^{\frac{1}{\alpha}}(v) y''(v) \int_u^v \frac{1}{r^{\frac{1}{\alpha}}(s)} ds,$$

for $v \geq u$. Integrating above inequality from u to v, we have

$$y(u) \geq r^{\frac{1}{\alpha}}(v)y''(v)\int_u^v \int_\zeta^v \frac{1}{r^{\frac{1}{\alpha}}(\zeta)}\mathrm{d}\zeta = r^{\frac{1}{\alpha}}(v)y''(v)R(v,u). \tag{38}$$

Integrating (6) from $\sigma(t)$ to t and using (38), we get (17)

$$\begin{aligned}\left(r(\sigma(t))(y''(\sigma(t)))^\alpha + \frac{p_0^\alpha}{\tau_0}r(\tau(\sigma(t)))(y''(\tau(\sigma(t))))^\alpha\right) &\geq \frac{k}{\mu}\int_{\sigma(t)}^t \hat{O}(s)y^\alpha(\sigma(s))\mathrm{d}s\\ &\geq \frac{k}{\mu}r(\sigma(t))(y''(\sigma(t)))^\alpha \int_{\sigma(t)}^t \hat{O}(s)R^\alpha(\sigma(t),\sigma(s))\mathrm{d}s.\end{aligned}$$

Using the fact that $(r(t)(y''(t))^\alpha)' < 0$ and (17), we obtain

$$r(\sigma(t))(y''(\sigma(t)))^\alpha < r(\tau(\sigma(t)))(y''(\tau(\sigma(t))))^\alpha,$$

and

$$\begin{aligned}r(\tau(\sigma(t)))(y''(\tau(\sigma(t))))^\alpha\left(1+\frac{p_0^\alpha}{\tau_0}\right) &\geq -\frac{k}{\mu}r(\sigma(t))(y''(\sigma(t)))^\alpha \int_{\sigma(t)}^t \hat{O}(s)R^\alpha(\sigma(t),\sigma(s))\mathrm{d}s\\ &\geq -\frac{k}{\mu}r(\tau(\sigma(t)))(y''(\tau(\sigma(t))))^\alpha \int_{\sigma(t)}^t \hat{O}(s)R^\alpha(\sigma(t),\sigma(s))\mathrm{d}s.\end{aligned}$$

Hence,

$$\left(1+\frac{p_0^\alpha}{\tau_0}\right) \geq -\frac{k}{\mu}\int_{\sigma(t)}^t \hat{O}(s)R^\alpha(\sigma(t),\sigma(s))\mathrm{d}s.$$

This led to a contradiction. The proof is complete. □

4. Applications

4.1. Asymptotic Properties

By combining Theorems 2–5 with Lemma 9, one can easily provide new criteria for the asymptotic properties of (1) as follows

Theorem 7. *Assume that (13) holds. Then, (1) is almost oscillatory.*

Theorem 8. *Assume that (19) and (35) or (36) hold. Then (1) is almost oscillatory.*

Theorem 9. *Assume that (11), (22) and either (35) or (36) hold. Then, (1) is almost oscillatory.*

Theorem 10. *Assume that (11) holds and, if there is a nondecreasing function $\rho \in C^1([t_0,\infty),(0,\infty))$ and $\sigma'(t) > 0$, such that (28) and either (35) or (36) hold, then (1) is almost oscillatory.*

Theorem 11. *Assume that (11) holds and if there is a nondecreasing function $\delta \in C^1([t_0,\infty),(0,\infty))$, such that (29) and either (35) or (36) hold, then (1) is almost oscillatory.*

4.2. Oscillation

In the following Theorem, we combine Theorems 2–5 with Theorem (37) to obtain new criteria for oscillation of (1)

Theorem 12. *If all assumptions of Theorem 1 or 2 or 3 or 4 or 5 and (37) hold, then (1) is oscillatory.*

Remark 2. *Compared to the existing results of [25,26], oscillation of (1) is attained by easier conditions.*

Example 1. Consider the third-order neutral delay differential equation

$$\left(t^2(x(t)+p_0x(\epsilon t))''\right)' + \frac{q_0}{t}y(0.5t) = 0, \ t \geq 1, \tag{39}$$

where $\epsilon \in (0,1)$ and $q_0 > 0$. We note that $r = t^2, \sigma(t) = 0.5t, \tau(t) = \epsilon t, p(t) = p_0$. It can easily be verified that $\hat{O}(t) = \frac{q_0}{t}$. By choosing $\delta(t) = \pi(t)\tau(t) = \epsilon$, Condition (19), (29), (28) and (37) become

$$q_0 > \frac{2(\tau_0 + p_0)}{\ln(2)\tau_0 e}, \tag{40}$$

$$q_0 > \frac{2}{(1-p_0)}, \tag{41}$$

$$q_0 > \frac{(\tau_0 + p_0)}{2\tau_0} \tag{42}$$

and

$$q_0 > \frac{\tau_0}{(\tau_0 + p_0)\left(0.5 + \ln 0.5 + \frac{1}{2}\ln^2 0.5\right)},$$

respectively. Using Theorems 8, 10 and 11, Equation (39) is almost oscillatory if (40) or (41) or (42) holds. Moreover, by Theorem 12, we see that (39) is oscillatory if

$$q_0 > \max\left\{\frac{\tau_0}{(\tau_0 + p_0)\left(0.5 + \ln 0.5 + \frac{1}{2}\ln^2 0.5\right)}, \frac{2(\tau_0 + p_0)}{\ln(2)\tau_0 e}\right\}.$$

Remark 3. *It is easy to verify that condition (13) fails; therefore, Theorem 1 does not apply.*

Remark 4. *If $p_0 = 1$ then our results are reduced to the results of Chatzarakis in [14].*

Example 2. Consider the third-order neutral delay differential equation

$$\left(t^2\left((x(t)+p_0x(\epsilon t))''\right)^\alpha\right)' + \frac{q_0}{t}y^\alpha(\lambda t) = 0, \ t \geq 1, \tag{43}$$

where $q_0 > 0$ and $\lambda, \epsilon \in (0,1)$. Condition (19), (28) and (37) reduse to

$$C1: q_0 > \frac{\epsilon + p_0}{\lambda \epsilon e \ln\left(\frac{1}{\lambda}\right)},$$

$$C2: q_0 > \frac{\mu\left(\epsilon + p_0^\alpha\right)^{1/\alpha}}{4\lambda \epsilon^{1/\alpha}}$$

and

$$C3: q_0 > \frac{\epsilon}{\left(\epsilon + p_0^\alpha\right)\left(1 - \lambda + \ln \lambda + \frac{1}{2}\ln^2 \lambda\right)},$$

respectively. Therefore, by Theorem 12, we see that (43) is oscillatory if

$$q_0 > \max\left\{\frac{\epsilon + p_0}{\lambda \epsilon e \ln\left(\frac{1}{\lambda}\right)}, \frac{\epsilon}{\left(\epsilon + p_0^\alpha\right)\left(1 - \lambda + \ln \lambda + \frac{1}{2}\ln^2 \lambda\right)}\right\} \tag{44}$$

or

$$q_0 > \max\left\{\frac{\mu\left(\epsilon + p_0^\alpha\right)^{1/\alpha}}{4\lambda \epsilon^{1/\alpha}}, \frac{\epsilon}{\left(\epsilon + p_0^\alpha\right)\left(1 - \lambda + \ln \lambda + \frac{1}{2}\ln^2 \lambda\right)}\right\}. \tag{45}$$

Remark 5. Consider a particular case of (43), namely,

$$\left(t^2\left(\left(x(t)+\frac{1}{4}x\left(\frac{t}{2}\right)\right)''\right)\right)' + \frac{q_0}{t}y(\lambda t) = 0, \tag{46}$$

Conditions (44) and (45) reduce to

$$q_0 > \max\left\{\frac{\frac{3}{4}}{\frac{1}{4}\lambda e \ln\left(\frac{1}{\lambda}\right)}, \frac{\frac{1}{2}}{\left(\frac{3}{4}\right)\left(1-\lambda+\ln\lambda+\frac{1}{2}\ln^2\lambda\right)}\right\} \tag{47}$$

and

$$q_0 > \max\left\{\frac{3}{8\lambda}, \frac{2}{3\left(1-\lambda+\ln\lambda+\frac{1}{2}\ln^2\lambda\right)}\right\}, \tag{48}$$

respectively; see Figure 1. Thus, by Theorem 12, Equation (46) is oscillatory if (47) or (48) satisfies. So, For a given $\lambda \in (0, 0.21)$, Condition (47) is sharp for oscillation, but in $\lambda \in (0.21, 1)$ Condition (48) is sharp for oscillation.

On the other hand, consider a particular case of (43), namely,

$$\left(t^2\left(\left(x(t)+\frac{1}{4}x\left(\frac{t}{2}\right)\right)''\right)^\alpha\right)' + \frac{q_0}{t}y\left(\frac{t}{4}\right) = 0, \tag{49}$$

where $\alpha > 1$. Conditions (44) and (45) reduce to

$$q_0 > \max\left\{\frac{3}{e\ln 2}, \frac{1}{2}\frac{2^{2\alpha+1}}{(2^{2\alpha}+2)\left(2\ln^2 2 - 2\ln 2 + \frac{3}{4}\right)}\right\}$$

or

$$q_0 > \max\left\{\frac{2}{17}2^\alpha\left(\frac{1}{2^{2\alpha}}\left(2^{2\alpha}+2\right)\right)^{\frac{1}{\alpha}}, \frac{1}{2}\frac{2^{2\alpha+1}}{(2^{2\alpha}+2)\left(2\ln^2 2 - 2\ln 2 + \frac{3}{4}\right)}\right\}.$$

Remark 6. It is easy to notice that the effect of the delay argument on the oscillation parameters varies from one example to another, and no consistent pattern can be found to determine this effect. Additionally, the oscillation test depends on two different conditions, so we notice the change in the effect of the delay argument on oscillation (from inverse to direct relationship). This also applies to the effect of α.

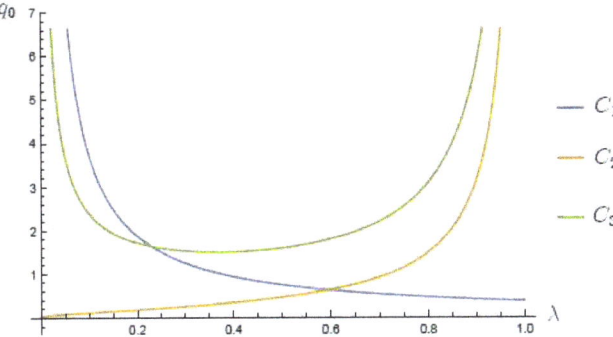

Figure 1. Test of the strength of criteria for (46).

5. Conclusions

In this paper, we introduced a simplified theorem for near oscillation; furthermore, we established oscillation criteria for (1). Using comparison theorems and the Riccati technique, we established criteria to check the oscillation under fewer restrictions, and compared this with some results published in the literature. Our results are an extension of and complement to existing results in some previous studies, such as [15,27,29].

The establishment of criteria for the oscillation of Equation (1) without the need for a condition $\sigma \circ \tau = \tau \circ \sigma$ and $\tau'(t) \geq \tau_0$ remains an open problem.

Author Contributions: Conceptualization, B.Q., O.M., S.S.S., S.N., D.S. and E.M.E.; methodology, B.Q., O.M., S.S.S., S.N., D.S. and E.M.E.; validation, B.Q., O.M., S.S.S., S.N., D.S. and E.M.E.; formal analysis, B.Q., O.M., S.S.S., S.N., D.S. and E.M.E.; investigation,B.Q., O.M., S.S.S., S.N., D.S. and E.M.E.; writing—review and editing, B.Q., O.M., S.S.S., S.N., D.S. and E.M.E.; supervision, B.Q., O.M., S.S.S., S.N., D.S. and E.M.E.; funding acquisition, D.S.; All authors have read and agreed to the published version of the manuscript.

Funding: D.S. was funded out under State Assignment Projects (No. FWEU-2021-0006, FWEU-2021-0001) of the Fundamental Research Program of Russian Federation 2021–2030 using the resources of the High-Temperature Circuit Multi-Access Research Center (Ministry of Science and Higher Education of the Russian Federation, project no 13.CKP.21.0038).

Conflicts of Interest: There are no competing interests.

References

1. Liu, M.; Dassios, I.; Tzounas, G.; Milano, F. Stability Analysis of Power Systems with Inclusion of Realistic-Modeling of WAMS Delays. *IEEE Trans. Power Syst.* **2019**, *34*, 627–636. [CrossRef]
2. Milano, F.; Dassios, I. Small-Signal Stability Analysis for Non-Index 1 Hessenberg Form Systems of Delay Differential-Algebraic Equations. *IEEE Trans. Circuits Syst. Regul. Pap.* **2016**, *63*, 1521–1530. [CrossRef]
3. Agarwal, R.P.; Berezansky, L.; Braverman, E.; Domoshnitsky, A. *Nonoscillation Theory of Functional Differential Equations with Applications*; Springer: Berlin/Heidelberg, Germany, 2012; 520p, ISBN978-1-4614-3454-2.
4. Hale, J.K. *Theory of Functional Differential Equations*; Springer: Berlin/Heidelberg, Germany, 1977.
5. Moaaz, O.; Elabbasy, E.M.; Qaraad, B. An improved approach for studying oscillation of generalized Emden–Fowler neutral differential equation. *J. Inequal. Appl.* **2020**, *2020*, 69. [CrossRef]
6. Chatzarakis, G.E.; Moaaz, O.; Li, T.; Qaraad, B. Some oscillation theorems for nonlinear second-order differential equations with an advanced argument. *Adv. Differ. Equ.* **2020**, 160. [CrossRef]
7. Bohner, M.; Grace, S.R.; Sager, Tunc, E. Oscillation of third-order nonlinear damped delay differential equations. *Appl. Math. Comput.* **2016**, *278*, 21–32. [CrossRef]
8. Chatzarakis, G.E.; Grace, S.R.; Jadlovska, I.; Li, T.; Tunc, E. Oscillation criteria for third-order Emden–Fowler differential equations with unbounded neutral coefficients. *Complexity* **2019**, *2019*, 5691758. [CrossRef]
9. Grace, S.R.; Graef, J.R.; Tunc, E. Oscillatory behaviour of third order nonlinear differential equations with a nonlinear nonpositive neutral term. *J. Taibah Univ. Sci.* **2019**, *13*, 704–710. [CrossRef]
10. Grace, S.R.; Jadlovska, I.; Tunc, E. Oscillatory and asymptotic behavior of third-order nonlinear differential equations with a superlinear neutral term. *Turk. J. Math.* **2020**, *44*, 1317–1329. [CrossRef]
11. Moaaz, O.; Qaraad, B.; El-Nabulsi, R.; Bazighifan, O. New Results for Kneser Solutions of Third-Order Nonlinear Neutral Differential Equations. *Mathematics* **2020**, *8*, 686. [CrossRef]
12. Elabbasy, E.M.; Qaraad, B.; Abdeljawad, T.; Moaaz, O. Oscillation Criteria for a Class of Third-Order Damped Neutral Differential Equations. *Symmetry* **2020**, *12*, 2020 [CrossRef]
13. Philos, C. On the existence of nonoscillatory solutions tending to zero at ∞ for differential equations with positive delays. *Arch. Math.* **1981**, *36*, 168–178. [CrossRef]
14. Chatzarakis, G.E.; Dzurina, J.; Jadlovsk, I. Oscillatory and asymptotic properties of third-order quasilinear delay differential equations. *J. Inequal. Appl.* **2019**, 23. [CrossRef]
15. Li, T.; Zhang, C.; Baculikova, B.; Dzurina, J. On the oscillation of third-order quasi-linear delay differential equations. *Tatra Mt. Math. Publ.* **2011**, *48*, 117–123. [CrossRef]
16. Dzurina, J.; Thapani, E.; Tamilvanan, S. Oscillation of solutions to third-order half-linear neutral differential equations. *Electron. J. Differ. Equ.* **2012**, *29*, 1–9.
17. Graef, J.; Tunc, E.; Grace, S. Oscillatory and asymptotic behavior of a third-order nonlinear neutral equation. *Opuscula Math.* **2017**, *37*, 839–852. [CrossRef]
18. Santra, S.S.; Ghosh, A.; Bazighifan, O.; Khedher, K.M.; Nofal, T.A. Second-order impulsive differential systems with mixed and several delays. *Adv. Differ. Equ.* **2021**, *1*, 1–12. [CrossRef]

19. Santra, S.S.; Baleanu, D.; Khedher, K.M.; Moaaz, O. First-order impulsive differential systems: Sufficient and necessary conditions for oscillatory or asymptotic behavior. *Adv. Differ. Equ.* **2021**, *1*, 1–20. [CrossRef]
20. Santra, S.S.; Tripathy, A.K. On oscillatory first order nonlinear neutral differential equations with nonlinear impulses. *J. Appl. Math. Comput.* **2019**, *59*, 257–270. [CrossRef]
21. Ruggieri, M.; Santra, S.S.; Scapellato, A. On nonlinear impulsive differential systems with canonical and non-canonical operators. *Appl. Anal.* **2021**. [CrossRef]
22. Moaaz, O. *Oscillation Theorems for Cartain Second-Order Differential Equations*; Lambert Academic Publishing: Saarbrücken, Germany, 2014.
23. Tunc, E. Oscillatory and asymptotic behavior of third-order neutral differential equations with distributed deviating arguments. *Electron. J. Differ. Equ.* **2017**. [CrossRef]
24. Tiryaki, A.; Aktas, M.F. Oscillation criteria of a certain class of third-order nonlinear delay differential equations with damping. *J. Math. Anal. Appl.* **2007**, *325*, 54–68. [CrossRef]
25. Baculikova, B.; Dzurina, J. Oscillation of third-order functional differential equations. *Electron. J. Qual. Theory of Diff. Equ.* **2010**, *43*, 1–10. [CrossRef]
26. Baculikova, B.; Dzurina, J. Oscillation of third-order nonlinear differential equations. *Appl. Math. Lett.* **2011**, *24*, 466–470. [CrossRef]
27. Grace, S.R.; Agarwal, R.P.; Pavani, R.; Thapani, E. On the oscillation of certain third order nonlinear functional differential equations. *Appl. Math. Comput.* **2008**, *202*, 102–112. [CrossRef]
28. Saker, S.H.; Dzurina, J. On the oscillation of certain class of third-order nonlinear delay differential equations. *Math. Bohem.* **2010**, *135*, 225–237. [CrossRef]
29. Ravi, P.; Agarwal, R.P.; Bohner, M.; Li, T.; Zhang, C. Oscillation of Third-Order Nonlinear Delay Differential Equations. *Taiwan. J. Math.* **2013**, *17*, 545–558.
30. Sidorov, N.A.; Trufanov, A.N. Nonlinear operator equations with a functional perturbation of the argument of neutral type. *Differ. Equ.* **2009**, *45*, 1840–1844. [CrossRef]
31. Zhang, S.Y.; Wang, Q. Oscillation of Second-Order Nonlinear Neutral Dynamic Equations on Time Scales. *Appl. Math. Comput.* **2010**, *216*, 2837–2848. [CrossRef]
32. Thapani, E.; Li, T. On the oscillation of third-order quasi-linear neutral functional differential equations. *Arch. Math.* **2011**, *47*, 181–199.

Article

Multi-Asset Barrier Options Pricing by Collocation BEM (with Matlab® Code)

Alessandra Aimi *,† and Chiara Guardasoni †

Department of Mathematical, Physical and Computer Sciences, Parco Area delle Scienze, 53/A, 43126 Parma, Italy; chiara.guardasoni@unipr.it
* Correspondence: alessandra.aimi@unipr.it; Tel.: +39-0521-906944
† Members of the INdAM-GNCS Research Group.

Abstract: In this paper, we extend the SABO technique (Semi-Analytical method for Barrier Options), based on collocation Boundary Element Method (BEM), to the pricing of Barrier Options with payoff dependent on more than one asset. The efficiency and accuracy already revealed in the case of a single asset is confirmed by the presented numerical results.

Keywords: boundary element method; barrier options; multi-asset options; basket options; spread options

MSC: 65M38; 91G60; 91G20; 65M80

Citation: Aimi, A.; Guardasoni, C. Multi-Asset Barrier Options Pricing by Collocation BEM (with Matlab® Code). *Axioms* **2021**, *10*, 301. https://doi.org/10.3390/axioms10040301

Academic Editors: Davron Aslonqulovich Juraev, Palle E. T. Jorgensen and Samad Noeiaghdam

Received: 29 September 2021
Accepted: 10 November 2021
Published: 12 November 2021

Publisher's Note: MDPI stays neutral with regard to jurisdictional claims in published maps and institutional affiliations.

Copyright: © 2021 by the authors. Licensee MDPI, Basel, Switzerland. This article is an open access article distributed under the terms and conditions of the Creative Commons Attribution (CC BY) license (https://creativecommons.org/licenses/by/4.0/).

1. Introduction to the Differential Model Problem

Partial Differential Equations (PDEs) of Mathematical Physics model a huge variety of real-life problems, from science to engineering. Since the work [1] of F. Black and M. Scholes, equations that model physical phenomena have been reconsidered to interpret financial phenomena. PDEs in space–time variables, modeling the price of the most evolved financial products, need efficient techniques to be numerically solved.

This work investigates the extensions of the so called SABO technique (Semi-Analytical method for Barrier Options), based on collocation Boundary Element Method (BEM) and applied so far for the numerical pricing of barrier options in a one dimensional asset framework [2].

The consideration of two dimensional partial differential problems can be suggested for several reasons: the desire to complicate the model to get closer to reality (such as, for example, introducing the dependence of option value on a stochastic volatility variable [3]) or the evaluation of options that depend not only on the asset value. In this last direction we have already approached Asian options whose payoff depends on the average of asset values [4,5] giving rise to a degenerate PDE based on two independent variables: the asset value and the average.

In this article, the extension is devoted to options whose payoff depends on more than one asset. The consequent differential model, described in the following, is set by a parabolic equation that, with suitable transformations, can be traced back to the heat equation [6].

From the computational point of view, the pricing of multi-asset options is recognized in the literature as a quite difficult issue and difficulties increase with the application of barriers [7]. The problem can be tackled starting from stochastic differential equations by Monte Carlo methods [8,9] or considering the problem in its partial differential formulation [10,11] with some limitations on the number of assets involved.

In principle our strategy can be applied to an undefined number of underlying assets, so theoretically the extension is straightforward, but the choice of approximation technique is crucial to face the *curse of dimensionality*: the application of collocation method ensures more efficiency than classical domain methods at low dimensions. This will be detailed in Section 4.

We assume the Black–Scholes–Merton scenario for the evaluation of an option $V(S_1, \ldots, S_n, t)$ based on n assets $S_1, \ldots, S_n$ during the time interval $[0, T]$, with no dividends: the behavior of the underlying assets is described by a geometric Brownian motion and $-1 \leq \rho_{ij} \leq 1$ is the correlation between the two assets i and j. The application of n-dimensional Itô Lemma and no-arbitrage arguments leads to the related backward parabolic differential model problem

$$\frac{\partial V}{\partial t} + \frac{1}{2}\sum_{i,j=1}^{n}\rho_{ij}\sigma_i\sigma_j S_i S_j \frac{\partial^2 V}{\partial S_i \partial S_j} + r\sum_{i=1}^{n} S_i \frac{\partial V}{\partial S_i} - rV = 0, \quad (S_1, \ldots, S_n, t) \in [\mathbb{R}^+]^n \times [0, T) \quad (1)$$

$$V(S_1, \ldots, S_n, T) = V_T(S_1, \ldots, S_n), \qquad (S_1, \ldots, S_n) \in [\mathbb{R}^+]^n$$

where r is the risk-free interest rate, σ_i is the volatility of underlying asset S_i and V_T is the option contract payoff at the expiry T that may depend on a strike price K and assume several expressions, for example, looking at call options in [12]:

- $\max(\sum_{i=1}^{n} c_i S_i - K, 0)$, for *Basket* options with weights c_i;
- for different kinds of *Rainbow* options,
  $\max(S_1, \ldots, S_n)$ n-color better-of option
  $\min(S_1, S_2)$ two-color worse-of option
  $\max(S_2 - S_1, 0)$ outperformance option
  $\max(\min(S_1 - K, \ldots, S_n - K), 0)$ min option
- $\max(S_1 - S_2 - K, 0)$, for *spread option*

The discussion in the following sections can be developed in the general dimension n [13]; however, for the sake of clarity and to simplify numerics, we will detail it considering two underlying assets S_1, S_2 only. Hence, the convenience and reliability of SABO will be highlighted with proofs and numerical examples for a two assets framework, and, in continuity with the previous article in this journal [4], we have inserted in Appendix A a ready to use Matlab® code.

2. Integral Representation Formula of the Solution for Options without Barriers

By the Green's theorem, we prove the integral representation formula for the solution of the differential model problem (1) describing the value of an option $V(S_1, S_2, t)$ based on two assets S_1, S_2 during the time interval $[0, T]$.

We recall the notion of joint transition probability density function $G(S_1, S_2, t; \tilde{S}_1, \tilde{S}_2, \tilde{t})$ (also known as Green's or fundamental solution) from the classical theory of PDEs [13]: denoting the space differential operator by

$$\mathcal{L}[V] := \frac{\sigma_1^2 S_1^2}{2}\frac{\partial^2 V}{\partial S_1^2} + rS_1\frac{\partial V}{\partial S_1} + \frac{\sigma_2^2 S_2^2}{2}\frac{\partial^2 V}{\partial S_2^2} + rS_2\frac{\partial V}{\partial S_2} + \rho\sigma_1\sigma_2 S_1 S_2 \frac{\partial^2 V}{\partial S_1 \partial S_2} - rV, \quad (2)$$

G is solution of (1), i.e.,

$$\frac{\partial G}{\partial t}(S_1, S_2, t) + \mathcal{L}[G](S_1, S_2, t) = 0 \quad (3)$$

w.r.t. the first tern of variables and of the adjoint model problem (4) and (5) w.r.t. the second tern of variables:

$$-\frac{\partial G}{\partial \tilde{t}}(\tilde{S}_1, \tilde{S}_2, \tilde{t}) + \mathcal{L}^*[G](\tilde{S}_1, \tilde{S}_2, \tilde{t}) = 0 \quad (\tilde{S}_1, \tilde{S}_2, \tilde{t}) \in \mathbb{R}^+ \times \mathbb{R}^+ \times (t, +\infty) \quad (4)$$

$$G(S_1, S_2, t; \tilde{S}_1, \tilde{S}_2, \tilde{t}) = \delta(S_1, \tilde{S}_1)\delta(S_2, \tilde{S}_2) \qquad (\tilde{S}_1, \tilde{S}_2) \in \mathbb{R}^+ \times \mathbb{R}^+; \quad (5)$$

moreover it is such that

$$(\mathcal{L}[V], G) = (V, \mathcal{L}^*[G]) \quad (6)$$

having set $(\cdot,\cdot)$ the standard L^2 scalar product. The adjoint operator $\mathcal{L}^*$ is defined as

$$\mathcal{L}^*[G] := \frac{\sigma_1^2 \tilde{S}_1^2}{2} \frac{\partial^2 G}{\partial \tilde{S}_1^2} + \frac{\partial G}{\partial \tilde{S}_1}(-r\tilde{S}_1 + 2\sigma_1^2 \tilde{S}_1 + \rho\sigma_1\sigma_2\tilde{S}_1)$$
$$+ \frac{\sigma_2^2 \tilde{S}_2^2}{2} \frac{\partial^2 G}{\partial \tilde{S}_2^2} + \frac{\partial G}{\partial \tilde{S}_2}(-r\tilde{S}_2 + 2\sigma_2^2 \tilde{S}_2 + \rho\sigma_1\sigma_2\tilde{S}_2) \quad (7)$$
$$+ \rho\sigma_1\sigma_2 \tilde{S}_1 \tilde{S}_2 \frac{\partial^2 G}{\partial \tilde{S}_1 \partial \tilde{S}_2} + G(\rho\sigma_1\sigma_2 - 3r + \sigma_1^2 + \sigma_2^2).$$

From [14] the fundamental solution is known to be

$$G(S_1, S_2, t; \tilde{S}_1, \tilde{S}_2, \tilde{t}) = \frac{e^{-r(\tilde{t}-t)}}{2\pi(\tilde{t}-t)} \frac{\exp\left(-\frac{\alpha'\Sigma^{-1}\alpha}{2}\right)}{\sigma_1\sigma_2\tilde{S}_1\tilde{S}_2\sqrt{\det\Sigma}} \quad (8)$$

with

$$\alpha_i = \frac{\log\frac{S_i}{\tilde{S}_i} + \left(r - \frac{\sigma_i^2}{2}\right)(\tilde{t}-t)}{\sigma_i\sqrt{\tilde{t}-t}}, \quad i = 1,2$$

the elements of the array α and

$$\Sigma = \begin{pmatrix} 1 & \rho_{12} \\ \rho_{12} & 1 \end{pmatrix}.$$

the correlation matrix.

Proposition 1. *The fundamental solution (8) of (3)–(5) verifies, $\forall \tilde{t}, t \in [0,T]$, the following identities*

$$\mathcal{I}_1 := \int_{\mathbb{R}^+ \times \mathbb{R}^+} G(S_1, S_2, t; \tilde{S}_1, \tilde{S}_2, \tilde{t}) d\tilde{S}_1 d\tilde{S}_2 = e^{-r(\tilde{t}-t)} \quad (9)$$

$$\mathcal{I}_2 := \int_{\mathbb{R}^+ \times \mathbb{R}^+} G(S_1, S_2, t; \tilde{S}_1, \tilde{S}_2, \tilde{t}) dS_1 dS_2 = e^{-(\tilde{t}-t)(3r - \sigma_1\sigma_2\rho_{12} - \sigma_1^2 - \sigma_2^2)} \quad (10)$$

Proof. From (8),

$$G(S_1, S_2, t; \tilde{S}_1, \tilde{S}_2, \tilde{t}) = \frac{e^{-r(\tilde{t}-t)}}{2\pi(\tilde{t}-t)} \frac{\exp\left(-\frac{\alpha_1^2 + \alpha_2^2 - 2\alpha_1\alpha_2\rho_{12}}{2(1-\rho_{12}^2)}\right)}{\sigma_1\sigma_2\tilde{S}_1\tilde{S}_2\sqrt{1-\rho_{12}^2}}$$

hence

$$\mathcal{I}_1 = \frac{e^{-r(\tilde{t}-t)}}{2\pi(\tilde{t}-t)\sigma_1\sigma_2\sqrt{1-\rho_{12}^2}} \int_{\mathbb{R}^+} \frac{\exp\left(-\frac{\alpha_1^2}{2(1-\rho_{12}^2)}\right)}{\tilde{S}_1} d\tilde{S}_1 \int_{\mathbb{R}^+} \frac{\exp\left(-\frac{\alpha_2^2 - 2\alpha_1\alpha_2\rho_{12}}{2(1-\rho_{12}^2)}\right)}{\tilde{S}_2} d\tilde{S}_2$$

by the changes of variables $\xi_i := \log(\tilde{S}_i)$ and exploiting the property $\int_0^{+\infty} e^{-\xi^2} d\xi = \sqrt{\pi}/2$, we obtain

$$= \frac{e^{-r(\tilde{t}-t)}}{2\pi(\tilde{t}-t)\sigma_1\sigma_2\sqrt{1-\rho_{12}^2}} \int_{\mathbb{R}^+} e^{-\frac{\alpha_1^2}{2}} \sqrt{2\pi\sigma_2^2(\tilde{t}-t)(1-\rho_{12}^2)} d\xi_1$$

$$= \frac{e^{-r(\tilde{t}-t)}}{\sqrt{2\pi(\tilde{t}-t)}\sigma_1} \int_{\mathbb{R}^+} e^{-\frac{\alpha_1^2}{2}} d\xi_1 = e^{-r(\tilde{t}-t)}.$$

Analogously

$$
\begin{aligned}
\mathcal{I}_2 &= \frac{e^{-r(\tilde{t}-t)}}{2\pi(\tilde{t}-t)\sigma_1\sigma_2\tilde{S}_1\tilde{S}_2\sqrt{1-\rho_{12}^2}} \int_{\mathbb{R}^+} e^{-\frac{a_1^2}{2(1-\rho_{12}^2)}} dS_1 \int_{\mathbb{R}^+} e^{-\frac{a_2^2-2a_1a_2\rho_{12}}{2(1-\rho_{12}^2)}} dS_2 \\
&= \frac{e^{(\tilde{t}-t)(-2r+\sigma_2^2-\frac{\rho_{12}^2\sigma_2^2}{2})}}{\sqrt{2\pi(\tilde{t}-t)}\sigma_1\tilde{S}_1} \int_{\mathbb{R}^+} e^{-\frac{a_1^2}{2}+a_1\rho_{12}(\tilde{t}-t)\sigma_2} dS_1 = e^{(\tilde{t}-t)(-3r+\sigma_2^2+\sigma_1^2+\sigma_1\sigma_2\rho_{12})}
\end{aligned}
$$

□

Remark 1. *Note that $\mathcal{I}_1$ and $\mathcal{I}_2$ turn out to coincide with the coefficients of the changes of variables $\tilde{G} = e^{r(\tilde{t}-t)}G$ and $\tilde{G} = e^{(3r-\sigma_1\sigma_2\rho_{12}-\sigma_1^2-\sigma_2^2)(\tilde{t}-t)}G$ allowing to obtain (3) and (4), respectively, without the last term.*

With the knowledge of the fundamental solution, we will derive, by mathematical analysis tools, the following integral representation formula:

Proposition 2. *The Feynman–Kac formula*

$$V(S_1, S_2, t) = \int_{\mathbb{R}^+ \times \mathbb{R}^+} V(\tilde{S}_1, \tilde{S}_2, T) G(S_1, S_2, t; \tilde{S}_1, \tilde{S}_2, T) d\tilde{S}_1 d\tilde{S}_2 \tag{11}$$

giving the option value as expected value of the payoff function, holds.

Proof. Multiply (1) by G and integrate in time-space domain

$$
\begin{aligned}
0 &= \int_t^T \int_{\mathbb{R}^+ \times \mathbb{R}^+} \frac{\partial V}{\partial \tilde{t}}(\tilde{S}_1, \tilde{S}_2, \tilde{t}) G(S_1, S_2, t; \tilde{S}_1, \tilde{S}_2, \tilde{t}) d\tilde{t} d\tilde{S}_1 d\tilde{S}_2 \\
&+ \int_t^T \int_{\mathbb{R}^+ \times \mathbb{R}^+} \mathcal{L}[V](\tilde{S}_1, \tilde{S}_2, \tilde{t}) G(S_1, S_2, t; \tilde{S}_1, \tilde{S}_2, \tilde{t}) d\tilde{t} d\tilde{S}_1 d\tilde{S}_2
\end{aligned}
$$

then, integrating by parts in time and in space and applying (6), one obtains

$$
\begin{aligned}
&= \int_{\mathbb{R}^+ \times \mathbb{R}^+} V(\tilde{S}_1, \tilde{S}_2, T) G(S_1, S_2, t; \tilde{S}_1, \tilde{S}_2, T) d\tilde{S}_1 d\tilde{S}_2 \\
&- \int_{\mathbb{R}^+ \times \mathbb{R}^+} V(\tilde{S}_1, \tilde{S}_2, t) G(S_1, S_2, t; \tilde{S}_1, \tilde{S}_2, t) d\tilde{S}_1 d\tilde{S}_2 \\
&- \int_t^T \int_{\mathbb{R}^+ \times \mathbb{R}^+} V(\tilde{S}_1, \tilde{S}_2, \tilde{t}) \frac{\partial G}{\partial \tilde{t}}(S_1, S_2, t; \tilde{S}_1, \tilde{S}_2, \tilde{t}) d\tilde{t} d\tilde{S}_1 d\tilde{S}_2 \\
&+ \int_t^T \int_{\mathbb{R}^+ \times \mathbb{R}^+} V(\tilde{S}_1, \tilde{S}_2, \tilde{t}) \mathcal{L}^*[G](S_1, S_2, t; \tilde{S}_1, \tilde{S}_2, \tilde{t}) d\tilde{t} d\tilde{S}_1 d\tilde{S}_2 \\
&= \int_{\mathbb{R}^+ \times \mathbb{R}^+} V(\tilde{S}_1, \tilde{S}_2, T) G(S_1, S_2, t; \tilde{S}_1, \tilde{S}_2, T) d\tilde{S}_1 d\tilde{S}_2 \\
&- \int_{\mathbb{R}^+ \times \mathbb{R}^+} V(\tilde{S}_1, \tilde{S}_2, t) \delta(S_1, \tilde{S}_1) \delta(S_2, \tilde{S}_2) d\tilde{S}_1 d\tilde{S}_2 \\
&- \int_t^T \int_{\mathbb{R}^+ \times \mathbb{R}^+} V(\tilde{S}_1, \tilde{S}_2, \tilde{t}) \left(\frac{\partial G}{\partial \tilde{t}} - \mathcal{L}^*[G]\right)(S_1, S_2, t; \tilde{S}_1, \tilde{S}_2, \tilde{t}) d\tilde{t} d\tilde{S}_1 d\tilde{S}_2
\end{aligned}
$$

and then, thanks to property (6), these steps lead to the Feynman-Kac formula. □

3. Integral Representation Formula of the Solution for Barrier Options

If we consider barrier options, sometimes analytical solutions are known (some examples are collected in [15]) but sometimes not. For example, in [12], the author considers the case of a European-style two-assets-basket double-barrier call option with payoff

$$V(S_1, S_2, T) = \max(S_1 + S_2 - K, 0) \tag{12}$$

and two knock-out barrier conditions

$$V(S_1, S_2, T) = 0 \quad S_1 + S_2 \leq B_1 \tag{13}$$
$$V(S_1, S_2, T) = 0 \quad S_1 + S_2 \geq B_2 \tag{14}$$

with down-and-out barrier B_1 and up-and-out barrier B_2. Then, the resulting domain Ω is partitioned as shown in Figure 1 in order to approximate the option value by a Finite Element Method.

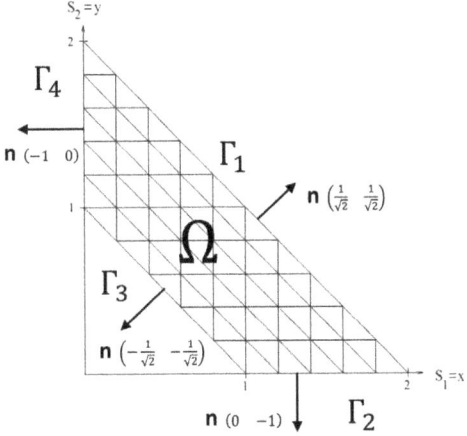

Figure 1. Triangulation of domain Ω in [12].

However, if we are interested in the computation of the option value only at a desired point (S_1, S_2) of the domain, it may certainly be convenient to apply the strategy already tested for various kind of options (based on one asset only) under several dynamics [2–5,16,17] and that we called SABO.

The method is based on a new analytical representation formula whose proof relies on Green's Theorem and retraces the proof of Feynman–Kac formula.

Proposition 3. *The solution of problem (1), (13) and (14), $\forall (S_1, S_2, t) \in \Omega \subseteq \mathbb{R}^2 \times [0, T)$, can be expressed by the following integral representation formula:*

$$V(S_1, S_2, t) = \int_\Omega V(\tilde{S}_1, \tilde{S}_2, T) G(S_1, S_2, t; \tilde{S}_1, \tilde{S}_2, T) d\tilde{S}_1 d\tilde{S}_2 \tag{15}$$
$$+ \int_t^T \int_{\partial \Omega} \varphi(\tilde{S}_1, \tilde{S}_2, \tilde{t}) \, d\tilde{t} d\tilde{S}_1 d\tilde{S}_2$$

with the functional φ defined in (17) below.

Proof. Multiply (1) by G and integrate in time and in the domain Ω

$$0 = \int_t^T \int_\Omega \frac{\partial V}{\partial \tilde{t}}(\tilde{S}_1, \tilde{S}_2, \tilde{t}) G(S_1, S_2, t; \tilde{S}_1, \tilde{S}_2, \tilde{t}) d\tilde{t} d\tilde{S}_1 d\tilde{S}_2$$
$$+ \int_t^T \int_\Omega \mathcal{L}[V](\tilde{S}_1, \tilde{S}_2, \tilde{t}) G(S_1, S_2, t; \tilde{S}_1, \tilde{S}_2, \tilde{t}) d\tilde{t} d\tilde{S}_1 d\tilde{S}_2$$

then, integrating by parts in time and applying Green's Theorem and (6), one obtains

$$= \int_\Omega V(\tilde{S}_1, \tilde{S}_2, T) G(S_1, S_2, t; \tilde{S}_1, \tilde{S}_2, T) d\tilde{S}_1 d\tilde{S}_2$$
$$- \int_\Omega V(\tilde{S}_1, \tilde{S}_2, t) G(S_1, S_2, t; \tilde{S}_1, \tilde{S}_2, t) d\tilde{S}_1 d\tilde{S}_2$$
$$- \int_t^T \int_\Omega V(\tilde{S}_1, \tilde{S}_2, \tilde{t}) \frac{\partial G}{\partial \tilde{t}}(S_1, S_2, t; \tilde{S}_1, \tilde{S}_2, \tilde{t}) d\tilde{t} d\tilde{S}_1 d\tilde{S}_2$$
$$+ \int_t^T \int_\Omega V(\tilde{S}_1, \tilde{S}_2, \tilde{t}) \mathcal{L}^*[G](S_1, S_2, t; \tilde{S}_1, \tilde{S}_2, \tilde{t}) d\tilde{t} d\tilde{S}_1 d\tilde{S}_2$$
$$+ \int_t^T \int_{\partial\Omega} \mathbf{p}(S_1, S_2, t; \tilde{S}_1, \tilde{S}_2, \tilde{t}) \cdot \mathbf{n} \, d\tilde{t} d\tilde{S}_1 d\tilde{S}_2$$

having defined $\partial\Omega$ as the boundary of the domain, $\mathbf{n}$ as its unit normal vector outwardly directed and $\mathbf{p} = (p_1, p_2)$ as

$$\begin{aligned} p_1 &= \frac{1}{2}\sigma_1^2 \tilde{S}_1^2 \left(G\frac{\partial V}{\partial \tilde{S}_1} - V\frac{\partial G}{\partial \tilde{S}_1}\right) + \frac{1}{2}\rho\sigma_1\sigma_2 \tilde{S}_1 \tilde{S}_2 \left(G\frac{\partial V}{\partial \tilde{S}_2} - V\frac{\partial G}{\partial \tilde{S}_2}\right) \\ &\quad - VG\left(\sigma_1^2 \tilde{S}_1 + \frac{1}{2}\rho\sigma_1\sigma_2 \tilde{S}_1 - r\tilde{S}_1\right) \\ p_2 &= \frac{1}{2}\sigma_2^2 \tilde{S}_2^2 \left(G\frac{\partial V}{\partial \tilde{S}_2} - V\frac{\partial G}{\partial \tilde{S}_2}\right) + \frac{1}{2}\rho\sigma_1\sigma_2 \tilde{S}_1 \tilde{S}_2 \left(G\frac{\partial V}{\partial \tilde{S}_1} - V\frac{\partial G}{\partial \tilde{S}_1}\right) \\ &\quad - VG\left(\sigma_2^2 \tilde{S}_2 + \frac{1}{2}\rho\sigma_1\sigma_2 \tilde{S}_2 - r\tilde{S}_2\right) \end{aligned} \quad (16)$$

Thus we can conclude that

$$0 = \int_\Omega V(\tilde{S}_1, \tilde{S}_2, T) G(S_1, S_2, t; \tilde{S}_1, \tilde{S}_2, T) d\tilde{S}_1 d\tilde{S}_2$$
$$- \int_\Omega V(\tilde{S}_1, \tilde{S}_2, t) \delta(S_1, \tilde{S}_1) \delta(S_2, \tilde{S}_2) d\tilde{S}_1 d\tilde{S}_2$$
$$- \int_t^T \int_\Omega V(\tilde{S}_1, \tilde{S}_2, \tilde{t}) \left(\frac{\partial G}{\partial \tilde{t}} - \mathcal{L}^*[G]\right)(S_1, S_2, t; \tilde{S}_1, \tilde{S}_2, \tilde{t}) d\tilde{t} d\tilde{S}_1 d\tilde{S}_2$$
$$+ \int_t^T \int_{\partial\Omega} \varphi(S_1, S_2, t; \tilde{S}_1, \tilde{S}_2, \tilde{t}) \, d\tilde{t} d\tilde{S}_1 d\tilde{S}_2$$

with

$$\varphi(S_1, S_2, t; \tilde{S}_1, \tilde{S}_2, \tilde{t}) = \mathbf{p}(S_1, S_2, t; \tilde{S}_1, \tilde{S}_2, \tilde{t}) \cdot \mathbf{n} \quad (17)$$

□

Remark 2. *The representation formula, and SABO in general, can be naturally adapted to barrier options configurations different from the proposed example. Note also that it is not really important to know exactly the expression of the functional φ because, as it depends on the unknown solution V, it is unknown in its turn and therefore it will be numerically determined.*

Remark 3. *The representation formula can be analytically derived in time or in space to obtain Greeks functions that can be straightforwardly evaluated by SABO without the need of evaluating option values (look at [2]).*

In the example (12)–(14), due to the configuration of the boundary $\partial\Omega = \Gamma_1 \cup \Gamma_2 \cup \Gamma_3 \cup \Gamma_4$ (see Figure 1), the integral of φ reduces to

$$\int_t^T \int_{\partial\Omega} \varphi(S_1, S_2, t; \tilde{S}_1, \tilde{S}_2, \tilde{t}) \, d\tilde{t} d\tilde{S}_1 d\tilde{S}_2 = \int_t^T \int_{\Gamma_1 \cup \Gamma_3} \varphi(S_1, S_2, t; \tilde{S}_1, \tilde{S}_2, \tilde{t}) \, d\tilde{t} d\tilde{S}_1 d\tilde{S}_2$$

because on $\Gamma_2 := \{(S_1, 0) : B_1 \leq S_1 \leq B_2\}$ the first component of the normal vector is null and every term of p_2 has the factor S_2, trivial on Γ_2; on $\Gamma_4 := \{(0, S_2) : B_1 \leq S_2 \leq B_2\}$ the second component of the normal vector is null and every term of p_1 has the factor S_1, trivial on Γ_4.

Moreover, considering double knock-out barriers on $\Gamma_1 := \{(S_1, S_2) : S_1 + S_2 = B_2\}$ and on $\Gamma_3 := \{(S_1, S_2) : S_1 + S_2 = B_1\}$, the option value V is equal to 0, justifying the representation formula: $\forall (S_1, S_2, t) \in \Omega \times [0, T)$

$$\begin{aligned} V(S_1, S_2, t) &= \int_\Omega V(\tilde{S}_1, \tilde{S}_2, T) G(S_1, S_2, t; \tilde{S}_1, \tilde{S}_2, T) d\tilde{S}_1 d\tilde{S}_2 \\ &+ \int_t^T \int_{\Gamma_1 \cup \Gamma_3} G(S_1, S_2, t; \tilde{S}_1, \tilde{S}_2, \tilde{t}) \phi(\tilde{S}_1, \tilde{S}_2, \tilde{t}) \, d\tilde{t} d\tilde{S}_1 d\tilde{S}_2 \end{aligned} \quad (18)$$

with ϕ depending only on $(\tilde{S}_1, \tilde{S}_2, \tilde{t}) \in \Gamma_1 \cup \Gamma_3 \times [0, T)$, i.e.,

$$\phi(\tilde{S}_1, \tilde{S}_2, \tilde{t}) = \left(\frac{1}{2} \sigma_1^2 \tilde{S}_1^2 \frac{\partial V}{\partial \tilde{S}_1} + \frac{1}{2} \rho \sigma_1 \sigma_2 \tilde{S}_1 \tilde{S}_2 \frac{\partial V}{\partial \tilde{S}_2} \right) n_1 + \left(\frac{1}{2} \sigma_2^2 \tilde{S}_2^2 \frac{\partial V}{\partial \tilde{S}_2} + \frac{1}{2} \rho \sigma_1 \sigma_2 \tilde{S}_1 \tilde{S}_2 \frac{\partial V}{\partial \tilde{S}_1} \right) n_2.$$

4. Boundary Integral Equation

The representation formula (18) of V holds in the whole domain and can be used once the unknown density ϕ is recovered. To this aim, the idea is to consider the application of the representation formula at the barriers, where $\forall (S_1, S_2, t) \in \Gamma_1 \cup \Gamma_3 \times [0, T)$ the option value is trivial, i.e., $V(S_1, S_2, t) = 0$ and therefore

$$\int_t^T \int_{\Gamma_1 \cup \Gamma_3} G(S_1, S_2, t; \tilde{S}_1, \tilde{S}_2, \tilde{t}) \phi(\tilde{S}_1, \tilde{S}_2, \tilde{t}) \, d\tilde{t} d\tilde{S}_1 d\tilde{S}_2 = \quad (19)$$
$$- \int_\Omega V(\tilde{S}_1, \tilde{S}_2, T) G(S_1, S_2, t; \tilde{S}_1, \tilde{S}_2, T) d\tilde{S}_1 d\tilde{S}_2.$$

This Boundary Integral Equation (BIE) in the unknown ϕ is a Volterra integral equation of first kind and can be numerically solved as described in the following section.

Remark 4. *This strategy can be extended to higher dimensions, denoting by Γ the hyperplane defined by the barrier conditions. The representation formula still holds: $\forall (S_1, \ldots, S_n, t) \in \Omega \times [0, T)$*

$$\begin{aligned} V(S_1, \ldots, S_n, t) &= \int_\Omega V(\tilde{S}_1, \ldots, \tilde{S}_n, T) G(S_1, \ldots, S_n, t; \tilde{S}_1, \ldots, \tilde{S}_n, T) d\tilde{S}_1 \ldots d\tilde{S}_n \\ &+ \int_t^T \int_\Gamma G(S_1, \ldots, S_n, t; \tilde{S}_1, \ldots, \tilde{S}_n, \tilde{t}) \phi(\tilde{S}_1, \ldots, \tilde{S}_n, \tilde{t}) \, d\tilde{t} d\tilde{S}_1 \ldots d\tilde{S}_n \end{aligned} \quad (20)$$

together with the BIE with knock-out condition: $\forall (S_1, S_2, t) \in \Gamma \times [0, T)$

$$\int_t^T \int_\Gamma G(S_1, \ldots, S_n, t; \tilde{S}_1, \ldots, \tilde{S}_n, \tilde{t}) \phi(\tilde{S}_1, \ldots, \tilde{S}_n, \tilde{t}) \, d\tilde{t} d\tilde{S}_1, \ldots, d\tilde{S}_n = \quad (21)$$
$$- \int_\Omega V(\tilde{S}_1, \ldots, \tilde{S}_n, T) G(S_1, \ldots, S_n, t; \tilde{S}_1, \ldots, \tilde{S}_n, T) d\tilde{S}_1 d\tilde{S}_2$$

and with the general expression of fundamental solution G written in [14].

Remark 5. *If the asset domain is either a 2 or 3 or 4 dimensional domain, to solve (21) means to reduce the dimensionality of the problem by one. In this case, the collocation method can be implemented, as proposed in Section 5 (introducing suitable mesh triangulations), since finite* element discretization can be easily applied up to three dimensions with presumably greater accuracy w.r.t. Monte Carlo simulations. However, of course, the collocation method suffers the curse of dimensionality. Therefore, first of all, SABO in this framework, is suggested as an alternative to deterministic methods (finite difference methods and finite element methods) for its higher efficiency.

With the increasing of the dimension $n > 4$, (21) is still valid and very general, not requiring special conditions as for example in [18], so the strategy of solving (21) can be anyway taken into account. An idea could be to involve Monte Carlo method, not as a method for path simulation [9], but considered as a quadrature method [19] directly applied to the integral terms in (21). Then unknown ϕ might be represented by linear combination of radial basis functions around some nodes spread out in the $n-1$-dimensional hyperplane Γ. This will be the subject of future investigation.

5. Numerical Approximation by Collocation BEM

We introduce

$$\tilde{\phi}(S_1, S_2, t) := \sum_{n=1}^{N_{\Delta t}} \sum_{m=1}^{M_{\Delta S}} \alpha_{nm} \bar{\phi}_m(S_1, S_2) \hat{\phi}_n(t) \tag{22}$$

for the numerical approximation of ϕ by:

- piecewise constant functions in time

$$\hat{\phi}_n(t) := H[t - t_{n-1}] - H[t - t_n] \qquad n = 1, \cdots, N_{\Delta t}$$

defined by the Heaviside functions $H[\cdot]$ on a uniform decomposition of the time interval $[0, T]$:

$$\Delta t = \frac{T}{N_{\Delta t}}, \qquad N_{\Delta t} \in \mathbb{N}^+ \qquad t_n = n N_{\Delta t}, \qquad n = 0, \cdots, N_{\Delta t},$$

- piecewise constant functions in space $\bar{\phi}_m(S) = \bar{\phi}_m(S_1, S_2)$, $m = 1, \ldots, \bar{M}_{\Delta S}, \bar{M}_{\Delta S} + 1, \ldots, M_{\Delta S}$ defined on a decomposition $\mathcal{T}_1 = \{e_1, \ldots, e_{\bar{M}_{\Delta S}}\}$ on Γ_1 and $\mathcal{T}_2 = \{e_{\bar{M}_{\Delta S}+1}, \ldots, e_{M_{\Delta S}}\}$ on Γ_2 constituted by $M_{\Delta S}$ segments such that $\text{length}(e_i) \leq \Delta S$ with $e_i \cap e_j = \emptyset$ if $i \neq j$.

The related unknown array

$$\boldsymbol{\alpha} = \left(\boldsymbol{\alpha}^{(1)} \boldsymbol{\alpha}^{(1)} \boldsymbol{\alpha}^{(2)} \ldots \boldsymbol{\alpha}^{(N_{\Delta t})}\right)^\top \qquad \boldsymbol{\alpha}^{(\ell)} = \left(\alpha_{\ell 1}, \cdots, \alpha_{\ell M_{\Delta S}}\right)^\top, \quad \ell = 1, \ldots, N_{\Delta t}$$

can be computed by solving a linear system

$$\mathbb{M} \boldsymbol{\alpha} = \boldsymbol{\beta}, \tag{23}$$

of $N_{\Delta t} \times M_{\Delta S}$ equations obtained by collocating the BIE (19) at the midpoints of time decomposition interval

$$\bar{t}_n = \frac{t_{n-1} + t_n}{2}, \quad n = 1, \ldots, N_{\Delta t} \tag{24}$$

and at the midpoints of space decomposition segments

$$\bar{S}_m = (\bar{S}_1^{(m)}, \bar{S}_2^{(m)}), \quad m = 1, \ldots, M_{\Delta S}. \tag{25}$$

Note that the matrix $\mathbb{M}$ has a block upper triangular Toeplitz structure

$$\begin{pmatrix} \mathbb{M}^{(1)} & \mathbb{M}^{(2)} & \mathbb{M}^{(3)} & \ldots & \mathbb{M}^{(N_{\Delta t})} \\ 0 & \mathbb{M}^{(1)} & \mathbb{M}^{(2)} & \ldots & \mathbb{M}^{(N_{\Delta t}-1)} \\ 0 & 0 & \mathbb{M}^{(1)} & \ldots & \mathbb{M}^{(N_{\Delta t}-2)} \\ \vdots & \vdots & \vdots & \ddots & \vdots \\ 0 & 0 & 0 & \ldots & \mathbb{M}^{(1)} \end{pmatrix} \tag{26}$$

and this is due to the form of the fundamental solution (8): the starting PDE Equation (1) has coefficients independent of time implying that the fundamental solution depends on time only through the difference $\tilde{t} - t$ and, by consequence, each block depends only on $t_n - \tilde{t}_n$. From the computational point of view, this means notable computational and memory savings because only the last column of blocks needs to be computed and the linear system (23) can be solved by block-backward substitution

$$\begin{cases} \mathbf{z}^{(\ell)} = \boldsymbol{\beta}^{(\ell)} - \sum_{j=2}^{N_{\Delta t}-\ell+1} \mathbb{M}^{(j)} \boldsymbol{\alpha}^{(\ell+j-1)} \\ \mathbb{M}^{(1)} \boldsymbol{\alpha}^{(\ell)} = \mathbf{z}^{(\ell)}. \end{cases} \quad \ell = N_{\Delta t}, \cdots, 1$$

where only the non singular block $\mathbb{M}^{(1)}$ needs to be inverted.

We can get the expression of a general entry of the matrix block $\mathbb{M}^{(\ell)}$, $\ell = 1, \ldots, N_{\Delta t}$, introducing, in the lhs of (19), a space-time piecewise constant approximation (22) and collocation points (24) and (25): $\forall \tilde{m}, m = 1 \ldots, M_{\Delta S}$, $\forall \tilde{n} = 1 \ldots, N_{\Delta t}, \forall n = \tilde{n} \ldots, N_{\Delta t}$

$$\mathbb{M}^{(\ell)}_{\tilde{m}m} = \mathbb{M}^{\tilde{n}n}_{\tilde{m}m} = \mathbb{M}^{(n-\tilde{n}+1)}_{\tilde{m}m} = \int_{\tilde{t}_{\tilde{n}}}^{T} \int_{e_m} G(\tilde{S}_1^{(\tilde{m})}, \tilde{S}_2^{(\tilde{m})}, \tilde{t}_{\tilde{n}}; \tilde{S}_1, \tilde{S}_2, \tilde{t}) \hat{\phi}_m(\tilde{S}_1, \tilde{S}_2) \hat{\phi}_n(\tilde{t}) \, d\tilde{t} \, d\tilde{S}_1 d\tilde{S}_2 \qquad (27)$$
$$= \int_{\max(t_{n-1}, \tilde{t}_{\tilde{n}})}^{t_n} \int_{e_m} G(\tilde{S}_1^{(\tilde{m})}, \tilde{S}_2^{(\tilde{m})}, \tilde{t}_{\tilde{n}}; \tilde{S}_1, \tilde{S}_2, \tilde{t}) \, d\tilde{t} \, d\tilde{S}_1 d\tilde{S}_2$$

and, collocating the rhs of (19) in the same way, we can write

$$\beta^{\ell}_{\tilde{m}} = \beta^{\tilde{n}}_{\tilde{m}} = -\int_{\Omega} V(\tilde{S}_1, \tilde{S}_2, T) G(\tilde{S}_1^{(\tilde{m})}, \tilde{S}_2^{(\tilde{m})}, \tilde{t}_{\tilde{n}}; \tilde{S}_1, \tilde{S}_2, T) d\tilde{S}. \qquad (28)$$

Once the array $\boldsymbol{\alpha}$ of coefficients in (22) are computed by solving the linear system (23), the unknown functional ϕ in (18) can be replaced by its approximation $\tilde{\phi}$, in order to get the approximation of the option price, at the desired time instant and assets values, without the need of introducing grids or triangulation over the domain Ω as in Figure 1.

Note that the evaluation of system entries can be simply performed by Matlab® "integral2" function as it presents only little troubles: the degeneration of the fundamental solution in time towards the Dirac delta distribution when $t \to \tilde{t}$ (look at (5)) and possible non-smoothness of the payoff function like, for example, in the case of Basket options.

6. Numerical Results and Discussion

- Consider the problem (1), (12), (13) and (14) of evaluating a double knock-out basket call option based on two assets

$$V(S_1, S_2, T) = \max(S_1 + S_2 - K, 0)$$

with data suggested in [12] and listed in Table 1. The domain is as depicted in Figure 1. Numerical results have been obtained by the code inserted in the Appendix A.

Table 1. Double knock-out basket call option data.

K	T	r	σ_1	σ_2	ρ	B_1	B_2
1	1	0.05	0.25	0.25	0.7	1	2

The approximated solution, represented in Figure 2, is evaluated over a rectangular grid of points, vertices of 16^2 simplices that subdivide the square $[0,2]^2$. It is obtained by SABO with $\Delta S = 0.125\sqrt{2}$ and $\Delta t = 0.1$. The behavior of the solution is in good agreement with that found in [12] ("Courtesy of A. Kvetnaia"), there approximated by Finite Element Method (FEM) after having transformed Equation (1) in a divergence-free form, not necessary or useful to apply SABO.

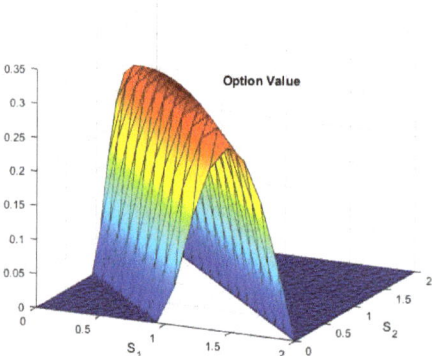

Figure 2. SABO approximation of a double knock-out basket call option value $V(S_1, S_2, 0)$ with data suggested in Table 1.

The greater accuracy, implying greater velocity, of SABO w.r.t. Finite Differences and Monte Carlo Methods has been already highlighted in our papers about barrier options, see for example [3,20]. Here, in Table 2, we compare the CPU time of computation (Laptop computer: CPU Intel i5, 4 Gb RAM.) of SABO w.r.t. FEM as applied in [12], i.e., linear finite elements in space coupled with Crank–Nicolson scheme in time: the chosen reference value is $V(0.5, 1, 0) = 0.306264$, obtained by stressing the discretization parameters in our SABO code ($M_{\Delta S} = N_{\Delta t} = 30$, tolerance on integrals computation equal to 10^{-12}) and both methods converge towards it. The starting space mesh for FEM, that will be denoted by R_0 from now on, is made by 48 uniform triangles as depicted in Figure 3, then any successive refinement, R_i $i = 1, 2, 3$, is obtained by dividing each triangle in four uniform triangles (by connecting the midpoints of the edges). Despite not having implemented a parallel code for the computation of independent blocks in (26), SABO allows to achieve the same accuracy of FEM with a computation time of at least one lower order of magnitude.

Table 2. Comparison between FEM and SABO: option value obtained at $S_1 = 0.5, S_2 = 1, t = 0$ (upper tables) and CPU time of computation in seconds (lower tables). Reference value: 0.306264.

		SABO					FEM		
$M_{\Delta S}$	$N_{\Delta t}$	2	4	8	16	$N_{\Delta t}$	25	50	100
2		0.304101	0.305635	0.305527	0.305501	R_0	0.293627	0.293663	0.293662
4		0.305242	0.306735	0.306631	0.306586	R_1	0.304109	0.304129	0.304134
8		0.304903	0.306424	0.306320	0.306274	R_2	0.305875	0.305891	0.305894
16		0.304916	0.306429	0.306329	0.306282	R_3	0.306171	0.306185	0.306188
						R_4	0.306229	0.306243	0.306246

		CPU time SABO					CPU time FEM		
$M_{\Delta S}$	$N_{\Delta t}$	2	4	8	16	$N_{\Delta t}$	25	50	100
2		2.1×10^{-1}	5.8×10^{-1}	8.6×10^{-1}	1.1×10^{0}	R_0	7.2×10^{-1}	1.4×10^{0}	3.4×10^{0}
4		5.1×10^{-1}	7.8×10^{-1}	1.2×10^{0}	2.1×10^{0}	R_1	4.3×10^{0}	8.3×10^{0}	1.5×10^{1}
8		9.2×10^{-1}	1.5×10^{0}	2.5×10^{0}	4.5×10^{0}	R_2	8.1×10^{0}	1.3×10^{1}	2.2×10^{1}
16		2.1×10^{0}	4.1×10^{0}	6.8×10^{0}	1.3×10^{1}	R_3	3.0×10^{1}	3.6×10^{1}	4.8×10^{1}
						R_4	6.4×10^{2}	6.7×10^{2}	7.1×10^{2}

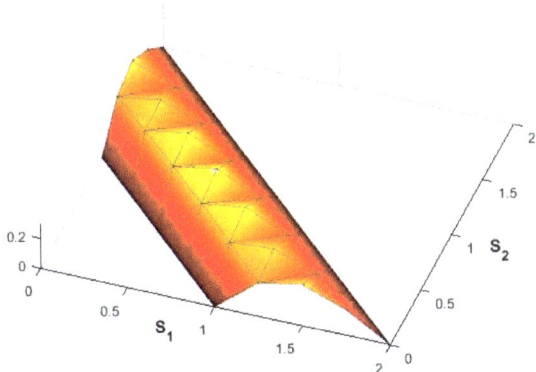

Figure 3. FEM approximation of a double knock-out basket call option value $V(S_1, S_2, 0)$ with data suggested in Table 1, mesh R_0 and $\Delta t = 0.02$.

In order to quantitatively check the reliability of numerical results, we show in Figure 4 that the approximated solution evaluated inside the domain but moving towards boundary Γ_4 (or equivalently Γ_2) converges to the exact solution of a double knock-out call option with the same financial parameters but based on one asset, whose explicit expression $C(S_1, t)$ can be found in [21] and is here reported for reader's convenience:

$$
\begin{aligned}
C(S, t) &= S \sum_{n=-\infty}^{\infty} \left\{ \left(\frac{B_2^n}{B_1^n}\right)^{\frac{2r}{\sigma^2}+1} (\mathcal{N}[d_1] - \mathcal{N}[d_2]) - \left(\frac{B_1^{n+1}}{B_2^n S}\right)^{\frac{2r}{\sigma^2}+1} (\mathcal{N}[d_3] - \mathcal{N}[d_4]) \right\} \\
&\quad - Ke^{-r(T-t)} \sum_{n=-\infty}^{\infty} \left\{ \left(\frac{B_2^n}{B_1^n}\right)^{\frac{2r}{\sigma^2}-1} (\mathcal{N}[d_1 - \sigma\sqrt{T-t}] - \mathcal{N}[d_2 - \sigma\sqrt{T-t}]) \right. \\
&\quad \left. - \left(\frac{B_1^{n+1}}{B_2^n S}\right)^{\frac{2r}{\sigma^2}-1} (\mathcal{N}[d_3 - \sigma\sqrt{T-t}] - \mathcal{N}[d_4 - \sigma\sqrt{T-t}]) \right\}
\end{aligned}
\quad (29)
$$

$$
d_1 = \frac{\ln\left(\frac{SB_2^{2n}}{KB_1^{2n}}\right) + \left(r + \frac{\sigma^2}{2}\right)(T-t)}{\sigma\sqrt{T-t}} \qquad d_2 = \frac{\ln\left(\frac{SB_2^{2n-1}}{B_1^{2n}}\right) + \left(r + \frac{\sigma^2}{2}\right)(T-t)}{\sigma\sqrt{T-t}}
$$

$$
d_3 = \frac{\ln\left(\frac{B_1^{2n+2}}{KSB_2^{2n}}\right) + \left(r + \frac{\sigma^2}{2}\right)(T-t)}{\sigma\sqrt{T-t}} \qquad d_4 = \frac{\ln\left(\frac{B_1^{2n+2}}{SB_2^{2n+1}}\right) + \left(r + \frac{\sigma^2}{2}\right)(T-t)}{\sigma\sqrt{T-t}}
$$

with $\mathcal{N}$ the standard normal cumulative distribution function.

These "boundary conditions" on Γ_2 and Γ_4 are necessary to apply Finite Element Method; on the contrary, they are not set during SABO implementation, but they are naturally matched by the approximated solution.

In order to show the stability of results we have changed also some financial parameters starting from Table 1: in Figure 5, the value of the strike price (on the top) and the volatility of one asset (on the bottom) and, in Figure 6, the correlation (poorly correlated assets with $\rho = 0.1$ on the top, highly correlated assets with $\rho = 0.9$ on the bottom).

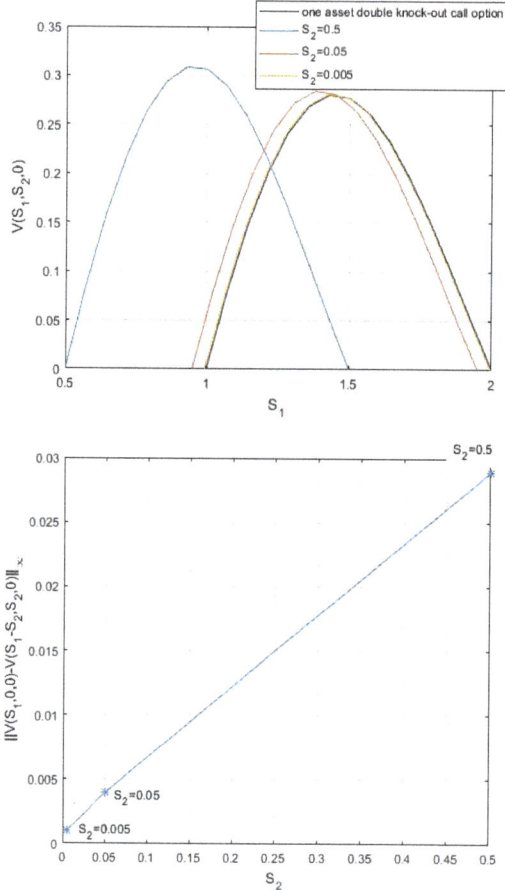

Figure 4. Convergence of the solution $V(S_1, S_2, 0)$ towards $V(S_1, 0, 0) = C(S_1, 0)$, exact option value on Γ_4: on the top, projection onto the planes perpendicular to the domain at fixed values of S_2; on the bottom, distance to $C(S_1, 0)$ in ∞−norm.

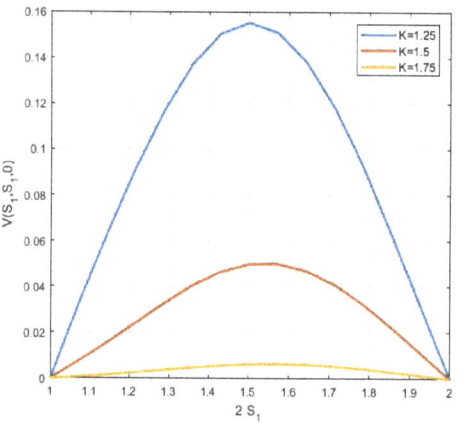

Figure 5. *Cont.*

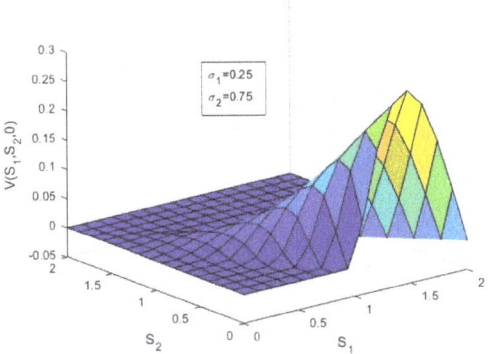

Figure 5. Variation of the option value profile $V(S_1, S_2, 0)$ along the line $S_1 = S_2$ in relation to the strike K (**top**); graph of $V(S_1, S_2, 0)$ for $\sigma_1 = 0.25$ and $\sigma_2 = 0.75$ (**bottom**).

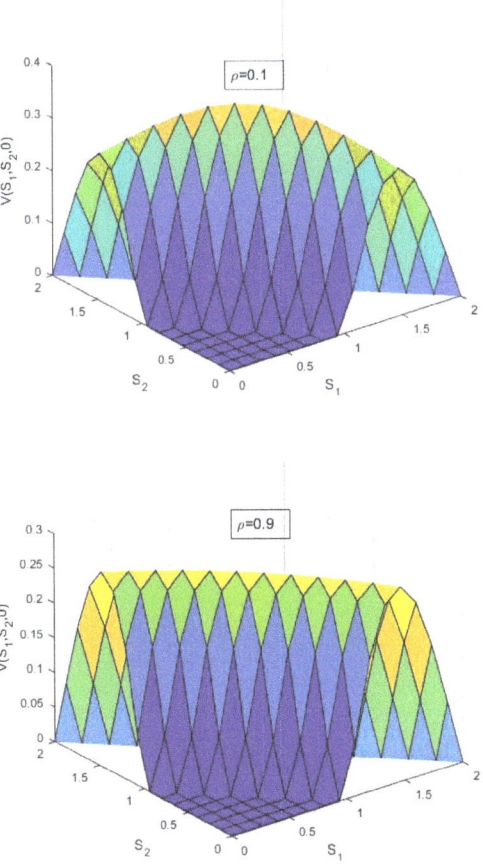

Figure 6. The different aspects of solution for different values of correlation.

- Consider a put down and out option based on two assets whose value is the solution of (1) provided with the payoff function

$$V(S_1, S_2, T) = \max(K - S_2, 0) \qquad \text{if } S_1 \in [B, +\infty]$$

meaning that there is a lower out barrier on the first asset only. The boundary and the domain are shown in Figure 7 top.

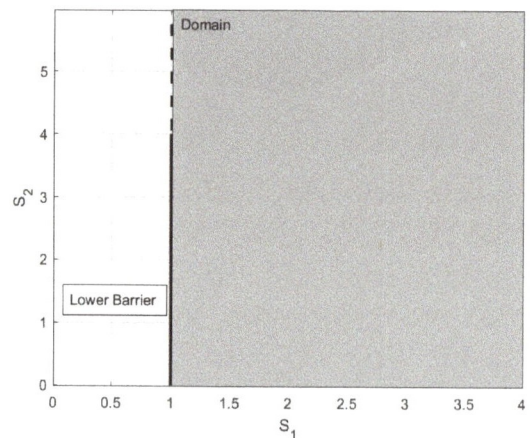

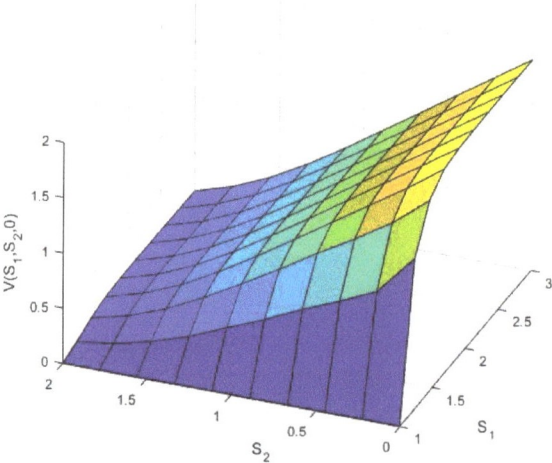

Figure 7. Domain related to second example and solution over a portion of the domain.

In Figure 7, bottom, there is the representation of the approximated solution at time $t = 0$, obtained with $M_{\Delta S} = N_{\Delta t} = 10$. In Table 3, we can observe the stabilization of digits in the numerical approximation of $V(S_1, S_2, 0)$ at $(S_1, S_2) = (2, 1)$ doubling the number of discretization nodes $M_{\Delta S}$ and $N_{\Delta t}$ in (22).

Table 3. Stabilization of value $V(S_1, S_2, 0)$ at $(S_1, S_2) = (2, 1)$ with the refinement of discretization parameters Δ_S and Δ_t.

$N_{\Delta t}$ \ $M_{\Delta S}$	5	10	20
5	0.89692	0.89731	0.89731
10	0.89679	0.89719	0.89719
20	0.89675	0.89715	0.89716
40	0.89673	0.89714	0.89714

7. Conclusions

In this paper, we have extended the SABO method, based on collocation BEM, for pricing barrier options depending on two assets. The good performance of this technique, largely employed in the past for options on a single asset, has been shown by means of numerical results. The implemented Matlab® code is enclosed in the Appendix A, in such a way that it can be directly used by any interested reader.

Author Contributions: All authors contributed equally to this work. All authors have read and agreed to the published version of the manuscript.

Funding: This work has been partially supported by INdAM-GNCS Research Projects.

Conflicts of Interest: The authors declare no conflict of interest.

Appendix A. Matlab® Code

All the above provided numerical results were obtained by codes developed with Matlab® Release 2007b running on a laptop computer (CPU Intel i5, 4 Gb RAM). The one implementing SABO algorithm applied to the first numerical example in this paper is below available.

```matlab
%Basket options
%Code for the example with data taken from the book of R. Seydel
close all
clear
clc
%%%%%%%%%%%%%%%%%%%%%%%%%%%%%%%%%%%%%%
global r sigma1 sigma2 rho
%financial parameters
B2=2; %50; %upper barrier
B1=1; %0.1; %lower barrier
T=1; %expiry
r=0.05; %interest rate
sigma1=0.25; %volatility of S1
sigma2=0.25; %volatility of S2
rho=0.7; %correlation
K=1; %1.5; %strike price
if K>=B2
    disp('strike > upper barrier -> option value = 0')
    return
end
%%%%%%%%%%%%%%%%%%%%%%%%%%%%%%%%%%%%%%
%discretization in space
tol=10^-12;
epsi=10^-10;
L1=B2*sqrt(2); %length of Gamma1
L3=B1*sqrt(2); %length of Gamma3
MS1=10; %n. of segments on Gamma1
dS1=B2/MS1;
MS3=ceil(B1/dS1); %n. of segments on Gamma3
dS3=B1/MS3;
L=[L1/MS1:L1 L3/MS3:L3]; %length of each segment
B=[B2*ones(1,MS1) B1*ones(1,MS3)]; %belonged barrier
```

```
x0=[B2:-dS1:dS1 0:dS3:B1-dS3]; %first abscissa of each segment
x0(MS1+1)=epsi;
x1=[B2-dS1:-dS1:0 dS3:dS3:B1]; %second abscissa of each segment
x1(MS1)=epsi;
y0=[0:dS1:B2-dS1 B1:-dS3:dS3]; %first ordinate of each segment
y0(1)=epsi;
y1=[dS1:dS1:B2 B1-dS3:-dS3:0]; %second ordinate of each segment
y1(end)=epsi;
xbar=[B2-dS1/2:-dS1:0 0+dS3/2:dS3:B1]; %abscissae midpoints
ybar=[0+dS1/2:dS1:B2 B1-dS3/2:-dS3:0]; %ordinates midpoints
%discretization in time
Nt=10; %8;
dt=T/Nt;
t=0:dt:T;
tbar=(t(1:end-1)+t(2:end))/2; %midpoints
%%%%%%%%%%%%%%%%%%%%%%%%%%%%%%%%%%%%%%
%Matrix entries
disp('Computation of the matrix entries')
M=zeros(MS1+MS3,MS1+MS3,Nt);
%Matrix diagonal block
for m=1:MS1+MS3
    for mbar=1:MS1+MS3
        M(mbar,m,1)=integral2(@(tau,l)...
            Solfond(xbar(mbar),ybar(mbar),tbar(1),l,B(m)-1,tau),...
            tbar(1)+epsi,t(2),x0(m),x1(m));%,'AbsTol',tol,'RelTol',tol);
    end
end

%Other matrix blocks of the first row
for ell=2:Nt
    for m=1:MS1+MS3
        for mbar=1:MS1+MS3
            M(mbar,m,ell)=integral2(@(tau,l)...
                Solfond(xbar(mbar),ybar(mbar),tbar(1),l,B(m)-1,tau),...
                t(ell),t(ell+1),x0(m),x1(m));
        end
    end
end

%%%%%%%%%%%%%%%%%%%%%%%%%%%%%%%%%%%%
%RHS
disp('Computation of the rhs entries')
MAX=max(B1,K);
Smin=@(S1) MAX-S1;
Smax=@(S1) B2-S1;
for ell=1:Nt
    for mbar=1:MS1+MS3
        Beta(mbar,ell)=-integral2(@(S1,S2) (S1+S2-K).*...
            Solfond(xbar(mbar),ybar(mbar),tbar(ell),S1,S2,T),...
            0+epsi,MAX-epsi,Smin,Smax);
        Beta(mbar,ell)=Beta(mbar,ell)-...
            integral2(@(S1,S2) (S1+S2-K).*...
            Solfond(xbar(mbar),ybar(mbar),tbar(ell),S1,S2,T),...
            MAX,B2-epsi,0+epsi,Smax);
    end
end

%%%%%%%%%%%%%%%%%%%%%%%%%%%%%%%%%%
%Linear System resolution
disp('Linear System resolution')
Alpha(:,Nt)=M(:,:,1)\Beta(:,Nt);
for ell=Nt-1:-1:1
    for j=2:Nt-ell+1
        Zeta(:,j-1)=M(:,:,j)*Alpha(:,ell+j-1);
    end
    Alpha(:,ell)=M(:,:,1)\(Beta(:,ell)-sum(Zeta,2));
end

%%%%%%%%%%%%%%%%%%%%%%%%%%%%%%%%%%%%%
```

```matlab
%Post processing at time t=0
disp('Post-processing: option value')
NXS1=25;
NXS2=25;
XS1=linspace(0,B2,NXS1);
XS2=linspace(0,B2,NXS2);
[X,Y]=meshgrid(XS1,XS2);
Value=zeros(NXS2,NXS1);
%Values of single asset double knock-out call on Gamma4
for j=1:NXS2
    Value(j,1)=doubleOUT_call(XS2(j),K,B1,B2,T,r,sigma2^2);
end
%Values of single asset double knock-out call on Gamma2
for i=1:NXS1
    Value(1,i)=doubleOUT_call(XS1(i),K,B1,B2,T,r,sigma1^2);
end
for i=2:NXS1
    for j=2:NXS2
        Value(j,i)=0;
        if(XS1(i)+XS2(j)>B1&&XS1(i)+XS2(j)<B2)
            Value(j,i)=integral2(@(S1,S2) (S1+S2-K).*...
                Solfond(XS1(i),XS2(j),0,S1,S2,T),...
                0+epsi,MAX-epsi,Smin,Smax);
            Value(j,i)=Value(j,i)+...
                integral2(@(S1,S2) (S1+S2-K).*...
                Solfond(XS1(i),XS2(j),0,S1,S2,T),...
                MAX,B2,0+epsi,Smax);
            for ell=1:1:Nt
                for m=1:MS1+MS3
                    SUM(m,ell)=Alpha(m,ell)*integral2(@(tau,l)...
                        Solfond(XS1(i),XS2(j),0,l,B(m)-l,tau),...
                        t(ell),t(ell+1),x0(m),x1(m));
                end
            end
            Value(j,i)=Value(j,i)+sum(sum(SUM,1),2);
        end
    end
end
surf(X,Y,Value)
xlabel('S1')
ylabel('S2')

%%%%%%%%%%%%%%%%%%%%%%%%%%%%%%%%%%%%%%%%%%%%%%%%%%%%%%%%
function G=Solfond(SS1,SS2,tt,S1,S2,t)
% fundamental solution
global r sigma1 sigma2 rho
alpha1=(log(SS1./S1)+(r-sigma1^2/2)*(t-tt))./(sigma1*sqrt(t-tt));
alpha2=(log(SS2./S2)+(r-sigma2^2/2)*(t-tt))./(sigma2*sqrt(t-tt));
G=exp(-r*(t-tt))./(2*pi*(t-tt))*sqrt(1-rho^2)/(sigma1*sigma2)./...
    (S1.*S2).*exp(-0.5*(alpha1.^2+alpha2.^2-2*alpha1.*alpha2*rho)/...
    (1-rho^2));
end

%%%%%%%%%%%%%%%%%%%%%%%%%%%%%%%%%%%%%%%%%%%%%%%%%%%%%%%%
function c=doubleOUT_call(S,K,L,U,t,r,sigma2)
% doubleOUT_call evaluates the call option with double out barrier
% according to the formula in section 4.17.3 in "The complete guide to
% option pricing formulas by E. G. Haug
%
% Inputs:
%   S: underlying asset value
%   K: strike price
%   L: lower barrier
%   U: upper barrier
%   t: time to maturity (0 not allowed!!!!!)
%   r: riskless interest rate
%   sigma2: variance of the assrt price
% Output:
%   c: value of the option at S at time t
```

```
%
if S<=L || S>=U
    c=0;
else
    N=5; % The sum whould be infinite, this is the approximation bound
    
    % Compute some parameters that do not change in the sum. it is therefore
    % convenient to compute them once for all
    sigma = sqrt(sigma2);
    m = (2*r)/sigma2 + 1;
    A = (r+0.5*sigma2)*t;
    B = sigma*sqrt(t);
    
    sum1=0;
    sum2=0;
    for ndx=-N:N
        % Define paramenters that change with ndx
        d1 = (log(S*U^(2*ndx)/(K*L^(2*ndx))) + A)./B;
        d2 = (log(S*U^(2*ndx-1)/(L^(2*ndx))) + A)./B;
        d3 = (log((L^(2*ndx+2))./(K*S*U^(2*ndx))) + A)./B;
        d4 = (log((L^(2*ndx+2))./(S*U^(2*ndx+1))) + A)./B;
        
        % Update the sums value
        sum1 = sum1 + (U/L)^(ndx*m) * (normcdf(d1)-normcdf(d2)) - ...
            ((L^(ndx+1))./(S*U^ndx)).^m .* (normcdf(d3)-normcdf(d4));
        sum2 = sum2 + (U/L)^(ndx*(m-2)) .*(normcdf(d1-B)-normcdf(d2-B)) - ...
            ((L^(ndx+1))./(U^ndx*S)).^(m-2).*(normcdf(d3-B)-normcdf(d4-B));
    end
    
    c = S .* sum1 - K*exp(-r*t) .* sum2;
end

end
```

References

1. Black, F.; Scholes, M. The Pricing of Options and Corporate Liabilities. *J. Political Econ.* **1973**, *81*, 637–654.
2. Guardasoni, C. Semi-Analytical method for the pricing of barrier options in case of time-dependent parameters (with Matlab® codes). *Commun. Appl. Ind. Math.* **2018**, *9*, 42–67.
3. Guardasoni, C.; Sanfelici, S. Fast numerical pricing of barrier options under stochastic volatility and jumps. *SIAM J. Appl. Math.* **2016**, *76*, 27–57.
4. Aimi, A.; Diazzi, L.; Guardasoni, C. Efficient BEM-based algorithm for pricing floating strike Asian barrier options (with MATLAB® code). *Axioms* **2018**, *7*, 40.
5. Aimi, A.; Guardasoni, C. Collocation Boundary Element Method for the pricing of Geometric Asian Options. *Eng. Anal. Bound. Elem.* **2018**, *92*, 90–100.
6. Guillaume, T. On the multidimensional Black Scholes partial differential equation. *Ann. Oper. Res.* **2019**, *281*, 229–251.
7. Glasserman, P.; Staum, J. Conditioning on one-step survival for barrier options simulations. *Oper. Res.* **2001**, *49*, 923–937.
8. Cuomo, S.; Di Lorenzo, E.; Di Somma, V.; G., T. A sequential Monte Carlo approach for the pricing of barrier option under a stochastic volatility model. *Electron. J. Appl. Stat. Anal.* **2020**, *13*, 128–145.
9. Giles, M.B. Multilevel Monte Carlo Path Simulation. *Oper. Res.* **2008**, *56*, 607–617.
10. Lars Kirkby, J.; Nguyen, D.; Nguyen, D. A general continuous time Markov chain approximation for multi-asset option pricing with systems of correlated diffusion. *Appl. Math. Comput.* **2020**, *386*, 125472.
11. Leentvaar, C.; Oosterlee, C. Multi-asset option pricing using a parallel Fourier-based technique. *J. Comput. Financ.* **2008**, *12*, 1–26.
12. Seydel, R.U. *Tools for Computational Finance*; Springer: Berlin, Germany, 2009.
13. Friedman, A. *Partial Differential Equations of Parabolic Type*; Prentice-Hall, Inc.: Englewood Cliffs, NJ, USA, 1964.
14. Wilmott, P. *Derivatives: The Theory and Practice of Financial Engineering*; John Wiley: Hoboken, NJ, USA, 1998.
15. Haug, E. *The Complete Guide to Option Pricing Formulas*; McGraw-Hill: New York, NY, USA, 2007.
16. Ballestra, L.; Pacelli, G. A boundary element method to price time-dependent double barrier options. *Appl. Math. Comput.* **2011**, *218*, 4192–4210.
17. Carr, P.; Itkin, A.; Muravey, D. Semi-closed form prices of barrier options in the time-dependent CEV and CIR models. *J. Deriv.* **2020**, *28*, 26–50.
18. Escobar, M.; Ferrando, S. Barrier options in three dimensions. *Int. J. Financ. Mark. Deriv.* **2014**, *3*, 260–292.
19. Todorov, V.; Dimov, I. Monte Carlo methods for multidimensional integration for European option pricing. *AIP Conf. Proc.* **2016**, *1773*, 100009. doi:10.1063/1.4965003.

20. Aimi, A.; Diazzi, L.; Guardasoni, C. Numerical pricing of geometric asian options with barriers. *Math. Methods Appl. Sci.* **2018**, *41*, 7510–7529.
21. Kunimoto, N.; Ikeda, M. Pricing options with curved boundaries. *Math. Financ.* **1992**, *2*, 275–298.

Review

Foundations of Engineering Mathematics Applied for Fluid Flows

Yuli D. Chashechkin

Laboratory of Fluid Mechanics, Ishlinsky Institute for Problems in Mechanics RAS, 119526 Moscow, Russia; yulidch@gmail.com; Tel.: +7-(495)-434-0192

Citation: Chashechkin, Y.D. Foundations of Engineering Mathematics Applied for Fluid Flows. *Axioms* **2021**, *10*, 286. https://doi.org/10.3390/axioms10040286

Academic Editors: Davron Aslonqulovich Juraev and Samad Noeiaghdam

Received: 23 September 2021
Accepted: 25 October 2021
Published: 29 October 2021

Publisher's Note: MDPI stays neutral with regard to jurisdictional claims in published maps and institutional affiliations.

Copyright: © 2021 by the author. Licensee MDPI, Basel, Switzerland. This article is an open access article distributed under the terms and conditions of the Creative Commons Attribution (CC BY) license (https:// creativecommons.org/licenses/by/ 4.0/).

Abstract: Based on a brief historical excursion, a list of principles is formulated which substantiates the choice of axioms and methods for studying nature. The axiomatics of fluid flows are based on conservation laws in the frames of engineering mathematics and technical physics. In the theory of fluid flows within the continuous medium model, a key role for the total energy is distinguished. To describe a fluid flow, a system of fundamental equations is chosen, supplemented by the equations of the state for the Gibbs potential and the medium density. The system is supplemented by the physically based initial and boundary conditions and analyzed, taking into account the compatibility condition. The complete solutions constructed describe both the structure and dynamics of non-stationary flows. The classification of structural components, including waves, ligaments, and vortices, is given on the basis of the complete solutions of the linearized system. The results of compatible theoretical and experimental studies are compared for the cases of potential and actual homogeneous and stratified fluid flow past an arbitrarily oriented plate. The importance of studying the transfer and transformation processes of energy components is illustrated by the description of the fine structures of flows formed by a free-falling drop coalescing with a target fluid at rest.

Keywords: fluid; flows; dynamic; structure; axiomatics; fundamental equations; dissipation; complete solution; ligaments; waves; vortices; plate; wake; drop; impact

MSC: 76A02; 76D50; 76G25

1. Introduction

By recognizing mathematics as "the language in which the book of the Universe is written..." in the polemical treatise "Assaying Master", G. Galilei ([1], 1623) opened a new epoch in the development of natural sciences. Unification of mathematics with applied disciplines enables introducing the concept of "accuracy" as an instrument for estimation of the conformity degree of conclusions to the basic axioms. Simultaneously, the dual concept of "error" value was introduced in practice as a variability measure for physical quantities and their differences from the "exact values" either calculated theoretically or prescribed. The concept of accuracy includes the estimation of adequacy, which is the mutual correspondence of basic ideas in various fields of knowledge. With the explanation of the physical meaning of "acceleration" and "inertia", introducing analogues of such quantities as "pressure" and "temperature", the new epoch came for physical quantity definitions which most fully characterized the physical state of matter and the measure of its variability. At the same time, attempts were made to find a measure for mechanical motion of macroscopic bodies (solid, liquid, or gaseous).

A crucial contribution to the development of the mathematical and physical sciences belongs to R. Descartes. His remarkable achievements in the context of this topic were the introduction of a coordinate system [2], which included the implicit definition of a dimension as belonging to a certain type of sets with given properties, in this case with a measure of the length and the numeric value for a quantity (i.e., the dimensionless ratio of the value to the unit of measure). Considering conservation laws as the basis for the definition of a measurement process, R. Descartes selected momentum, being the product of the body mass multiplied by the velocity, as a measure for body motion ([3], 1641). J.C. Maxwell recognized the vector nature of the momentum and velocity two centuries later [4] based on Hamilton's theory of quaternions [5]. By introducing a coordinate system, Descartes not only unified algebra and geometry but also created a basis for searching for connections between the dynamic and geometric parameters of physical phenomena, the properties of which are conserved with the transition to a new coordinate system.

In a short polemic note, G.W. Leibniz introduced "vis viva" ("live force"), equal to double the kinetic energy of a body's movement as a new measure of motion ([6], 1686). "Vis viva" is a part of the energy as a scalar parameter with a more general nature, which includes both potential and other types of energy in general cases. The universality degree of the concept of energy goes far beyond the framework of mechanical processes. Energy is used to describe all physical processes in the widest range of time and space scales.

Almost synchronously with Leibniz's note, I. Newton published his fundamental treatise ([7], 1687) with an alternative approach to mechanics based on the concept of force and the postulation of "laws of motion". The introduction of the "material point", force, and acceleration concepts exerted and still continue to have a crucial influence on the development of theoretical mechanics [8] and many related sciences. The works by Descartes, Leibniz, and Newton contributed to the successful joint development of both hydrodynamics and mathematics as consensual tools for describing complex phenomena within the framework of the "continuous medium" concept.

Based on the concept of "infinitesimal quantities", the theory of fluid flows began to take its modern form in the middle of the 18th century, when J.-le R. d'Alembert was the first to construct a solution to the partial differential equation describing string oscillations. Later, J.-le R. d'Alembert formulated the continuity equation which is the differential form of the matter conservation law for both incompressible fluids and compressible gases [9]. He thereby established the most general form of the local conservation law as a connection of the volume temporal variation value with flux through the covering's surface. A series of experiments on the body motion's drag in fluids carried out by d'Alembert together with de Condorcet and l'abbe Bossut [10] showed the effectiveness of the coordinated efforts of mathematicians and experimenters in solving practical important problems that ensure the protection of public finances from "indomitable inventors".

Based on the Newton's laws of motion [7], L. Euler applied "solidification of a liquid particle" and obtained the first closed system of equations describing the motion of "ideal" compressible and incompressible fluids. In the current interpretation, Euler's system written for density, velocity, pressure, and gravity acceleration [11] is a form of the representation of mass (continuity of the medium) and momentum conservation equations. The conclusion in [11] is as follows: "Everything that the Theory of liquids is contained is held the two above equations (§ 34), so that for continuation of these studies, we lack not the laws of mechanics, but only the Analysis, which is not yet sufficiently developed for this purpose", stimulated the search for solutions to the Euler equations. This search still continues successfully and brings new approximate and exact solutions to particular problems, including the description of traveling gravitational waves in water [12].

The tool needed to correct the deficiency of the Euler equations—the absence of viscous friction, which became more and more apparent—was developed by J. J. Fourier [13] at the beginning of the 19th century. The value of the new differential equation of a parabolic type found by J.J. Fourier in the analysis of heat conduction processes and the method of its solution can hardly be overestimated. In particular, the developed operator was included by his follower C. Navier into the equation to describe a viscous dissipation [14]. It is interesting to note that C. Navier's explanation of the equation's derivation for a viscous continuous medium was based on P.-S. Laplace's idea of the discrete (atomic) structure of matter. A. Fick also used Fourier's representations in deriving the equations of simple diffusion [15].

Working on improvement of the accuracy of pendulum gravimeters, G.G. Stokes reinterpreted Navier's equations within the framework of the theory of continuous medium motion, making a number of reasonable assumptions, particularly stating the independence of viscous forces on pressure, and giving them their modern form [16]. Stokes's works for many years remained invisible in the shadow of the active, practically important research of linear and nonlinear waves (Russell, Rayleigh, Boussinesq, Thomson, Airy, and many others) and vortices (Helmholtz, Thomson, Kirchhoff, and others), while H. Lamb, in his extensive treatise [17], did not emphasize their fundamental nature. Basically, Stokes, like most scientists in the 19th century, studied the flows of a homogeneous and two-layer (multilayer) fluid, although the very fact of density variability was well known since the 18th century [18].

Russian scientists traditionally paid attention to the study of the density variability and its impact on fluid flows. In famous articles, M.V. Lomonosov described the atomic nature of air elasticity [19] and indicated the influence of density inhomogeneity on the air flow in mines [20]. The extended article by M.V. Lomonosov on the significance of the Arctic Ocean [21], which was presented in 1763 and first published a hundred years later, has been analyzed and widely cited until now.

The initiator of the development of a heavy apparatus for flights in the air and the creator of the scientific foundations of the theory of aeronautics [22], the great encyclopedist D.I. Mendeleev investigated the state equation for gases [23], pure liquids [24], and solutions [25]. The title of one of the Mendeleev's fundamental monographs [23] reproduced the name of the article [20], emphasizing the continuation and relations of the ideas.

However, active accounting for the continuous density variability in fluid flows was not realized due to the supposed small effect of its insignificant relative changes both in natural conditions and in many industrial technologies. Actually, the impurities in the fluid itself were presented as certain "passive substances", with the density given by an empirical expression [26] independent on the condition of fluid and gas motions. Furthermore, in fact, the density was really excluded from most parts of the theoretical investigations by the assumption of its constancy, omitting it as an overall multiplier from dynamic equations together with an equation of state closing the system. In this approximation, it was sufficient for describing the flows to calculate the fields of the velocity components and pressure only [27].

The situation began to change significantly at the end of the 19th century, when J. Gibbs discovered the relations between the thermodynamic potentials and the physical properties of fluids or gases, which are density, pressure, entropy, and temperature, among others [28]. Moreover, Gibbs revealed available potential surface energy, which is an additional form of internal energy in fluids with the surface tension's specified existence of the free surface of a fluid. Now, when the thermodynamic parameters of fluids and gases are defined as derivatives of the Gibbs potential [29,30], the energy concept is used for description of both the static properties of fluids and the dynamics of their variations in a flow. However, practical implementation of the dual nature of fluid parameters, such as density, pressure, and enthalpy, which have mechanical and thermodynamic senses in the description of fluid flows, still remained very limited.

Thus, by the end of the 19th century, the equations representing all the conservation laws, which were necessary for the description of a fluid flow, were written by D'Alembert (continuity) as well as Navier, Stokes, Fourier, and Fick for the transport of momentum, temperature, and dissolved matter. Moreover, the energetic basis for descriptions of the medium state (introduced thermodynamic potentials and their derivatives) was constructed and later implemented in the form of the state equations. However, the idea of considering all governing equations together as a self-consistent system and performing analysis while taking into account the compatibility condition was not expressed in general and was not practically implemented in the form of particular examples. As one of the reasons, one can indicate the "subconscious" influence of the smallness of the ratios between the variations in energy (and density) and the value of the total energy, as well as the lack of scrutinizing analysis of the energy and density distributions in motionless and flowing fluid.

At the same time, due to the insufficient development of methods for analyzing complex systems of nonlinear differential equations for solving practical problems, a variety of semi-empirical and purely constitutive theories were created. There were proposed theories of linear and nonlinear waves [31,32], turbulence, which was being actively developed in the works of W. Thomson, J. Boussinesq, and especially O. Reynolds [33,34], boundary layers [35,36], vortices [37], and others.

In the middle of the last century, stratification effects, which ensure the existence of internal waves [38] and multilayer convective flows [39], began to be studied. However, rare and separate works did not change the essence of the general approach, although the scientific community as a whole recognized the need to use the fundamental equations, which were differential analogs of the conservation laws of matter, momentum, and energy all together [27,40,41], and the importance of the density variability effects. However, complete solutions were not constructed, and only partial solutions describing different structural components of flows such as separate waves, vortices, jets, and wakes were practically studied. For the most part, large-scale (wave) components were investigated, and only a few fine components were studied (e.g., Stokes's solution of flow produced by an oscillating plate along its surface [27]).

Since even the linearized system of fundamental equations has a high order, the complete solution includes several functions of different types [42]. However, in practice, following the work of Stokes and Rayleigh, only one "main solution" is searched for. The existence of additional functions is discussed quite rarely. One of the few examples of an exception is the solution to the problem of sound reflection from a solid wall [27].

Another undiscussed feature of modern hydrodynamics is the freedom of choice of incompatible model equations. Special reduced models and families of constitutive ones remain the main tools to study linear and non-linear waves, boundary layers, jets, wakes, vortices, and vortex systems. For some models, the results of their calculations are of good consistency between each other and with the experiments, albeit in a rather narrow range of parameters.

As the calculations of the continuous groups show, every system of equations within common models of fluid flows is characterized by its own set of infinitesimal symmetries [43] with a corresponding limited number of conserved quantities. The differences in the physical sense of the quantities used, denoted by the same symbols in different systems of equations, make it difficult and even impossible to compare the results and bring them to their general forms, which is necessary to unify the data. Differences in the properties of the included quantities do not allow for creating common requirements for the numerical and experimental techniques or the rules for comparing the data obtained. As a result, demands for indicating experimental measurement error and the temporal and spatial resolutions of the instruments have practically disappeared.

The main object of studies in hydrodynamics is still the limited system of equations, including continuity and momentum transfer in a fluid with a constant density, which are Euler equations (EEs) for an ideal fluid and Navier–Stokes equations (NSEs) for a viscous one. The solvability of a 3D NSE in the constant density approximation has not been proven yet ("6th Millennium Problem" [44]).

Calculations of the flow velocity in a homogeneous fluid within the EE and the NSE cannot be directly compared with the experimental data due to the impossibility of identifying a "liquid particle", which has no distinguishable boundaries. All the indirect and associative methods for the fluid velocity measurements are based on explicit or implicit assumptions, including "passivity" of impurities, the applicability of the Bernoulli equation, and independence of the parameters of the diffusion and heat transfer processes of the experimental conditions, making the implementation degree for real flows difficult to estimate.

Extensive observations show that all flows of real fluids are characterized by a fine structure. The structure is expressed more or less clearly depending on the conditions of the experiment, the quality, and the completeness of the measuring systems. The structure parameters depend on a large number of impacting factors characterizing the properties of the medium, boundary conditions, and external perturbations. The structures of a flow include high-gradient boundaries separating the regions with a more uniform distribution of the parameters of flow pattern change under the impact of non-equilibrium processes, chemical reactions, and the transformation of matter (e.g., ionization, radiation, and absorption of radiation energy).

When studying flows, the impact of the internal energy transfers and conversion processes was not practically analyzed. Transformation of internal energy from its latent potential form into active perturbations of pressure and temperature has not been explored yet, nor has the impact of the form complexity of the state equation, which in modern physics is determined on the basis of thermodynamic potential. The traditional thermodynamic quantities, such as density, entropy, pressure, and concentration, which are defined as derivatives of free enthalpy (Gibbs potential [28]), can change very quickly in a flow.

The introduction of thermodynamic potentials allows for expanding the number of energy transfer mechanisms in hydrodynamics. Thus, the traditional energy transfer by a flow with a velocity U is supplemented with transport by waves of various types with a group velocity c_g, the impact of fast thermodynamically non-equilibrium processes of fast local energy release or absorption, and slow dissipative processes. Fast processes of internal energy conversion manifest themselves in flows induced by a freely falling drop in a liquid at rest [45].

At the same time, the need to improve the theory of flows is growing. The density and total amount of energy in natural and technological processes are increasing due to a number of factors, and as a sequence, is growing the value of the damage caused by natural disasters both local (e.g., fires, floods, and heavy storms) and global ones are associated with natural weather variability and climate change, supplemented by the uncontrolled anthropogenic impact on the environment.

The density and total amount of energy in natural and technological processes are increasing due to a number of factors, and as a sequence, are growing the value of the damage caused by natural disasters both local (e.g., fires, floods, and heavy storms) and global ones are associated with natural weather variability and climate change, supplemented by the uncontrolled anthropogenic impact on the environment.

The highlighted role of mathematics in the description of hydrodynamic processes forces us to return once again to the analysis of the interaction of two key branches of natural science, such as mathematics and hydrodynamics, in order to clarify the content of the terms used, the rules for choosing systems of equations, the methods for their analytical or numerical solutions, as well as the requirements for experimental techniques following from the theory.

In the absence of a canonical definition of mathematics, in practice, different representations are used. These include definite ones, such as "Mathematics is a science of quantity: discrete quantities were studied by arithmetic, continuous ones were done by geometry" by Aristotle, "Mathematics includes only those sciences in which either order or measure is considered" by R. Descartes, "Mathematics... is a science of quantitative relations and spatial forms of the real world" (F. Engels—A.N. Kolmogorov), figurative ones, such as "Mathematics is a set of abstract forms—mathematical structures" by N. Bourbaki, "Mathematics is a language" by G. Galilei and J. Gibbs, "Mathematics is a millstone that grinds what is poured into them" by T.G. Huxley, and pragmatic ones, such as "The science on indirect measurements" by O. Comte.

The abundance of definitions reflects the variety of methods and tools for this branch of science, which is applied to an increasing number of disciplines. Among them, engineering sciences occupy a special place, being focused on the practically used description of a research object (natural system or technology) and the prognosis of its future behavior.

Engineering sciences, which are focused on optimizing the description of the physical object, predicting its long-term variability, defining criteria for identify catastrophic scenarios of events and signs (precursors) for their realization, and assessing the object's response to anthropogenic factors, including directed efforts for controlling the state, contain such disciplines as engineering mathematics and technical mechanics.

Applied to descriptions of fluid and gas flows, *Engineering mathematics* is defined as

"An axiomatic science about the principles of choosing the content of symbols, rules of operations and criteria for assessing accuracy".

The goals of engineering mathematics are to describe the current physical status of fluids and gases as well as the dynamics and structure of a flow and to predict its natural evolution and responses to additional external influences, including targeted control.

Connected by the general principles with engineering mathematics, Technical mechanics is defined as *"An empirio-axiomatic science on the rules for choosing physical quantities, and methods for measuring and evaluating errors in the determination of their values"*.

Technical mechanics is aimed at the selection of physical quantities corresponding to the principles of engineering mathematics which can be measured with a guaranteed estimation of accuracy in the frame of current metrology sciences or by introducing a new procedure, which gives room to measure the physical quantities characterizing fluid flow dynamics and structure. The selected measurable physical parameters of flows must allow for performing a qualitative and quantitative comparison with the results of engineering mathematics applied for the description of a phenomenon under study.

A special place in the definition of dual disciplines belongs to the concept of "accuracy" and "errors or inaccuracy" in the theoretical and practical description of phenomena, respectively.

In mathematics, the internal criteria for assessing accuracy are naturally formulated in the arithmetic and algebra of number fields on the basis of the distinguished properties of two numbers: "zero" and "one". In the mathematical analysis of continuous quantities, the analogues procedure for the comparison of infinitely and regularly decreasing variations of the primary variable and functions (i.e., the Cauchy–Weierstrass algorithm) is applied. In modern applied mathematics, where calculations with the use of conditionally converging series and diverging (singular) functions are widely used, the introduction of a universal criterion at this stage is difficult and requires individual analysis of the problem under study. One of the tools for determining the accuracy is the procedure for comparing the calculations with the experimental data, which needs to include the identity proof of the compared quantities defined in different branches of sciences. In a number of countries, for determining the measurement error, the standards of physical quantities and the procedures for their application are used, being recommended by the relevant international or national organizations.

The opportunity for directly comparing these dual engineering disciplines of mathematics and mechanics is ensured by the implementation of the general scientific (logical and philosophical) principles underlying modern natural science, as well as by the unity in defining the content of the concepts used, which characterize the physical quantities and laws of their changes.

In physics, as the basis for describing the dynamics of nature, a set of conservation laws for the basic physical quantities or their differential analogues is chosen. Their implementation takes into account the conventional logic principles, forming a basement for formal rules of scientific studies. The modern set of these principles is given below.

2. General Principles ("Laws", Demands, and Regulations) of the Science Philosophy

In developing the theory of knowledge, Aristotle [46] formulated the first group of philosophy laws, which included the following principles:

- **Identities**: the concept should be used in the same meaning in the course of reasoning;

- **Internal consistency**: "it is impossible to exist and not to exist at the same time", or "it is impossible to speak correctly, simultaneously affirming and denying something" (i.e., binary logic);
- **Excluded third**: "A or not-A is true, there is no third".

When analyzing the nature of objectivity in describing nature, G.V. Leibniz, who believed that "The great foundation of mathematics is the principle of consistency, that is the statement that a judgment cannot be true and false at the same time", following Aristotle, Descartes, and a number of other predecessors, expanded a number of principles of the philosophy of science. Later, Leibniz's list was repeatedly supplemented, shortened, modified, extended, and today includes the following principles:

- **Meaningfulness**: definability of the essence of a studied subject, object, concept, method, or position in the considered and independent categories. The basis of mechanics is formed by the axiomatically defined mathematical and physical concepts of number, set, space and time, motion, matter (constituents of fluid or gas and their properties), and flow;
- **Identities**: complete conservation of the meaning of an object at operation;
- **Consistency**: two opposed properties cannot be simultaneously true or false;
- **Uniqueness**: excluded third. From the conflicting judgments, one is true, the other is false, and the third is not given. At the same time, this permits describing independent properties in different categories, like the dualism of a "point particle" and a wave distributed in space;
- **Sufficient reason (raison d'etre)**: the presence of history and confirmation of meaning;
- **Minimum sufficiency**: "You should not multiply things unnecessarily", "Do not multiply the number of entities beyond measure", "It is useless to do less with more", "Blessed is the Lord who made everything difficult unnecessary and everything necessary easy!" (G. Skovoroda), and "Of all the explanations, the best is the simplest";
- **Causality**: changes are a consequence of the previous and the cause of the future;
- **Completeness**: description of the known properties of an object with an error estimation and the potential to include newly discovered properties.

In accordance with the outlined principles, the main problem of fluid mechanics is description of the self-consistent temporal change of the fluid status, spatial position, and interaction of the studied medium with outer solids, fluids, and gases.

For solving the main problem of fluid mechanics, which is describing the structure and dynamics of a fluid or gas flow, the following is necessary:

- Indicate the principles for the research object's definition and select the physical quantities characterizing the object;
- Choose the methods for studying the properties of the studied object and processes in the course of their change;
- Give examples of studying the selected phenomena by various independent methods and show the consistency of the results obtained with the estimation of the calculation accuracy and the measurement errors. The general basis for the construction of the theory and methods of experimental studies of a fluid and gas flow in engineering mechanics is the laws of physical quantity conservation within the continuous medium model, which admits infinitesimal representations of physical quantities.

In mathematics, the condition for the conservation of distance defines one of the types of space transformation into itself, which is the transformation of motion [47], which coincides with the concept of an ideal fluid flow [48,49].

In mechanics, the conservation laws have generalized the historical experience of describing phenomena, reflecting the fundamental properties of the existence and immutability of matter (its possible transformations, such as radioactive transformations in nuclear physics, will not be considered here), as well as the parameters of its motion. They are based on the concept of homogeneity of space and time and the isotropy of space.

Traditionally, the fluid mechanics are developed in the "continuous medium" approximation. However, the "continuous medium" methodology, with continuous values of the physical quantities themselves and their derivatives on arbitrary scales, does not match with the concepts of discrete structure of matter. Atomic–molecular properties are expressed in scales to the order of $10^{-8}\ldots10^{-7}$ cm, and nuclear properties are expressed in scales to the order of 10^{-13} cm.

At intermediate scales to the order of $10^{-7}\ldots10^{-6}$ cm (i.e., the size of an atomic-molecular cluster [50]), the influence of both the atomic–molecular interactions [51,52] and macroscopic properties, for example, in the form of a latent potential part of the internal energy's corresponding surface tension [53], are significant.

The distribution of internal energy and pressure in the near-surface layer determines the state of the medium that can be liquid or gaseous and actively influences the structure and dynamics of the ongoing processes, particularly the dynamics of the ocean and atmosphere interaction [40,41]. The choice of a description based on a scale-invariant set of conservation laws allows for the passing of a discrete medium from the model to a continuous one while preserving the meaning of the characterizing quantities. The sizes of the microstructural components establish the natural limits of applicability of the continuous medium model. Namely, the minimal sizes of the studied macroscopic phenomena should exceed the scale of the molecular cluster such that $l_f > \delta_c \sim 10^{-6}\ldots10^{-5}$ cm.

The engineering sciences under consideration are based on the universal conservation laws of matter (total mass or density), as well as the measures of motion, which are momentum and total energy. The mobility of atoms, molecules, and their associations—clusters or macroscopic "liquid particles" with larger structural components—leads to a continuous change of the distribution of matter and the tensor of inertia in space. Due to the independent mobility of small components changing in the moment of inertia and the inhomogeneous dissipation rate of the momentum or realizing latent internal energy, the angular momentum cannot be used as an invariant parameter of the flow and thus is not considered further.

The main parameter of the state and dynamics of fluid and gas is the total energy E_t, including the mechanical part (kinetic and potential) and internal energy, which is determined by the equilibrium thermodynamic parameters (Gibbs potential) [28–30]. The internal energy contains the available potential surface, chemical, electromagnetic, and other types of energy.

Taking into account the total energy gives room to consider all the mechanisms for its transfer in a flow, including transfer with a local flow velocity and a group wave one, slow diffusion processes with a characteristic rate, and rather fast ones in the course of localized direct atomic–molecular interactions (for example, at the size of a molecular cluster upon free surface elimination in merging fluids [45,53]).

The slow transfer of invariant quantities by atomic–molecular processes is described by its own laws, including the corresponding dissipative coefficients, such as the kinematic viscosity ν, thermal diffusivity κ_T, and diffusion κ_S for momentum, heat, and substance transfer, respectively. Although the potentials characterize thermodynamically equilibrium states, their application to the description of the characteristics of non-equilibrium processes is justified by small deviations of the state of the systems from the equilibrium one. Large deviations from the equilibrium values are taken into account by introducing fast localized sources, which determine the energy changes in the course of direct atomic–molecular processes, such as in the fast release of the available surface potential energy contained in the eliminated free surface of merging fluids [45].

The rules for choosing the quantities and methods for comparing their values in relation to the description of a fluid or gas flow will be considered below.

3. Elementary Mathematics in the Theory of a Fluid Flow

The mathematical basis for the theory of a fluid flow, describing changes in the position, dynamic state, and physical properties of a medium under study, includes *real* numbers, with the properties defined here a priori (Note: In hydrodynamics, *real numbers* are used both to mark points in space when introducing a coordinate system and to describe physical quantities. The use of *complex numbers*, which facilitate calculations in the study of dissipating media (i.e., the immersion of the configuration space into the algebra of complex numbers) leads to an extension of the feasible solution space and requires performing special analysis to select among the solutions the physically justified ones). The rules for classifying the constituent elements of sets, being "a collection of certain and distinguishable objects, conceived as a single whole", for performing operations with them are given in a number of monographs and reference books [47]. In mechanics, one of the main types of sets, which is a mathematically defined vector space, was selected as the basis for describing physical phenomena.

The concepts of "space" and "time" in classical mechanics have been introduced axiomatically as primary quantities and are considered as two independent continuous sets existing independently for matter and material processes.

By introducing a coordinate frame, each point in space is associated with an element from the set of real numbers. The minimum quantity of the set elements, which are coordinates characterizing a selected point in the space, is called its dimension.

The axiomatic of a vector space allows the operations of summation and multiplication; internal composition (summation of vectors); associativity; commutativity with the rule of algebraic summation (subtraction) of vectors; associativity of the product of factors; multiplication by one; and distributivity. An important property of the available operations is an outer composition, which is the conservation of a scalar by a vector product in the initial vector space [47].

A space admitting the introduction of a distance $\rho(1, 2) = \sqrt{(x_i^1 - x_i^2)^2}$, between elements x_i^1, x_i^2, with the properties $\rho \geq 0$, $\rho(1, 2) + \rho(2, 3) \geq \rho(1, 3)$, $\rho(1,1) = 0$, is called metric [47]. The space of basic variables expressed with length coordinates is named the configuration space.

The space permits deformations and transformations into itself, both discrete (affine or projective) and continuous, with a transformation parameter, which can have its own dimension or be reduced to the dimension of the base space by introducing a dimensional coefficient (in the case of classical space–time using the world constant (i.e., the light velocity)). By introducing a new variable, the transformation parameter, the configuration space is extended up to a four-dimensional space.

Observable invariants of the spaces, which are the distances between the points or time intervals between the events, are used to define the coordinate values. In mechanics, the space is considered to be a three-dimensional metric (Euclidean) with a standard basis given by orts e_1, e_2, e_3 or (x, y, z).

The mathematical definition of motion is based on the introduction of an absolute coordinate system centered at the point where the radius vector specifies the position of a material point with mass M in the configuration space $\mathbb{R}^3$ at the initial and subsequent time instants, which is in a 4D unified configuration and time space.

In the aggregate of the configuration space (x, y, z) transformations, the orthogonal mapping into themselves (into space (x', y', z')) with the conservation of the distance between the elements is distinguished. This transformation in the Cartesian coordinate system is given by the formulas $x'_i = a_{ik}x_k$, $a_{ji}a_{jk} = 0$ at $i \neq k$ and $a_{ji}a_{jk} = 1$ at $i = k$.

The *motion* in a geometric sense is defined as a continuous orthogonal transformation of metric space into itself, with the time t as an independent continuous parameter of transformation, which conserves the distances between points and the relative positions of objects [47]. In this case, the determinant composed of the coefficients of the matrix a_{ik} equals $\|a_{ik}\| = +1$. An orthogonal transformation with the determinant $\|a_{ik}\| = -1$, which does not conserve the orientation of the figures, specifies a reflection about some axis.

The introduction of the concept of motion, which is characterized by its own laws (functions) of changing the positions of objects, leads to a further expansion of the space and the introduction of the functional space of the problem.

A motion in Euclidean space $\mathbb{R}^3$ is characterized by a group of transformations, which includes independent subgroups of rectilinear shifts $\delta r = v_t \delta t$ with a velocity v_τ and rotations around the instantaneous center $r = 0$ with an angular velocity Ω:

$$\delta \mathbf{r} = (\mathbf{v}_t + \Omega \times \delta \mathbf{r})\delta t. \tag{1}$$

The transformations given by the group of motions are studied by elementary geometry.

The motion in the prescribed coordinate frame with the center at a point $\mathbf{r} = 0$ is characterized by the trajectory $S_t(x, y, z)$ and the velocity $\mathbf{v} = (v_x, v_y, v_z) = \frac{d\mathbf{R}}{dS}\frac{dS_t}{dt} = \tau\frac{dS_t}{dt}$, where $\mathbf{R}$ is the radius vector to the body and τ is the unit vector specifying the local direction of the tangent to the trajectory of the body.

The transition from geometry to mechanics includes the introduction of the concept of mass M, which is an additional independent quantity with its own dimension. Mass is a positively defined scalar quantity which, hereinafter, is considered constant for a mechanical body with a finite size and infinitesimal material points. In mechanics, the mass M is a measure of inertia and the gravitational interaction of bodies. In a general case, a body mass M is a variable, and the nature of its change must be determined independently. The introduction of a new quantity (i.e., mass) leads to a further extension of the dimension of the problem's functional space, which becomes five-dimensional.

The **physical definition** of mechanical motion is based on recording the distances between bodies, each of them being characterized by its own mass and changes in distance over time. In this case, some bodies with fixed distances between them form a basis in which the position of a moving body (material point) is recorded. Distances are invariants, as their values do not depend on a choice of the coordinate system, and their values when passing from one system to another are transformed in accordance with the systems of measurement.

The length, time, and mass standards are accurate enough to enable the functioning of such sophisticated instruments as global positioning systems. Based on the external composition rule in the list of properties of vector space, it follows the equivalence of the vector spaces of the momentum $\mathbf{p}$ and the velocity of the body $\mathbf{v}$. Consequently, the definition of the physical motion of a material point relative to the system of bodies, which form the coordinate frame, is equivalent to the operation of transforming space into itself (i.e., the geometric definition of motion). It is the unity of the different forms of motion definition which connects the invariant properties of space (homogeneity and isotropy) with the conservation laws (Noether's theorem). The invariants of motion of a material point are the momentum $\mathbf{p}$, kinetic energy $E_a^k = \frac{M_a v_a^2}{2} = \frac{p^2}{2M_a}$, and angular momentum $\mathbf{M}_a = [\mathbf{r}_a \mathbf{p}_a]$.

The description of body motions is carried out on the basis of Newton's laws [7] in algebras of real and complex numbers or quaternions (the latter representation is preferable for symbolic programming in navigation). All parameters of solid motion of a mathematical (kinematic) nature, which are based on definitions of the coordinates, velocities, and accelerations, or a physical nature (momentum and energy) are observable (i.e., their value can be determined (measured) by various independent methods with objective control of the error). The methods of their definitions ultimately come down to measurements of the distances, time intervals, and mass. Precision samples and procedures (international standards) were developed for that purpose.

Descriptions of the dynamics of bodies are also carried out in phase space (formed by the components of velocities, momentums, or wave number vectors). An extended six-dimensional configuration space, which unites the spaces of coordinates and velocities, is used for a complete description of the motion as well. The complete functional space of the problem of body movement is eight-dimensional and includes three spatial coordinates, three speeds, time, and mass.

Since the physical state of solids does not change in the course of motion, the set of such parameters as mass, velocity or momentum, angular momentum, and energy turns out to be complete and sufficient to describe all types of motion.

4. Parameters of a Fluid Flow

The description of substance flows in liquid, gaseous, and plasma states is based on the concept of a "continuous medium", which allows the use of continuous functions over the entire range of scales. The key property of a fluid is fluidity (i.e., the ability to move under the action of any small perturbations). This property manifests itself in the decomposition of the flow velocity which, according to the definition of Cauchy–Helmholtz [54], has the following form:

$$v_i(r_r + \delta r_k) = v_i(r_k) + \varepsilon_{ijk}\Omega_j \delta r_k + \frac{\partial v_i}{\partial x_l}\delta r_l, \qquad (2)$$

where ε_{ijk} is a unit antisymmetric tensor of the third rank. In addition to the terms describing displacement and rotation in Equation (1), Equation (2) takes into account the change in the "liquid particle" shape. An additional term describing the shear of the velocity destroys the independence of the action of the rectilinear shift and rotation operators and changes the group properties of the motion operator as a whole.

The difference between the representations of "motion" (Equation (1)) and "decomposition" (Equation (2)) reflects the existence of two independent continuums, those being a metric space (whose motions (Equation (1)) are characterized by the group of motion) and a medium immersed in it (decomposition (Equation (2)) includes the shift operator). To describe the body motion, it is sufficient to use a three-dimensional Euclidean space $\mathbb{R}^3$, supplemented by a point body with constant mass M which acts as a parameter connecting the kinematically determined velocity **v**, with the momentum $\mathbf{p} = M\mathbf{v}$ serving the measure of motion.

The demand for a uniqueness rule requires identification of the difference between the properties of the metric space and the physical space of the continuous medium submerged in it. For an independent description of the flows of a continuous medium, the dimension of the physical space of a problem must have a higher dimension than that of the configuration space in order to save their independent identities. A natural extension of the problem space is introduction into the analysis of the density inhomogeneity in the initial state and its further variability $\rho = \rho(x, y, z, t)$, which corresponds to the properties of real fluids.

The transition to a space with a higher dimension significantly changes the technique of mathematical description of flows and the physical content of the mathematical quantities, particularly the kinematic vorticity $\omega = \mathrm{rot}\mathbf{v}$. Now, the rate of vorticity generation is defined not only by the spatiotemporal variability of the flow velocity but also by the gradients of the thermodynamic quantities $\frac{d\omega}{dt} = \nabla P \times \nabla \rho^{-1}$, which are gradients of the pressure P and density (Bjerknes's theorem).

In the Helmholtz interpretation, the fluid vorticity is identified through rotation of the elements of the medium [54]. The difference between the concepts of "rotation of a part of a continuous medium" and vorticity as a measure of deformation of a fluid particle was noted by J. Bertrand [55–57] and S. Lee [58] in the 19th century. However, due to the insufficient development of some branches of mathematics and the technique of hydrodynamic experiments, their ideas and objections were not supplemented by a constructive development.

One of the hidden difficulties in describing flows is associated with the assumptions of homogeneity, continuity, and deformability of a continuous medium, which give no room for identifying an individual "particle" [11] having no physically distinguishable boundaries. The mass of a "particle" decreases indefinitely when its size tends toward zero, and the object of study disappears.

In the experiment, to measure the flow velocity, the tracking of markers is used, including solid particles, gas bubbles, and droplets of immiscible liquids, which have stable individual characteristics. However, the introduction of a marker, which is an additional physical object, complicates the behavior of a carrying fluid, as a new and more complex multi-component medium is formed. It was Descartes who first noted that a free macroscopic solid is not only carried by a flow but also twisted by the flow around its own axis [3]. The rotating marker additionally perturbs the fluid environment. A small marker becomes involved in the Brownian motion. The transport of soluble impurities is influenced by diffusion effects. The dynamics of droplets of immiscible liquids are perturbed by the effects of non-uniform surface tension.

The combined action of many factors leads to an uncontrolled difference between the motion of markers and the flow of the carrier fluid in which they are immersed. The generally accepted hypothesis of "passivity" of impurities is not confirmed in fine-resolution experimental studies on the redistribution of impurities both in waves, where an initially homogeneous solution or suspension is redistributed and forms a fine structure [53], or in vortices, where a patch of dyed fluid transforms into spiral arms and helical lines and splits into individual fibers.

The absence of criteria for identifying a "liquid particle" means that the "fluid velocity", having a mathematical meaning, is not a physically observable flow parameter, for which one can specify a method for evaluating the error in the course of an experiment and a procedure for reducing it to a given value. The need to meet the criterion of observability of physical quantities for the fluid flow parameters was noted by Stokes, Maxwell, Reynolds, and many others leading hydrodynamics researchers, but analysis of the conditions for observability of a "liquid particle" has not previously been carried out. From the conditions for the conservation laws' application, it follows that the observable (measured with the error control) flow parameters are the specific momentum $\mathbf{p}$, which can be determined by measuring the interaction of the flow impact on a standard body, or the flow rate, as well as the complete energy E_t. Both universal and specific methods to determine their values for concrete experimental conditions have not been developed yet.

At the beginning of the last century, it was found that the property of liquid fluidity (i.e., the ability to move under arbitrary small external influences), was due to the mobility of atoms and molecules, their associations (i.e., clusters [50–52]), and individual structural components [53], each of which is characterized by its own energy. Large-scale components (waves, flows, vortices, and ligaments) interact with each other and all other multi-scale components [59] and provide action of various mechanisms for energy transfer. The transitions of kinetic, potential (gravitational, chemical, and concentration), and internal energy into other forms complicate the descriptions of flows.

In modern fluid mechanics, the main parameter of an equilibrium continuous medium at rest is the internal energy, which is described by the Gibbs potential G [30]. Derivatives of the thermodynamic potential define the traditional parameters of a continuous medium, such as the density $\rho(x_1, x_2, x_3)$, pressure $P(x_1, x_2, x_3)$, temperature $T(x_1, x_2, x_3)$, and concentration of dissolved or suspended particles $S_i(x_1, x_2, x_3)$, which have a clear physical meaning and are available for observation. In modern fluid mechanics, the parameters of the medium, such as the density and pressure, are considered to be quantities with a double nature, namely mechanical and thermodynamic.

The main flow parameters are the total specific energy E_t and momentum $\mathbf{p}$, which manifests itself in the forceful action in dynamics and continuous variations in the flow structure (i.e., in the evolution of distinguished spatial patterns of different physical quantities).

Within the classical hydrodynamics, in the systems of equations of motion for ideal (Euler equations (EEs) [11]) and viscous fluids (Navier–Stokes equations (NSEs) [27,40,41]), the constant density, as a coefficient in all terms of the equations, can be omitted. Then, the systems of equations for fluid motion [27,40,41] are transformed into algorithms for a special transformation of the Euclidean space into itself. In this case, the difference between the two various continuums, which are metric space and the functional space of submerged fluid, is lost. In the three-dimensional formulation, the sets of EE and NSE equations degenerate on singular components and become insoluble [42].

The axiomatic definition of the medium state (fluid and gas) and its flow is further carried out, taking into account the need to preserve the meaning of the physical quantities used and the condition for resolving the governing system of equations. A unified description of the dynamics and structure of flows of fluids and gases is carried out in the absolute Cartesian coordinate system of the metric (Euclidean) space based on the fundamental conservation laws, following the basic methodology of physics [27,40,41].

5. Definition of Fluid Flow

A medium which has the property of fluidity (i.e., the ability to change position in space under the influence of arbitrary small perturbations of physical quantities), characterized by thermodynamic potentials and their derivatives such as thermodynamic quantities and kinetic as well as other physical coefficients (in particular, those determining the propagation of electromagnetic or acoustic waves), is named weakly compressible fluid if occupying a finite volume (or gas or plasma) if it fills all available space.

The definition of a fluid flow is as follows:

The transfer of momentum, energy, and matter, accompanied by self-consistent changes in physical quantities which determine the state of a continuous medium.

The redistribution of matter and energy without a transfer of momentum is called a process (for example, the diffusion transport of matter). Flows are characterized by dynamics (changes in values of physical quantities and the magnitude of the fluid impact on solids) and structure (spatial pattern of the physical quantities' distribution).

The measurable quantities describing a fluid flow are the density ρ, scalar total energy E_t, including the specific kinetic $E_M = \frac{\rho v^2}{2}$, potential E_p and internal E_i energy described by the Gibbs potential G ($E_t = E_M + E_p + E_i$), and vector momentum $\mathbf{p} = (p_x, p_y, p_z) = \rho \mathbf{v}$, which are supplemented with the parameters of the fluid state. The fluid velocity is defined as the instantaneous ratio of two invariant quantities, which are the momentum and density ($\mathbf{v} = \mathbf{p}/\rho$). Furthermore, it is assumed that the fluid velocity $\mathbf{v}$ is identical to the velocity of transformation of the Euclidean space $\mathbb{R}^3$ into itself.

Taking into account different mechanisms of energy transfer with intrinsic spatiotemporal parameters as the base value, characterizing the medium equilibrium state, the Gibbs potential is chosen, as well as its derivatives, determining the density, pressure, temperature, entropy, and other parameters [28–30]. The set of the state quantities includes

$$G(x, y, z, t) = G(\rho(x, y, z, t), P(x, y, z, t), T(x, y, z, t), S_i)(x, y, z, t), \quad (3)$$
$$\rho(x, y, z, t) = \rho(P((x, y, z, t), T(x, y, z, t), S_i(x, y, z, t)).$$

To improve the accuracy of determining the values of physical quantities in practice, additional functional relations between the physical quantities characterizing a fluid are also used as equations of state. Among them are the dependences of the velocity of sound, electrical conductivity, the optical refraction index on pressure, temperature, and salinity [26].

The *axiomatically* introduced system of equations for fluid motion, taking into account the general principles of the choice of physical quantities and fundamental conservation laws [27,40,41], has the following form:

$$\begin{cases} \rho = \rho(P, T, S_n), \quad G = G(x,y,z), \\ \frac{\partial \rho}{\partial t} + \nabla \cdot (\mathbf{p}) = Q_m, \\ \frac{\partial S_i}{\partial t} + \nabla \cdot (S_i \mathbf{v} + \mathbf{I}_i) = Q(S_i), \\ \frac{\partial (p^i)}{\partial t} + \nabla_j \Pi^{ij} = \rho g^i + 2\rho \, \varepsilon^{ijk} v_j \Omega_k + Q^i(f), \\ \frac{\partial E}{\partial t} + \nabla_i (E \, v u^i) + \nabla_i \left(q^i + P v^i - \sigma^{ij} v_j + \frac{\partial w}{\partial S_n} I_n^i \right) = Q(e), \end{cases} \quad (4)$$

where ρ is the density, the ratio of two invariant quantities $\mathbf{v} = \mathbf{p}/\rho$ is the fluid velocity, $\mathbf{I}_n$ is the concentration and density of the diffusion flux of the nth impurity, $\Pi^{ij} = \rho \, u^i u^j + P \delta^{ij} - \sigma^{ij}$ is the momentum flux density tensor, $\sigma^{ij} = \mu \left(\frac{\partial v^i}{\partial x^j} + \frac{\partial v^j}{\partial x^i} - \frac{2}{3} \delta^{ij} \frac{\partial v^k}{\partial x^k} \right) + \zeta \delta^{ij} \frac{\partial v^k}{\partial x^k}$ is the symmetric viscous stress tensor, δ^{ij} is the fundamental metric tensor μ, ζ is the first and second dynamic viscosities, Ω is the global rotation angular velocity, $E = \rho\left(\frac{v^2}{2} + \varepsilon + U\right)$ is the total energy density, ε is the specific internal energy, $w = \varepsilon + \frac{P}{\rho}$ is the specific enthalpy, U is the specific gravity potential, $\mathbf{g} = -\nabla U$ is the acceleration of gravity, and $\mathbf{q}$ is the vector of the heat flux density. The acceded sources of mass for the nth impurity, force, and energy Q_m, $Q(S_n)$, $Q^i(f)$, $Q(e)$, respectively, describe the impact of non-hydrodynamic processes, which take place in fluid flows and can be equal to zero. The introduction of sources reflects the possible impact on the fluid flow structure and dynamics of external factors with non-hydrodynamic natures and the uncertainties of phenomena on the smallest length scale of the order of atomic sizes. Sources in the basic system (Equation (4)) give room to analyze the transformation of latent potential energy into its active form and estimate its impact on the flow dynamics and structure. In the absence of external disturbances and internal sources, they vanish. To the author's knowledge, the set of Equation (1) has not been analyzed up to now, taking into consideration the compatibility conditions necessary for construction of the complete set of solutions.

Under the assumption that the gradients of potentials, physical quantities, and the intensity of external sources are small, the system in Equation (1) is transformed into a system of equations for describing the transfer of matter, concentration of individual components, temperature, and momentum [27], which is widely used in environmental and technical fluid mechanics [40,41]:

$$\begin{cases} G = G(P,S,T) = G(\mathbf{x},t),\ \rho = \rho(P,S,T) = \rho(\mathbf{x},t), \\ \frac{\partial \rho}{\partial t} + \nabla_j(p^j) = Q_\rho, \\ \frac{\partial(p^i)}{\partial t} + \left(\nabla_j \frac{p^j}{\rho}\right)p^i = -\nabla^i P + \rho\, g^i + \nu\Delta(p^i) + 2\varepsilon^{ijk}p_j\Omega_k + Q^i, \\ \frac{\partial \rho T}{\partial t} + \nabla_j \cdot (p^j T) = \Delta(\kappa_T \rho T) + Q_T, \\ \frac{\partial \rho S_i}{\partial t} + \nabla_j \cdot (p^j S_i) = \Delta(\kappa_S \rho S_i) + Q_{S_i}. \end{cases} \quad (5)$$

The system in Equation (5), which includes the quantities of mechanical, thermodynamic, and kinematic natures, takes into account the dissipation of momentum and the influence of internal or external sources of matter, temperature, and substances on a flow structure and dynamics.

The system in Equation (5) is supplemented with the initial and boundary conditions, including impermeability for the density and the components of the substance (impurity concentration) in the fluid, the values of the temperature or its flux, the no-slip condition for momentum or velocity on solid boundaries, the equality of momentum fluxes on the contact surfaces of two fluids, and the damping of all disturbances at infinity. It should be noted the high dimension of the physical problem space, which is determined by all independent quantities of the system (i.e., the dimensions of the coordinate's space, time, density, pressure, temperature, concentration of solutes or suspended particles, as well as kinetic coefficients) included in the equations.

Connecting Equation (5) forms a system of coupled algebraic differential equations, with the solution being constructed by taking into account the compatibility condition [42]. The rank of the system (the order of the highest derivative, if it is possible to reduce the system to one equation), as well as the order of its linearized version and the degree of the characteristic (dispersion) equation, determines the minimal number of eigenfunctions which constitutes a complete solution. The complete system with the diffusion equation for one impurity has the tenth rank [42]. Accordingly, the flow pattern for this set is formed by composing ten functions with intrinsic spatiotemporal scales. The abundance of flow components differing in scales and structure is manifested in continuous changes in the observed flow pattern. Over time, the number of structural components can increase due to the processes of nonlinear interaction of the flow components [59].

Due to the independence (individual behavior) of physical quantities, the fields for each of them, which are characterized by their own geometry, spatial, and temporal scales, must be simultaneously determined in the experiment. The accuracy of the state equations, which is the difference in the values of the density, speed of sound, refractive index, and other reliably determined quantities calculated by solving the system in Equation (5) and obtained experimentally by measuring the values of the temperature, pressure, salinity, and other quantities, determines the error of analytical and numerical calculations.

Consistency of the infinitesimal symmetries of the system in Equation (5) with the basic principles of physics [43] testifies to the validity of its choice as a basis for studying fluid flows. The practical recommendations for the selection of parameters for numerical simulation and experimental techniques are based on the analysis of the intrinsic spatial-time scales of the system in Equation (5) with the initial and boundary conditions of the problem.

The unity of the content of the physical parameters—in theory, numerical modeling and experiments based on the system Equation (5)—allows one to estimate the accuracy of the solutions based on the physical properties of fluids and experimental conditions. Equation (5) contains no additional parameters and does not require their introduction for the development of numerical codes.

Equation (5) is parametrically and length-scale invariant [42,43]. The emergence of "new flow regimes" is usually explained by the influence of ignored energy conversion from its potential form to an active one and back, a change in forms of the state equation, or limitations of techniques, particularly insufficient or excessive sensitivity, temporal or spatial resolution, limitations of the dynamic range of an instrument, or a calculation method necessary to identify all structural components.

Since the symmetries of the fundamental system and equations of other flow models (numerous versions of turbulence theories, theories of waves, vortices, jets, wakes, and others) differ significantly [42,43], the same symbols in different systems have various physical meanings. To check the consistency of the results, it is necessary to calculate the identically defined fields of parameters or to find, for example, the forces and torques acting on a selected obstacle in its own coordinate frame or the flow rate in a selected section.

Until now, the general properties of Equation (2) and even its linearized version have hardly been studied. Periodic solutions of the linearized system have been constructed by methods of the theory of singular perturbations [60] in the approximation of weak dissipation [42].

6. Classification of Infinitesimal Periodic Flow Components

The fluid in natural and industrial conditions, where the density $\rho(z)$ is specified by the distributions of pressure, temperature, and concentration of impurities, under the action of buoyancy effects becomes stably stratified. The total density variability in a stratified fluid and partial contributions due to temperature and salinity variations are described by the buoyancy length scales $\Lambda = \left|\frac{1}{\rho}\frac{d\rho}{dz}\right|^{-1}$, $\Lambda_T = \left|\frac{1}{\rho(T)}\frac{d\rho(T)}{dz}\right|^{-1}$, and $\Lambda_S = \left|\frac{1}{\rho(S)}\frac{d\rho(S)}{dz}\right|^{-1}$, frequency $N_\rho = \sqrt{g/\Lambda}$, $N_T = \sqrt{g/\Lambda_T}$, and $N_S = \sqrt{g/\Lambda_S}$, and period $T_b = 2\pi/N$, T_T, T_S (the axis z is directed vertically upward, g is the gravity acceleration, and the effect of compressibility is neglected). In strongly stratified fluid, which is typical for laboratory conditions, $N \sim 1\,\text{s}^{-1}$ in the environment, $N \sim 0.01\,\text{s}^{-1}$ in potentially homogeneous fluid, and $N \sim 10^{-5}\text{s}^{-1}$ for actually homogeneous fluid, which is generally used in theory ($N \equiv 0$). Furthermore, values for the buoyancy frequency, where N_ρ, N_T, and N_S are supposed to be constant at all depths.

Taking into account that the stratification effects allows one to construct complete solutions of the linearized system in Equation (5) using the compatibility condition and give a classification of the flow structural components, the set of Equation (5) for stratified fluids contains small dissipative coefficients and can be treated by singular perturbation theory methods [60]. Substitution of the solution in the form of plane waves with a positive frequency $\omega > 0$ and a complex wave number $\mathbf{k} = \mathbf{k}_1 + i\mathbf{k}_2$ for the density $\rho' = A_\rho \exp i(\mathbf{k}x - \omega t)$ and similar treatment for perturbations of other physical quantities into the linearized system in Equation (5) gives a dispersion equation of the tenth degree [42]:

$$D_V(k,\omega) \cdot F(k,\omega) = 0$$
$$F(k,\omega) = -D_V(k,\omega)D_{\kappa_T}(k,\omega)D_{\kappa_S}(k,\omega)\left(k^2 + i\frac{k_z(\Lambda_T+\Lambda_S)}{\Lambda_T\Lambda_S}\right) +$$
$$D_{\kappa_T}(k,\omega)\left(\frac{\omega\, k_z}{\Lambda_S}D_V(k,\omega) - N_S^2 k_\perp^2\right) + D_{\kappa_S}(k,\omega)\left(\frac{\omega\, k_z}{\Lambda_T}D_V(k,\omega) - N_T^2 k_\perp^2\right) \quad (6)$$
$$D_V(k,\omega) = -i\omega + \nu\, k^2, D_{\kappa_T}(k,\omega) = -i\omega + \kappa_T\, k^2, D_{\kappa_S}(k,\omega) = -i\omega + \kappa_S\, k^2,$$
$$k_\perp^2 = k_x^2 + k_y^2.$$

The multiplicative structure of Equation (6) is composed of three singular operators of the viscous boundary layer type $D_V(k,\omega) = -i\omega + \nu\, k^2$ and is similar for the temperature and concentration components. The main wave operator reflecting the action of all factors, which are buoyancy, compressibility, and dissipation, clearly demonstrate a multiscale structure of a periodic flow, simultaneously containing both large wave and fine components [42].

The regular roots of Equation (6), with the imaginary part being small compared with the real one, describe infinitesimal waves of various types, which are inertial, gravitational, acoustic, or hybrid ones. A rich family of roots with a singularly perturbed type characterizes *ligaments*. The real and imaginary parts of these roots are in the same order.

Waves are defined as components of a flow where the parameters of local temporal variability (frequency ω) and an instantaneous spatial structure (wavenumber **k** or wavelength λ) are related by a dispersion relation $\omega = \omega(\mathbf{k}, \mathbf{kA}, \ldots)$, where **A** is the amplitude.

Ligaments are thin and extended components of flows described by the set of singular solutions of the complete and linearized system in Equation (5) and the corresponding algebraic dispersion relation in Equation (6) for fluid with density, temperature, and salinity stratification. The number of ligaments depends on the system rank (Equation (5)). The transverse scales of periodic ligaments $\delta_\omega^\nu = \sqrt{\nu/\omega}$, $\delta_\omega^{\kappa_T} = \sqrt{\kappa_T/\omega}$, and $\delta_\omega^{\kappa_S} = \sqrt{\kappa_S/\omega}$, as well as for transient ligaments $\delta_\tau^\nu = \sqrt{\nu \cdot \tau}$, $\delta_\tau^{\kappa_T} = \sqrt{\kappa_T \cdot \tau}$, and $\delta_\tau^{\kappa_S} = \sqrt{\kappa_S \cdot \tau}$, where τ is the duration of the flow formation, or ligaments in a stationary flow with velocity U are $\delta_U^\nu = \nu/U$, $\delta_U^{\kappa_T} = \kappa_T/U$, and $\delta_U^{\kappa_S} = \kappa_S/U$. The length scales are defined by the kinematic coefficients, ν, κ_T, and κ_S, as well as the wave frequency ω, the time interval τ of the flow formation duration, and the velocity U. The length of the ligaments depends on the lifetime of the process under study. Ligaments are distinguished by a high level for the vorticity and mechanical energy dissipation rate. All the flow components co-exit, transfer, and disappear simultaneously, despite the difference in characteristic scales. Each of the flow components provides the transfer of energy, matter, and vorticity.

The total numbers and type of ligaments are determined by the rank of the set in Equation (5) and the order of its linearized version. The minimum number of ligaments is four. Their existence is provided by the variability in density and action of the fluid viscosity. There are six ligaments when heat conduction effects are included. The eight ligaments exist for the case of including the equation for salinity diffusion's presence into the set of governing equations [27]. With time, the number of ligaments and their locations are changed as a result of the non-linear interactions of all the flow components, both for large waves and fine ligaments [59].

It is unsteady ligaments with transverse scales $\delta_\tau^\nu = \sqrt{\nu \cdot \tau}$, $\delta_\tau^{\kappa_T} = \sqrt{\kappa_T \cdot \tau}$, and $\delta_\tau^{\kappa_S} = \sqrt{\kappa_S \cdot \tau}$, and a length $l_l = u \cdot \tau$ (u is the local velocity) which provide the connection of atomic–molecular processes with macroscopic structural components, such as waves (intrinsic time is the inverse period ω), vortices, and others.

As the kinetic coefficients tend toward zero, the thickness of the ligaments decreases uniformly. The vanishing of the coefficient lowers the rank of the system in Equation (5) and discretely reduces the number of ligaments. In the three-dimensional formulation, there are up to six ligaments if diffusion is taken into account, and there are four if only the viscosity is kept.

In the case of a really homogeneous fluid in infinite space, different roots of the dispersion equation become identical for an incompressible fluid:

$$\mathbf{k}^2\left(\omega + i\nu \mathbf{k}^2\right)^2 = 0 \tag{7}$$

This is also the case for a compressible gas [42]:

$$\left(\mathbf{k}^2\left(1 - \frac{i\omega\,\tilde{\nu}}{c_s^2}\right) - \frac{\omega^2}{c_s^2}\right)\left(\omega + i\nu\,\mathbf{k}^2\right)^2 = 0. \tag{8}$$

where $\tilde{\nu} = \zeta + 4\nu/3$, the shear (first) and convergence (second) kinematic viscosity are ν and ζ, respectively, and the sound velocity is c_s. The multiplicity of the roots in Equations (7) and (8) indicates the degeneration of the Navier–Stokes equations for a homogeneous and barotropic fluid [26] in singular components. In this case, two different spaces, which are the metric $\mathbb{R}^3$ that permits motion with the operator in Equation (1) and the space of a submerged homogeneous fluid with the flow decomposition in Equation (2), become indistinguishable. The violation of the uniqueness principle in the determination of the same object manifests here in the degeneration of the problem.

A consequence of the high rank of the system of fundamental equations, solutions of which include several superposed regular and singular functions with incommensurable spatial and temporal properties, is the non-stationarity of all types of flows. The patterns of such flows are constantly being self-transformed.

In some experiments and numerical simulations, ligaments cannot be identified due to insufficient sensitivity or resolution of the instruments, as well as shadowing by high-level perturbations generated by other flow components.

Vortices are complex unsteady flow components with relatively high vorticities $\omega = \text{rot}\mathbf{u}$, which are composed of a set of ligaments. In a vortex, free solids are transported by the flow and simultaneously twist around their own axis, which was noted by Descartes [7] and confirmed later in many laboratory experiments. The uniform miscible fluid volume is split by ligaments, dividing the vortex into individual fibers. The solutions of the Navier–Stokes system characterizing the velocity and pressure fields do not only admit experimental verification with control of the accuracy in an actually homogeneous incompressible fluid where "Eulerian liquid particles" become unidentifiable, and the fluid velocity is a non-observable quantity.

Using all solutions for the system in Equation (5) and the dispersion in Equation (6) allows for solving the linear problem of periodic internal wave generation by an oscillating body in complete 2D and 3D formulations with physically justified initial and boundary conditions [61].

Calculating the ligaments generated by periodic internal waves incident on an inclined wall or a critical level, at which the wave and buoyancy frequencies coincide in a medium with a variable density gradient, gives room to completely solve the wave problem [62]. The differences between the calculations of evanescent waves infiltrating into the supercritical region where their frequency exceeds the local buoyancy frequency, which was published in 1998 [62], from the results of experiments [63] published in 2012 do not exceed a few percent.

Since both waves and ligaments are described by functions of the same type in a linear formulation, they all directly interact with each other, despite differences in their own length scales. Thin high-gradient interfaces (ligaments) are formed in the regions of intersections of internal wave beams [42,64], and as a result of ligament interactions, new ligaments and internal waves are generated [59].

7. Theoretical and Laboratory Studies of Flows around an Obstacle Based on a Reduced Fundamental System

The atmosphere and hydrosphere of the Earth with their densities set by the distributions of all thermodynamic quantities, such as the pressure, temperature, concentration of dissolved substances, and suspended fine particles, are generally stably stratified due to gravity action. The transfer of matter by diffusion processes is being studied quite actively. Less attention is paid to the study of the accompanying energy transfer, although it is a reserve of available potential energy that gives room to produce the mechanical motion of a stratified fluid, particularly diffusion-induced flows on topography [65,66].

The stratification effects on the dynamics, structure, and geometry of flows are noticeable, despite the smallness of the relative changes in density values and are actively studied in environmental and laboratory conditions [38–42]. Due to the limited vertical dimensions of the working volumes in the laboratory and the sensitivity of the density measurement methods, differential heating and regulation of the concentration of the stratifying additive are used to simulate stratification. Differentially heated atmospheric air and water, various gases, and aqueous solutions of metal salts are used as working media. Due to the smallness of the ratio of kinetic coefficients (both for seawater and an aqueous solution of common table salt, where the Lewis number is Le $= \kappa_T/\kappa_S \sim 100$ [26]), the salt stratification decays more slowly than the temperature one and is used more widely.

Due to the high heat capacity of aqueous solutions, the dissipation effects practically do not heat the fluid, and their influence remains unnoticed when studying the flow patterns around bodies in laboratory conditions. In this regard, when performing calculations and laboratory modeling of flows in a stratified basin, the fluid is considered isothermal and incompressible, with the state equation defined by the distribution of the dissolved salt concentration. The initial stable density distribution is often chosen to be $\rho_0(z) = \rho_{00} s(z) = \rho_{00} \exp(-z/\Lambda)$, where ρ_{00} is the density on the reference level, with constant values for the buoyancy scale Λ, frequency N, and period T_b.

The reduced system of fundamental Equation (6) for a one-component incompressible stratified medium in the Boussinesq approximation, when density variations are neglected everywhere except for the term with gravity, takes the following form:

$$\begin{cases} \rho = \rho_0 + \rho_{00} \cdot s, & \text{div } \mathbf{v} = 0, \\ \frac{\partial \mathbf{v}}{\partial t} + \nabla \cdot (\mathbf{v}\ \mathbf{v}) = -\frac{1}{\rho_{00}} \nabla P + \nabla \cdot (\nu \nabla \mathbf{v}) - s \cdot \mathbf{g}, \\ \frac{\partial s}{\partial t} + \nabla \cdot (s\ \mathbf{v}) = \nabla \cdot (\kappa_S \nabla s) + \frac{v_z}{\Lambda}, \end{cases} \quad (9)$$

where the fluid velocity is $\mathbf{v} = \mathbf{p}/\rho$, $P(x, z, t)$ is the pressure except for the hydrostatic one, and s is the salinity perturbation including the salt contraction coefficient.

Physically valid no-slip and no-flux boundary conditions on the surface of a solid impermeable body with geometric dimensions of a length L, width W, and height h which can move with a constant velocity U starting at $t = 0$ have the following form:

$$\begin{aligned} \mathbf{u}|_{t \le 0} &= 0, & s|_{t \le 0} &= 0, & P|_{t \le 0} &= 0, \\ u_x|_\Sigma &= u_z|_\Sigma = 0, & \left[\frac{\partial s}{\partial n}\right]\Big|_\Sigma &= \frac{1}{\Lambda} \frac{\partial z}{\partial n}, \\ u_x|_{x,\ z \to \infty} &= U, & u_z|_{x,\ z \to \infty} &= 0. \end{aligned} \quad (10)$$

The governing system of Equation (9), together with the initial and boundary conditions in Equation (10), is characterized by a set of temporal and spatial scales having significantly different values. Among the large linear scales are the buoyancy scale $\Lambda = |d \ln \rho/dz|^{-1}$, which characterizes the manifestation level of the initial stratification, the geometric dimensions of the body h, L, and W, the attached internal wave length $\lambda_a = UT_b$, and the viscous wave scale. The small scales characterizing basic ligaments accompanying stratification in a viscous fluid and the basic flow with velocity U are $\delta_U^\nu = \sqrt{\nu/U}$, $\delta_N^{\kappa_S} = \sqrt{\kappa_S/N}$, and $\delta_U^\nu = \nu/U$, $\delta_U^{\kappa_S} = \kappa_S/U$, respectively. The system in Equations (9) and (10) was selected as a basis for design of the stands and development of a technique for laboratory studies of flows around an obstacle.

The laboratory experiments based on the system in Equation (9) were conducted at the stands of the Unique Science Facility's "Hydrophysical Complex for modeling hydrodynamic processes in the environment and their impact on underwater technical objects, as well as distribution of impurities in the ocean and atmosphere (USF HPC IPMech

RAS)" [67]. The laboratory set-up included transparent tanks with different dimensions, additional devices for filling with a stratified fluid, towing models, generated surface and internal waves, placing and moving sensors, sonar, hydrophones and microphones, schlieren and optic flow visualization instruments, an experiment control, and data processing.

Numerical simulation for the formulated mathematical problem is constructed on the basis of the finite volume method using the computational utility OpenFOAM [68]. The open source of this computational package enabled the creation of original program codes in the C++ environment for modifying and improving the existing approaches and standard solvers for numerical simulation of stratified flows in order to be able to perform direct numerical simulations in a wide range of flow parameters. More detailed information on the modifications made to the standard OpenFOAM solvers can be found in other papers by the authors indicated in the references in [68].

7.1. Slow Diffusion-Induced Flow on a Sloping Plate

Stratified fluids in the field of mass forces (in particular, gravity) are examples of thermodynamically non-equilibrium systems. They have a reserve of available potential gravitational energy, since the center of mass lies below the geometric center of the fluid volume. At the impermeable boundaries, the molecular flux of a stratified substance is interrupted. The violation of the flux leads to an accumulation of diffusing material in some places and a deficit in others. An arising hydrostatic pressure gradient in the field of mass forces (gravitation) accelerates the fluids and forms a flow, which exists even in the absence of destabilizing external factors [65,66]. Due to the easy attainability of the formation conditions, such flows are formed in a stratified fluid or gas near any inclined boundaries. They are common in the atmosphere ("mountain and valley winds"), the world ocean, where their formation is associated with the stratification and global rotation effects, and in thermally inhomogeneous lakes.

First, a stationary solution of the linearized system in Equation (9) with the boundary conditions of Equation (10) describing the flow on an inclined plane was constructed. Similar salinity and velocity profiles of the flow were characterized by a single combination scale $\delta = \sqrt[4]{\nu \kappa_S / N^2 \sin^2 \alpha}$, where α was the inclination angle of the plane to the horizon [65,66]. The thickness of the induced flow tended toward infinity as the angle α tended toward zero, and the solution itself became divergent.

The profiles of the salinity perturbations along the normal to the plate surface in the asymptotic solution for the transient flow formation problem and the small-time approximation of the exact solution of the creeping flow formation problem are characterized by the length scale $\delta_N^{\kappa_S} = \sqrt{\kappa_S/N}$, where the time t is normalized by the buoyancy period, $\tau = t/T_b$, (ξ, ζ) is the local coordinate frame, and the axis ζ is normal to the plane:

$$s' = -2\frac{\delta_N^{\kappa_S}\sqrt{\tau}}{\Lambda}\text{ierfc}\left(\frac{\zeta}{2\delta_N^{\kappa_S}\sqrt{\tau}}\right) \tag{11}$$

Meanwhile, the induced velocity is described by both diffusion scales (i.e., for the density $\delta_N^{\kappa_S}$ and the velocity $\delta_N^\nu = \sqrt{\nu/N}$):

$$u(\zeta) = \frac{N^2 \delta_s \tau^{3/2}}{\nu - \kappa_s}\left[i^3\text{erfc}\left(\frac{\zeta}{2\delta_\nu\sqrt{\tau}}\right) - i^3\text{erfc}\left(\frac{\zeta}{2\delta_s\sqrt{\tau}}\right)\right]\sin 2\alpha,$$
$$i^n\text{erfc}(z) = \int_z^\infty i^{n-1}\text{erfc}(x)dx, \quad i^0\text{erfc}(z) = \frac{2}{\sqrt{\pi}}\int_z^\infty e^{-x^2}dx, \quad i^{-1}\text{erfc}(z) = \frac{2}{\sqrt{\pi}}e^{-z^2}. \tag{12}$$

The structural flow components with incommensurable values of the length scale evidence the total unsteadiness of the phenomena under study.

The calculations of the flow pattern in the complete non-linear formulation show a system of cells which was formed near a plate with a length l (Figure 1) within the whole range of the inclination angle of the plate $0 \leq \alpha < 90°$. In this case, the flow pattern around the center of the plate agreed with the exact and asymptotic solutions for the sloping infinite plane.

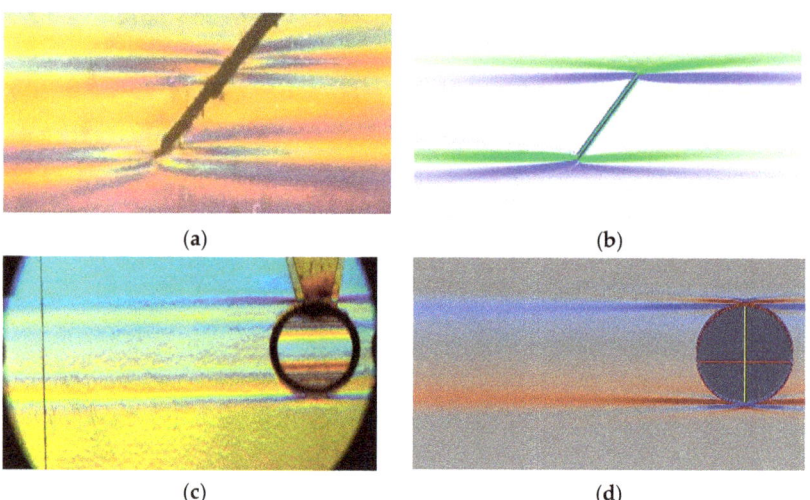

Figure 1. Schlieren and numerical visualization of the diffusion-induced flow: (**a**,**b**) on a fixed inclined plate, $L = 2.5$ cm, $N = 0.84$ s^{-1}, and $T_b = 7.5$ s; (**c**,**d**) cylinder, $D = 5$ cm, and $T_b = 10.5$ s.

Diffusion-induced flows on topography have been actively studied for the past 70 years. However, due to the internal multiscale nature, a complete calculation of such a flow even for the case of bodies with a simple shape, such as a cylinder, a plate of a finite length, or a wedge, is carried out only with the use of supercomputer technologies.

The calculated patterns of the fields of physical quantities, which were consistent with the data of a high-resolution schlieren visualization of diffusion-induced flows, are presented in Figure 1. The fields of the horizontal component of the refractive index gradient, which were visualized by schlieren instruments, were compared with the calculated density gradient fields (these parameters were connected by a linear relation for a solution of sodium chloride). The experiments were conducted in a transparent tank filled with a continuously stratified solution of common table salt [69].

The structure of the visualized perturbation fields shows that some flow components with the parameters from Equation (12) belonged to the class of ligaments. The corresponding attached internal waves (at $U = \sqrt{\nu N} \sim 0.1$ cm/s and $\lambda = UT_b \sim 0.5$ cm) were too small in amplitude and could not be detected by existing instruments. The creeping diffusion-induced flows, which are formed on asymmetric bodies, provide their self-propulsion at the horizons of neutral buoyancy [69].

When calculating the formation of a flow pattern around a body starting to move, the diffusion-induced flow was chosen as the initial condition. The solvability of the problem of flow around a plate in a two-dimensional formulation for both stratified and homogeneous fluids was used for comparison of the new results with the previously obtained ones and to note good agreement in terms of the drag estimate [68].

The complete solution of the system in Equation (6) enables calculating the fields of all physical parameters of the flow, including the vorticity and its baroclinic generation rate, as well as the energy and its dissipation rate. The condition for the solvability of the minimum scales of the flow structural components should be taken into account when choosing the mesh parameters and experimental techniques [68].

7.2. Pattern of Flow around a Moving Plate in Wave and Vortex Flow Regimes

The problem of flow around a moving body with a velocity U based on the system in Equation (9) with the boundary conditions in Equation (10) includes a set of length scales of a geometric and dynamic nature. The set includes the buoyancy scale Λ, body sizes L, W, H, and length of the attached internal wave $\lambda_a = 2\pi U T_b$. The group of fine scales contains the thicknesses of the ligaments associated with the natural oscillation of the medium δ_N^ν, $\delta_N^{\kappa_S}$ and those accompanying the internal attached waves δ_U^ν, $\delta_U^{\kappa_S}$. For reliable registration of all the flow components, the dimensions of the observation or calculation area should noticeably exceed the macroscale of the problem L, W, H, λ, and the dimensions of the resolution cells should be several times smaller than the microscales δ_N^ν, $\delta_N^{\kappa_S}$ and δ_U^ν, $\delta_U^{\kappa_S}$. The time for observing the flow should noticeably exceed the buoyancy period T_b. The time step Δt must satisfy the Courant criterion $Co = |\mathbf{v}|\Delta t/\Delta r \leq 1$, which is determined by the mesh cell size Δr, values of the microscales δ_N^ν, $\delta_N^{\kappa_S}$, δ_U^ν, $\delta_U^{\kappa_S}$, and local flow velocity $\mathbf{v}$.

The flow parameters depend on the medium's properties, such as the density, stratification, coefficients of viscosity and diffusion of the stratifying component, and the size, shape, position, surface quality, direction, and velocity of movement of a body. We consider the motion of a thin rectangular sharp-edged plate with its plane oriented at an arbitrary angle α to the horizon (with an angle of attack) which starts a uniform motion with a velocity U in the horizontal direction.

When studying the flow pattern formed during the motion of a plate with a length L and velocity U oriented along the trajectory, the system of equations for incompressible stratified fluid motion in the two-dimensional formulation and local coordinate system associated with the body ξ, ζ is transformed into a single equation of internal waves for the stream function Ψ (fluid velocity $v_x = \frac{\partial \Psi}{\partial z}, v_z = \frac{\partial \Psi}{\partial x}$):

$$\left[\frac{\partial^2}{\partial t^2}\left(\frac{\partial^2}{\partial \xi^2} + \frac{\partial^2}{\partial \zeta^2}\right) + N^2\left(\cos\varphi \frac{\partial}{\partial \xi} - \sin\varphi \frac{\partial}{\partial \zeta}\right)^2 - \nu\frac{\partial}{\partial t}\left(\frac{\partial^2}{\partial \xi^2} + \frac{\partial^2}{\partial \zeta^2}\right)^2\right]\Psi = 0 \qquad (13)$$

where the boundary conditions for the stream function on the plate surface are

$$\left.\frac{\partial \Psi}{\partial \zeta}\right|_{\zeta=0} = U\vartheta\left(\xi + \frac{L}{2} - Ut\right)\vartheta\left(\frac{L}{2} + Ut - \xi\right), \quad \left.\frac{\partial \Psi}{\partial \xi}\right|_{\zeta=0} = 0, \qquad (14)$$

This includes the attenuation of all disturbances at infinity, where φ is the inclination angle of the trajectory to the horizon and ϑ is the Heaviside function.

Substitution of the solution in the form of plane wave transforms (Equation (13)) into the dispersion equation is expressed as

$$\omega^2\left(k^2 + k_z^2\right) - N^2(k\cos\varphi - k_z\sin\varphi)^2 + i\omega\nu\left(k^2 + k_z^2\right)^2 = 0, \qquad (15)$$

This takes the simplest form when the plate moves along a horizontal surface:

$$\omega^2\left(k^2 + k_z^2\right) - N^2 k^2 + i\omega\nu\left(k^2 + k_z^2\right)^2 = 0 \qquad (16)$$

The roots of Equation (16), which significantly differ in the relations between the real and imaginary parts, are denoted by the indices (w) and (l). They characterize the internal waves and thin ligaments:

$$k_{l,w}(\omega, k) = \pm\sqrt{-k^2 + \frac{i\omega}{2\nu}\left[1 \pm \sqrt{1 + \frac{4i\nu k^2 N^2}{\omega^3}}\right]}. \qquad (17)$$

By substituting Equation (17) into the boundary conditions and solving the linear system of equations, one can find the stream function and then calculate the velocity components and all other physical quantities of the problem. However, even in the simplest case, the integrals cannot be calculated analytically, and in order to illustrate the flow properties, they are calculated numerically, as in more complex cases, including direct numerical solving of Equation (9) in a complete nonlinear formulation. Below, we present the results of numerical visualization of the analytical solutions of the linearized system in Equation (9), as well as the problem in the complete formulation.

At relatively low Reynolds numbers, a typical stratified flow pattern around a moving horizontal plate, which is shown in Figure 2, consists of specifically arranged groups of upstream perturbations, attached internal waves, and a thin density wake formed by extended ligaments in the form of thin and long interfaces.

(a)
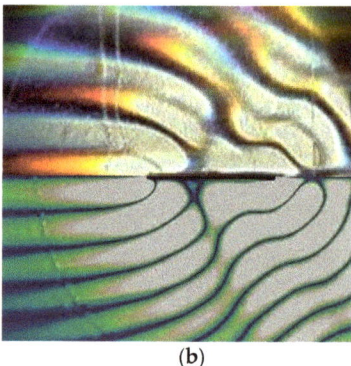
(b)

Figure 2. Schlieren (upper part of the images) and calculated (lower one) stratified flow patterns around a uniformly moving horizontal plate: (**a**,**b**) $T_b = 7.6$ s, $L = 7.5$ cm, $U = 0.27, 0.39$ cm/s, where the thin wavy lines in the Schlieren images are density markers which visualize the profile of the horizontal velocity component in the fields of upstream perturbations.

The phase surfaces of the internal waves separate the half-waves of crests and troughs which bind toward horizons in front of the body and close up behind it. For all the cases considered, perturbations were pronounced at the edges of a uniformly moving plate.

In the case of the long plate with respect to the attached internal wavelength $\lambda_a < L$, the phase surfaces of the internal waves were broken above and beneath the plate. The comparisons in Figure 2 show that the numerical and experimental data were in good qualitative agreement for the calculated and visualized internal wave fields [68].

With the increase in the velocity, the general flow structure undergoes some modifications, such as decreased declination of the internal wave phase surfaces toward the direction of the body motion, a change in geometry of the fine-structural interfaces, and the degree of manifestation of separate flow components (Figure 3a). The strongest structural changes were revealed in the wake past the plate, where a system of short transverse interfaces in the form of tilted ligaments (streaky structure) was observed (Figure 3b).

With a further increase in the velocity, the ligaments in the wake became more and more pronounced and actively interacted. Short interfaces were gradually elongated and transformed into a vortex system where typical vortex elements, such as vortex dipoles outlined by thin interfaces, were observed (Figure 4).

In the vortex flow regime, when the vortex elements become a dominant flow component, with the internal wavelength being comparable to the observation area size, the most contrasting structural changes are manifested in the wake flow past the plate (Figure 5).

Both the laboratory and numerical simulations show that the wake flow structure past the tilted plate consisted of a typical vortex street in the form of a sequence of mushroom-like elements. In the strongly stratified medium (Figure 5a), the wake vortices gradually collapsed downstream, being broken up into a set of fine structural elements, while in the homogeneous fluid (Figure 5b), the vortex street expanded while evolving downstream in the vertical direction within the observation area. A variety of multilayer fine structural flow elements was formed on the vortex shells and in the regions of interaction of the vortex flow multiscale components between themselves and with the plate surface.

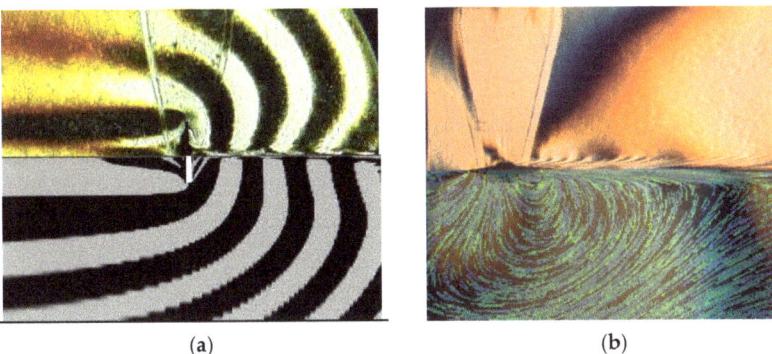

Figure 3. Internal wave field patterns around a uniformly moving horizontal plate in a stratified fluid, with schlieren visualization by the "vertical slit–filament" method and numerical visualization of the complete solution of the linearized problem $T_b = 7.6$ s: (**a**) $L = 7.5$ cm and $U = 0.40$ cm/s; (**b**) $L = 2.5$ cm and $U = 2.3$ cm/s. The upper parts of the images correspond to the schlieren visualization, and the lower one shows the calculation results of the density gradient field and streamline patterns.

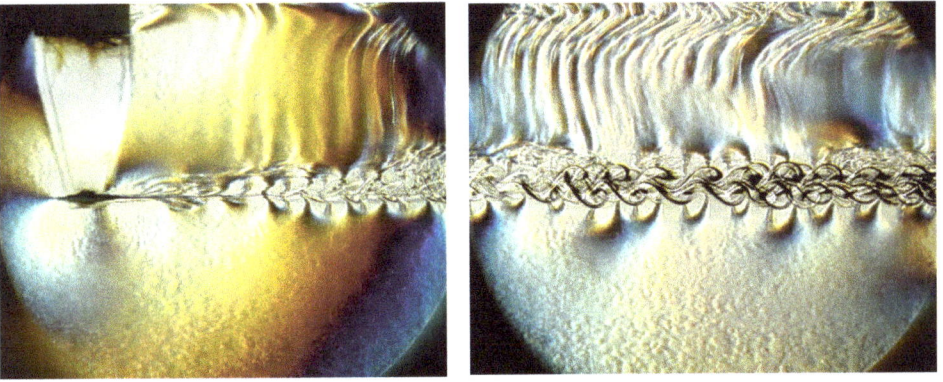

Figure 4. Schlieren images of the flow past a uniformly moving plate, with gradual transformation of the ligaments into a sequence of vortex dipoles inside the density wake; $T_b = 7.5$ s, $L = 2.5$ cm, $U = 5.25$ cm/s.

All the flow components evolved and actively interacted with each other and with the free stream. In the unsteady flow regime, one can distinguish slowly evolving components, such as upstream and attached wave fields, rapidly changing ones, including fine-structured layers or ligaments, and their sets, which are vortices. The calculations and observations of the flow patterns were in good qualitative agreement with each other in all the flow regions, including the upstream perturbations, a system of internal waves, the wake with fine structures, and the vortices.

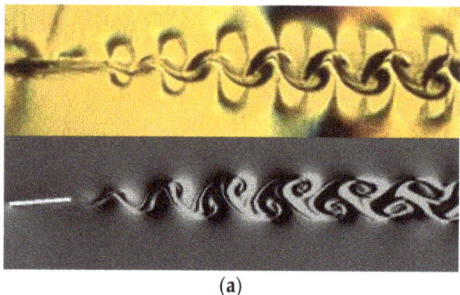 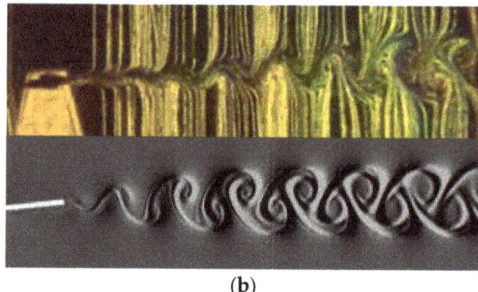

Figure 5. Schlieren (upper part of the images) and calculated (lower one) vortex flow patterns past a uniformly moving tilted plate ($L = 2.5$ cm, $U = 4.3$ cm/s): (**a**,**b**) $T_b = 7.6$; $6.3 \cdot 10^5$ s, $N = 0.83$; 10^{-5} s^{-1}.

The fine structure of the flow patterns, which is a consequence of atomic–molecular interactions in moving matter and the intense action of energy transformations in the atomic–molecular processes at the boundary of fluid (gas) with a submerged solid body, as well as inside of an inhomogeneous fluid (or gas), exists in all flows at any phase states of matter. As an illustration, Figure 6 presents the schlieren visualization data of the flow pattern around the wing in the wind tunnel (photographs were kindly provided by Professor V.G. Sudakov, TsAGI) and the plate towed in the stratified basin (Figure 6b,d). As can be seen in the photographs presented, the families of transversely located fine interfaces or ligaments are visualized both in compressible gas flows with transonic velocities and near a slowly moving body in a weakly compressible stratified fluid. The structural similarity of the gas and fluid flow images indicates the parametric invariance of the fact of the ligaments' existence.

The presented results of theoretical and experimental studies of the flow pattern around the strip show that the reduced system of fundamental Equation (9) with physically justified boundary conditions described all the details of the observed flow pattern, which were upstream disturbances, internal waves, wakes, vortices, and ligaments in a wide range of flow parameters. Complete solutions of the system in Equation (9) in a unified formulation described both the creeping diffusion-induced flows on a fixed obstacle and the wave and vortex structures at sufficiently large Reynolds numbers to the order of $10^4 \div 10^5$ without involving additional parameters [68].

At high velocities, the flow pattern becomes more complicated, and the flow, which contains a large number of structural components with its own parameters, is continuously transformed. The algorithms and criteria for constructing the complete solutions of Equations (9) and (10) allow for calculating all the physical quantities of the flows, which are the fields of density, velocity, pressure, vorticity, its baroclinic generation rate, mechanical energy dissipation rate, forces, and torques acting on a 2D body in a flow without involving additional hypotheses and constants in either stratified or homogeneous (potential or actual) environments [68]. The transition to a 3D formulation requires the development of new algorithms and numerical codes, which take into account the dimension expansion of the complete space of a problem and the appearance of new groups of ligaments. In this case, solving problems in a 2D formulation can be used to control the quality of the developed algorithms and methods for comparing the theoretical and experimental results.

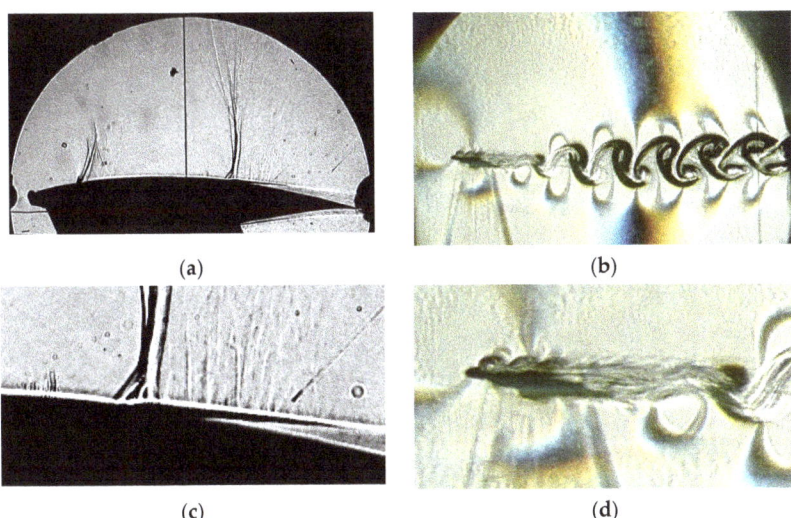

Figure 6. Schlieren images of flow around bodies. (**a**,**c**) Wing in the TsAGI wind tunnel at Ma = 0.77 and plate in the stratified basin of the LMT IPMech RAS (T_b = 7.55 s, L = 2.5 cm, angle of attack α = 12.5°, U = 3.6 cm/s, Re = 900, Fr = 1.73). (**b**,**d**) Enlarged sections of the figures with fine structures.

Returning to the definition of mathematics, in the conclusion of this section, one can say that the physical variables of the problem are the density, momentum (in the Boussinesq approximation, the velocity $\mathbf{v} = \mathbf{p}/\rho_{00}$ differs from the momentum by a constant coefficient ρ_{00}^{-1}), pressure, and concentration of the stratifying component, which are chosen in accordance with the conservation laws. In this work, the principles of accuracy (error) control can be implemented by comparing the fields of the selected physical quantities obtained by independent methods, including analytical (in this case, non-uniform asymptotic expansions do not allow an estimation of accuracy), numerical (obtained while taking into account the solvability condition of the ligaments), and experimental methods, with a high spatial resolution. Here, the objects of comparison are the components of the density gradient fields, calculated theoretically and reconstructed from the schlieren visualization data of flows of an aqueous solution of table salt. The density and refractive index of the working fluid are related by an almost constant coefficient.

8. Influence of Fast Energy Transfer Processes on the Dynamics and Structure of Impact Flows Produced in a Motionless Target Fluid of a Coalescing Free-Falling Drop

Attention has been paid in recent decades to the study of the hydrodynamics and acoustics of a drop's impact and is explained by the fundamental nature of the topic, the growth of technical applications, the improvement of instruments, and programs of data collection and processing. The compactness of the process allows for carrying out research even in small laboratories, and the diversity and reproducibility of the ongoing processes (in general) provide the potential to obtain new experimental data, clarifying and expanding the existing representations on the physical nature of flows. The complexity and outlines of elements as well as the reproducibility of the fast changeable flow pattern allow one to trace the action of various mechanisms for energy transfer, including both small-scale fast and slow diffusion mechanisms. Attention is paid to the study of the relationships between hydrodynamic and acoustic processes, and the search for mechanisms for the excitation of gravitational capillary waves or soundwaves.

In experiments, new groups of capillary waves appeared sequentially when the flow structure changed. Short waves appeared first at the boundary of the parch of a droplet coalescing with the target fluid [45] and even on the surface of the coalescing drop, followed by around the growing crown, later at the edge of the descending crown and inside the cavity, and further around the basement of the growing splash and at its top around a detachment zone of an escaping drop. The formation of a streamer and cavity after its submerging was accompanied by the generation of new groups of circular capillary waves [70,71].

At the initial contact of a freely falling drop with the target fluid, a high-frequency sound packet was formed, and with some delay, a new lower-frequency packet or group of packets was observed [72]. The main source of sound in the droplet impact flow was considered to be an oscillating gas volume, excited by the compression during the primary contact of the merging liquids and by the rapid retraction of the remainder of the air bridge during the pinch-off of a large bubble [73].

The emission processes of waves during the coalescence of a freely falling drop visualize the action of different mechanisms for transfer of the total energy, including the kinetic energy of motion, potential, and internal energy. To describe the internal energy, the free enthalpy (the Gibbs potential G) was chosen, with its derivatives determining the traditional thermodynamic quantities [29,30,71].

The thermodynamic and kinetic quantities of a drop's impact flow included in the equations are the density of the air ρ_a and water ρ_d (further $\rho_{a,d}$); kinematic $\nu_{a,d}$ and dynamic $\mu_{a,d}$ viscosities of the media; conventional σ_d^a and normalized $\gamma = \sigma_d^a/\rho_d$ cm^3/s^2 coefficients of the surface tension; the acceleration of gravity g; the diameter D, surface area S_d, volume V_d, mass M, and velocity U of the droplet's contact with a target fluid, and the duration of its complete coalescence $\tau_D = D/U \sim 10^{-3}$ s. The ratios of these parameters set the characteristic dimensionless parameters, such as the numbers of Reynolds Re $= UD/\nu$; Froude Fr $= U^2/gD$; Bond Bo $= gD^2/\gamma$; Onezorge Oh $= \nu/\sqrt{\gamma D}$; and Weber We $= U^2D/\gamma$. The kinetic energy of a freely falling liquid droplet is equal to $E_k = \frac{MU^2}{2}$, and the available surface potential energy (ASPE) is $E_\sigma = \sigma S_d$. The potential energy E_p is determined by the position of the fluid in the gravity field and particularly by the shape and area of the free surface.

The Gibbs potential is distributed non-uniformly inside a fluid with a free surface. Far from the boundaries inside of a homogeneous fluid, the value of the potential is determined by the entropy s, temperature T, specific volume $V = 1/\rho$, and pressure P [28,29]:

$$G_f = -sT + VP \tag{18}$$

The differential of the Gibbs potential dG_S depends on the concentration of the impurity components S_i and the chemical potential μ_i:

$$dG_i = -sdT + VdP + \mu_i dS_i. \tag{19}$$

The anisotropy of the atomic–molecular interactions near the free surface forms large gradients of physical quantities. It was found by optical and X-ray reflectometry and atomic force microscopy that the density, dielectric constant, and dipole moment in the bulk of the fluid and in the structurally distinguished near-surface layer with a thickness of the order of the molecular cluster size ($\delta_\sigma \sim 10^{-6}$ cm) differed markedly [50–53]. Taking into account the differences of the physical properties near the free surface of a fluid, the additional term $\Delta G = -S_\sigma d\sigma$ is introduced in the Gibbs potential, which has the meaning of the ASPE:

$$dG_\sigma = -sdT + VdP - S_\sigma d\sigma. \tag{20}$$

For drop with diameter $D \sim 0.5$ cm falling with a velocity $U \sim 1$ m/s, the ratio of the ASPE E_σ to the kinetic energy of the falling drop E_k does not exceed several percent, but the ASPE density $W_\sigma = \frac{E_\sigma}{V_\sigma}$ is three orders of magnitude higher than the kinetic energy density $W_k = \frac{E_k}{V}$. When fluids coalesce, free surfaces are eliminated, and potential energy is quickly transformed into an active form, including pressure, thermal, chemical, and mechanical energy perturbations.

The ASPE transformation process plays a significant role in the formation of the fine structure of the impurity distribution in the impact flows of a freely falling drop. The elimination of the near-surface layers of a droplet and target fluid with a thickness of the order of the size of a molecular cluster ($\delta_\sigma \sim 10^{-6}$ cm) for the falling drop velocity U of the order of several meters per second occurs rather rapidly ($\tau_\sigma = \delta_\sigma/U \sim 10^{-8}$ s), while the entire droplet coalesces out in a time of the order of $\tau_D = D/U \sim 10^{-3}$ s. The released energy is stored in a thin layer on the outer contour of the confluence region and accelerates thin jets of coalesced fluids, which form typical line structures at the bottom of the cavity and the walls of the crown [45]. In turn, the energy of mechanical motion is partially converted into ASPE with an increase in the area of the deformed free surface.

A photograph of the flow pattern of a coalescing drop of a ferric chloride solution and an ink drop decaying into separate fibers is shown in Figure 7. Upon reaching the surface of the fluid, the jets going out from the drop–target fluid confluence region form thin spikes. Small droplets (splashes) fly out from the tops of the spikes, with the velocity being an order higher in magnitude than the contact velocity of the droplet. Over time, as the line of contact of the fluids advances, due to the effects of molecular diffusion of momentum (viscous damping of flows), the thickness of the jets increases, and the velocity decreases. At the same time, the size of the droplets, flying more and more slowly, grows as well.

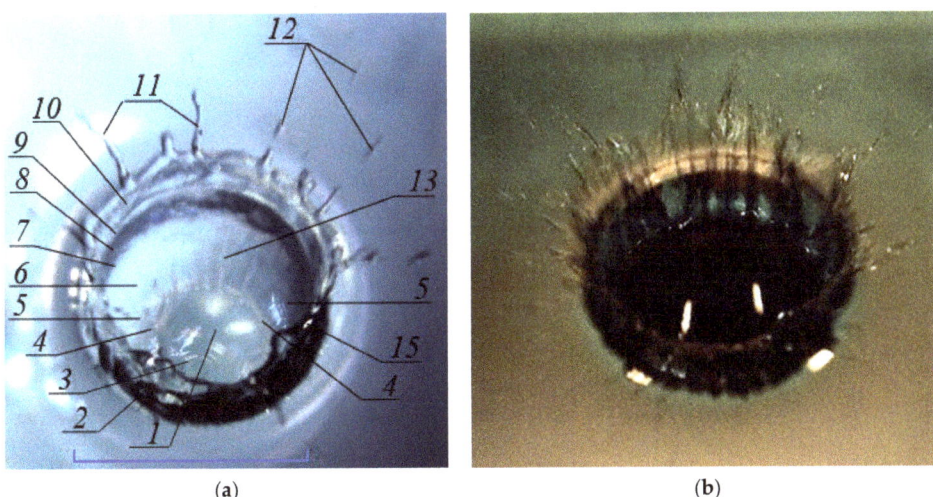

Figure 7. Decay of a freely falling droplet into individual jets at the boundary of the confluence region with the target fluid. (**a**) A drop of a saturated solution of ferrous sulfate merges with water; $D = 0.5$ cm, $U = 3.5$ m/s. (**b**) A drop of ink merges with water; $D = 0.5$ cm, $U = 3.2$ m/s, with the weak color background of the cavity and crown wall and dark fast jets (ligaments) reaching the tips of the spikes. Denotation: 1 = drop, 2 = edge of a crown with teeth, 3 = a wake of an emitted droplet's impact on the surface of a coalescing drop, 4 = the boundary of the region of the drop and target fluid confluence, 5 = annular capillary waves at the bottom of the cavity covering the region of confluence, 6 = bottom of the cavity, 7 = the border of the cavity and the crown, 8 and 9 = the wall and the upper edge of the crown, respectively 10 = veil, 11 = spikes, 12 = small droplets (splashes), 13 = fast jets (ligaments) at the bottom of the cavity, 14 = 3D texture of the crown wall, and 15 = teeth on the crown edge; the length of the tag is 1 cm.

As the flow pattern further evolves, the kinetic energy of the formed flow transforms into potential energy and the ASPE of the distorted surface of the fluid, and this is backwards when the crown and splashes sink and the ejected drops return. The reverse transformation of the energy of motion in the ASPE of the newly formed free surface occurs much more slowly ($\tau \sim 10^{-3} \ldots 10^{-1}$ s).

After the end of the active phase of evolution, the heavier colored fluid forms a toroidal vortex in the body of the target fluid, consisting of colored fibers separated by transparent sections. The vortex slowly sinks, breaking up into new rings and forming an expanding cascade [74]. The movement of the entire structure and its constituent individual elements gradually slows down, as well as the diffuse spreading of the fibers and the decrease in the density difference between the fibers and the receiving fluid as well as the viscous dissipation of the remnants of the initial momentum of the drop. In the final phase, it is the dissipative processes which determine the geometry and dynamics of the flows.

The dynamics of the processes of total energy conversion and their influence on the evolution and structure of a fluid flow requires further study.

9. Discussion of Results

Based on the results of a brief excursion into the history of the development of nature research, the following principles for constructing the methodology of a scientific investigation are formulated:

- **Meaningfulness**: definability of the essence of a subject under study;
- **Identities**: immutability of the content of a subject;
- **Consistency**: internal unity of operations;
- **Uniqueness**: binary logic;
- **Sufficient reason**: prehistory and a consistent environment;
- **Minimum sufficiency**: as simple as possible;
- **Causality**: the directed temporal evolution of events;
- **Completeness**: description of the known properties without involving additional concepts and the openness to accept new facts.

The implementation of the listed principles is carried out within the framework of engineering mathematics, which is defined as "Axiomatic science for the principles of choosing the content of symbols, rules of operations and criteria for assessing accuracy", as well as supplementing technical mechanics, or "Empirio-axiomatic science for the criteria of choosing physical quantities, measurement techniques and procedures for assessing the error".

The unity of the scientific basis allows direct comparison of the results of theoretical and experimental studies with independent and mutual control of the accuracy of calculations and measurement errors.

The general basis for consistent engineering mathematics and experimental mechanics of fluid flows within the framework of a continuous medium model is the axiomatically introduced system of fundamental equations representing differential analogues of the conservation laws for matter, concentration of dissolved components, momentum, and energy (including latent energy and the mechanisms for its conversion into active forms). The system of equations determines the physical quantities characterizing the equilibria and flows of fluids or gases.

The energy basis for describing the state and flows of a fluid involves all forms of energy, which are mechanical kinetic, potential, and internal, including chemical, surface, and electromagnetic energies. The mechanisms for reciprocal energy transfer include direct and reverse transition processes from one form to another (for example, available potential surface or chemical energy due to fluid stratification can be converted into mechanical energy of fluid flows and vice versa). The energy is transferred with the flow velocity, group velocity of waves, and dissipative processes and is rapidly transformed from one form to another as the result of direct atomic–molecular interactions (for example, during the conversion of ASPE into other forms).

Taking into account the distinguished role of energy in the structure formation and the impact on the dynamics of flows, the internal energy was selected as the first physical quantity used for describing the equilibrium state of fluids and gases in the form of a thermodynamic potential (the Gibbs potential was nominated as the basic one). The derivatives of the Gibbs potential and their combinations define traditional thermodynamic quantities, such as the density, pressure, temperature, concentration of components, and surface tension coefficient. The inescapable molecular processes of matter, momentum, heat, and transfer are characterized by kinetic coefficients. Additional coefficients describe the processes of propagation of electrical current and acoustic and electromagnetic waves with various lengths. All the coefficients really are functions of thermodynamic variables and can be involved in complementary equations of state.

Fluid or gas flow is defined as an inseparable transfer of the independent measures of fluid motion, such as momentum, energy, and matter, accompanied by self-consistent changes in the thermodynamic (density, pressure, temperature, and concentration of components), kinetic (dissipative coefficients), and additional coefficients characterizing the propagation of acoustic or electromagnetic waves, electric current, and other phenomena.

Within the framework of the "continuous medium" model, the description of fluid flows is carried out by continuous functions based on the solutions of the system of axiomatically introduced fundamental equations describing the transfer of momentum, energy, and matter with physically substantiated initial and boundary conditions.

The analysis of the fundamental system of equations was carried out while taking into account the compatibility condition, which determined the rank of the complete nonlinear system, the order of its linearized version, and the degree of the characteristic (dispersion) equation. The rank of the system, which defines the minimum number of independent functions making up the complete solution, is 6 for a one-component medium without a thermodynamic state equation (NSE for stratified and homogeneous fluids), 8 when energy (temperature) transfer is taken into account, and 10 for the case of introducing the additional equation of matter transfer into the system. Due to the nonlinearity of the equations, all components of the flows interact with each other. The superposition of a large number of inseparable eigenfunctions with independent space–time properties, which form the fields of registered physical quantities, is manifested in the unsteadiness and evolution of the flow structure.

The classification of the flow structural components for a fluid with weak dissipation, carried out on the basis of the complete solution of the linearized system of fundamental equations, includes large-scale regular components, such as waves or vortices, jets or wakes, as well as singular components, including families of fine ligaments. The transverse scales of ligaments are determined by the dissipative properties of the medium and the characteristic time of the process (i.e., by the duration of flow formation, frequency of periodic flow, or velocity of a uniform free stream). Due to the dual nature of ligaments, they reflect the impact of the atomic–molecular properties which are presented by the kinetic and mechanical coefficients defined by the flow type. Ligaments connect processes on micro- and macroscales. The nonlinear interactions of the structural components of all solutions of the linearized system of fundamental equations generate new components, including vortices which are unsteady localized perturbations with a high level of vorticity formed by embedded ligaments.

The construction of programs for numerical simulations and experimental techniques, taking into account the fluid properties based on the system of fundamental equations, enables carrying out coordinated theoretical and experimental studies of flows and estimating the accuracy of calculations and the error of experiments without involving additional hypotheses, equations, and parameters.

As an illustration, the 2D problem on the uniform flow of (strongly and weakly) stratified and (potentially and actually) homogeneous fluids around an arbitrarily oriented plate was considered within the framework of a reduced system of fundamental equations when the heat conductivity and compressibility effects were neglected. The fields of various physical variables were calculated for a wide range of flow parameters, including diffusion-induced creeping flows and pronounced wave and unsteady vortex regimes, within a unified formulation without involving additional equations or constants. The computational and experimental results were in good agreement as a whole and in their individual details.

In the observations of the patterns of droplet flows, a number of effects have been identified which show the influence of the processes of rapid conversion of the available potential surface energy on the pattern of substance distribution. These include the disintegration of a drop into separated filaments in the vicinity of the confluence line, the formation of fast, small droplets moving faster than falling droplets, the formation of linear and mesh structures of filaments containing drop matter on the walls of the cavity and crown, and the filament structure of vortex elements at all the stages of subsequent flow evolution.

The development of numerical simulation codes based on the system of fundamental equations in a full 3D formulation, together with consistent experimental research techniques which allow registering the large-scale flow components and resolving all the fine-structural ones with the estimation of measurement errors, will help to significantly improve the accuracy of describing the dynamics and structure of a fluid and gas flow and develop substantiated estimates for predicting their evolution, effective flow control methods, as well as motion control techniques for a free vehicle in both gaseous and liquid media.

10. Conclusions

In the context of compatible axiomatic engineering mathematics and empirio-axiomatic technical mechanics as branches of sciences having in-definition demands to control the theoretical accuracy and experimental errors, the description of the fluid or gas status and flows is based on the conserved mathematical and physical quantities. Motion is defined as the transformation of metric space into itself with the conservation of distances. A fluid flow is defined as a transfer of momentum, matter, and energy.

The state and flow of a fluid or gas are defined by the complete set of fundamental equations, describing the transport of matter, constituents, momentum, and complete energy, including kinetic, potential, and internal energy. The thermodynamic parameters of fluid are defined as the derivatives of thermodynamic potentials. The transport of physical quantities is characterized by kinetic and material coefficients. The conventional system of fundamental equations supplementing empirical equations of state for thermodynamic potentials (the Gibbs potential is selected as basic) and density, as well as the boundary and initial conditions, is closed, well-posed, and resolvable. The system, which is characterized by a high rank and high dimension of the functional space, becomes degenerated on singular components in an approximation of a constant density.

The complete solutions of the fundamental equation system, which was analyzed while taking into account the owed modern version of conventional rules by Archimedes–Leibnitz (Meaningfulness, Identities, Consistency, Uniqueness, Sufficient reason, Minimum sufficiency, Causality, and Completeness) and the compatibility conditions, describe large-scale flow components (e.g., waves, vortices, jets, and wakes) and a rich family of ligaments (singularly perturbed components in an approximation of weak dissipation). The fluid flows, as a result of the superposition of many functions with different spatiotemporal parameters, are unsteady.

The nonlinear interactions of flow components result in the permanent evolution of the dynamical and structural parameters of a flow.

A common fundamental basis allows for providing a direct comparison of experimental and theoretical results with an estimation of the accuracy and errors. To increase the accuracy of the system state description and the prognostic potential of solutions, new technical instruments and codes allowing the definition of all flow parameters have to be developed.

Funding: This research was funded by the Russian Science Foundation (RSF), grant number 19-19-00598 "Hydrodynamics and energetics of drops and droplet jets: formation, motion, break-up, interaction with the contact surface".

Institutional Review Board Statement: Not applicable.

Informed Consent Statement: Not applicable.

Data Availability Statement: Data sharing not applicable.

Acknowledgments: This paper is dedicated to the memory of my dear friend and colleague Yuri Vasil'evich Kistovich (1955–2001), who made an invaluable contribution to the development of the theory of waves of various types and stratified flows.

Conflicts of Interest: The author declares no conflict of interest. The funders had no role in the design of the study; in the collection, analyses, or interpretation of data; in the writing of the manuscript, or in the decision to publish the results.

References

1. Galilei, G. *Il Saggiatore*; Appresso, G., Ed.; Mascardi: Rome, Italy, 1623; p. 236.
2. Descartes, R. *A Discourse on the Method, Optics, Geometry and Meteorology*; Olscamp, P.J., Ed.; Hackett Publishing: Indianapolis, IN, USA, 2001; p. 424.
3. Descartes, R. *Principia Philosophiae. Apud Ludovicum Elzevirium*; Springer Science: Berlin, Germany, 1984; p. 328.
4. Maxwell, J.C. Remarks on the Mathematical Classification of Physical Quantities. *Proc. L. Math. Soc.* **1871**, *3*, 224–233.
5. Sir, W.R.H.; Hamilton, W.E. (Eds.) *Elements of Quaternions*; Longmans, Green, & Co.: London, UK, 1866; p. 800.
6. Leibniz, G.W. Brevis demonstration erroris memorabilis Cartesii et al.lorum circa legem naturalem, secundum quam volunt a Deo eandem semper quatitatem motus conservari, quia et re mechanica abuntur. *Acta Erud.* **1786**, *3*, 161–163.
7. Newton, I. *Philosophiæ Naturalis Principia Mathematica*; J.Streater: London, UK, 1687; p. 320.
8. Zhuravlev, V.F. *Fundamentals of Theoretical Mechanics*; Izdatel'stvo Fiziko-Matematicheskoj Literatury: Moscow, Russia, 2001; p. 320. (In Russian)
9. D'Alembert, J.-L.R. *Réflexions sur la Cause Générale des Vents*; David: Paris, France, 1747; p. 372.
10. D'Alembert, J.-L.R.; la Marquis de Condorcet, J.M.A.; l'abbe Bossut, C. *Nouvelles Expériences sur la Résistance des Fluids*; C.-A. Jombert: Paris, France, 1777; p. 232.
11. Euler, L. Principes généraux du mouvement des fluides. *Mémoires L'académie Des. Sci. Berl.* **1757**, *11*, 274–315.
12. Kistovich, A.V.; Chashechkin, Y.D. Propagating stationary surface potential waves in a deep ideal fluid. *Water Res.* **2018**, *45*, 719–727. [CrossRef]
13. Fourier, J.B.J. *Théorie Analytique de la Chaleur*; Firmin Didot Père et Fils: Paris, France, 1822; p. 639.
14. Navier, C.-L.-M.-H. Mémoire sur les Lois du Mouvement des Fluids. *Mém. l'Acad. Sci.* **1822**, *6*, 389–417.
15. Fick, A.E. On liquid diffusion. *Philos. Mag.* **1855**, *10*, 30–39. [CrossRef]
16. Stokes, G.G. On the theories of the internal friction of fluids in motion, and of the equilibrium and motion of elastic bodies. *Trans. Camb. Philos. Soc.* **1845**, *8*, 287–305.
17. Lamb, H. *Treatise on the Motion of Fluids*, 6th ed.; CUP: Cambridge, UK, 1879; p. 258.
18. Franklin, B. Behavior of oil on water. Letter to J. Pringle. In *Experiments and Observations on Electricity*; R. Cole: London, UK, 1769; pp. 142–144.
19. Lomonosov, M. De motu aeris in fodinis observato. *Novi Comm. Acad. Scie. Petropolit.* **1750**, *1*, 267–275.
20. Lomonosov, M. Tentamen theoriae de vi aëris elastic. *Novi Comm. Acad. Scie. Petropolit.* **1750**, *1*, 230–244.
21. Lomonosov, M.V. *A Brief Description of Various Voyages in the North. Seas and an Indication of the Possible Passage from the Siberian Ocean. to East. India*; Marine Techn. Comm.: St. Petersburg, Russia, 1763; p. 34.
22. Mendeleev, D.I. *On Drag of Fluids and on Aeronautics*; Typo. V. Demakova: St. Petersburg, Russia, 1880; p. 80. (In Russian)
23. Mendeleev, D.I. *On the Elasticity of Gases*; Typo. V. Demakova: St. Petersburg, Russia, 1875; p. 262.
24. Mendeleev, D.I. *Studies of Water Solutions on Specific Gravity*; Typo. V. Demakova: St. Petersburg, Russia, 1877; p. 536.
25. Mendeleeff, D. The variation in density of water with temperature. *Philos. Mag.* **1892**, *33*, 99–132. [CrossRef]
26. Popov, N.I.; Fedorov, K.N.; Orlov, V.M. *Sea Water*; Nauka: Moscow, Russia, 1979; p. 330. (In Russian)
27. Landau, L.D.; Lifshitz, E.M. *Fluid Mechanics. V.6. Course of Theoretical Physics*; Pergamon Press: Oxford, UK, 1987; p. 560.
28. Gibbs, J.W. *Elementary Principles in Statistical Mechanics*; Scribner's and Sons: New York, NY, USA, 1902; p. 207.
29. Feistel, R.; Harvey, A.H.; Pawlowicz, R. International Association for the Properties of Water and Steam. In Proceedings of the Advisory Note No. 6: Relationship between Various IAPWS Documents and the International Thermodynamic Equation of Seawater—2010 (TEOS-10), Dresden, Germany, 1–5 September 2016.
30. Feistel, R. Thermodynamic properties of seawater, ice and humid air: TEOS-10, before and beyond. *Ocean. Sci.* **2018**, *14*, 471–502. [CrossRef]
31. Russell, J.S. Report on Waves. In Proceedings of the 14th Meeting of the British Association for the Advancement of Science, York, UK, 2 September 1844; pp. 311–390.
32. Rayleigh, L.; Strutt, J.W. *Theory of Sound. V.1*; CUP: Cambridge, UK, 1877.
33. Reynolds, O. An experimental investigation of the circumstances which determine whether the motion of water shall be direct or sinuous, and of the law of resistance in parallel channels. *Proc. R. Soc. Lond.* **1883**, *35*, 84–99.
34. Reynolds, O. On the dynamical theory of incompressible viscous fluids and the determination of the criterion. *Philos. Trans.* **1895**, *186*, 123–164.

35. Prandtl, L. Über Flüssigkeitsbewegung bei sehr kleiner Reibung. In *Proceedings of the Verhandlungen des Dritten Internationalen Mathematiker-Kongresses, Heidelberg, Germany, 8–13 August 1904*; Teubner: Leipzig, Germany, 1905; pp. 485–491.
36. Schlichting, H. *Boundary Layer Theory*; McGraw Hill Co.: New York, NY, USA, 1955; p. 812.
37. Alekseenko, S.V.; Kuibin, P.A.; Okulov, V.L. *Theory of Concentrated Vortices*; Springer: Berlin, Germany, 2007; p. 494.
38. Lighthill, J. *Waves in Fluids*; CUP: Cambridge, UK, 1978; p. 504.
39. Turner, J.S. *Buoyancy Effects in Fluids*; CUP: Cambridge, UK, 1979; p. 368.
40. Müller, P. *The Equations of Oceanic Motions*; CUP: Cambridge, UK, 2006; p. 302.
41. Vallis, G.K. *Atmospheric and Oceanic Fluid Dynamics*; CUP: Cambridge, UK, 2017; p. 745.
42. Chashechkin, Y.D. Singularly perturbed components of flows—Linear precursors of shock waves. *Math. Model. Nat. Phenom.* **2018**, *13*, 1–29. [CrossRef]
43. Baidulov, V.G.; Chashechkin, Y.D. Comparative analysis of symmetries for the models of mechanics of nonuniform fluids. *Dok. Phys.* **2012**, *57*, 192–196. [CrossRef]
44. Ladyzhenskaya, O.A. Sixth problem of the millennium: Navier-Stokes equations, existence and smoothness. *Russ. Math. Surv.* **2003**, *58*, 251–286. [CrossRef]
45. Chashechkin, Y.D.; Ilinykh, A.Y. Drop decay into individual fibers at the boundary of the contact area with a target fluid. *Dokl. Phys.* **2021**, *66*, 101–105.
46. Aristoteles_Metaphysics. Book IV. Available online: https://www.documentacatholicaomnia.eu/03d/-384_-322,_Aristoteles,_Metaphysics,_EN.pdf (accessed on 15 September 2021).
47. Manturov, O.V.; Solntsev, Y.K.; Sorkin, Y.I.; Fedin, N.G. *Explanatory Dictionary of Mathematical Terms*; Education: Moscow, Russia, 1964; p. 539. (In Russian)
48. Serrin, J. *Mathematical Principles of Classical Fluid Mechanics*; Handbuch der Physik, Band VIII/1: Berlin, Germany, 1959; pp. 125–263.
49. Mase, G.E. *Theory and Problems of Continuum Mechanics*; McGraw-Hill: New York, NY, USA, 1964; p. 270.
50. Eisenberg, D.; Kauzmann, W. *The Structure and Properties of Water*; Oxford University Press: Oxford, UK, 2005; p. 308.
51. Bunkin, N.F.; Suyazov, N.V.; Shkirin, A.V.; Ignat'ev, P.S.; Indukaev, K.V. Study of Nanostructure of highly purified water by measuring scattering matrix elements of laser radiation. *Phys. Wave Phenom.* **2008**, *16*, 243–260. [CrossRef]
52. Teschke, O.; de Souza, E. Water molecule clusters measured at water/air interfaces using atomic force microscopy. *Phys. Chem.–Chem. Phys.* **2005**, *7*, 3856–3865. [CrossRef]
53. Chashechkin, Y.D. Evolution of the fine structure of the matter distribution of a free-falling droplet in mixing liquids. *Izv., Atmos. Ocean. Phys.* **2019**, *55*, 285–294. [CrossRef]
54. Helmholtz, H. Über Integrale der hydrodynamischen Gleichungen, welche den Wirbelbewegungen entsprechen. *J. Reine Angewandte. Dieprechen* **1858**, *55*, 25–55.
55. Bertrand, J. Théorème relative au mouvement le plus général d'un fluide. *Comp. Rend.* **1868**, *66*, 1227–1330.
56. Bertrand, J. Note relative à la théorie des fluides. Résponse à lacommunication de M. Helmholtz. *Comp. Rend.* **1868**, *67*, 267–269.
57. Bertrand, J. Observations nouvelles sur un mémoire de M. Helmholtz. *Comp. Rend.* **1868**, *67*, 469–472.
58. Lie, S. *Zur Allgemeinen Theorie der Partiellen Differentialgleichungen Beliebiger Ordnung*; Berichte Sächs. Ges.: Leipzig, Germany, 1895; pp. 53–128.
59. Chashechkin, Y.D. Conventional partial and new complete solutions of the fundamental equations of fluid mechanics in the problem of periodic internal waves with accompanying ligaments generation. *Mathematics* **2021**, *9*, 586. [CrossRef]
60. Nayfeh, A.H. *Introduction to Perturbation Techniques*; John Wiley & Sons: New York, NY, USA, 2011; p. 533.
61. Vasiliev, A.Y.; Chashechkin, Y.D. Generation of beams of three-dimensional periodic internal waves by sources of various types. *J. Appl. Mech. Tech. Phys.* **2006**, *47*, 314–323. [CrossRef]
62. Kistovich, Y.V.; Chashechkin, Y.D. Linear theory of beams internal wave propagation an arbitrarily stratified liquid. *J. Appl. Mech. Tech. Phys.* **1998**, *39*, 302–309. [CrossRef]
63. Paoletti, M.S.; Swinney, H.L. Propagating and evanescent internal waves in a deep ocean model. *J. Fluid Mech.* **2012**, *706*, 571–583. [CrossRef]
64. Teoh, S.G.; Ivey, G.N.; Imberger, J. Laboratory study of the interaction between two internal wave rays. *J. Fluid Mech.* **1997**, *336*, 91–122. [CrossRef]
65. Prandtl, L. *Führer Durch die Strömungslehre*; Verlagskatalog Von Friedr. Vieweg & Sohn in Braunschweig: Berlin, Germany, 1942; p. 337.
66. Phillips, O.M. On flows induced by diffusion in a stably stratified fluid. *Deep-Sea Res.* **1970**, *17*, 435–443. [CrossRef]
67. USF "HPC IPMech RAS"—Hydrophysical Complex for Modeling Hydrodynamic Processes in the Environment and Their Impact on Underwater Technical Objects, as Well as the Distribution of Impurities in the Ocean and Atmosphere, Institute for Problems in Mechanics RAS. Available online: https://ipmnet.ru/uniqequip/gfk (accessed on 28 October 2021).
68. Chashechkin, Y.D.; Zagumennyi, I.V. 2D hydrodynamics of a plate: From creeping flow to transient vortex regimes. *Fluids* **2021**, *6*, 310. [CrossRef]
69. Levitsky, V.V.; Dimitrieva, N.F.; Chashechkin, Y.D. Visualization of the self-motion of a free wedge of neutral buoyancy in a tank filled with a continuously stratified fluid and calculation of perturbations of the fields of physical quantities putting the body in motion. *Fluid Dynam.* **2019**, *54*, 948–957. [CrossRef]

70. GuoZhen, Z.; ZhaoHui, L.; DeYong, F. Experiments on ring wave packet generated by water drop. *Chin. Sci. Bull.* **2008**, *53*, 1634–1638. [CrossRef]
71. Chashechkin, Y.D. Packets of capillary and acoustic waves of drop impact. *Her. Bauman Mosc. State Tech. Univ. Ser. Nat. Sci.* **2021**, *1*, 73–91. (In Russian)
72. Prosperetti, A.; Oguz, H.N. The impact of drops on liquid surfaces and the underwater noise of rain. *Annu. Rev. Fluid Mech.* **1993**, *25*, 577–602. [CrossRef]
73. Prokhorov, V.E. Acoustics of oscillating bubbles when a drop hits the water surface. *Phys. Fluids* **2021**, *33*, 083314. [CrossRef]
74. D'Arcy, W.T. *On Growth and Forms*; Dover Publications: New York, NY, USA, 1992; p. 1116.

Article

Modeling Multi-Dimensional Public Opinion Process Based on Complex Network Dynamics Model in the Context of Derived Topics

Tinggui Chen [1,2,*], Xiaohua Yin [1], Jianjun Yang [3], Guodong Cong [4] and Guoping Li [5]

1. School of Statistics and Mathematics, Zhejiang Gongshang University, Hangzhou 310018, China; yinxh0213@163.com
2. Academy of Zhejiang Culture Industry Innovation & Development, Zhejiang Gongshang University, Hangzhou 310018, China
3. Department of Computer Science and Information Systems, University of North Georgia, Oakwood, GA 30566, USA; Jianjun.Yang@ung.edu
4. School of Tourism and Urban-Rural Planning, Zhejiang Gongshang University, Hangzhou 310018, China; cgd@mail.zjgsu.edu.cn
5. Zhejiang Liziyuan Food Co., Ltd., Hangzhou 310018, China; winnerwon@163.com
* Correspondence: ctgsimon@mail.zjgsu.edu.cn

Citation: Chen, T.; Yin, X.; Yang, J.; Cong, G.; Li, G. Modeling Multi-Dimensional Public Opinion Process Based on Complex Network Dynamics Model in the Context of Derived Topics. *Axioms* **2021**, *10*, 270. https://doi.org/10.3390/axioms10040270

Academic Editor: Palle E.T. Jorgensen

Received: 1 September 2021
Accepted: 19 October 2021
Published: 21 October 2021

Publisher's Note: MDPI stays neutral with regard to jurisdictional claims in published maps and institutional affiliations.

Copyright: © 2021 by the authors. Licensee MDPI, Basel, Switzerland. This article is an open access article distributed under the terms and conditions of the Creative Commons Attribution (CC BY) license (https://creativecommons.org/licenses/by/4.0/).

Abstract: With the rapid development of the Internet, the speed with which information can be updated and propagated has accelerated, resulting in wide variations in public opinion. Usually, after the occurrence of some newsworthy event, discussion topics are generated in networks that influence the formation of initial public opinion. After a period of propagation, some of these topics are further derived into new subtopics, which intertwine with the initial public opinion to form a multidimensional public opinion. This paper is concerned with the formation process of multi-dimensional public opinion in the context of derived topics. Firstly, the initial public opinion variation mechanism is introduced to reveal the formation process of derived subtopics, then Brownian motion is used to determine the subtopic propagation parameters and their propagation is studied based on complex network dynamics according to the principle of evolution. The formula of basic reproductive number is introduced to determine whether derived subtopics can form derived public opinion, thereby revealing the whole process of multi-dimensional public opinion formation. Secondly, through simulation experiments, the influences of various factors, such as the degree of information alienation, environmental forces, topic correlation coefficients, the amount of information contained in subtopics, and network topology on the formation of multi-dimensional public opinion are studied. The simulation results show that: (1) Environmental forces and the amount of information contained in subtopics are key factors affecting the formation of multi-dimensional public opinion. Among them, environmental forces have a greater impact on the number of subtopics, and the amount of information contained in subtopics determines whether the subtopic can be the key factor that forms the derived public opinion. (2) Only when the degree of information alienation reaches a certain level, will derived subtopics emerge. At the same time, the degree of information alienation has a greater impact on the number of derived subtopics, but it has a small impact on the dimensions of the final public opinion. (3) The network topology does not have much impact on the number of derived subtopics but has a greater impact on the number of individuals participating in the discussion of subtopics. The multidimensional public opinion dimension formed by the network topology with a high aggregation coefficient and small average path length is higher. Finally, a practical case verifies the rationality and effectiveness of the model proposed in this paper.

Keywords: multi-dimensional public opinion; topic derivation; complex network dynamics model; online comments; hot events

1. Introduction

In the era of big data, massive amounts of information are generated on social platforms. Each netizen can make comments on a hot topic based on the information the individual has obtained. When the topic is sufficiently popular, an initial public opinion will form about the event in question. At the same time, with the further disclosure of information, related derived subtopics may be generated and more netizens may join the discussion. Along with the evolution of initial public opinion and the influence of various external factors, some derived subtopics will form a second stage of public opinion [1]. This new derived public opinion and the initial public opinion are intertwined to form a multi-dimensional public opinion. In real life, with the evolution of emergency events and the change of related information, more initial public opinion will generate derived public opinion. Taking the topic "The COVID-19 incident at the end of 2019", for instance, the online opinion derived from "Wuhan epidemic" to other derived topics, such as "Conspiracy theory of the epidemic origin", "Li Wenliang, the first man who discovered the epidemic", and "U.S. congressmen concealed the epidemic". The discussions that form this multi-dimensional public opinion make people nervous and scary and this has a significant impact on social harmony and stability. Based on this, analyzing the formation mechanism of derived public opinion, as well as the formation process of multi-dimensional public opinion, has important theoretical and practical significance for studying public opinion.

At present, there is relatively little research into the formation process of multi-dimensional public opinion in the context of topic derivation. The main research methods are qualitative analysis and quantitative modeling [2]. In terms of qualitative analysis, scholars mostly use specific cases to discuss the definition, potential harm, and common characteristics of network-derived public opinion. In terms of quantitative modeling, researchers try to discover the rules of public opinion derivation by establishing models to better study its evolution so as to effectively prevent possible public opinion crises. Among them, the SIR infectious disease model is often used to analyze network derivation effects. However, currently, only a single-dimensional initial public opinion is analyzed, and multi-dimensional public opinion is rarely considered. In fact, under the influences of multiple information, initial public opinion often derives public opinion in multiple dimensions. Based on this, after the initial topic is discussed and the initial public opinion is formed, this paper first introduces the initial public opinion derivation mechanism to reveal the formation process of derived subtopics, and then analyzes its evolution law based on the SIR disease model and maps the propagation process of multiple subtopics into multiple layers. Basic reproduction number is introduced to determine whether derived subtopics can form derived public opinion, thereby revealing the entire process of multi-dimensional public opinion formation.

The structure of this paper is organized as follows: Section 2 is literature review; Section 3 builds a model for the formation of multi-dimensional public opinion in a topic-derived context; Section 4 uses simulation experiments to analyze the influences of some of the main factors on the formation of multi-dimensional public opinion; Section 5 uses actual cases to verify the model proposed in this paper; Section 6 contains the conclusions of the study and considers the prospects for future work.

2. Literature Review

This paper defines multi-dimensional public opinion as public opinion formed by combining initial public opinion and derived public opinion. When single or multiple derived public opinions are generated, they together go to constitute multi-dimensional public opinion. Derived public opinion is the basis of multi-dimensional public opinion and the main subject of current research. Therefore, the discussion in this paper mainly focuses on the formation of derived public opinion. At present, many scholars have conducted research on the formation of derived public opinion, which mainly includes two aspects: one is the study of the propagation and evolutionary mechanism of derived public opinion; the other is the study of the reasons for the formation of derived public opinion and countermeasures to it.

At present, the research on the propagation and evolutionary mechanism of derived public opinion is mostly based on infectious disease model, but the focus is on specific public opinion topics, such as the research on rumors and fake news events. The representative literatures are as follows: Lan et al. [3] set up a mathematic model of the derivative effect of network public opinion based on a logistic model from the viewpoint of information alienation. Zhang and Feng [4] put forward a two-layer coupled SEIR public opinion propagation model for derived topics that applied when news was circulated. Korobeinikov [5] studied global properties of SIR and SEIR epidemic models with multiple parallel infectious stages and verified that these systems possessed the only globally stable equilibrium state. Arenas et al. [6] adopted a Microscopic Markov Chain Approach (MMCA) meta population mobility model to study the cases of COVID-19. Estrada [7] modeled the Singapore COVID-19 pandemic with an SEIR multiplex network model. Yang et al. [8] proposed a competitive diffusion model, namely, the Linear Threshold model with One Direction state Transition (LT1DT), in order to explore the problem of minimizing the spread of rumor in social networks. Zanette and Damian [9] explored the dynamics of an epidemic-like model for the spread of a rumor on a small-world network. Moreno et al. [10] deduced the mean-field equations used to describe the dynamics of a rumor process occurring on top of complex heterogeneous networks. Zhou et al. [11] considered the influence of network topological structure and the unequal footings of neighbors of an infected node in propagating the rumor and found that the number of final infected nodes depended on the topology of the network. Tan et al. [12] organically combined the analytic hierarchy process and wavelet neural network to develop an effective and feasible network public opinion monitoring system, and analyzed the Theater High Altitude Area Defense (THAAD) incident as a case, verifying that the system had good evaluation performance and estimation accuracy. You et al. [13] proposed a social network-oriented public opinion monitoring platform based on ElasticSearch (SNES), and proved with a large body of empirical evidence that the platform could well adapt to social networks with high real-time data and good performance in public opinion monitoring. Chen et al. [14] proposed a monitoring and identification method for high-risk users of enterprise public opinion combined with user portrait technology and a random forest algorithm. The proposed scheme helped enterprises identify high-risk users with inadequate experience who may trigger negative public opinion. Wang et al. [15] proposed an improved energy model to characterize the propagation of rumors on social networks quantitatively and used experiments to evaluate the influences of model parameters, network structures, and effective linkage rates. Askarizadeh et al. [16] presented an evolutionary game model to analyze the rumor process in social networks and the analysis results showed that propagation of convincing anti-rumor messages and the location of rumor control centers had an important effect on debunking rumors. Although the aforementioned literature presents a propagation model of derived public opinion, it mostly uses topic derivation rate parameters to quantify the topic derivation process, and only explores the influence of the change of topic derivation rates on initial public opinion along a single dimension and does not consider multi-dimensional public opinion situations with the combination of derived public opinion and initial public opinion.

Some scholars have conducted research on the reasons for opinion formation and the response strategies of derived public opinion, and the representative literatures are as follows: From the viewpoint of spreading factors, Zhang [17] studied causes of network public opinion derivation triggered by public emergencies and concluded that the government, the media, and the public were important factors in the formation of derived public opinion, their different behavioral logics based on their respective interests in spreading information being the fundamental reason for the generation of derived public opinion. Duncan and Peter [18] found that large cascades of influence were driven not by influential individuals but by a critical mass of easily influenced individuals. Li [19] considered that the fundamental reason for the formation of derived network public opinion lies in information alienation. Snyder et al. [20] held that many large-scale phenomena, such as rapid changes in public opinion and the outbreak of disease epidemics, could be fruitfully modeled as cascades of activation on networks. In addition, according to the general law of network derived public opinion, Wang and Dai [21] summed up three basic derived chain structure types and corresponding probability algorithms, and finally listed the operational steps of a network derived public opinion chain. Moreover, by introducing social preference theory, Chen et al. [22] revealed the micro-interaction mechanism of public opinion polarization and their simulation results showed that different social preferences held by individuals had different influences on public opinion polarization effects. Although the above-mentioned literature reveals the formation and propagation mechanisms of derived public opinion to a certain extent, its research methods are mostly inductive, and are mostly based on single-dimensional network public opinion development. Few scholars adopt the perspective of multi-dimensional public opinion.

In addition, a number of scholars have conducted preliminary discussions of the multi-dimensional public opinion evolution model. By combining social judgment theory with the multi-agent model, Li and Xiao [23] proposed a multidimensional opinion evolution model for studying the dynamics of opinion polarization and the simulation results demonstrated that the polarization process was affected by assimilation effect parameters and contrast effect parameters. Parsegov et al. [24] proposed a significant extension of the classical Friedkin–Johnsen model so as to describe the evolution of agents' opinions on several topics. In addition, based on the real processes by which multiple topics concerning the same event were generated and disseminated, Sun and Chai [25] divided the portraits of online learners into three dimensions and constructed a labeling system for the portraits of learners based on the data fields of an online learning platform. Wang et al. [26] designed a topic detection algorithm that worked on these multidimensional public opinion networks and their simulation results demonstrated that this model could be used to effectively characterize the communication characteristics of multiple topics on "We the Media" networks. Although the above literature considers the multi-dimensional characteristics of online public opinion and reveals its evolution mechanism to a certain extent, most scholars directly discuss its nature and characteristics from the perspective of multi-dimensional public opinion. Few scholars focus on its formation mechanisms. In the propagation of subtopics, few scholars consider the interweaving and the mutual influence of multi-dimensional derived public opinion and initial public opinion. This has resulted in an incomplete understanding of the phenomena, and further research is needed.

In summary, current scholars mostly use single-dimensional online public opinion models to conduct research. Few scholars combine derived public opinion with initial public opinion and do not discuss the formation process of multi-dimensional public opinion from the perspective of derived topics. In reality, after the outbreak of an initial public opinion on a topic, multiple subtopics are often derived from the network. Some of these subtopics form derived public opinion in the propagation process and they are intertwined with the initial public opinion to form a more influential multi-dimensional public opinion. Based on this, from the perspective of derived topics, the formation mechanism of multi-dimensional public opinion can be studied more clearly. This paper discusses the formation process of multi-dimensional public opinion in the context of derived topics, in order to provide a reasonable guide.

3. Model Construction

The entire process of modeling multi-dimensional public opinion formation is based on complex network dynamics. First, the complex network simulation is used to generate the intricate relationship among netizens in the real world. In addition, the agent is used to represent the individual nodes in the network and the network scale is set to N, meaning it is assumed that there are N nodes in the network, which are divided into three states:

susceptible, infective, and recovered. At the same time, the SIR model is introduced to analyze the evolution of the three types.

Generally, for the initial topic formed by the emergent news worthy event, the perception of netizens and the propagation of events will prompt formation of an initial public opinion on the Internet. However, as time goes by, netizens' subjective understanding of the initial public opinion information will gradually become biased. With the gradual accumulation of deviations, the initial public opinion yield derivations, resulting in multiple subtopics. Some derived subtopics may form derived public opinions due to intense discussion and enthusiasm and a large amount of information. The intertwining of these with the earlier state of public opinion forms a multi-dimensional public opinion. Based on this, the specific research ideas of this paper are shown in Figure 1.

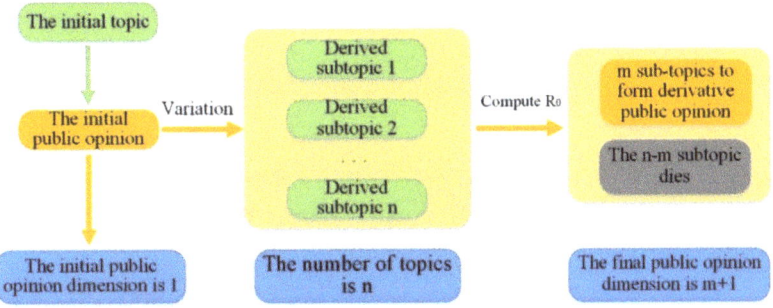

Figure 1. Research framework.

According to the Figure 1, this paper divides the formation process of multi-dimensional public opinion into four stages, as follows:

Stage 1: The formation and propagation stage of initial public opinion. For the initial topic formed by the emergent hot events, a large number of netizens participate in the discussion and spread the topic, prompting the initial topic to form an initial public opinion. The individual netizens participating in the initial public opinion discussion are divided into susceptible, infected and recovered status, and the SIR model is introduced to analyze their propagation process.

Stage 2: The generation stage of derived subtopics. As the initial public opinion spreads and the external environment is stimulated, the probability of the initial public opinion variation gradually increases. The variation degree P is introduced to describe the variation degree of the initial public opinion. If the variation degree at time T_i is at a high interval, it will generate a derived subtopic.

Stage 3: The information propagation stage of derived subtopics. After the derived subtopic is generated, netizens accept and pay attention to this source of information. When the individual's attention to the derived subtopic is greater than the threshold g_0 at time T_i, the individual will participate in the discussion of the sub-topic. Infection rate and immunity rate change based on Brownian motion. The subtopic information evolves according to the SIR model of the new propagation parameters.

Stage 4: The multi-dimensional public opinion formation stage. Although the initial public opinion may derive multiple subtopics, not all subtopics can form public opinion. Only when the propagation reproduction number R_0 of a derived subtopic reaches a certain threshold is it considered that the subtopic forms a derived public opinion, and the newly emerging derived public opinion is intertwined with the initial public opinion to form a multi-dimensional public opinion.

The parameters and variables involved in the formation of multi-dimensional public opinion are shown in Tables 1 and 2.

Table 1. Relevant parameters.

Parameters	Description
α	Infection rate
β	Immunity rate
ρ	Information alienation rate
δ	Environmental forces
θ_i	The topic correlation between the ith derived subtopic and the initial public opinion
σ_i	The amount of information contained in the ith derived subtopic
P_U	The parameter of highly variable degree threshold
g_0	The attention threshold

Table 2. Relevant variables.

Variable	Description
$S(t)$	The numberof susceptible individuals in the network at time t
$I(t)$	The number of infective individuals in the network at time t
$R(t)$	The number of recovered individuals in the network at time t
P	The degree of variation of initial public opinion
g_i	Individual i's attention to subtopics
R_0	Basic reproduction number

3.1. Initial Public Opinion Propagation Model

After the initial public opinion is formed, individuals in the social network begin to receive information and spread public opinion. The propagation process of initial public opinion is analyzed through the SIR model. First of all, individuals in the network are divided into three states in proportion, namely, susceptible state (S), infective state (I), and recovered state (R). S is the class of netizens who have not received relevant public opinion information, I that of netizens who have received public opinion information and who actively spread it, R that of netizens who are not interested in public opinion information. The transformation relationship of the three is shown in Figure 2.

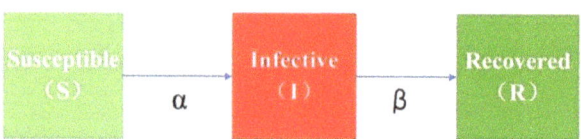

Figure 2. State transformation.

According to the Figure 2, when the public opinion information of a hot event is formed, susceptible individuals (S) will receive relevant information under the influence of surrounding infected individuals and thus become infective individuals (I). Assuming that there are r susceptible individuals among the neighbors of the infective person at this time, the susceptible person will become infective with probability α after contact with the infective person, while infective individuals will gradually lose interest as time goes by and recover with probability β [27].

Suppose that $S(t)$, $I(t)$, and $R(t)$, respectively, represent the number of individuals in a susceptible state, an infective state, and a recovered state in the network at time t, and meet the following Formula (1):

$$S(t) + I(t) + R(t) = N \qquad (1)$$

System dynamics equations are introduced to simulate the evolution of individuals in these three states in the network. Since dynamical equations are used only as a tool for public opinion evolution in this paper, the simple SIR model of rumor propagation in uniform network established by Lu [28] is used as the evolutionary model for initial public opinion. The dynamic equations are shown in the following Formula (2):

$$\begin{cases} \frac{dS(t)}{dt} = -\bar{k} * \alpha * I(t) * \frac{S(t)}{N}; \\ \frac{dI(t)}{dt} = \bar{k} * \alpha * I(t) * \frac{S(t)}{N} - \beta * I(t) \; ; \\ \frac{dR(t)}{dt} = \beta * I(t). \end{cases} \quad (2)$$

where $\bar{k} = \frac{1}{N} * \sum_i^N k(i)$ means the average degree of all nodes in the network. After the initial public opinion of the sudden hot event is formed, the infected node contacts with $\bar{k}$ neighbor nodes. Among them, the neighboring node in a susceptible state becomes infective with probability α, and at the same time the infective node recovers with probability β.

3.2. The Formation Process of Derived Subtopics

Compared with traditional media, the spread of public opinion on the Internet is more sudden and unpredictable. After the initial public opinion of a newsworthy event is formed, related information will be viewed, commented on, and reposted by a large number of netizens in a relatively short period of time, its scope gradually expanding over time. With the continuous release of initial public opinion information, the netizens' subjective understanding of initial public opinion will gradually generate deviations. With the gradual accumulation of these deviations, the information will be distorted in the process of propagation and then magnified by the influence of social network environments. In this process, the initial public opinion will change [29]. Generally speaking, the variation of public opinion events is multi-directional, and people's views on the same event tend to become diversified. Therefore, after public opinion becomes variable, multiple derived subtopics will usually be generated and the subtopics may be further derived to form a public opinion which is consistent with the initial public opinion and constitutes a multi-dimensional public opinion situation.

3.2.1. The Degree of Variation P

Generally, the process of public opinion variation is relatively slow, and the state of variation gradually deepens after a long period of accumulation. The degree of public opinion variation P is introduced to reflect the state of initial public opinion variation. With the evolution of public opinion, P slowly increases, and when P reaches a certain threshold, the initial public opinion will be in a highly variable state, resulting in derived subtopics. As public opinion further develops and fades, P also decreases to a smaller value. Generally speaking, the variation degree of public opinion is affected by many factors. The most important, however, is the degree of information alienation in the propagation of public opinion, along with the environmental forces of public opinion. The degree of information alienation ρ refers to the contradiction between information produced and consumed and is due to the influence of various factors during the formation, propagation, and use of information, such that a subject loses the ability to handle incoming information, losing sight of the initial accounts, and becomes enslaved to and dominated by this new information [30]. For example, after the "Death of He Hongshen in Macao" event, due to changes in netizens' gossip psychology and attention perspective, the information reflecting the event itself lost its original significance and attention was displaced on to his family property, resulting in a series of subtopics such as "He's Original Wife" and "He's Property". Environmental force refers to the influence of the general environment, such as the propagation of events or network communication channels on public opinion, which is often affected by two aspects: firstly, there are the characteristics of the event itself, such as the degree of social disputes in different events; secondly, there is the impact of the network environment on events. δ is used to describe the strength of environmental forces. At present, with the prevalence of new media communication methods, information communication channels have become more diverse, and the network environment has played a guiding role in the opinions, attitudes and communication behaviors of netizens. The scope of the incident may gradually expand, and it is more likely to form public opinion. The incident has been distorted, leading to deviations from the original development trend, and given rise to derivations.

Combining the above two factors, according to the existing literature [31], the evolution process of public opinion variation is divided into a budding period, an outbreak period, a diffusion period, and a dissipative period. The variation process of public opinion is analyzed and the initial public opinion variation degree is described with the following Formula (3):

$$P = \rho * e^{-(\frac{T-9}{\delta})^2} \tag{3}$$

where T represents the evolution time of public opinion, ρ is the degree of information alienation, δ is the environmental forces, and the degree of variation P first increases and then decreases with the development of public opinion. Generally, a deeper degree of information alienation means a higher degree of variation, a greater environmental force, and a higher degree of variation.

3.2.2. The Formation of Subtopics

The initial public opinion variation degree is divided into two intervals, with P_U as the threshold. When $0 < P < P_U$, the degree of variation is lower; when $P_U < P < 1$, the degree of variation is higher. The specific process is shown in Figure 3.

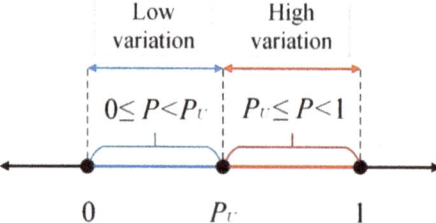

Figure 3. Variation range.

According to Figure 3, when $P_U < P < 1$ at time T_i and the degree of variation P_U exceeds the variation threshold, a derived subtopic is generated. Here, the variation threshold P_U is 0.8.

3.3. The Propagation Process of Derived Subtopics

After the derived subtopics are generated, the popularity of the initial public opinion has not completely disappeared. Therefore, the information covered by the derived subtopics and the initial public opinion information are disseminated throughout the network at the same time. In turn, the multi-layer SIR model is used to simulate the propagation process. It is assumed that when an individual's attention to a subtopic is greater than a certain threshold, individuals participate in the discussion of the subtopic at that level. At the same time, the information propagation parameters of the new subtopics and the number of individuals participating in the discussion have changed, so this section defines these two parameters.

3.3.1. The Propagation Parameter of Derived Subtopics

When derived subtopics are generated, it means that a new topic begins to spread, and whether it can form derived public opinion is closely related to its propagation parameters and the number of individuals participating in the discussion. Consequently, Brownian motion is introduced to allow for random disturbance to the original public opinion parameters, which determines the information propagation parameters of the derived subtopics.

Definition 1 [32]. (Ω, F, P) is a probability space. If an adaptation process B_t satisfies the following conditions in this probability space, B_t is called Brownian motion or the Wiener process. For almost all $(\omega \in \Omega)$, the sample B_t is continuous, and $B_0 = 0$.

(1) For all real numbers s, t satisfies $0 \leq s \leq t$, $B_t - B_s$ and F_s are independent;
(2) When $0 \leq s \leq t$, $B_t - B_s$ obeys the normal distribution $N(0, t - s)$, the normal distribution satisfies the mean value of 0 and the variance is $t - s$.

Since the subtopic is closely related to the initial public opinion, its propagation parameters are related to the initial public opinion propagation parameters. Here, the infection rate and recovery rate parameters of the ith derived subtopic are defined by the following Formulas (4) and (5):

$$\alpha_i = \alpha_0 + \sigma_i \dot{B}_i(t), i = 2, 3, \ldots \quad (4)$$

$$\beta_i = \beta_0 + \sigma_i \dot{B}_i(t), i = 2, 3, \ldots \quad (5)$$

where $B_i(t)$ is the independent Brownian motion, α_0 and β_0 is the infection rate and immunity rate of the initial public opinion, respectively, and $\sigma_i > 0$ ($0 < \sigma_i < 1$) is a constant, which represents the amount of information of the current subtopic, that is, the intensity of the disturbance. The greater the intensity, the greater the change in infection rate.

3.3.2. The Propagation Model of Derived Subtopics

After the derived subtopics are generated, different people will show different degrees of attention when facing the same hot topic. This difference in attention is a manifestation of the heterogeneous characteristics of individual [33] and it will affect whether the individual participates in the discussion of subtopics. Individuals' attention to derived subtopics is often affected by two factors. The first of these is the degree of relevance between the subtopic and the topic that formed the initial public opinion. Generally speaking, the higher the degree of relevance between the subtopic and its correlation, the more attention the individual pays to the subtopic. The second factor is the external influence from the individual's surrounding environment. If more people discuss the subtopic in the individual's surrounding environment, it is easier for the individual to pay attention to the subtopic.

The topic correlation and the number of infected neighbors k_{inf} are used to describe the degree of individual attention to derived subtopics. Setting k_{inf} as the number of neighboring nodes infected around a node with degree k, then the attention degree function of a node with degree k in the network [34] is expressed by the following Formula (6):

$$g_i = 1 - (1 - \theta)^{k_{inf}} \quad (6)$$

where the topic correlation $\theta \in (0, 1)$. Generally speaking, when the correlation between the subtopic and the initial topic is 0, it means that the subtopic is completely irrelevant to the initial topic. At this time, the subtopic cannot be regarded as a derived subtopic of the initial topic, so the individual attention is set to 0 at this time. When the relevance is 1, it means that the subtopic is completely related to the initial topic. At this time, the relevant content of the subtopic is equivalent to the initial topic, so the individual attention is set to 1. In addition, a greater degree of correlation between the subtopic and the initial topic means more individual attention given to the subtopic. The higher number of infections in the neighboring nodes around the individual means that the individual receives more information and consequently pays more attention to subtopics.

Since the number of infected neighbors around each individual is different and their number will change over time, the attention of the individual is also dynamically changing, that is, when a derived subtopic is generated at time T_i, the individuals with high attention to that subtopic will continue to participate in the propagation of the subtopic, while individuals with lower attention will lose interest in the subtopic and will no longer participate in the discussion of the subtopic, resulting in reduced topic spread. It is defined that when $g_i > g_0$ at the time T_i, individuals in the network will participate in the discussion of the subtopic and carry out the propagation of the subtopic. These individuals propagate according to the SIR model with the new propagation parameters and are intertwined with the initial public opinion to form a multi-layer SIR propagation network. The schematic diagram of their propagation transformation is shown in Figure 4.

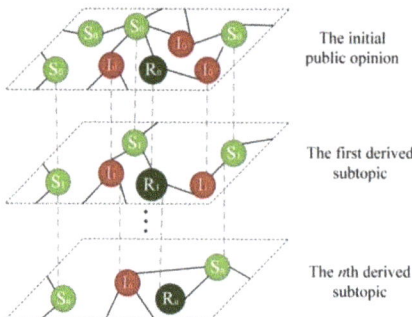

Figure 4. Propagation transformation of derived subtopics.

It can be seen from Figure 4 that the spread of initial public opinion corresponds to the first-level SIR communication network. After the first derived subtopic is generated, individuals in the network choose whether to participate in the discussion of the first derived subtopic according to their degree of attention. A second-layer SIR propagation network emerges, and the process goes on until n subtopics are derived, when an $(n + 1)$-layer SIR propagation network is formed, corresponding to the propagation process of $(n + 1)$ topics.

3.4. The Formation of Multi-Dimensional Derived Public Opinion

Though derived subtopics have been generated and spread, not all subtopics will form derived public opinion. Therefore, this paper introduces the basic reproduction number from infectious disease modeling as the criterion for judging whether a given subtopic forms public opinion. The basic reproduction number refers to the number of people who can be infected by an infective person on average in an environment where all people are susceptible without intervention [35], and its expression is given in the following Formula (7) [36]:

$$R_0 = \left(1 + \frac{r}{\alpha}\right)\left(1 + \frac{r}{\beta}\right) \quad (7)$$

Formula (7) is calculated by the new infection number data and transmission parameters at each time point, these being based on the above transmission dynamics model, where r refers to the growth rate of the number of new infections at time T_i, and α and β refer to the infection rate and immunity rate, respectively. In the infectious disease model, an important critical point of R_0 is $R_0 = 1$. The larger the value of R_0 is, the more difficult it is to control the epidemic. When $R_0 < 1$, the infectious disease will gradually disappear. When $R_0 = 1$, the infectious disease will become endemic. When $R_0 > 1$, infectious diseases will spread exponentially. According to the propagation of derived subtopics, the basic reproduction number R_0 of each derived subtopic is dynamically changing. It is defined here that when the basic reproduction number R_0 of the subtopic is larger than 1 and lasts for a period of time, the subtopic forms a network public opinion, i.e., a dimension is added to the initial public opinion.

According to the above analysis, the specific simulation process of multi-dimensional public opinion generation is shown in Figure 5.

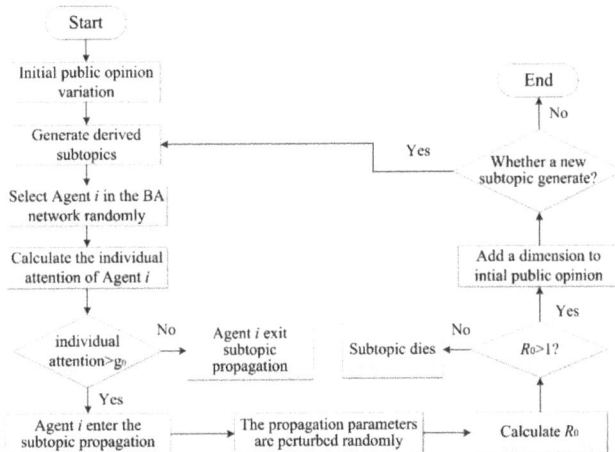

Figure 5. Multi-dimensional public opinion generation.

It can be seen from Figure 5 that when a derived subtopic is generated, the attention of an individual in the network to the subtopic is calculated at that moment. If it is greater than the attention threshold, the individual participates in the discussion of the subtopic. At the same time, the propagation parameters of the subtopic change, and the basic regeneration number also changes accordingly. When the basic reproduction number $R_0 > 1$, it is considered that the subtopic forms a derived public opinion. When multiple derived subtopics form a derived public opinion, this and the initial public opinion form a multi-dimensional public opinion.

4. Simulation Experiments

In this section, combined with the multi-dimensional public opinion formation model built above, we discuss the influence of the degree of information alienation, environmental force, topic correlation, and the amount of information contained in subtopics on the formation process of multi-dimensional public opinion. The degree of information alienation and environmental force are the initial influencing factors on derived subtopics, and the generation of derived subtopics is the basis for the formation of multi-dimensional public opinion. Therefore, when discussing these two factors, they are divided into two parts. The first is the influence on the number of derived subtopics and the propagation process. The second is the impact on the dimensions of multi-dimensional public opinion. In order to find the key factors that affect the dimensions of multi-dimensional public opinion, the above four factors are combined and compared, and finally the influence of different network topologies on the formation of multi-dimensional public opinion is further simulated.

4.1. The Impact of Information Alienation on the Formation of Multi-Dimensional Public Opinion

Based on the multi-dimensional public opinion formation model established in the previous section, initial parameters are set first. Assuming that in the initial state, only the initial public opinion is spread in the network, and the initial degree of variation is not 0, at this time the derived subtopic has not yet been generated. Set the initial public opinion propagation parameters as: $\alpha = 0.17$, $\beta = 0.5$. The complex network chooses the BA (proposed by Barabasi and Albert) scale-free network [37]; the node size is set to $N = 1000$. The relevant threshold is set to: $P_U = 0.8$, $g_0 = 0.6$. In order to facilitate the observation of the formation of multi-dimensional public opinion, the subtopics are set to be derived from the variation of the initial public opinion information. At this time, each derived subtopic presents a positive correlation with the initial public opinion. Taking the comprehensive visualization into consideration, other parameters are set as: $\delta = 9$, $\theta \sim N(0.9, 0.3)$, $\sigma \sim N(0.9, 0.3)$, and the evolution time is set to $T = 40$.

4.1.1. The Impact of Information Alienation on the Amount of Derived Subtopics and Propagation Process

When the information of the initial public opinion is varied, the degree of variation is higher and the degree of variation of the initial public opinion is greater. When the degree of variation reaches the threshold, derived subtopics are generated, and the longer the duration of the high variation state is, the more times the degree of variation reaches the threshold and the number of subtopics also increases. Based on this, this section will simulate and analyze the changes of variation under different information alienation settings. The results are shown in Figures 6 and 7.

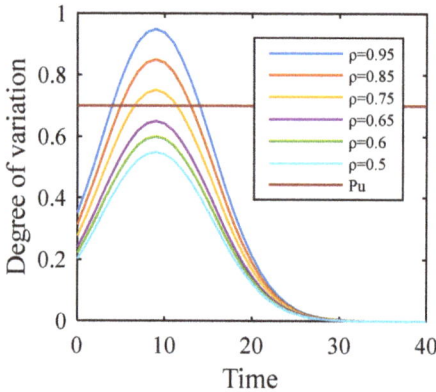

Figure 6. The changes of degree of variation under different information alienation settings.

It can be seen from Figure 6 that no matter how the degree of information alienation changes, the degree of initial public opinion variation always follows a similar trend, that is, as time goes by, the degree of variation will always reach a peak at the same time, and the overall trend will increase first and then decrease. In addition, when time = 0, the initial public opinion variation degree is between 0.2 and 0.4, indicating that the degree of information alienation has little effect on the degree of variation at the initial moment. As the degree of information alienation increases, the peak value of the degree of variation gradually increases. The degree of variation at the initial moment is almost unchanged. This shows that as the degree of information alienation increases, the degree of initial public opinion variation will reach the threshold of variation in a shorter time, resulting in the generation of new derived subtopics. Secondly, after the same evolution time, the stronger the degree of information alienation, the lower the initial public opinion is and the higher is the degree of variation. It can be seen from Figure 7 that when the degree of information alienation is lower than 0.75, the number of derived subtopics generated is 0, which means that only when the degree of information alienation reaches a certain value, there will be derived subtopics, and the number of derived subtopics increases with the deepening of the degree of information alienation.

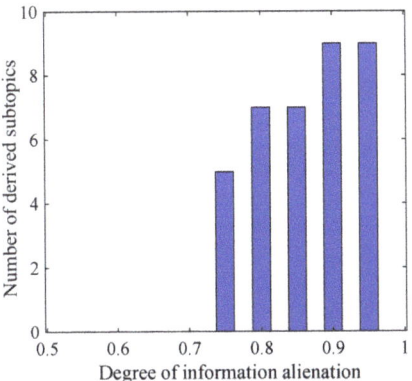

Figure 7. The number of derived subtopics under different information alienation settings.

In summary, the information alienation degree must reach a certain value before derived subtopics are generated. Therefore, the information alienation degree ρ is set to 0.75, 0.85, and 0.95 respectively, and the derived subtopics generated are numbered according to the formation time, and the first is set as 1. The influence of different information alienation settings on the propagation process of the subtopics is analyzed by analogy.

It can be seen from Figure 8 that with the increase in the degree of information alienation, the number of individuals discussing the derived subtopic 1 has not changed significantly, but the number of individuals discussing the derived subtopics 2, 3, and 4 has greatly increased. At $\rho = 0.95$, the number of individuals discussing derived subtopics 2 and 3 is greater than that of derived subtopic 1. The reason may be that the deviation caused by information alienation requires time to accumulate. Since the derived subtopic 1 is generated earlier and has a higher degree of overlap with the initial public opinion information, even if the degree of information alienation increases, the influence on the derived subtopic 1 will be small. At the same time, it shows that when the degree of information alienation is not high, most individuals tend to pay attention to the earliest derived subtopics and participate in the spread of subtopics. When the degree of information alienation is high, the number of derived subtopics generated is also large enough. The derived subtopic with the highest individual participation may not be the one that broke out first. In addition, it can be seen from Figure 8 that as the degree of information alienation increases, the number of derived subtopics further increases.

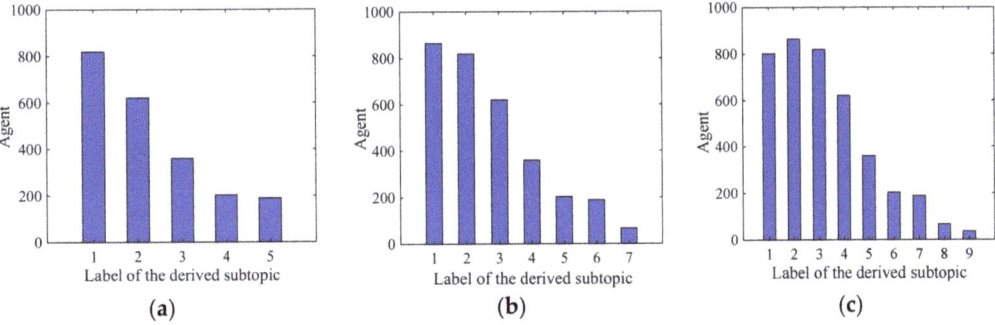

Figure 8. The number of agents participating in subtopic communication under different information alienation. (**a**) $\rho = 0.75$. (**b**) $\rho = 0.85$. (**c**) $\rho = 0.95$.

It can be seen from Figure 9 that the blue, red, and green lines represent the number of agents who are susceptible, infective, and recovered, respectively. Comparing the curve of the number of susceptible persons and the number of infective persons with respect to the initial public opinion and the subtopics in Figure 9, it can be seen that although the initial number of susceptible agents discussing subtopics is less than the number of those discussing the initial topics, the number of infective agents discussing some subtopics is the same as that discussing the initial topics. This indicates that the popularity is the same as the initial public opinion. With the deepening of information alienation, the number of such subtopics further increases.

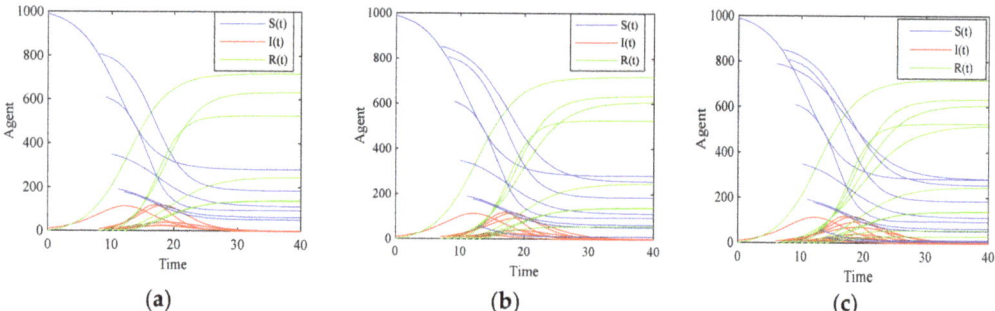

Figure 9. Evolution of initial public opinion and subtopics under different information alienation settings. (**a**) $\rho = 0.75$. (**b**) $\rho = 0.85$. (**c**) $\rho = 0.95$.

4.1.2. The Impact of Information Alienation on the Degree of Multi-Dimensional Public Opinion

The different degree of information alienation affects the time for the generation of subtopics, the number of subtopics and the number of individuals participating in the discussion of the subtopics, which will affect the formation of subsequent multi-dimensional public opinions. Based on this, this section simulates the formation process of multi-dimensional public opinion under different degrees of information alienation and the results are shown in Figures 10 and 11.

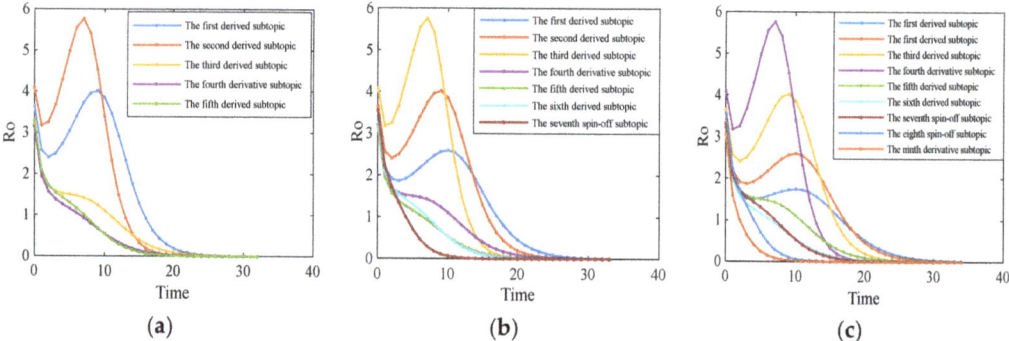

Figure 10. Changes of subtopics' R_0 values under different information alienation settings. (**a**) $\rho = 0.75$. (**b**) $\rho = 0.85$. (**c**) $\rho = 0.95$.

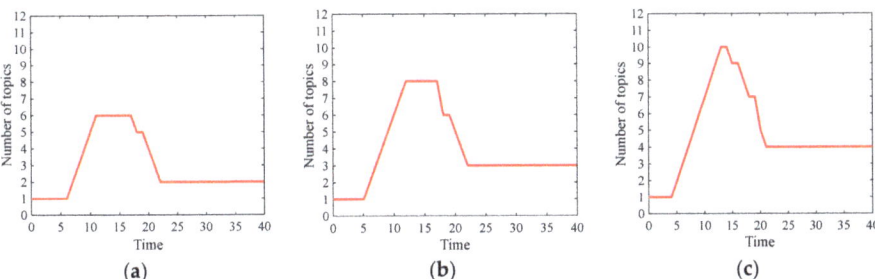

Figure 11. Changes of the number of topics under different information alienation settings. (**a**) $\rho = 0.75$. (**b**) $\rho = 0.85$. (**c**) $\rho = 0.95$.

It can be seen from Figure 10 that the changes in the basic reproduction numbers of each derived subtopic under different information alienation settings shows two trends: one is a short-term decline in the initial stage, then a rise to a peak, and finally a decline and stabilization at 0; the other is a gradual or slow decrease and final stabilization at zero. This shows that after the emergence of subtopics, there has not yet been a large-scale discussion in the initial stage, so the basic reproduction number declines. With the increase of netizens' participation in discussions, some subtopics may become hot topics due to the large amount of information or the influence of factors such as the network environment. Therefore, the basic reproduction number rapidly increases. At the same time, some subtopics may disappear as the initial basic reproduction number declines. In addition, with the increase of the degree of information alienation, the explosive growth of the basic reproduction number of the first derived subtopic becomes weaker and weaker. This shows that when the degree of information alienation is high, the earliest derived subtopics may not necessarily become explosive hot topics, but the derived subtopics generated in the middle and later stages are more likely to become hot topics.

In addition, it can be seen from Figure 11 that the number of topics are all 1 in the initial state. From time = 5, derived subtopics are successively generated, and the number of topics also rapidly increases. With further propagation, some subtopics disappeared, and the number of topics also gradually decreases. In the end, some derived subtopics form derived public opinion, and the dimensions of multi-dimensional public opinion also stabilize at a certain value. Comparing Figure 10a–c, it can be seen that as the degree of information alienation increases, the number of topics starts to rise earlier, and as the number of subtopics increases, the peak of the number of topics rises from six to ten. The increase is large, but the final dimensions of the stable state has changed slightly, rising from two to four, which shows that the degree of information alienation has an impact on the number of derived subtopics, but it has a greater impact on the final public opinion dimension. The degree of influence is relatively small.

4.2. The Impact of Environmental Forces on the Formation of Multi-Dimensional Public Opinion

Environmental force δ refers to the influence of the spread of events or network communication channels on public opinion. Different social hot events receive different environmental forces, and the degree of information variation will be different. The number of derived subtopics generated, along with the formation process of multi-dimensional public opinion, will change accordingly, so the formation process of multi-dimensional public opinion under different environmental forces is analyzed. In order to describe the impact of different environmental forces on social hot events, set $\delta = 3, \delta = 5, \delta = 7, \delta = 9$, and the other parameters as: $\alpha = 0.17, \beta = 0.5, R = 0.9, P_U = 0.7, g_0 = 0.6, \rho = 0.85, T = 40$.

4.2.1. The Impact of Environmental Forces on the Number of Derived Subtopics and the Propagation Process

Generally speaking, the greater the environmental impact of the initial public opinion, the higher the degree of variability, and the number of derived subtopics will increase accordingly. Based on this, the generation process of derived subtopics under different environmental forces is analyzed and the specific results are shown in Figures 12–15.

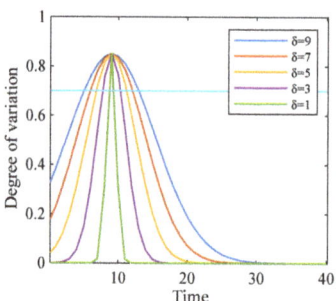

Figure 12. Changes of degree of variation under different information alienation settings.

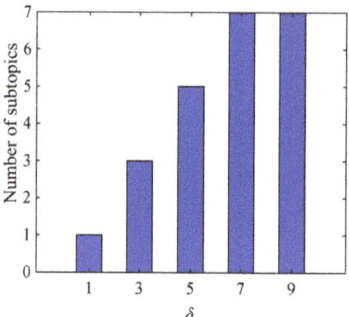

Figure 13. The number of derived subtopics under different information alienation settings.

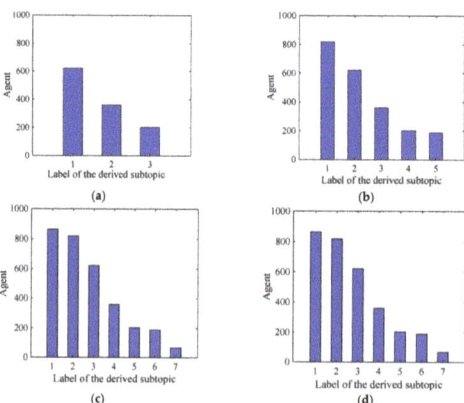

Figure 14. The number of agents participating in subtopic communication under different environmental forces. (**a**) $\delta = 3$. (**b**) $\delta = 5$. (**c**) $\delta = 7$. (**d**) $\delta = 9$.

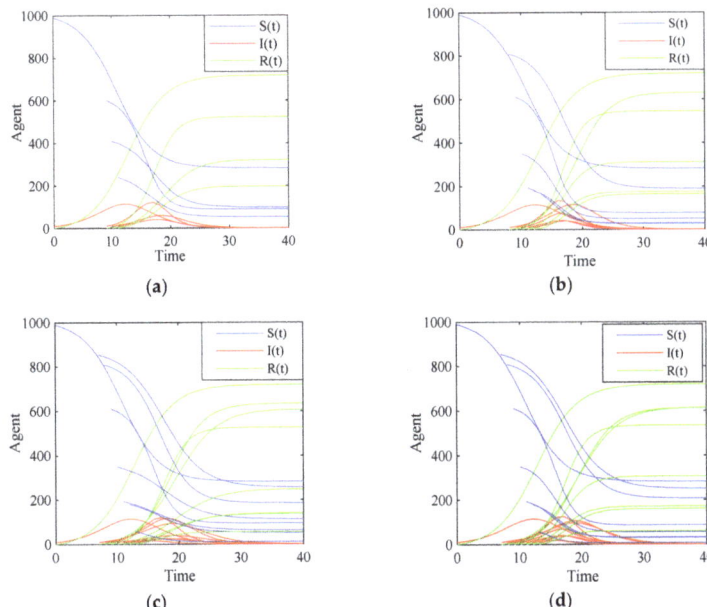

Figure 15. Evolution of initial public opinion and subtopics under different environmental forces. (**a**) $\delta = 3$. (**b**) $\delta = 5$. (**c**) $\delta = 7$. (**d**) $\delta = 9$.

It can be seen from Figure 12 that when the environmental force changes, the variation at the initial moment gradually increases but the peak value of the variation is basically the same; and when the environmental force is very small, the variation can still reach the threshold at a certain moment. Environmental forces have great impact on the initial degree of variation of social hot events, but little impact on the degree of variation in the evolution of initial public opinion. It can be seen from Figure 13 that as the environmental force increases, the number of derived subtopics increases. Even if the environmental force is only 1, there are still derived subtopics. However, when the value reaches a certain level, the number of derived subtopics reaches a stable level. In addition, it can be seen from Figure 14 that with the increase of environmental forces, the number of individuals discussing derived subtopics does not change significantly, and, the earlier derived subtopics are generated, the more individuals participate in the discussion. With the increase of environmental forces, the number of individuals participating in each derived subtopic increases. In addition, it can be seen from Figure 15 that when the environmental force expands from 1 to 3, the number of susceptible agents participating in the communication of the subtopic increases significantly.

However, when the environmental force continues to increase, the numbers of susceptible and infective people do not change significantly. This indicates that when the environmental force reaches a certain value, its influence on the spread of subtopics is small.

4.2.2. The Impact of Environmental Forces on the Dimensions of Multi-Dimensional Public Opinion

Different environmental forces change the time point, the number of subtopics, and the number of agents participating in the discussion of the subtopics, which further affect the subsequent formation of derived public opinion. Based on this, the basic reproduction number R_0 and the multidimensional public opinion dimensions of the subtopics under different environmental forces are simulated here and the results are shown in Figures 16 and 17.

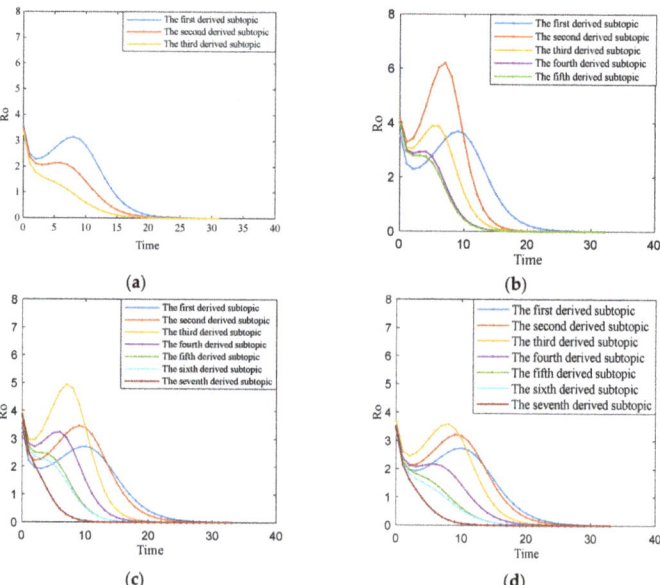

Figure 16. The change of R_0 under different environmental forces. (**a**) $\delta = 3$. (**b**) $\delta = 5$. (**c**) $\delta = 7$. (**d**) $\delta = 9$.

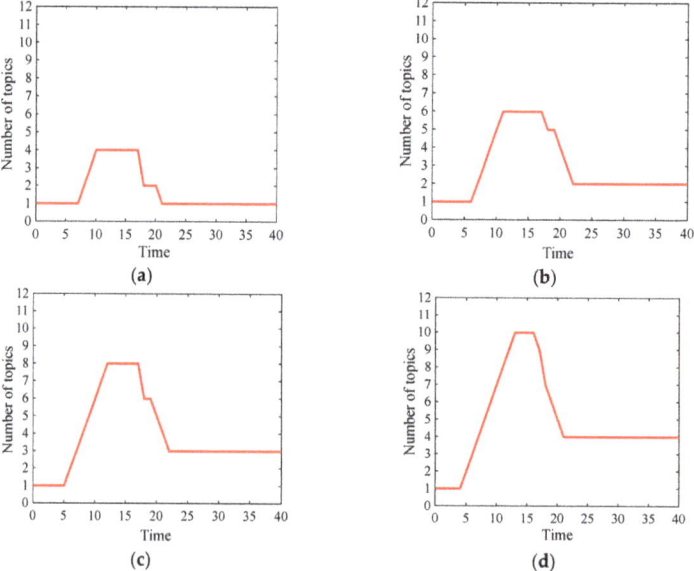

Figure 17. The change of the number of subtopics under different environmental forces. (**a**) $\delta = 3$. (**b**) $\delta = 5$. (**c**) $\delta = 7$. (**d**) $\delta = 9$.

It can be seen from Figure 16a–d that with the increase of environmental forces, the evolution trend of the basic regeneration number curve corresponding to derived subtopic 1 hardly changes, while the basic regeneration number of the second subtopic changes from the continuous decrease to growth, and the peak of R_0 reaches 6. However, due to the increasing environmental force, the peak of the second derived topic is 3. This shows that the environmental force has a relatively small impact on the first derived subtopic, and it has a greater impact on the second and third derived subtopics, and the basic regeneration number of the second subtopic increases with the increase in environmental force. However, when the environmental force increases to a certain level, the increase in the basic regeneration rate begins to slow down. From Figure 17, it can be seen that when the environmental force is 3, although three derived subtopics are generated, the final public opinion dimension is still 1, and no multi-dimensional public opinion is formed. At the same time, the increase of environmental forces has a positive correlation with the increase of the final public opinion dimension.

4.2.3. Analysis of the Combination of Factors Influencing the Number of Derived Subtopics

Through the above simulation analysis, it can be seen that the main factors affecting the number of derived subtopics are the degree of information alienation and the environmental forces. The derived subtopic as a transitional link in the formation of public opinion has a greater impact on its dimension, so these factors are combined for analysis. The result is shown in Figure 18.

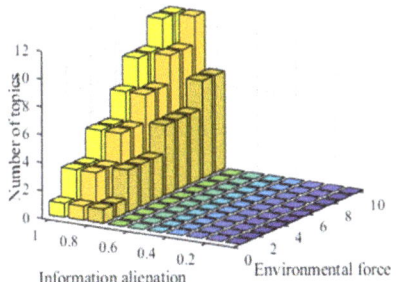

Figure 18. Analysis of the combination of factors influencing the number of derived subtopics.

It can be seen from Figure 18 that only when the degree of information alienation reaches a certain threshold will derived subtopics be generated. When the degrees of information alienation and environmental forces both increase, the number of derived subtopics increases to a certain extent, but the environmental force has a greater impact on the number of derived subtopics. When the degree of information alienation is 0.8 and the environmental force is 1, the number of derived subtopics is 1. When the degree of information alienation is unchanged and the environmental force is 9, the number of derived subtopics is 7, and the range of change is large. This shows that the degree of information alienation has an important impact on the formation of derived subtopics, and when there are derived subtopics, environmental forces will have more influence on the number of derived subtopics.

4.3. The Impact of Topic Correlation on the Formation Process of Multi-Dimensional Public Opinion

When the ith derived subtopic is generated at time T_i, individuals in the network will pay attention to certain subtopics and participate in the discussion of that subtopic according to their own characteristics and environment. The degree of individual attention will be affected by the degree of correlation between the current subtopic and the initial public opinion and the number of surrounding infected neighbors. Therefore, by analyzing the degree of correlation between the subtopic and the initial public opinion, we can further explain which subtopics can form derived public opinions. The simulation results are as follows.

In order to explore whether subtopics that are highly related to the initial public opinion are more likely to form derived public opinions, a comparative analysis of Figures 19–21 is carried out. When $\theta \sim N(0.3, 0.3)$, it can be seen from Figure 19a that the first derived subtopic has the greatest correlation with the initial public opinion and the correlation with other subtopics is small. From Figures 20a and 21a, we can see that the peak value of the basic reproduction number R_0 of the first subtopic is also the highest, and the final public opinion dimension is 2. When $\theta \sim N(0.5, 0.3)$, the first and third derived subtopics have a relatively high relevance. From Figures 20b and 21b, we can see that the peak value of the basic reproduction number R_0 of the first, second, and third subtopics is higher, and the final public opinion dimension is 4. When $\theta \sim N(0.9, 0.3)$, the correlation degree of each derived subtopic with the initial public opinion is relatively large, and the correlation degree of the second, fourth, sixth, and seventh subtopics even reaches 1. From Figures 20c and 21c, the peak value of the basic reproduction number R_0 of the first, second, third, and fourth subtopics is higher, but the peak value reached by the curve has decreased, and the final public opinion dimension is 5. When $\theta \sim U[0, 1]$, the second, sixth, and seventh derived subtopics have a higher degree of correlation. From Figures 20d and 21d, it can be seen that the peak value of the basic reproduction number R_0 of the second subtopic is higher. The final public opinion dimension is 4. It can be seen from the above phenomenon that the peak of the basic reproduction number of topics with a high degree of correlation is correspondingly higher, which means that it is easier to form derived public opinion. When all the subtopics have a relatively high degree of correlation with the initial public opinion, the subtopic with the highest degree of correlation is relatively high. Topics may not be able to form derived public opinion, but the subtopics generated in the early stage are more likely to become hot topics and form derived public opinions. The reason may be that when the correlation of subtopics is relatively high, an individual will have a sense of freshness in the early subtopics, and as time goes by, they will gradually lose interest in the initial public opinion, even if there is a degree of correlation with initial public opinion in the later period. The subtopics with higher correlation are not very popular, leading to the failure to form a new public opinion.

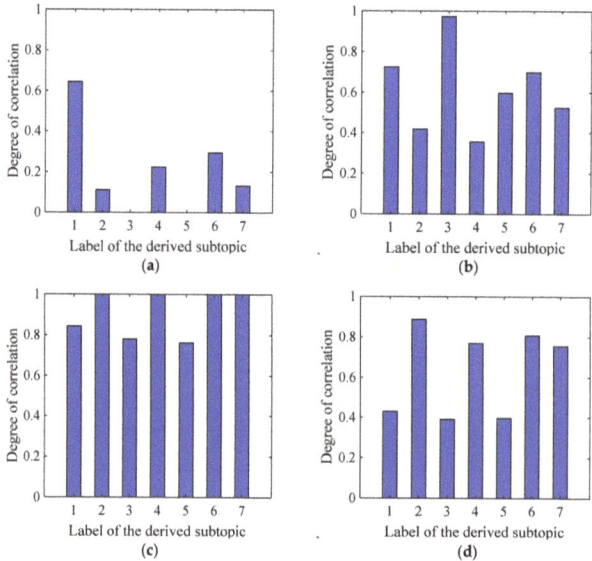

Figure 19. The correlation between derived subtopic number and initial public opinion. (a) $\theta \sim N(0.3, 0.3)$. (b) $\theta \sim N(0.5, 0.3)$. (c) $\theta \sim N(0.9, 0.3)$. (d) $\theta \sim U[0, 1]$.

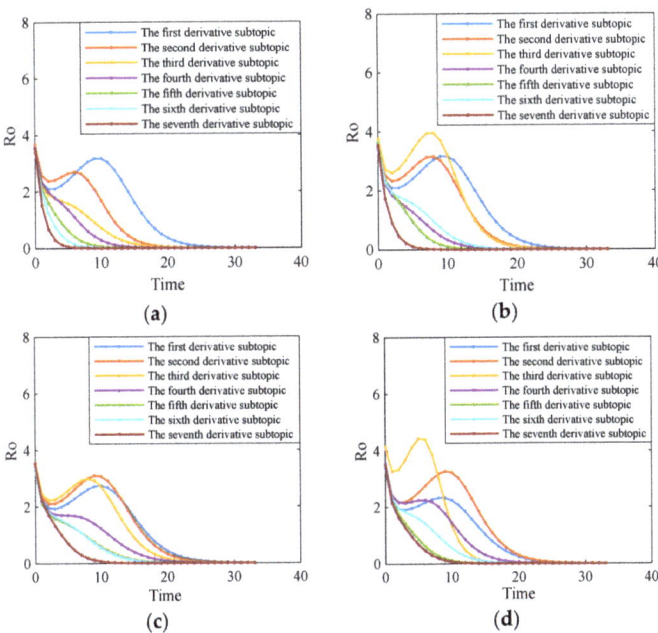

Figure 20. The change of R_0 under different distributions of correlation. (a) $\theta \sim N(0.3, 0.3)$. (b) $\theta \sim N(0.5\ 0.3)$. (c) $\theta \sim N(0.9, 0.3)$. (d) $\theta \sim U[0, 1]$.

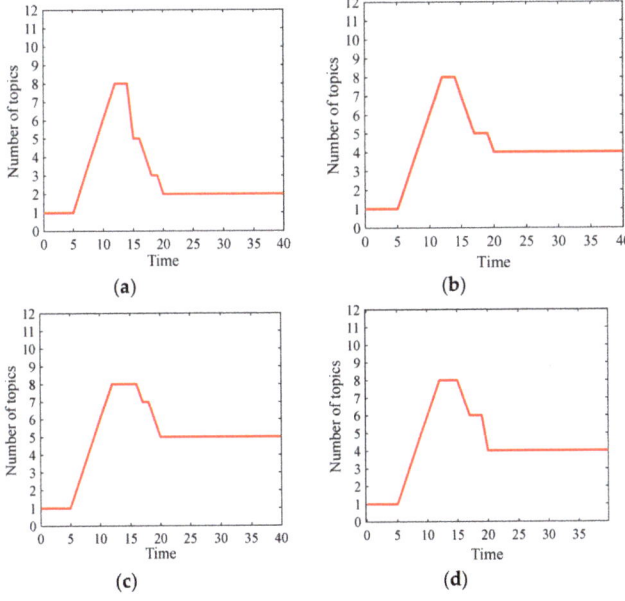

Figure 21. The change of the number of topics under different distributions of correlation. (a) $\theta \sim N(0.3, 0.3)$. (b) $\theta \sim N(0.5, 0.3)$. (c) $\theta \sim N(0.9, 0.3)$. (d) $\theta \sim U[0, 1]$.

4.4. The Impact of the Amount of Information Contained in Subtopics on the Formation Process of Multi-Dimensional Public Opinion

After different derived subtopics are generated, the content posted by individuals participating in the topic discussion is different, and the amount of information contained is also different. In order to explore whether derived subtopics with a large amount of information are more likely to form derived public opinion, simulation analysis is used. The impact of the amount of information contained in subtopics on the formation of multi-dimensional public opinion is analyzed, and the specific results are shown in Figures 22–24.

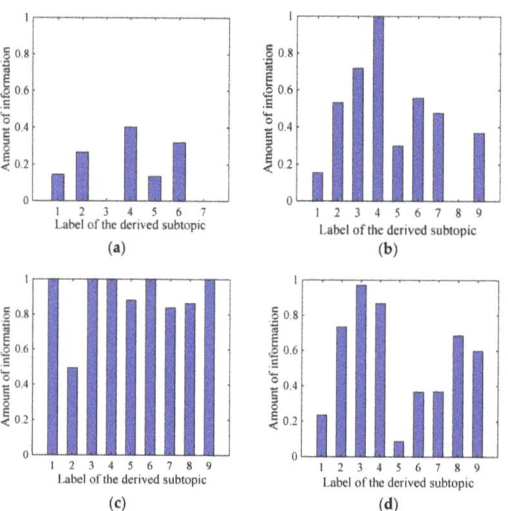

Figure 22. The amount of information of each subtopic under different distributions. (a) $\sigma_i \sim N(0.1, 0.3)$. (b) $\sigma_i \sim N(0.5, 0.3)$. (c) $\sigma_i \sim N(0.9, 0.3)$. (d) $\sigma_i \sim U[0, 1]$.

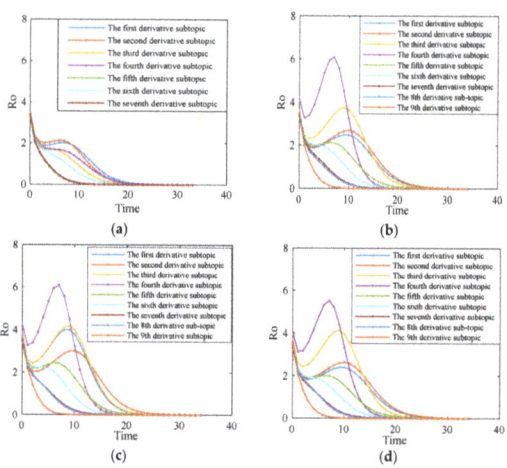

Figure 23. The change of R_0 under different distributions of the amount of information. (a) $\sigma_i \sim N(0.1, 0.3)$. (b) $\sigma_i \sim N(0.5, 0.3)$. (c) $\sigma_i \sim N(0.9, 0.3)$. (d) $\sigma_i \sim U[0, 1]$.

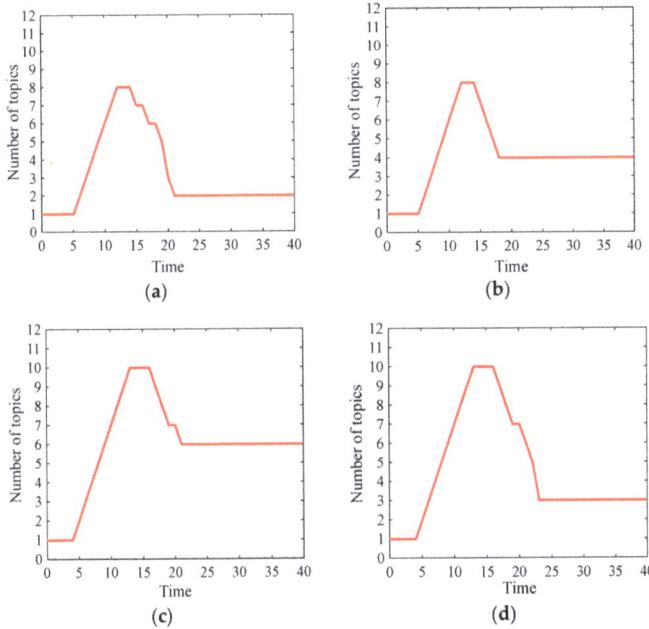

Figure 24. The change of the number of topics under different distributions of the amount of information. (**a**) $\sigma_i \sim N(0.1, 0.3)$. (**b**) $\sigma_i \sim N(0.5, 0.3)$. (**c**) $\sigma_i \sim N(0.9, 0.3)$. (**d**) $\sigma_i \sim U[0, 1]$.

In order to analyze whether subtopics with a large amount of information are more likely to form derived public opinion, a comparative analysis is carried out. When $\sigma \sim N(0.3, 0.3)$, it can be seen from Figure 22a that the amount of information contained in all derived subtopics is relatively small. From Figures 23a and 24a, we can see that the peak of R_0 is low, and the final public opinion dimension is only 2. When $\sigma \sim N(0.5, 0.3)$, the second and third derived subtopics contain more information. From Figures 23b and 24b, it can be seen that the peak value of the basic reproduction number R_0 of the second and third subtopics is also the highest, and the final public opinion dimension is 4. When $\sigma \sim N(0.9, 0.3)$, each derived subtopic contains a large amount of information. Among them, the amount of information contained in the first, third, fourth, and ninth subtopics even reaches 1. From Figures 23c and 24c, we can see that the peak value of the basic reproduction number R_0 of the first, third, fourth, and fourth subtopics is also higher, and the final public opinion dimension is 6. When $\sigma \sim U[0, 1]$, the third and fourth subtopics contain more information. From Figures 23d and 24d, it can be seen that the peak of the basic reproduction number R_0 of the third and fourth subtopics is also higher, and the final public opinion dimension is 3. From the above phenomenon, it can be seen that the peak value of the basic reproduction number R_0 of the subtopic with a large amount of information is correspondingly higher, and it is easier to form a new one-dimensional derived topic.

4.5. Combination Analysis of Various Factors Affecting the Multi-Dimensional Public Opinion

Through the above simulation analysis, it can be known that there are many factors that affect the dimensions of multi-dimensional public opinion. In the actual evolution of public opinion, due to the urgency of online public opinion control, it is usually necessary to focus on key links in order to reduce the cost of public opinion prevention and control. Therefore, finding the most critical factor from the many factors that affect the formation of multi-dimensional public opinion is more in line with realistic requirements. Therefore, we will find out the key factors through a combination analysis of the factors that affect the formation of multi-dimensional public opinion. The results are shown in Figures 25–28.

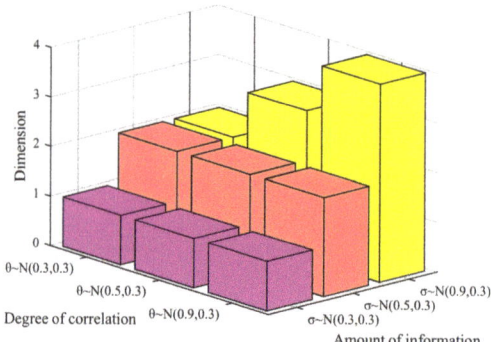

Figure 25. Relationship among degree of correlation, amount of information and dimension.

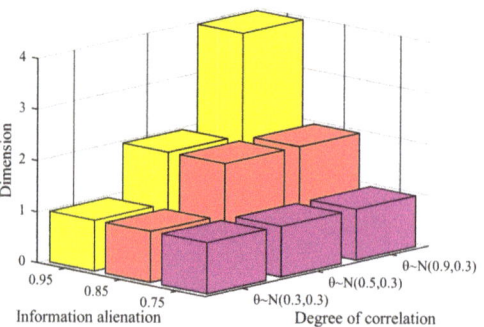

Figure 26. Relationship among degree of correlation, information alienation and dimension.

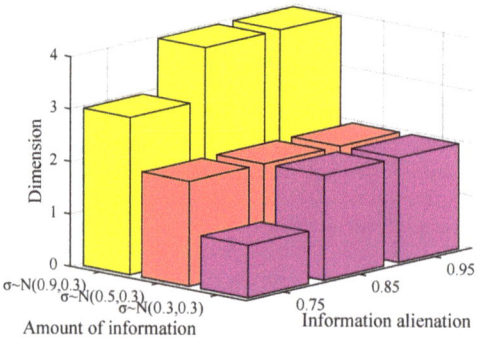

Figure 27. Relationship among amount of information, information alienation and dimension.

Figures 25–27 respectively show the relationship among degree of correlation, amount of information, and dimension, the relationship among degree of correlation, information alienation and dimension, the relationship among amount of information, information alienation and dimension. It can be seen from Figure 25 that when the amount of information contained in a subtopic is small, its correlation does not impact on the dimension.

When the amount of information contained is large, as the correlation increases, the dimension increases. No matter how the degree of correlation changes, the dimensions of public opinion increase when the amount of information increases. Therefore, comparing the effect of correlation, the amount of information contained in subtopics has a greater impact on the dimension. It can be seen from Figure 26 that with the increase in the degree of correlation and information alienation, the dimension increases, and the increase in the two is not much different. It can be seen from this that: the degree of correlation and the degree of information alienation have little effect on the dimensions of public opinion. It can be seen from Figure 27 that as the degree of information alienation and the amount of information contained in subtopics increase, the dimensions of public opinion increase to a certain extent. When the amount of information is $\sigma \sim N(0.5, 0.3)$, as the degree of information alienation increases, the dimensions of public opinion do not change. When the degree of information alienation increases with the amount of information contained in the subtopics, the dimensions of public opinion change. It can be seen that, compared to the degree of information alienation, the amount of information contained in subtopics has a greater impact on the dimensions of public opinion.

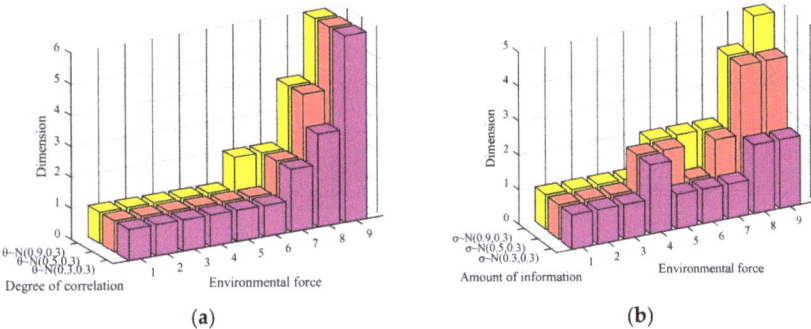

Figure 28. Combination analysis of environmental forces, correlation and the amount of information contained in subtopics. (**a**) The relationship among dimension, environmental forces and correlation. (**b**) The relationship among dimension, environmental forces and amount of information.

In addition, Figure 28 shows the relationship among environmental forces, correlation and the amount of information contained in subtopics. From Figure 28a, it can be seen that when the environmental force is small, it has little effect on the dimensions of public opinion. When the environmental force is large, the dimensions of public opinion increase sharply with the increase of environmental force. With the increase of the degree of correlation, the dimensions of public opinion have not changed or only increased slightly. Therefore, compared with the degree of correlation, the influence of environmental forces on the dimensions of public opinion is greater. From Figure 28b, it can be seen that as the amount of information contained in the subtopics and the environmental forces increase, the dimensions of public opinion increase significantly. Therefore, the amount of information and environmental forces contained in the subtopics impact more on the dimensions of public opinion. Based on the above combination analysis, the amount of information contained in subtopics and environmental forces are the key factors that affect the dimensions of public opinion.

4.6. The Influence of Network Topology on the Formation Process of Public Opinion Dimensions

Differences in network structure have an important influence on the formation and evolution of public opinion. Therefore, it is necessary to simulate and analyze the formation process of multi-dimensional public opinion under different network topologies. Since the previous simulation analysis is based on the BA network, the influence of different network topologies on the dimensions of public opinion is studied. The BA network, a fully connected network and a WS small world network are selected for comparative analysis. The network topology parameters are as follows.

It can be seen from Table 3 that the aggregation coefficient and average of the fully connected network are larger than other networks, but the average path length is smaller. Other parameters are set as follows: $\rho = 0.85$, $\delta = 8$, $\sigma \sim N(0.5, 0.3)$, $\theta \sim N(0.9, 0.3)$, $P_U = 0.7$, $g_0 = 0.6$, $N = 1000$. The simulation results are shown in Figures 29 and 30.

Table 3. Different network topology parameters.

Network Name	Average Path Length	Aggregation Coefficient	Average
WS small world network	5.3719	0.0088	4
BA network	4.0282	0.034	3.964
Fully connected network	1	1	999

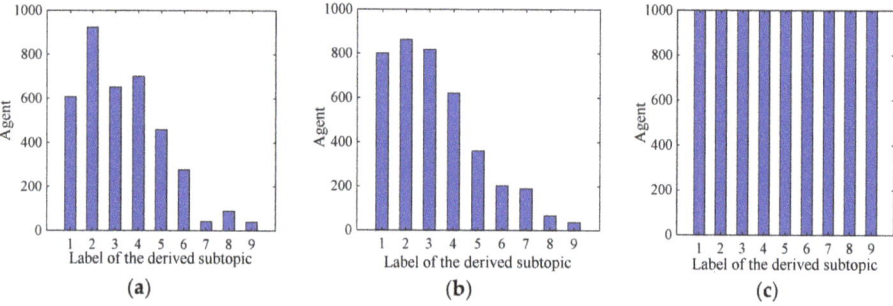

Figure 29. The number of agents participating in the discussion of derived subtopics under different network topologies. (a) WS small world network. (b) BA network. (c) Fully connected network.

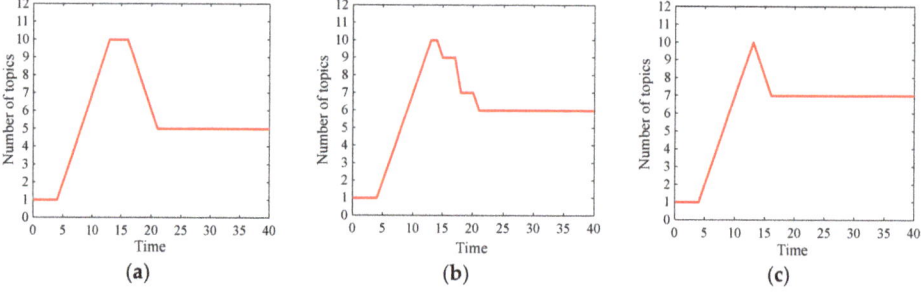

Figure 30. The change of number of topics under different network topologies. (a) WS small world network. (b) BA network. (c) Fully connected network.

It can be seen from Figure 29 that 9 derived subtopics are generated under different network structures. In addition, in the fully connected network, the number of agents participating in the discussion of each derived subtopic is 1000, indicating that the participation of each subtopic is very high. While in the WS small world network and the BA network, only part of the subtopics' participation is high, which means that the network topology does not have much impact on the number of subtopics, it has a greater impact on the number of agents participating in the discussion of the subtopics. In a fully connected network, the participation of each derived subtopic is up to the highest. As can be seen from Figure 30, in different network structures, the dimensions ultimately formed by multi-dimensional public opinion are different. Fully connected networks will eventually form the highest dimensions, and WS small world networks will eventually form the lowest dimensions. This change is related to network topology parameters. The change of the aggregation coefficient is the same, which is opposite to the change of the average path length. Therefore, it can be seen that the multi-dimensional public opinion formed by the network topology with a high aggregation coefficient and short average path length has a higher dimension.

5. Empirical Analysis

The news of COVID-19 that broke out in late 2019 swept across China and the world. At the same time, hot events related to the epidemic are fermenting on the Internet, and the spread of some rumors has caused many netizens to panic. After a long period of anti-epidemic work, the epidemic situation in China has basically stabilized and people's lives have gradually returned to normal. However, on 9 November 2020, China's CCTV news broadcast "One new local confirmed case of COVID-19 in Shanghai" ignited netizens' panic. With the development of the Shanghai epidemic, the incident has spawned many topics, and even the discussion on the derived topics has surpassed that of the original topics. The original topic of "One new local confirmed case of COVID-19 in Shanghai" derived further subtopics, such as "Shanghai Epidemic Prevention and Control Work Conference", "One Community in Pudong New Area will be Upgraded to Medium Risk Degree", and "4015 people in Shanghai Pudong Hospital have been quarantined". Discussions on these derived subtopics have even pushed the public's panic to a climax. Here, we searched the Baidu index with "Shanghai epidemic" as the key word and obtained the search index from 18 November to 9 December 2020, as shown in Figure 31.

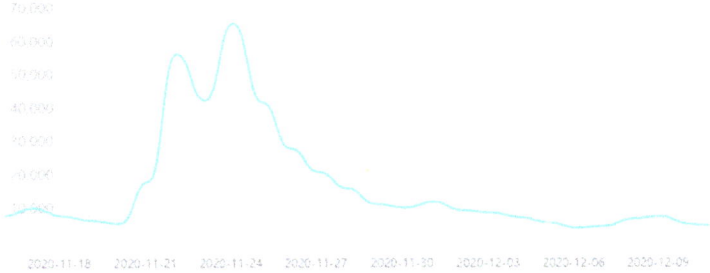

Figure 31. "Shanghai epidemic" Baidu index.

It can be seen from Figure 31 that the "One new local confirmed case of COVID-19 in Shanghai" reached the highest search interest degree from November 21 to 24, and then stabilized, and most derived subtopics were generated during this time period, indicating that network public opinion has an important influence on the formation and propagation of derived subtopics.

In order to study the phenomenon of the formation of multi-dimensional public opinion for the above-mentioned events, event-related microblogs released by media outlets such as "Tou tiao", "The Paper News", and "CCTV News" between 9 November and 29 November 2020 are collected. The results are shown in Table 4.

It can be seen from Table 4 that after the outbreak of the original topic "One new local confirmed case of COVID-19 in Shanghai", with the development of the incident, 11 derived subtopics were successively formed and the amount of reading and discussion of each derivative subtopic was different. The distribution map of the number of individuals participating in the discussion is simulated based on the time when the derived subtopics are generated and the number of discussions among netizens. It also defines that if the total number of readings of a derived subtopic is greater than 100 million and the total number of discussions by netizens is greater than 5000, the subtopic is regarded as a derived public opinion, that is, a dimension is added to the initial public opinion, and the time cut-off point is set by December 9th. Based on this simulation, the public opinion dimension change map of the "One new local confirmed case of COVID-19 in Shanghai" event was generated and the result is shown in Figure 32.

Table 4. Topic related to "Shanghai epidemic" public opinion.

Release Time	Topic	Reading Volume	Discussion Volume	Topic Number
11.09	#One new local confirmed case of COVID-19 in Shanghai #	240 m	6607	0
11.10	#Shanghai Epidemic Prevention and Control Work Conference#	300 m	18,000	1
11.21	#One Community in Pudong New Area Was Upgraded to Medium Risk Degree#	23.868 m	594	2
11.21	#One Community in Pudong New Area Will Be Upgraded to Medium Risk Degree Tomorrow#	130 m	3452	3
11.21	#4015 people in Shanghai Pudong Hospital have been quarantined#	450 m	17,000	4
11.21	#83 people that once contacted with infected person were tracked#	100 m	6383	5
11.21	#1 new COVID-19 case confirmed among15,416 people in Shanghai#	32.914 m	1319	6
11.23	#2 new local confirmed cases of COVID-19 in Shanghai#	710 m	28,000	7
11.23	#One COVID-19 patient once exposed toan aviation container#	310 m	8109	8
11.29	#Shanghai Songjiang #	12.469 m	3106	9
11.29	#No COVID-19 in Shanghai Songjiang #	6.118 m	431	10
1129	#Reasults of 6 local confirmed case of COVID-19 in Shanghai#	190 m	3670	11

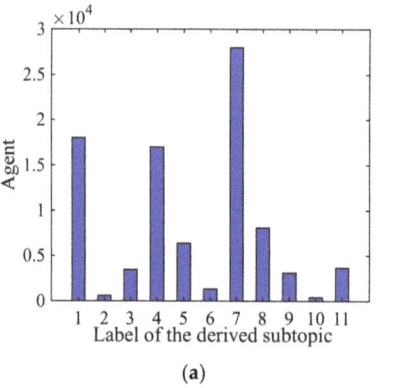
(a)

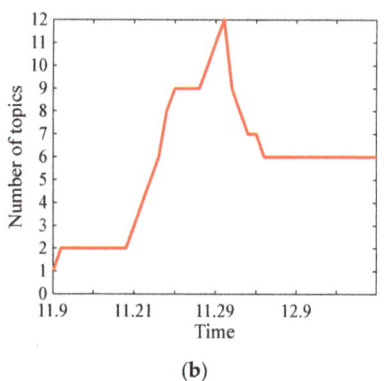
(b)

Figure 32. Simulation of derived subtopics for this public opinion. (**a**)The numberof agents discussing derived subtopics. (**b**) Change inthe number of topics.

It can be seen from Figure 32a that the number of agents associated with the first, fourth, and seventh derived subtopics is relatively large, which means that participation is high. It can be seen from Figure 32b that the dimensions of public opinion began to rise on November 21, reached a peak on the 29th, and finally stabilized at six dimensions, and formed a six-dimensional public opinion. This six-dimensional public opinion is the original public opinion.

The following is a simulation of the event based on the multi-dimensional public opinion formation model mentioned in this paper. Due to the large amount of case data and comprehensive visualization considerations, the simulation network scale is set to 1000. Since there are many derived subtopics generated, settings were at $\rho = 0.95$, $\delta = 9$, and the amount of information contained in the subtopics $\sigma \sim N(0.9, 0.3)$, the initial public opinion infection rate set at $\alpha = 0.2$, $\beta = 0.5$. The result is shown in Figure 33.

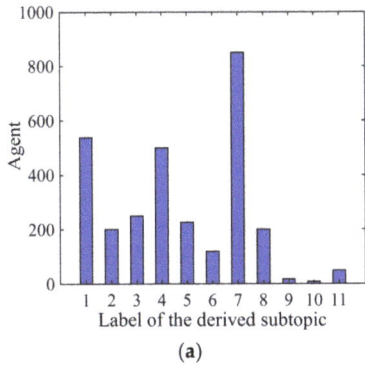
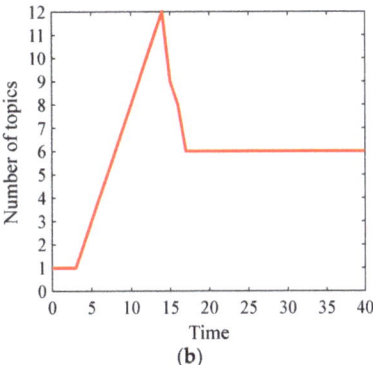

Figure 33. Simulation of derived subtopics for this public opinion used the model proposed in this paper. (**a**) The number of agents discussing derived subtopics. (**b**) Change in the number of topics.

It can be seen from Figure 33 that the simulation of the multi-dimensional public opinion using the model proposed in this paper is not much different from the actual results. First of all, it can be seen from Figure 33a that the number of individuals participating in the first, fourth, and seventh derived subtopics is the largest, which is consistent with reality. It can be seen from Figure 33b that the dimensions of public opinion reached up to twelve, then began to decline, and finally stabilized at six dimensions. Although the rise and decline process is slightly different from the reality, the overall trend is basically the same. In general, the multi-dimensional public opinion formation model proposed in this paper can better simulate social hot events in reality and has important guiding significance for analyzing the causes of multi-dimensional public opinion and predicting the evolutionary trends of multi-dimensional public opinion.

6. Conclusions

This paper discusses the formation process of multi-dimensional public opinion based on derived topics. Usually, after a hot news story breaks out, the social network platforms generate the initial topic of the event and form the initial public opinion as the discussion heats up. As time goes by, netizens' subjective understanding of the initial public opinion information will gradually become biased and gradually accumulate, leading to variation in the initial public opinion so that multiple subtopics are derived. After the subtopics are generated, the SIR model is used to analyze the propagation process and calculate the basic reproduction number to determine whether the derived public opinion can be formed, and this derived new public opinion is intertwined with the initial public opinion to form a multi-dimensional public opinion together. Finally, simulation experiments are conducted to explore the influence of factors, such as the degree of information alienation, environmental forces, topic correlation, and the amount of information contained in subtopics on the formation of multi-dimensional public opinion.

The following conclusions are obtained through simulation experiments:

(1) When the degree of information alienation reaches a certain threshold, derived subtopics will be generated. In addition, when the degree of information alienation is high, the earliest derived subtopics may not necessarily form derived public opinion but later derived subtopics may also be generated. Derived public opinion is formed, and the degree of information alienation has a greater impact on the number of derived subtopics but has small impact on the dimensions of the final state of public opinion.

(2) Environmental forces and the amount of information contained in subtopics are the key factors that affect the formation of multi-dimensional public opinion. Among them, environmental forces have a greater impact on the early subtopics, and the amount of information contained in subtopics is key to forming derived public opinion.

(3) Subtopics that are highly related to the initial public opinion topic are more likely to form derived public opinions. When all subtopics are highly correlated with the initial public opinion, the subtopic with the highest degree of correlation may not necessarily form a derived public opinion, but subtopics generated in the early stage are more likely to form a derived public opinion.

(4) The network topology does not have much impact on the number of subtopics, but it has a greater impact on the number of individuals participating in the discussion of the subtopics, and the dimensions of multidimensional public opinion formed by the network topology with a high aggregation coefficient and short average path length are greater.

However, this paper still has the following shortcomings, which will require further study:

(1) This paper does not consider the influence of external information intervention in the study of the formation process of multi-dimensional public opinion and subsequent intervention mechanisms which can be introduced to study the influence of external information on initial public opinion and derived public opinion.

(2) The paper considers the situation of static nodes without considering the increase or withdrawal of Internet users' nodes [38]. In reality, individuals participating in discussion of derived subtopics often increase and decrease. Therefore, the multidimensional public opinion evolution mechanism under the dynamic network can be considered in the follow-up research.

(3) In this paper, the dynamic equations for complex networks only considers the node average degree of uniform networks and cannot reflect the connections between each node and its neighbor. Therefore, more complex dynamic equations should be considered to simulate the infection process of each node in future.

Author Contributions: T.C. described the proposed framework and wrote the whole manuscript; X.Y. implemented the simulation experiments; J.Y. and G.L. collected the data; G.C. revised the manuscript. All authors have read and agreed to the published version of the manuscript.

Funding: This research is supported by the National Social Science Foundation of China (Grant No. 20BTQ059), the Project of China (Hangzhou) Cross-border E-commerce College (No. 2021KXYJ07), the Contemporary Business and Trade Research Center and Center for Collaborative Innovation Studies of Modern Business of Zhejiang Gongshang University of China (Grant No. 14SMXY05YB), as well as the Characteristic & Preponderant Discipline of Key Construction Universities in Zhejiang Province (Zhejiang Gongshang University-Statistics).

Institutional Review Board Statement: Not applicable.

Informed Consent Statement: Not applicable.

Data Availability Statement: The data used to support the findings of this study are available from the corresponding author upon request.

Conflicts of Interest: The authors declare that they have no conflicting interest.

References

1. An, L.; Dai, Y.; Zhou, Y. Research on the formation and evolution of public opinion derived from public safety events-based on topic and time series analysis. *Public Secur. Stud.* **2020**, *12*, 18–35.
2. Chen, T.; Rong, J.; Yang, J.; Cong, G.; Li, G. Combining Public Opinion Dissemination with Polarization Process Considering Individual Heterogeneity. *Healthcare* **2021**, *9*, 176. [CrossRef]
3. Lan, Y.; Dong, X.; Zeng, R.; Qi, Z. Research on effects model of derived of network public opinion from the perspective of information alienation. *J. Intell.* **2015**, *34*, 139–143.
4. Zhang, Y.; Feng, Y. Two-layer coupled network model for topic derivation in public opinion propagation. *China Commun.* **2020**, *17*, 176–187. [CrossRef]
5. Korobeinikov, A. Global properties of SIR and SEIR epidemic models with multiple parallel infectious stages. *Bull. Math. Biol.* **2009**, *71*, 75. [CrossRef] [PubMed]
6. Arenas, A.; Cota, W.; Gomez-Gardenes, J.; Gomez, S.; Steinegger, B. A mathematical model for the spatiotemporal epidemic spreading of COVID19. *MedRxiv* **2020**, *10*, 1101.
7. Estrada, E. COVID-19 and SARS-CoV-2. Modeling the present, looking at the future. *Phys. Rep.* **2020**, *869*, 1–51. [CrossRef]
8. Yang, L.; Li, Z.; Giua, A. Containment of rumor spread in complex social networks. *Inf. Sci.* **2020**, *506*, 113–130. [CrossRef]
9. Zanette, D.H. Dynamics of rumor propagation on small-world networks. *Phys. Rev. E* **2002**, *65*, 041908. [CrossRef]
10. Moreno, Y.; Nekovee, M.; Pacheco, A.F. Dynamics of rumor spreading in complex networks. *Phys. Rev. E* **2004**, *69*, 066130. [CrossRef]
11. Zhou, J.; Liu, Z.; Li, B. Influence of network structure on rumor propagation. *Phys. Lett. A* **2007**, *368*, 458–463. [CrossRef]
12. Tan, Y.; Lin, Q.; Luan, Y.; Chen, T.; Qiao, Y.; Luan, Y. Campus Network Public Opinion Monitoring System Based on Reptile Technology. *IOP Conf. Ser. Earth Environ. Sci.* **2019**, *252*, 21. [CrossRef]
13. You, C.; Zhu, D.; Sun, Y.; Ye, A. A Social-Network-Oriented Public Opinion Monitoring Platform Based on ElasticSearch. *Comput. Mater. Contin.* **2019**, *61*, 1271–1283. [CrossRef]
14. Chen, T.; Yin, X.; Peng, L.; Rong, J.; Yang, J.; Cong, G. Monitoring and Recognizing Enterprise Public Opinion from High-Risk Users Based on User Portrait and Random Forest Algorithm. *Axioms* **2021**, *10*, 106. [CrossRef]
15. Wang, C.; Wang, G.; Luo, X.; Li, H. Modeling rumor propagation and mitigation across multiple social networks. *Phys. A Stat. Mech. Its Appl.* **2019**, *535*, 122240. [CrossRef]
16. Askarizadeh, M.; TorkLadani, B.; Manshaei, M.H. An evolutionary game model for analysis of rumor propagation and control in social networks. *Phys. A Stat. Mech. Its Appl.* **2019**, *523*, 21–39. [CrossRef]
17. Zhang, S. Multi-Dimensional Analysis of causes of network-derived public opinion from the perspective of spreading factors. *J. Beijing Adm. Inst.* **2019**, *4*, 56–63.
18. Duncan, J.; Peter, S. Influentials, networks, and public opinion formation. *J. Consum. Res.* **2007**, *34*, 441–458.
19. Li, H. Research on evolution law and countermeasure of the derived network public opinion based on information alienation theory-taking internet rumors governance for example. *J. Mod. Inf.* **2015**, *35*, 4–8.
20. Snyder, J.; Cai, W.; D'Souza, R. Degree-targeted cascades in modular, degree-heterogeneous networks. *Phys. Soc.* **2020**, *1*, 09316.
21. Wang, L.; Dai, J. Research on the quantitative analytical method of network public opinion derived chain. *Inf. Sci.* **2016**, *34*, 59–63.
22. Chen, T.; Li, Q.; Fu, P.; Yang, J.; Xu, C.; Cong, G.; Li, G. Public opinion polarization by individual revenue from the social preference theory. *Int. J. Environ. Res. Public Health* **2020**, *17*, 946. [CrossRef]
23. Li, J.; Xiao, R. Agent-based modelling approach for multidimensional opinion polarization in collective behaviour. *J. Artif. Soc. Soc. Simul.* **2017**, *20*, 14. [CrossRef]
24. Parsegov, S.E.; Proskurnikov, A.V.; Tempo, R.; Friedkin, N.E. Novel multidimensional models of opinion dynamics in social networks. *IEEE Trans. Autom. Control* **2016**, *62*, 2270–2285. [CrossRef]
25. Sun, Y.; Chai, R. An Early-Warning Model for Online Learners Based on User Portrait. *Ingénierie Des Systèmesd' Inf.* **2020**, *25*, 535–541. [CrossRef]
26. Wang, G.; Chi, Y.; Liu, Y.; Wang, Y. Studies on a multidimensional public opinion network model and its topic detection algorithm. *Inf. Process. Manag.* **2019**, *56*, 584–608. [CrossRef]
27. Wang, Y.; Wang, J. SIR rumor spreading model considering the effect of difference in nodes' identification capabilities. *Int. J. Mod. Phys. C* **2017**, *28*, 15. [CrossRef]
28. Lu, Y. Study on the Spread of Complex Network Virus Based on Human Behavior. Ph.D. Thesis, Nanjing University of Posts and Telecommunications, Nanjing, China, 2015.
29. Liu, D.; Wang, W.; Li, H. Evolutionary mechanism and information supervision of public opinions in internet emergency. *Procedia Comput. Sci.* **2013**, *17*, 973–980. [CrossRef]
30. Wang, H.; Lan, Y.; Pan, Y. Research on the causes of internet derived public opinion from the perspective of information alienation dynamics. *Mod. Intell.* **2013**, *33*, 59–63. [CrossRef]
31. Guo, Y. Analysis of Variation and Evolution of Network Public Opinion Based on Cellular Automata. Ph.D. Thesis, Beijing Jiaotong University, Beijing, China, 2016.
32. Liu, C. *Random Process*; Huazhong University of Science and Technology Press: Wuhan, China, 2008.

33. Chen, T.; Peng, L.; Yin, X.; Jing, B.; Yang, J.; Cong, G.; Li, G. A Policy Category Analysis Model for Tourism Promotion in China During the COVID-19 Pandemic Based on Data Mining and Binary Regression. *Risk Manag. Healthc. Policy* **2020**, *13*, 3211–3233. [CrossRef]
34. Chen, T.; Peng, L.; Yang, J.; Cong, G. Analysis of User Needs on Downloading Behavior of English Vocabulary APPs Based on Data Mining for Online Comments. *Mathematics* **2021**, *9*, 1341. [CrossRef]
35. Adnerson, R.M.; May, R.M. Infectious diseases of humans: Dynamics and control. *Science* **1991**, *254*, 591–592.
36. Wallinga, J.; Lipsitch, M. How generation intervals shape the relationship between growth rates and reproductive numbers. *Proc. Biol. Sci.* **2007**, *274*, 599–604. [CrossRef] [PubMed]
37. Goh, K.I.; Kahng, B.; Kim, D. Universal behavior of load distribution in scale-free networks. *Phys. Rev. Lett.* **2001**, *87*, 278701. [CrossRef]
38. Chen, T.; Rong, J.; Peng, L.; Yang, J.; Cong, G.; Fang, J. Analysis of Social Effects on Employment Promotion Policies for College Graduates Based on Data Mining for Online Use Review in China during the COVID-19 Pandemic. *Healthcare* **2021**, *9*, 846. [CrossRef] [PubMed]

Article

Existence of Mild Solutions for Multi-Term Time-Fractional Random Integro-Differential Equations with Random Carathéodory Conditions

Amadou Diop [1] and Wei-Shih Du [2,*]

1 Department of Mathematics, Université Gaston Berger de Saint-Louis, UFR SAT, Saint-Louis B.P. 234, Senegal; diop.amadou@ugb.edu.sn
2 Department of Mathematics, National Kaohsiung Normal University, Kaohsiung 82444, Taiwan
* Correspondence: wsdu@mail.nknu.edu.tw

Abstract: In this paper, we investigate the existence of mild solutions to a multi-term fractional integro-differential equation with random effects. Our results are mainly relied upon stochastic analysis, Mönch's fixed point theorem combined with a random fixed point theorem with stochastic domain, measure of noncompactness and resolvent family theory. Under the condition that the nonlinear term is of Carathéodory type and satisfies some weakly compactness condition, we establish the existence of random mild solutions. A nontrivial example illustrating our main result is also given.

Keywords: measure of noncompactness; random effect; random operator; Mönch's fixed point theorem; multi-term fractional differential equation; Carathéodory condition; resolvent family theory

MSC: 34A08; 34K30; 47G20; 60H25

1. Introduction

For more than three decades, fractional calculus has played an important role in the study of linear and nonlinear fractional integro-differential equations that arise from the modeling of nonlinear phenomena, optimal control of complex systems and other scientific research (see, e.g., [1,2]). Multi-term time-fractional differential systems have also attracted a great interest in recent years, see for instance [3–6] and references cited therein. As inherently deterministic extensions, random fractional differential equations exist in many applications and have been studied by many authors and more details from historical points of view and recent developments of such equations are reported to the monographs [7,8], papers [9–11], and the references cited therein. To be more precise, the existence results and qualitative properties for fractional differential equations with random effects are examined in [12,13] and references cited therein. Very recently, considerable attention has been given to multi-term time-fractional differential systems. For instance, Pardo and Lizama [6] studied the existence of mild solutions under Carathéodory type conditions by using measure of noncompactness techniques, Singh and Pandey [14] have established the existence and uniqueness of mild solutions for multi-term time-fractional differential systems with non-instantaneous impulses and finite delay by using Banach fixed point theorem whereas Chang and Ponce [15], with the help of the theory of fractional resolvent families, established the existence of mild solutions to a multi-term fractional differential equation. It may be noted here that the mentioned works are confined to deterministic systems. Inspired by the aforementioned papers [6,12–14], this work focuses on the existence of mild solution of problem (1) with multi-term time-fractional integro-differential equations with random effects in the form

$$\begin{cases} {}^C D^{1+\beta}\vartheta(t,\omega) + \sum_{k=1}^{n} \alpha_k\, {}^C D^{\gamma_k}\vartheta(t,\omega) = \mathcal{A}\vartheta(t,\omega) + F\left(t,\vartheta(t,\omega),\int_0^t \mathcal{B}(t,s)\vartheta(s,\omega)ds,\omega\right), \\ \vartheta(0,\omega) = \vartheta_0(\omega), \\ \vartheta'(0,\omega) = \vartheta_1(\omega), \end{cases} \quad (1)$$

for $0 < t \leq b < \infty$ and $\omega \in \Omega$, where the state $\vartheta(\cdot,\cdot)$ takes values in a separable Banach space $\mathbb{X}$ with norm $\|\cdot\|$, $(\Omega,\mathcal{F},\mathbb{P})$ is a complete probability space, ${}^C D^u$ stand for the Caputo fractional derive of order $u > 0$, $0 < \beta \leq \gamma_n \leq \cdots \leq \gamma_1 \leq 1$ and $\alpha_k \geq 0$, $k = 1,2,\cdots n$ be given and $\mathcal{A} : \mathcal{D}(\mathcal{A}) \subset \mathbb{X} \to \mathbb{X}$ is the infinitesimal generator of a bounded and strongly continuous cosine family. $F : [0,b] \times \mathbb{X} \times \mathbb{X} \times \Omega \to \mathbb{X}$ is a random nonlinear function to be specified later. The operator $\mathcal{B}(\cdot,\cdot) : \Delta \to \mathbb{R}^+$ is a continuous operator satisfying

$$\zeta_B = \sup_{s,t \in [0,b]} \int_0^b \mathcal{B}(t,s)ds < \infty,$$

where $\Delta = \{(t,s) \in \mathbb{R}^2 : 0 \leq s \leq t < b\}$ and $\vartheta_0(\cdot)$ and $\vartheta_1(\cdot)$ are given random functions.

To the best of our knowledge, the study of the existence of mild solutions of multi-term time-fractional integro-differential equations with random effects by the abstract form (1) has not yet been treat in the literature. The main contributions of this paper are: Firstly, the study of existence of random multi-term time-fractional integro-differential equations of the form (1) via measure of noncompactness is an untreated topic in the literature. Secondly, the nonlinear term satisfies a weak compactness condition that does not require the compactness of the resolvent family and sufficient conditions for the existence of mild solutions where the solution operators are only equicontinuous, are established by means of Mönch fixed point theorem and a random fixed point theorem with stochastic domain via the noncompactness measure. At last, our theorems guarantee the effectiveness of existence results under some weakly compactness condition and the work can considered as a supplemented for the case that the corresponding (β,γ_k)-resolvent operator is compact and deterministic one. The results are established using of the (β,γ_k)-resolvent operators developed in [6].

This paper is organized as follows. Section 2 contains preliminary details. In Section 3, we show the existence of random mild solutions by Mönch's fixed point theorem combined with a fixed point theorem with stochastic domain and (β,γ_k)-resolvent family. A nontrivial example illustrating our main result (Theorem 2; see below) is also given.

2. Preliminaries

In this section, we recall some basic concepts, notations, definitions, lemmas, and preliminary facts, which are used throughout this article. We set $(\Omega,\mathcal{F},P)$ be a complete probability space. Let $\mathbb{X}$ be a separable Banach space and denote $\mathcal{C}([0,b],\mathbb{X})$ be the Banach space of all continuous $\mathbb{X}$-valued functions on interval $[0,b]$ equipped with the supremum norm $\|\vartheta\| = \sup\{\|\vartheta(t)\| : t \in [0,b]\}$. In the sequel, a mapping $\vartheta : [0,b] \times \Omega \to \mathbb{X}$ is said to be a stochastic process if for each $t \in [0,b]$, $\vartheta(t,\cdot) = \vartheta(t)(\cdot)$ is measurable. First, we recall some basics definitions and properties related to random operators which are used in this paper.

Definition 1. *A mapping* $F : I \times \mathbb{X} \times \mathbb{X} \times \Omega \to \mathbb{X}$ *is said to be random Carathéodory if the following hold:*

(a) *The mapping* $(t,\omega) \to F(t,,x,y,\omega)$ *is jointly measurable for all* $x \in \mathbb{X}$ *and for all* $y \in \mathbb{X}$;
(b) *The mapping* $(x,y) \to F(t,,x,y,\omega)$ *is jointly continuous for almost each* $t \in [0,b]$ *and for all* $\omega \in \Omega$;

Definition 2 (see [16]). *Let* $\mathbb{X}$ *be a separable Banach space with Borel σ-algebra* $\mathcal{B}$. *A mapping* $Y : \Omega \times \mathbb{X} \to \mathbb{X}$ *is called a random operator if* $Y(.,y)$ *is measurable for each* $y \in \mathbb{X}$.

It is generally expressed as $Y(\omega, y) = Y(\omega)y$. We will use these two expressions interchangeably in this paper.

Definition 3 (see [16]). *Let* $D : \Omega \to 2^{\mathbb{X}}$ *be a mapping and*

$$U = \{(\omega, y) : \omega \in \Omega \text{ and } y \in D(\omega)\}.$$

(i) *A mapping* $Y : U \to \mathbb{X}$ *is called a random operator with stochastic domain* D *if*

    (a) D *is measurable (i.e,* $\{\omega \in \Omega : D(\omega) \cap A \neq \emptyset\} \in \mathcal{F}$ *for all* $A \subseteq \mathbb{X}$*);*
    (b)   *for every open set* $\mathcal{O} \subseteq \mathbb{X}$ *and any* $y \in \mathbb{X}$,

$$\{\omega \in \Omega : y \in D(\omega) \text{ and } Y(\omega, y) \in \mathcal{O}\} \in \mathcal{F}.$$

(ii) *We say that* Y *is continuous if every* $Y(\omega)$ *is continuous.*

Definition 4 (see [16]). *For a random operator* Y, *a mapping* $y : \Omega \to \mathbb{X}$ *is called a random (stochastic) fixed point of* Y *if for* $\mathbb{P}$-*almost all* $\omega \in \Omega$, *we have*

$$y(\omega) \in D(\omega),$$

$$Y(\omega)y(\omega) = y(\omega)$$

and

$$\{\omega \in \Omega : y(\omega) \in \mathcal{O}\} \subset \mathcal{F}$$

for every open set $\mathcal{O} \subseteq \mathbb{X}$ *(i.e., y is measurable).*

Lemma 1 (see [16]). *Let* $(\Omega, \mathcal{F}, \mathbb{P})$ *be complete and let* $y_0 : \Omega \to \mathbb{X}$ *and* $r : \Omega \to \mathbb{R}_+^*$ *be measurable. Then* $D : \Omega \to 2^{\mathbb{X}}$ *defined by*

$$D(\omega) = \{y \in \mathbb{X} : \|y - y_0(\omega)\| \leq r(\omega)\}$$

is a measurable multivalued mapping.

Lemma 2 (see [16]). *Let* $D : \Omega \to 2^{\mathbb{X}}$ *be measurable with* $D(\omega)$ *closed, convex and solid (i.e.,* $\operatorname{int}(D(\omega)) \neq \emptyset$*) for all* $\omega \in \Omega$. *Assume there exists a measurable random variable* $y_0 : \Omega \to \mathbb{X}$ *with* $y_0(\omega) \in \operatorname{int}(D(\omega))$ *for all* $\omega \in \Omega$. *Let* Y *be a continuous random operator with stochastic domain* D *such that for every* $\omega \in \Omega$,

$$\{y \in D(\omega) : Y(\omega)y = y\} \neq \emptyset.$$

then Y *has a stochastic fixed point.*

In this work, the existence of a mild solution to problem (1) is related to the existence of resolvent family introduce by Pardo and Lizama [6].

2.1. Resolvent Family

Now, we recall some definitions and basic results on fractional calculus. Let $\Gamma(\cdot)$ denote the gamma function and define g_x for $x > 0$ by

$$g_x(t) = \begin{cases} \dfrac{t^{x-1}}{\Gamma(x)}, & t > 0; \\ 0, & t \leq 0. \end{cases}$$

It is known that g_x satisfies the following properties:

(i)   for any $a, b > 0$, $(g_a \star g_b)(t) = g_{a+b}(t)$;

(ii) for $a, \lambda > 0$ and $\text{Re}(\lambda) > 0$, $\hat{g}_a(\lambda) = 1/\lambda^a$, where $\widehat{(\cdot)}$ and $(\cdot \star \cdot)(\cdot)$ denote the Laplace transformation and convolution, respectively.

The most frequently encountered tools in the theory of fractional calculus are provided by the Riemann–Liouville and Caputo fractional differential operators.

Definition 5. *The Riemann–Liouvulle fractional integral of a function $f \in L^1_{loc}([0, \infty), \mathbb{X})$ of order $\eta > 0$ with lower limit zero is defined as follows*

$$\mathbb{I}^\eta f(t) = (g_\eta \star f)(t) = \int_0^t g_\eta(t-s) f(s) \, ds \quad \text{for } t > 0$$

and $\mathbb{I}^0(t) = f(t)$, provided that side integral is point-wise defined in $[0, \infty)$.

Definition 6. *Let $\eta > 0$ be given and denote $m = \lceil \eta \rceil$. The Caputo fractional derivative of order $\eta > 0$ of a function $f \in C^m([0, \infty), \mathbb{X})$ with lower limit zero is given by*

$$^cD^\eta f(t) = \mathbb{I}^{m-\eta} D^m f(t) = \int_0^t g_{m-\eta}(t-s) D^m f(s) \, ds,$$

and $^cD^0 f(t) = f(t)$, where $D^m = d^m/dt^m$ and $\lceil \cdot \rceil$ is ceiling function.

For more progress and important properties about fractional calculus and its applications, we refer the reader to [1,2] and references therein. The following definition was introduced by Pardo and Lizama [6] and provides a suitable representation of a mild solution for Problem (1) in terms of a specific family of bounded and linear operators.

Definition 7 (see [6]). *Let A be a closed linear operator on a Banach space $\mathbb{X}$ with domain $\mathcal{D}(A)$ and let $\beta > 0$, γ_k, α_k, $k = 1, 2, \cdots n$ be real positive numbers. Then A is called the generator of a (β, γ_k)-resolvent family if there exists $\kappa \geq 0$ and a strongly continuous function $\mathcal{R}_{\beta, \gamma_k} : \mathbb{R}^+ \to \mathcal{L}(\mathbb{X})$ such that*

$$\left\{ \lambda^{\beta+1} + \sum_{k=1}^n \alpha_k \lambda^{\gamma_k} : \text{Re}(\lambda) > \kappa \right\} \subset \rho(A)$$

and

$$\lambda^\beta \left(\lambda^{\beta+1} + \sum_{k=1}^n \alpha_k \lambda^{\gamma_k} - A \right)^{-1} \vartheta = \int_0^\infty e^{-\lambda t} \mathcal{R}_{\beta, \gamma_k}(t) \vartheta \, dt,$$

where $\text{Re}(\lambda) > \kappa$ and $\vartheta \in \mathbb{X}$.

Theorem 1 (see [6]). *Let $0 < \beta \leq \gamma_n \leq \gamma_{n-1} \leq \cdots \leq \gamma_1 \leq 1$ and $\alpha_k \geq 0$, $k = 1, 2, \cdots, n$ be given and let A be a generator of a bounded and strongly continuous cosine family $\{C(t)\}_{t \in \mathbb{R}}$. Then, A generates a bounded (β, γ_k)-resolvent family $(\mathcal{R}_{\beta, \gamma_k}(t))_{t \geq 0}$.*

Motivated by Pardo and Lizama [6], we introduce the concept of random mild solution for Equation (1) as follows.

Definition 8. *Let $0 < \beta \leq \gamma_n \leq \cdots \leq \gamma_1 \leq 1$ and $\alpha_k \geq 0$, $k = 1, 2, \cdots n$ be given and $\mathcal{A}$ be a generator of a bounded (β, γ_k)-resolvent family $\{\mathcal{R}_{\beta, \gamma_k}(t)\}_{t \geq 0}$. Then, a stochastic process $\vartheta : [0, b] \times \Omega \to \mathbb{X}$ is said to be random mild solution of Equation (1) if $\vartheta(\cdot, \omega) \in \mathcal{C}([0, b], \mathbb{X})$, $\vartheta(0, \omega) = \vartheta_0(\omega)$, $\vartheta'(0, \omega) = \vartheta_1(\omega)$ and satifies the following integral Equation*

$$\vartheta(t,\omega) = \mathcal{R}_{\beta,\gamma_k}(t)\vartheta_0(\omega) + (g_1 \star \mathcal{R}_{\beta,\gamma_k})(t)\vartheta_1(\omega) + \sum_{k=1}^{n}\alpha_k \int_0^t \frac{(t-s)^{\beta-\gamma_k}}{\Gamma(1+\beta-\gamma_k)}\mathcal{R}_{\beta,\gamma_k}(s)\vartheta_0(\omega)ds$$

$$+ \int_0^t \mathcal{T}_{\beta,\gamma_k}(t-s)F\left(s,\vartheta(s,\omega),\int_0^s B(s,\tau)\vartheta(\tau,\omega)d\tau,\omega\right)ds$$

for $(t,\omega) \in [0,b] \times \Omega$, where $\mathcal{T}_{\beta,\gamma_k}(t) = (g_\beta \star \mathcal{R}_{\beta,\gamma_k})(t)$.

2.2. Measures of Noncompactness

We recall some fundamental definitions and lemmas related to the measure of noncompactness. We introduce first the definition for Hausdorff's measure of noncompactness and its properties.

Definition 9 (see [17]). *The Hausdorff measure of noncompactness $\chi(\cdot)$ defined on bounded set E of Banach space $\mathbb{X}$ is*

$$\chi(E) = \inf\{\epsilon > 0 : E \text{ can be covered by finite number of balls of radii smaller then } \epsilon\}.$$

More details on the Hausdorff's measure of noncompacness can be found in Goebel [17] and Deimling [18].

The notations $\chi(\cdot)$ and $\chi_C(\cdot)$ stand for the Hausdorff measure of noncompactness on the bounded set of $\mathbb{X}$ and $\mathcal{C}([0,b],\mathbb{X})$, respectively. For any $V \subset \mathcal{C}([0,b],\mathbb{X})$ and $t \in [0,b]$, set $V(t) = \{\vartheta(t) : \vartheta \in V\}$. Then $V(t) \subset \mathbb{X}$.

The next results play an important role in demonstrating our main result.

Lemma 3 (see [17]). *Let $V \subset \mathcal{C}([0,b],\mathbb{X})$ be bounded, then*

$$\chi(V(t)) \leq \chi_C(V) \quad \text{for all } t \in [0,b],$$

where $V(t) = \{v(t); v \in V\}$. Furthermore, if V is equicontinuous on $[0,b]$, then $\chi(V(t))$ is continuous on $[0,b]$, and $\chi_C(V) = \sup_{t \in [0,b]} \alpha(V(t))$.

Lemma 4 (see [19]). *Let $\{Z_n : n \in \mathbb{N}\}$ be a sequence of Bochner integrable functions from $[0,b]$ into $\mathbb{X}$ such that $\|Z_n(t)\| \leq f(t)$ for every $n \geq 1$ and almost all $t \in [0,b]$, where $f \in \mathbb{L}^1([0,b],\mathbb{R}^+)$, then the function $Z(t) = \chi\{Z_n(t) : n \geq 1\} \in \mathbb{L}^1([0,b],\mathbb{R}^+)$ and satisfy*

$$\chi\left(\int_0^t Z_n(s)\,ds : n \geq 1\right) \leq 2\int_0^t Z(s)ds.$$

Definition 10 (see [20]). *A countable set $\{Z_n : n \geq 1\} \subset \mathbb{L}^1([0,b],\mathbb{X})$ is called semicompact if there exists a function $f \in \mathbb{L}^1([0,b],\mathbb{R}^+)$ satisfying $\sup_{n \geq 1} Z_n(t) \leq f(t)$ for a.e. $t \in [0,b]$ and the sequence $\{Z_n : n \geq 1\}$ is relatively compact in $\mathbb{X}$.*

Lemma 5. *[20] Let $(\mathbb{Q}Z)(t) = \int_0^t \mathcal{T}_{\beta,\gamma_k}(t-s)Z(s)ds$ and the sequence $\{Z_n : n \geq 1\} \subset \mathbb{L}^1([0,b],\mathbb{X})$ be semicompact. Then the following statements hold:*
(i) *The set $\{\mathbb{Q}Z_n : n \geq 1\} \subset \mathbb{L}^1([0,b],\mathbb{X})$ is relatively compact in $\mathcal{C}([0,b],\mathbb{X})$.*
(ii) *If $Z_n \rightharpoonup Z$, then $(\mathbb{Q}Z_n)(t) \to (\mathbb{Q}Z)(t)$, as $n \to \infty$, for all $t \in [0,b]$.*

To prove our existence results, we shall use the following Mönch's fixed point theorem combined with the stochastic fixed point theorem (i.e., Lemma 2).

Lemma 6 (see [21]). *Let $\mathbb{V}$ be a closed and convex subset of $\mathbb{X}$ and $0 \in \mathbb{V}$. Then a continuous mapping $\mathcal{Q} : \mathbb{V} \to \mathbb{X}$ which satisies Mönch condition (i.e., $\mathbb{W} \subseteq \mathbb{V}$ is countable and $\mathbb{W} \subseteq \overline{con}(\{0\} \cup \mathcal{Q}(\mathbb{W})) \Longrightarrow \overline{\mathbb{W}}$ is compact) has a fixed point in $\mathbb{V}$.*

3. Some Existence Results

In this section, we investigate the existence of random mild solution for Equation (1). The following conditions will be used in our main theorem.

(H1) $\mathcal{A}$ be a generator of a bounded (β, γ_k)-resolvent family $\{\mathcal{R}_{\beta,\gamma_k}(t)\}_{t \geq 0}$ which is equicontinous. Let $M = \sup_{t \in [0,b]} \|\mathcal{R}_{\beta,\gamma_k}(t)\|_{\mathcal{L}(\mathbb{X})}$.

(H2) Then function $F : [0,b] \times \mathbb{X} \times \mathbb{X} \times \Omega \to \mathbb{X}$ is random Caratheodory and the functions $\vartheta_0(\cdot) : \Omega \to \mathbb{X}$ and $\vartheta_1(\cdot) : \Omega \to \mathbb{X}$ are measurables and essentialy bounded. Let

$$\max \left\{ \operatorname{ess\,sup}_{\omega \in \Omega} \|\vartheta_0(\omega)\| \, ; \operatorname{ess\,sup}_{\omega \in \Omega} \|\vartheta_1(\omega)\| \right\} \leq c_0$$

for some constant $c_0 \in \mathbb{R}^+$.

(H3) There exist two functions $f : [0,b] \times \Omega \to \mathbb{R}^+$ and $G : \mathbb{R}^+ \times \Omega \to \mathbb{R}^+$ such that for each $\omega \in \Omega$, $f(\cdot, \omega) \in L^{1/p_1}([0,b], \mathbb{R}^+)$ for a constant $p_1 \in (0,1)$ and $G(\cdot, \omega)$ a nondeacresing continuous function with

$$\|F(t,x,y,\omega)\|_\mathbb{X} \leq f(t,\omega) G(\|x\| + \|y\|, \omega), \text{ for a.e } t \in [0,b] \text{ and each } x \in \mathbb{X}, y \in \mathbb{X}.$$

(H4) For a constant $p_2 \in (0,1)$ and bounded subsets $\mathbb{V}_1, \mathbb{V}_2 \subset \mathbb{X}$, there exists a function $h : [0,b] \times \Omega \to \mathbb{R}^+$ such that for each $\omega \in \Omega$, $h(\cdot, \omega) \in \mathbb{L}^{1/p_2}([0,b], \mathbb{R}^+)$ with

$$\chi(F(t, \mathbb{V}_1, \mathbb{V}_2, \omega)) \leq h(t, \omega)[\chi(\mathbb{V}_1) + \chi(\mathbb{V}_2)].$$

(H5) There exists a random function $r : \Omega \to \mathbb{R} \setminus \{0\}$ such that

$$M c_0 \left[1 + b + \sum_{k=1}^{n} \frac{\alpha_k \, b^{(1+\beta-\gamma_k)}}{\Gamma(2+\beta-\gamma_k)} \right] + \frac{M G((1+\zeta_B)r(\omega), \omega)}{\Gamma(1+\beta)} \widehat{P}_1 \|f(\cdot,\omega)\|_{\mathbb{L}^{1/p_1}} \leq r(\omega)$$

where $\widehat{P}_1 = \left[\left(\frac{1-p_1}{1+\beta-p_1} \right) b^{(1+\beta-p_1)/(1-p_1)} \right]^{1-p_1}$.

The following existence theorem is one of the main results of this paper.

Theorem 2. *Assume that the assumptions (H1)–(H5) are valid, then the multi-term time-fractional integro-differential problem (1) has at least one mild random solution on $[0,b]$ provided that*

$$\frac{M \widehat{P}_2 (1+\zeta_B)}{\Gamma(1+\beta)} \|h(\cdot,\omega)\|_{\mathbb{L}^{1/p_2}} < \frac{1}{2}, \tag{2}$$

where $\widehat{P}_2 = \left[\left(\frac{1-p_2}{1+\beta-p_2} \right) b^{(1+\beta-p_2)/(1-p_2)} \right]^{1-p_2}$.

Proof. It is noted that

$$\|\mathcal{T}_{\beta,\gamma_k}(t)\|_{\mathcal{L}(\mathbb{X})} \leq \frac{M t^\beta}{\Gamma(1+\beta)}.$$

Consider the random operator $\mathcal{N}: \Omega \times \mathcal{C}([0,b], \mathbb{X}) \to \mathcal{C}([0,b], \mathbb{X})$ defined by

$$(\mathcal{N}(\omega)\vartheta)(t) = \mathcal{R}_{\beta,\gamma_k}(t)\vartheta_0(\omega) + (g_1 \star \mathcal{R}_{\beta,\gamma_k})(t)\vartheta_1(\omega)$$
$$+ \sum_{k=1}^{n} \alpha_k \int_0^t \frac{(t-s)^{\beta-\gamma_k}}{\Gamma(1+\beta-\gamma_k)} \mathcal{R}_{\beta,\gamma_k}(s)\vartheta_0(\omega)ds \quad (3)$$
$$+ \int_0^t \mathcal{T}_{\beta,\gamma_k}(t-s) F\left(s, \vartheta(s,\omega), \int_0^s \mathcal{B}(s,\tau)\vartheta(\tau,\omega)d\tau, \omega\right)ds$$

for $t \in [0,b]$. We divide the proof into a sequence of steps.

Step 1. We show that the mapping $\mathcal{N}$ is a random operator with stochastic domain.
By assumption (**H2**), we know that $F(t, x, y, \cdot)$ for $t \in [0,b]$, $x, y \in \mathbb{X}$, $\vartheta_0(\cdot)$ and $\vartheta_1(\cdot)$ are measurables. Then $\mathcal{N}(\cdot)\vartheta : \Omega \to \mathcal{C}([0,b], \mathbb{X})$ is a random varriable. Let $D : \Omega \to 2^{\mathcal{C}([0,b], \mathbb{X})}$ be defined by

$$D(\omega) = \{\vartheta \in \mathcal{C}([0,b], \mathbb{X}) : \|\vartheta\| \leq r(\omega)\}.$$

Thus, the set $D(\omega)$ is bounded, closed, convex and solid for all $\omega \in \Omega$. So D is measurable by Lemma 1. For each $\vartheta \in D(\omega)$, using (**H2**), (**H3**), (**H4**), and Hölder's inequality, we have

$\|(\mathcal{N}(\omega)\vartheta)(t)\|$

$\leq \|\mathcal{R}_{\beta,\gamma_k}(t)\vartheta_0(\omega)\| + \|(g_1 \star \mathcal{R}_{\beta,\gamma_k})(t)\vartheta_1(\omega)\| + \|\sum_{k=1}^{n} \alpha_k \int_0^t \frac{(t-s)^{\beta-\gamma_k}}{\Gamma(1+\beta-\gamma_k)} \mathcal{R}_{\beta,\gamma_k}(s)\vartheta_0(\omega)ds\|$

$+ \|\int_0^t \mathcal{T}_{\beta,\gamma_k}(t-s) F\left(s, \vartheta(s,\omega), \int_0^s \mathcal{B}(s,\tau)\vartheta(\tau,\omega)d\tau, \omega\right)ds\|$

$\leq M c_0 + b M c_0 + \sum_{k=1}^{n} \frac{\alpha_k M c_0 b^{(1+\beta-\gamma_k)}}{\Gamma(2+\beta-\gamma_k)}$

$+ \int_0^t \|\mathcal{T}_{\beta,\gamma_k}(t-s)\|_{\mathcal{L}(\mathbb{X})} \|F\left(s, \vartheta(s,\omega), \int_0^s \mathcal{B}(s,\tau)\vartheta(\tau,\omega)d\tau, \omega\right)\|ds$

$\leq M c_0 + b M c_0 + \sum_{k=1}^{n} \frac{\alpha_k M c_0 b^{(1+\beta-\gamma_k)}}{\Gamma(2+\beta-\gamma_k)}$

$+ \frac{M}{\Gamma(1+\beta)} \int_0^t (t-s)^{\beta} f(s,\omega) G\left(\|\vartheta(t,\omega)\| + \|\int_0^s \mathcal{B}(s,\tau)\vartheta(\tau,\omega)d\tau\|, \omega\right)$

$\leq M c_0 + b M c_0 + \sum_{k=1}^{n} \frac{\alpha_k M c_0 b^{(1+\beta-\gamma_k)}}{\Gamma(2+\beta-\gamma_k)} + \frac{M}{\Gamma(1+\beta)} \int_0^b (b-s)^{\beta} f(s,\omega) G((1+\zeta_B)r(\omega), \omega)ds$

$\leq M c_0 + b M c_0 + \sum_{k=1}^{n} \frac{\alpha_k M c_0 b^{(1+\beta-\gamma_k)}}{\Gamma(2+\beta-\gamma_k)}$

$+ \frac{M G((1+\zeta_B)r(\omega), \omega)}{\Gamma(1+\beta)} \left(\int_0^b (b-s)^{\beta/(1-p_1)} ds\right)^{1-p_1} \left(\int_0^b (f(s,\omega))^{1/p_1} ds\right)^{p_1}$

$$\leq M c_0 + b M c_0 + \sum_{k=1}^{n} \frac{\alpha_k M c_0 b^{(1+\beta-\gamma_k)}}{\Gamma(2+\beta-\gamma_k)}$$

$$+ \frac{M G((1+\zeta_B)r(\omega), \omega)}{\Gamma(1+\beta)} \left[\left(\frac{1-p_1}{1+\beta-p_1} \right) b^{(1+\beta-p_1)/(1-p_1)} \right]^{1-p_1} \|f(\cdot, \omega)\|_{\mathbb{L}^{1/p_1}}$$

$$\leq M c_0 \left[1 + b + \sum_{k=1}^{n} \frac{\alpha_k b^{(1+\beta-\gamma_k)}}{\Gamma(2+\beta-\gamma_k)} \right] + \frac{M G((1+\zeta_B)r(\omega), \omega)}{\Gamma(1+\beta)} \widehat{P}_1 \|f(\cdot, \omega)\|_{\mathbb{L}^{1/p_1}}$$

$$\leq r(\omega),$$

which implies that $\mathcal{N}$ is a random operator with stochastic domain D and $\mathcal{N}(\omega) : D(\omega) \to D(\omega)$ for each $\omega \in \Omega$.

Step 2. Show that $\mathcal{N}$ is continuous on $D(\omega)$.

Let $(\vartheta^{(m)})_{m \in \mathbb{N}}$ be a sequence in $D(\omega)$ satisfying $\vartheta^m \to \vartheta$ in $D(\omega)$ and define

$$\widetilde{F}_m(s) = F\left(s, \vartheta^{(m)}(s, \omega), \int_0^s \mathcal{B}(s, \tau) \vartheta^{(m)}(\tau, \omega) d\tau, \omega\right)$$

and

$$\widetilde{F}(s) = F\left(s, \vartheta(s, \omega), \int_0^s \mathcal{B}(s, \tau) \vartheta(\tau, \omega) d\tau, \omega\right).$$

then

$$\|(\mathcal{N}(\omega) \vartheta^{(m)})(t) - (\mathcal{N}(\omega) \vartheta)(t)\| \leq \frac{M}{\Gamma(1+\beta)} \int_0^t (t-s)^\beta \|\widetilde{F}_m(s) - \widetilde{F}(s)\| ds.$$

by assumption (**H2**) and combining with Lebesgue dominated convergence theorem, we get

$$\int_0^t (t-s)^\beta \|\widetilde{F}_m(s) - \widetilde{F}(s)\| ds \to 0 \text{ as } m \to \infty, \text{ for } t \in [0, b].$$

consequently, we obtain

$$\|(\mathcal{N}(\omega) \vartheta^{(m)})(t) - (\mathcal{N}(\omega) \vartheta)(t)\| \to 0 \text{ as } m \to \infty.$$

hence $\mathcal{N}$ is continuous.

Step 3. Show that for every $\omega \in \Omega$, $\{\vartheta \in D(\omega) : \mathcal{N}(\omega) \vartheta = \vartheta\} \neq \emptyset$.

To achieve this, we going to demontrate that the Mönch condition holds. Let $\omega \in \Omega$ be arbitrary fixed. First, let show that $\mathcal{N}$ maps bounded sets into equicontinuous sets of $D(\omega)$. Let $t_1, t_2 \in [0, b]$ with $t_2 > t_1$ and $\vartheta \in D(\omega)$. By the equicontinuouty of $\mathcal{R}_{\beta, \gamma_k}(t)$, we have

$$\|(\mathcal{N}(\omega)\vartheta)(t_2) - (\mathcal{N}(\omega)\vartheta)(t_1)\|$$

$$\leq \|\mathcal{R}_{\beta,\gamma_k}(t_2) - \mathcal{R}_{\beta,\gamma_k}(t_1)\|_{\mathcal{L}(\mathbb{X})} \|\vartheta_0(\omega)\|$$

$$+ \|(g_1 \star \mathcal{R}_{\beta,\gamma_k})(t_2)\vartheta_1(\omega) - (g_1 \star \mathcal{R}_{\beta,\gamma_k})(t_1)\vartheta_1(\omega)\|$$

$$+ \|\sum_{k=1}^{n} \alpha_k \int_0^{t_2} \frac{(t_2-s)^{\beta-\gamma_k}}{\Gamma(1+\beta-\gamma_k)} \mathcal{R}_{\beta,\gamma_k}(s)\vartheta_0(\omega) ds - \sum_{k=1}^{n} \alpha_k \int_0^{t_1} \frac{(t_1-s)^{\beta-\gamma_k}}{\Gamma(1+\beta-\gamma_k)} \mathcal{R}_{\beta,\gamma_k}(s)\vartheta_0(\omega) ds\|$$

$$+ \|\int_0^{t_2} \mathcal{T}_{\beta,\gamma_k}(t_2-s)\widetilde{F}(s) ds - \int_0^{t_1} \mathcal{T}_{\beta,\gamma_k}(t_1-s)\widetilde{F}(s) ds\|$$

$$\leq \|\mathcal{R}_{\beta,\gamma_k}(t_2) - \mathcal{R}_{\beta,\gamma_k}(t_1)\|_{\mathcal{L}(\mathbb{X}} \|\vartheta_0(\omega)\|$$

$$+ \int_{t_1}^{t_2} \|\mathcal{R}_{\beta,\gamma_k}(s)\|_{\mathcal{L}(\mathbb{X}} \|\vartheta_1(\omega)\| ds$$

$$+ \sum_{k=1}^{n} \alpha_k \left\| \int_0^{t_2} \frac{(t_2-s)^{\beta-\gamma_k}}{\Gamma(1+\beta-\gamma_k)} \mathcal{R}_{\beta,\gamma_k}(s) ds - \int_0^{t_1} \frac{(t_1-s)^{\beta-\gamma_k}}{\Gamma(1+\beta-\gamma_k)} \mathcal{R}_{\beta,\gamma_k}(s) ds \right\| \|\vartheta_0(\omega)\|$$

$$+ \int_0^{t_1} \|\mathcal{T}_{\beta,\gamma_k}(t_2-s) - \mathcal{T}_{\beta,\gamma_k}(t_1-s)\| \|\widetilde{F}(s)\| ds - \int_{t_1}^{t_2} \|\mathcal{T}_{\beta,\gamma_k}(t_1-s)\| \|\widetilde{F}(s)\| ds$$

$$\leq c_0 \|\mathcal{R}_{\beta,\gamma_k}(t_2) - \mathcal{R}_{\beta,\gamma_k}(t_1)\|_{\mathcal{L}(\mathbb{X})} + Mc_0(t_2 - t_1)$$

$$+ \sum_{k=1}^{n} \frac{\alpha_k M}{\Gamma(1+\beta-\gamma_k)} \left[\int_0^{t_1} |(t_2-s)^{\beta-\gamma_k} - (t_1-s)^{\beta-\gamma_k}| ds - \int_{t_1}^{t_2} |(t_2-s)|^{\beta-\gamma_k} ds \right] c_0$$

$$+ \int_0^{t_1} \|\mathcal{T}_{\beta,\gamma_k}(t_2-s) - \mathcal{T}_{\beta,\gamma_k}(t_1-s)\| \|\widetilde{F}(s)\| ds + \frac{M}{\Gamma(1+\beta)} \int_{t_1}^{t_2} |(t_1-s)|^{\beta} \|\widetilde{F}(s)\| ds$$

$$\leq c_0 \|\mathcal{R}_{\beta,\gamma_k}(t_2) - \mathcal{R}_{\beta,\gamma_k}(t_1)\|_{\mathcal{L}(\mathbb{X})} + Mc_0(t_2 - t_1)$$

$$+ \sum_{k=1}^{n} \frac{\alpha_k M}{\Gamma(1+\beta-\gamma_k)} \left[\int_0^{t_1} |(t_2-s)^{\beta-\gamma_k} - (t_1-s)^{\beta-\gamma_k}| ds - \int_{t_1}^{t_2} |(t_2-s)|^{\beta-\gamma_k} ds \right] c_0$$

$$+ \int_0^{t_1} \|\mathcal{T}_{\beta,\gamma_k}(t_2-t_1+s) - \mathcal{T}_{\beta,\gamma_k}(s)\| \|\widetilde{F}(s)\| ds + \frac{M}{\Gamma(1+\beta)} \int_{t_1}^{t_2} |(t_1-s)|^{\beta} \|\widetilde{F}(s)\| ds.$$

By the equicontinuity of $\mathcal{R}_{\beta,\gamma_k}$ and Lebesgue dominated convergence theorem, we conclude that the right side of the above inequality tends to zero (independently of ϑ) as $t_1 \to t_2$. Thus, $\mathcal{N}(\omega)(\mathbb{V})$ is equicontinuous on $[0, b]$.

Now, let us assume that $\mathbb{V} = \{\vartheta^{(m)}\}_{m=1}^{\infty}$ be a countable subset of $D(\omega)$ and $\mathbb{V} \subseteq \overline{con}(\{0\} \cup \mathcal{N}(\omega)(\mathbb{V}))$. Since $\mathcal{N}(\omega)(\mathbb{V})$ is is bounded and equicontinuous, we have $\mathbb{V} = \{\vartheta^{(m)}\}_{m=1}^{\infty}$ is bounded and equicontinuous and therefore by Lemma 3, the function $t \to \chi(\mathbb{V}(t))$ is continuous on $[0, b]$. By Lemma 4, we get

$$\chi\big(\{(\mathcal{N}(\omega)\vartheta^{(m)})(t)\}_{m=1}^{\infty}\big)$$

$$\leq \chi\bigg(\bigg\{\int_0^t \mathcal{T}_{\beta,\gamma_k}(t-s)F\bigg(s,\vartheta^{(m)}(s,\omega),\int_0^s \mathcal{B}(s,\tau)\vartheta^{(m)}(\tau,\omega)d\tau,\omega\bigg)ds\bigg\}_{m=1}^{\infty}\bigg)$$

$$\leq 2\frac{M(1+\zeta_B)}{\Gamma(1+\beta)}\int_0^t (t-s)^{\beta}h(s,\omega)\chi\big(\{\vartheta^{(m)}(s)\}_{m=1}^{\infty}\big)ds$$

$$\leq 2\frac{M(1+\zeta_B)}{\Gamma(1+\beta)}\int_0^b (b-s)^{\beta}h(s,\omega)\sup_{s\in[0,b]}\chi\big(\{\vartheta^{(m)}(s)\}_{m=1}^{\infty}\big)ds \tag{4}$$

$$\leq 2\frac{M(1+\zeta_B)}{\Gamma(1+\beta)}\sup_{s\in[0,b]}\chi\big(\{\vartheta^{(m)}(s)\}_{m=1}^{\infty}\big)\bigg(\int_0^b (b-s)^{\beta/(1-p_2)}ds\bigg)^{1-p_2}\bigg(\int_0^b (h(s,\omega))^{1/p_2}ds\bigg)^{p_2}$$

$$\leq 2\frac{M(1+\zeta_B)}{\Gamma(1+\beta)}\sup_{s\in[0,b]}\chi\big(\{\vartheta^{(m)}(s)\}_{m=1}^{\infty}\big)K_2\|h(\cdot,\omega)\|_{\mathbb{L}^{1/p_2}}.$$

By Lemma 3, we obtain

$$\chi(\mathcal{N}(\omega)\mathbb{V}) \leq \bigg[2\frac{MK_2(1+\zeta_B)}{\Gamma(1+\beta)}\|h(\cdot,\omega)\|_{\mathbb{L}^{1/p_2}}\bigg]\chi(\mathbb{V}).$$

From Mönch condition, we get

$$\chi(\mathbb{V}) \leq \chi(\overline{con}(\{0\}\cup\mathcal{N}(\omega)\mathbb{V})) = \chi(\mathcal{N}(\omega)\mathbb{V}) \leq \bigg[2\frac{MK_2(1+\zeta_B)}{\Gamma(1+\beta)}\|h(\cdot,\omega)\|_{\mathbb{L}^{1/p_2}}\bigg]\chi(\mathbb{V}).$$

From inequality (2), we deduce that $\chi(\mathbb{V}) = 0$. As a consequence of Theorem 6, we show that $\mathcal{N}$ has a fixed point $\vartheta(\omega) \in D(\omega)$. Since $\cap_{\omega\in\Omega}D(\omega) \neq \emptyset$, the hypothesis that a measurable selector of $int(D)$ holds. By Lemma 2, the random operator $\mathcal{N}$ has a stochastic fixed point $\vartheta^{\star}(\omega)$, which is a mild solution of (1). This complete the proof. □

With another growth condition of $F(\cdot,\cdot,\cdot)$, the condition (**H4**) can be deleted. Precisely, we replace the assumption (**H3**) with (**H3'**), where

(**H3'**) There exist two functions $Z_1, Z_2 : [0,b]\times\Omega \to \mathbb{R}^+$ such that for each $\omega \in \Omega$, $Z_1(\cdot,\omega), Z_2(\cdot,\omega) \in L^{1/p_1}([0,b],\mathbb{R}^+)$ for a constant $p_1 \in (0,1)$ with

$$\|F(t,x,y,\omega)\|_{\mathbb{X}} \leq Z_1(t,\omega) + Z_2(t,\omega)(\|x\|+\|y\|)$$

for $t \in [0,b]$ and $x,y \in \mathbb{X}$.

Theorem 3. *Under the assumptions (**H1**), (**H2**), (**H3'**), (**H4**) and condition (2), the multi-term time-fractional integro-differential problem (1) has at least one mild random solution on $[0,b]$.*

Proof. Using (**H1**), (**H2**), (**H3'**), and Hölder's inequality, we have

$$\|(\mathcal{N}(\omega)\vartheta)(t)\|$$

$$\leq \|\mathcal{R}_{\beta,\gamma_k}(t)\vartheta_0(\omega)\| + \|(g_1 \star \mathcal{R}_{\beta,\gamma_k})(t)\vartheta_1(\omega)\| + \|\sum_{k=1}^{n} \alpha_k \int_0^t \frac{(t-s)^{\beta-\gamma_k}}{\Gamma(1+\beta-\gamma_k)} \mathcal{R}_{\beta,\gamma_k}(s)\vartheta_0(\omega)ds\|$$

$$+ \|\int_0^t \mathcal{T}_{\beta,\gamma_k}(t-s)F\left(s,\vartheta(s,\omega),\int_0^s B(s,\tau)y(\tau,\omega)d\tau,\omega\right)ds\|$$

$$\leq M c_0 + b M c_0 + \sum_{k=1}^{n} \frac{\alpha_k M c_0 b^{(1+\beta-\gamma_k)}}{\Gamma(2+\beta-\gamma_k)}$$

$$+ \int_0^t \|\mathcal{T}_{\beta,\gamma_k}(t-s)\|_{\mathcal{L}(\mathbb{X})} \|F\left(s,\vartheta(s,\omega),\int_0^s \mathcal{B}(s,\tau)\vartheta(\tau,\omega)d\tau,\omega\right)\|ds$$

$$\leq M c_0 + b M c_0 + \sum_{k=1}^{n} \frac{\alpha_k M c_0 b^{(1+\beta-\gamma_k)}}{\Gamma(2+\beta-\gamma_k)}$$

$$+ \frac{M}{\Gamma(1+\beta)} \int_0^t (t-s)^{\beta} \left[Z_1(s,\omega) + Z_2(s,\omega)\left(\|\vartheta(t,\omega)\| + \|\int_0^s \mathcal{B}(s,\tau)\vartheta(\tau,\omega)d\tau\|\right)\right]ds$$

$$\leq M c_0 + b M c_0 + \sum_{k=1}^{n} \frac{\alpha_k M c_0 b^{(1+\beta-\gamma_k)}}{\Gamma(2+\beta-\gamma_k)}$$

$$+ \frac{M}{\Gamma(1+\beta)} \int_0^b (b-s)^{\beta} [Z_1(s,\omega) + Z_2(s,\omega)(1+\zeta_B)\|\vartheta\|]ds$$

$$\leq M c_0 + b M c_0 + \sum_{k=1}^{n} \frac{\alpha_k M c_0 b^{(1+\beta-\gamma_k)}}{\Gamma(2+\beta-\gamma_k)} + \frac{M}{\Gamma(1+\beta)} \left(\int_0^b (b-s)^{\beta/(1-p_1)}ds\right)^{1-p_1} \left(\int_0^b (Z_1(s,\omega))^{1/p_1}ds\right)^{p_1}$$

$$+ \frac{M(1+\zeta_B)}{\Gamma(1+\beta)}\|\vartheta\| \left(\int_0^b (b-s)^{\beta/(1-p_1)}ds\right)^{1-p_1} \left(\int_0^b (Z_2(s,\omega))^{1/p_1}ds\right)^{p_1}$$

$$\leq M\left[c_0 + b c_0 + \sum_{k=1}^{n} \frac{\alpha_k c_0 b^{(1+\beta-\gamma_k)}}{\Gamma(2+\beta-\gamma_k)} + \frac{1}{\Gamma(1+\beta)} \widehat{P}_1 \left(\int_0^b (Z_1(s,\omega))^{1/p_1}ds\right)^{p_1}\right]$$

$$+ M \frac{(1+\zeta_B)}{\Gamma(1+\beta)} \widehat{P}_1 \left(\int_0^b (Z_2(s,\omega))^{1/p_1}ds\right)^{p_1} \|\vartheta\|$$

$$\leq M K_1(\omega) + M K_2(\omega) \|\vartheta\|,$$

where

$$K_1(\omega) = c_0 + b c_0 + \sum_{k=1}^{n} \frac{\alpha_k c_0 b^{(1+\beta-\gamma_k)}}{\Gamma(2+\beta-\gamma_k)} + \frac{1}{\Gamma(1+\beta)} \widehat{P}_1 \left(\int_0^b (Z_1(s,\omega))^{1/p_1}ds\right)^{p_1}$$

and

$$K_2(\omega) = M\frac{(1+\zeta_B)}{\Gamma(1+\beta)} \widehat{P}_1 \left(\int_0^b (Z_2(s,\omega))^{1/p_1}ds\right)^{p_1}.$$

choosing $r(\omega) = \dfrac{MK_1(\omega)}{1 - MK_2(\omega)}$ and

$$D(\omega) = \{\vartheta \in \mathcal{C}([0,b], \mathbb{X}) : \|\vartheta\| \leq r(\omega)\} \text{ for all } \vartheta \in D(\omega),$$

we get

$$\|(\mathcal{N}(\omega)\vartheta))(t)\| \leq r(\omega).$$

The proof can be complete similarly to Theorem 2. □

Remark 1. *Consider the measure of noncompactness χ_C and ν defined in $\mathcal{C}([0,b], \mathbb{X})$ by*

$$\chi_C(\mathbb{K}) = \max_{\mathbb{O} \in \Delta(\mathbb{K})} \{\nu(\mathbb{O}), mod_c(\mathbb{O})\} \quad \text{and} \quad \nu(\mathbb{O}) = \sup_{t \in [0,b]} e^{-Lt}\chi(\mathbb{O}(t)), \qquad (5)$$

for all bounded subsets $\mathbb{K}$ of $\mathcal{C}([0,b], \mathbb{X})$, where $\Delta(\mathbb{K})$ stand for the collection of all countable subsets of $\mathbb{K}$, $\mathbb{O}(t) = \{x(t)$, $x \in \mathbb{O}$, $t \in [0,b]\}$ and L is an appropriate constant to be defined later. mod_C is the modulus of equicontinuity of the function set $\mathbb{O}$ defined by

$$mod_C(\mathbb{O}) = \lim_{\delta \to 0} \sup_{x \in \mathbb{O}} \max \Big\{ \|x(t_2) - x(t_1)\| \ : \ t_1, t_2 \in [0,b], \ |t_2 - t_1| \leq \delta \Big\}.$$

From [20], we know that there exists a $\mathbb{O}^\star$ which achieves the maximun in (5). Furthermore, the measure χ_C is nonsingular, monotone and regular.

Applying the abaove regular measure of noncompactness, we obtain the following result.

Theorem 4. *Assume that the assumptions (H1)–(H5) are valid, then the multi-term time-fractional integro-differential problem (1) has at least one mild random solution on $[0,b]$.*

Proof. As in the proof of Theorem 2, we show that the operator $\mathcal{N} : D(\omega) \to D(\omega)$ has a stochastic fixed point. We know that $\mathcal{N}$ is a random operator with stochastic domain which is continuous and maps bounded sets into equicontinuous sets of $D(\omega)$. So, in order to finish the proof it is sufficient to show that $\mathcal{N}$ satisfies the Mönch's condition.

Let $\omega \in \Omega$ and $\mathbb{V}$ be a countable subset of $D(\omega)$ and $\mathbb{V} \subseteq \overline{con}(\{0\} \cup \mathcal{N}(\omega)(\mathbb{V}))$. Since $\chi_C((\mathcal{N}(\omega)\mathbb{V}))$ has a maximum, let $\{z^{(m)}(\cdot, \omega)\}_{m=1}^\infty \subset (\mathcal{N}(\omega)\mathbb{V})$ be a denumerable set which attain its maximum. Then there exists a set $\{\vartheta^{(m)}(\cdot, \omega)\}_{m=1}^\infty \subset \mathbb{V}$ such that

$$(\mathcal{N}(\omega)\vartheta^{(m)})(t) = z^m(\cdot, \omega) \quad \text{for all } t \in [0,b]. \qquad (6)$$

We choose $L > 0$ such that

$$K = \dfrac{2M(1 + \zeta_B)}{\Gamma(1 + \beta)} \sup_{t \in [0,b]} \int_0^b (b-s)^\beta h(s,\omega) e^{L(s-t)} ds < 1. \qquad (7)$$

Now, by (4), we derive that

$$\chi(\{(\mathcal{N}(\omega)\vartheta^{(m)})(t)\}_{m=1}^{\infty})$$

$$\leq 2\frac{M(1+\zeta_B)}{\Gamma(1+\beta)} \int_0^t (t-s)^\beta h(s,\omega) \chi\left(\{\vartheta^{(m)}(s)\}_{m=1}^{\infty}\right) ds$$

$$\leq 2\frac{M(1+\zeta_B)}{\Gamma(1+\beta)} \int_0^b (b-s)^\beta h(s,\omega) e^{Ls} \sup_{s\in[0,b]} \left[e^{-Ls}\chi\left(\{\vartheta^{(m)}(s)\}_{m=1}^{\infty}\right)\right] ds \quad (8)$$

$$\leq 2\frac{M(1+\zeta_B)}{\Gamma(1+\beta)} \nu\left(\{\vartheta^{(m)}\}_{m=1}^{\infty}\right) \int_0^b (b-s)^\beta h(s,\omega) e^{Ls} ds.$$

By using (5) and (8), we get

$$\nu(\{z^{(m)}\}_{m=1}^{\infty}) \leq \sup_{t\in[0,b]} e^{-Lt}\chi(\{(\mathcal{N}(\omega)\vartheta^{(m)})(t)\}_{m=1}^{\infty})$$

$$\leq \sup_{t\in[0,b]} e^{-Lt} \frac{2M(1+\zeta_B)}{\Gamma(1+\beta)} \nu\left(\{\vartheta^{(m)}\}_{m=1}^{\infty}\right) \int_0^b (b-s)^\beta h(s,\omega) e^{Ls} ds \quad (9)$$

$$\leq \frac{2M(1+\zeta_B)}{\Gamma(1+\beta)} \nu\left(\{\vartheta^{(m)}\}_{m=1}^{\infty}\right) \sup_{t\in[0,b]} \int_0^b (b-s)^\beta h(s,\omega) e^{L(s-t)} ds$$

$$\leq K \nu\left(\{\vartheta^{(m)}\}_{m=1}^{\infty}\right).$$

Thus, we have

$$\nu\left(\{\vartheta^{(m)}(\cdot,\omega)\}_{m=1}^{\infty}\right) = \nu(\mathbb{V}) \leq \nu(\overline{con}(\{0\} \cup \mathcal{N}(\omega)(\mathbb{V})))$$

$$= \nu\left(\{z^{(m)}(\cdot,\omega)\}_{m=1}^{\infty}\right) \leq K \nu\left(\{\vartheta^{(m)}(\cdot,\omega)\}_{m=1}^{\infty}\right).$$

From (7), we obtain

$$\nu\left(\{\vartheta^{(m)}(\cdot,\omega)\}_{m=1}^{\infty}\right) = \nu(\mathbb{V}) = \nu\left(\{z^{(m)}(\cdot,\omega)\}_{m=1}^{\infty}\right) = 0.$$

By the definition of ν, we get

$$\chi\left(\{\vartheta^{(m)}(t,\omega)\}_{m=1}^{\infty}\right) = \chi\left(\{z^{(m)}(t,\omega)\}_{m=1}^{\infty}\right) = 0 \text{ for every } t \in [0,b].$$

From (**H4**), we obtain

$$\chi\left(\left\{F\left(s, \vartheta^{(m)}(s,\omega), \int_0^s \mathcal{B}(s,\tau)\vartheta^{(m)}(\tau,\omega)d\tau, \omega\right)\right\}_{m=1}^{\infty}\right)$$

$$\leq h(t,\omega)(1+\zeta_B)\chi\left(\{\vartheta^{(m)}(t,\omega)\}\right) = 0 \quad (10)$$

which implies that

$$\mathcal{S} := \left\{F\left(s, \vartheta^{(m)}(s,\omega), \int_0^s \mathcal{B}(s,\tau)\vartheta^{(m)}(\tau,\omega)d\tau, \omega\right)\right\}_{m=1}^{\infty}$$

is relatively compact for almost all $t \in [0, b]$ in $\mathbb{X}$. Since $\{\vartheta^{(m)}(\cdot, \omega)\}_{m=1}^{\infty} \in D(\omega)$, by **(H3)**, we derive that $\mathcal{S}$ is uniformly integrable for a.e $t \in [0, b]$. From Definition 10, we conclude that $\mathcal{S}$ is semicompact and by Lemma 4, the set

$$\left\{ \int_0^t \mathcal{T}_{\beta, \gamma_k}(t-s) F\left(s, \vartheta^{(m)}(s, \omega), \int_0^s \mathcal{B}(s, \tau) \vartheta^{(m)}(\tau, \omega) d\tau, \omega \right) ds \right\}_{m=1}^{\infty}$$

is relatively compact in $\mathbb{X}$. Hence, by (6), $\{z^{(m)}(\cdot, \omega)\}_{m=1}^{\infty}$ is also relatively compact in $\mathcal{C}([0, b], \mathbb{X})$. Since $\chi_{\mathcal{C}}$ is a nonsingular, monotone and regular measure of noncompactness, then by Mönch condition, we have

$$\chi_{\mathcal{C}}(\mathbb{V}) \leq \chi_{\mathcal{C}}(\overline{con}(\{0\} \cup \mathcal{N}(\omega)(\mathbb{V}))) = \chi_{\mathcal{C}}\left(\{z^{(m)}(\cdot, \omega)\}_{m=1}^{\infty}\right) = 0,$$

which shows that $\mathbb{V}$ is relatively compact in $\mathcal{C}([0, b], \mathbb{X})$. The proof is completed. □

Remark 2. *In comparison to Theorem 2, the result obtain in Theorem 3 is more general and interesting. Due to the choice of the measure of noncompactness, we can notice that inequality (2) in Theorem 2 is not necessary in Theorem 3.*

In Theorem 3, when we replace the condition **(H4)** by **(H3′)**, we obtain the following result where the condition **(H5)** is released.

Theorem 5. *Under the assumptions (H1), (H2), (H3′) and (H4), the multi-term time-fractional integro-differential problem (1) has at least one mild random solution on $[0, b]$.*

Proof. As in the proof of Theorem 3, there exists a random function $r : \Omega \to \mathbb{R} - \{0\}$ such that the operator $\mathcal{N} : D(\omega) \to D(\omega)$ is a random operator with stochastic domaine $D(\omega) = \{\vartheta \in \mathcal{C}([0, b], \mathbb{X}) : \|\vartheta\| \leq r(\omega)\}$. Furthermore, we know that $\mathcal{N}$ is a random operator with stochastic domain which is continuous and maps bounded sets into equicontinuous sets of $D(\omega)$. So, we only need to check that $\mathcal{N}$ satisfies the Mönch's condition. Following a similar argument as in the proof of Theorem 3, one can verify the conclusion. □

Remark 3. *The random differential equation with delay is a special type of random functional differential equations. The random functional differential equations with state-dependent delay have many important applications in mathematical models of real phenomena. By applying the ideas and techniques as in this article and making some appropriate conditions, one can obtain the existence results for a class of multi-term time-fractional random integro-differential equations with state-dependent delay.*

4. A Nontrivial Example of Application of Theorem 2

In this section, we give a nontrivial example to illustrate our main results.

Example 1. *Let $\beta, \gamma_k > 0 \ k = 1, 2, \cdots, m$ be such that $0 < \beta \leq \gamma_m \leq \cdots \leq \gamma_1 \leq 1$. We consider the following problem with random effects:*

$$\begin{cases} {}^C D^{1+\beta} u(t,x,\omega) + \sum_{k=1}^{m} \alpha_k \, {}^C D^{\gamma_k} u(t,x,\omega) = \dfrac{\partial^2 u(t,x,\omega)}{\partial^2 x} \\[4pt] \qquad \dfrac{(1+t)^{-1/3}}{20} Q(\omega)\left[\sin(u(t,x,\omega)) + \displaystyle\int_0^t \cos(t-s) u(s,x,\omega)\, ds \right],\ x \in [0,\pi],\ t \in [0,1],\ \omega \in \Omega \\[4pt] u(t,0,\omega) = u(t,\pi,\omega) = 0,\ t \in [0,1],\ \omega \in \Omega, \\[4pt] u(0,x,\omega) = u_0(x,\omega),\ x \in [0,\pi],\ \omega \in \Omega, \\[4pt] \left.\dfrac{\partial u(t,x,\omega)}{\partial t}\right|_{t=0} = u_1(x,\omega),\ x \in [0,\pi],\ \omega \in \Omega, \end{cases} \quad (11)$$

where Q is a real-valued random variable, $u_0, u_1 : [0,\pi] \times \Omega \to \mathbb{R}$ are given functions and $(\Omega, \mathcal{F}, \mathbb{P})$ a complete probability space. Let $\mathbb{X} = \mathbb{L}^2([0,\pi], \mathbb{R})$ and define the operator $\mathcal{A} : D(\mathcal{A}) \subset \mathbb{X} \to \mathbb{X}$ by $\mathcal{A}z = z''$, where

$$D(\mathcal{A}) = \{ z \in \mathbb{X} : z, z' \text{ are absolutely continuous}, z'' \in \mathbb{X}, z(0) = z(\pi) = 0 \}.$$

then

$$\mathcal{A}z = \sum_{n=1}^{\infty} n^2 (z, w_n) w_n,\ z \in D(\mathcal{A}),$$

where $w_n(\theta) = \sqrt{\dfrac{2}{\pi}} \sin(n\theta)$, $n = 1, 2, \cdots$ is the orthogonal set of eigenvectors of $\mathcal{A}$. Thus, $\mathcal{A}$ generates a strongly continuous cosine family $\{C(t)\}_{t \in \mathbb{R}}$ given by

$$C(t)z = \sum_{n=1}^{\infty} \cos(nt)(z, w_n) w_n\ \text{ for } z \in \mathbb{X}.$$

Since $\beta, \gamma_k > 0$ $k = 1, 2, \cdots, m$ be such that $0 < \beta \le \gamma_m \le \cdots \le \gamma_1 \le 1$, by Theorem 1, we deduce that $\mathcal{A}$ generates a bounded (β, γ_k)-resolvent family

$$\mathcal{R}_{\beta,\gamma_k}(t)z = \int_0^{\infty} \dfrac{1}{t^{(1+\beta)/2}} \Phi_{(1+\beta)/2}(s t^{-(1+\beta)/2}) C(s) z\, ds,\ z \in \mathbb{X},\ t \in [0,1],$$

where

$$\Phi_{(1+\beta)/2}(y) = \sum_{n=0}^{\infty} \dfrac{(-y)^n}{n!\, \Gamma(-(\beta(n+1)) - n)},\ y \in \mathbb{C},$$

is the Wright functions. Furthermore, we define

$$\vartheta(t,\omega)(x) = u(t,x,\omega) \text{ for } t \in [0,1],\ x \in [0,\pi] \text{ and } \omega \in \Omega$$

and

$$\vartheta_0(\omega)(x) = u_0(x,\omega),\quad \vartheta_1(\omega)(x) = u_1(x,\omega) \text{ for } x \in [0,\pi] \text{ and } \omega \in \Omega.$$

For every $t \in [0,1]$, $x \in [0,\pi]$ and $\omega \in \Omega$, define

$$F\left(t, \vartheta(t,\omega), \int_0^t \mathcal{B}(t,s)\vartheta(s,\omega)\,ds, \omega\right)(x) = f(t,\omega)\left[\sin(u(t,x,\omega)) + \int_0^t \cos(t-s)u(s,x,\omega)\,ds\right]$$

where $f(t,\omega) = \dfrac{(1+t)^{-1/3}Q(\omega)}{20}$. Then Equation (11) can be rewritten in the abtract form of Equation (1). From the definition of the nonlinear term F, we have

$$\left\| F\left(t,\vartheta(t,\omega),\int_0^t \mathcal{B}(t,s)z_1(s,\omega)ds,\omega\right) - F\left(t,\tilde{\vartheta}(t,\omega),\int_0^t \mathcal{B}(t,s)z_2(s,\omega)ds,\omega\right) \right\|$$

$$\leq f(t,\omega)\Big[\|\vartheta(\cdot,\omega)-\tilde{\vartheta}(\cdot,\omega)\| + \zeta_B\|z_1(\cdot,\omega)-z_2(\cdot,\omega)\|\Big],$$

where $\zeta_B = \sup\limits_{t\in[0,1]} \int_0^t \cos(t-s)ds \leq 1$. Therefore

$$\left\| F\left(t,\vartheta(t,\omega),\int_0^t \mathcal{B}(t,s)z(s,\omega)ds,\omega\right) \right\| \leq f(t,\omega)\Big[\|\vartheta(\cdot,\omega)\| + \zeta_B\|z(\cdot,\omega)\|\Big]$$

and for any bounded and contable set $\mathbb{V}$ of $\mathbb{X}$, we obtain

$$\chi\left(F\left(t,\mathbb{V}(t),\int_0^t \mathcal{B}(t,s)\mathbb{V}(s)ds,\omega\right)\right) \leq h(t,\omega)[\chi(\mathbb{V}) + \zeta_B\chi(\mathbb{V})]$$

where $h(t,\omega) = f(t,\omega)$. Taking $p_1 = p_2 = \dfrac{1}{2}$ and $\beta = \dfrac{3}{4}$, we have

$$P_1 = P_2 = \sqrt{\dfrac{2}{5}} \text{ and } \|f(\cdot,\omega)\| = \dfrac{3}{50}\left[(2)^{5/6} - 1\right]^{1/2}|Q(\omega)| \leq 0.06 \times |Q(\omega)|.$$

If $\dfrac{2M\sqrt{2/5}}{\Gamma(7/4)}0.06 \times |Q(\omega)| < \dfrac{1}{2}$ then by Theorem 2, the random system (11) has at least one random mild solution.

5. Conclusions

Random fractional integro-differential equations are one of the most important research topics in the past thirty years. In this paper, we investigate the existence of mild solutions to a multi-term fractional integro-differential Equation (1) with random effects (see Theorems 2–5). A nontrivial example illustrating Theorem 2 is also given.

Author Contributions: Writing original draft, A.D. and W.-S.D. Both authors have read and agreed to the published version of the manuscript.

Funding: The second author is partially supported by Grant No. MOST 110-2115-M-017-001 of the Ministry of Science and Technology of the Republic of China.

Institutional Review Board Statement: Not applicable.

Informed Consent Statement: Not applicable.

Data Availability Statement: Not applicable.

Acknowledgments: The authors wish to express their hearty thanks to the anonymous referees for their valuable suggestions and comments.

Conflicts of Interest: The authors declare no conflict of interest.

References

1. Podlubny, I. *Fractional Differential Equations*; Academic Press: New York, NY, USA, 1999.
2. Kilbas, A.A.; Srivastava, H.M.; Trujillo, J.J. *Theory and Applications of Fractional Differential Equations*, North-Holland Mathematics Studies 204; Elsevier Science: Amsterdam, The Netherlands, 2006.
3. Keyantuo, V.; Lizama, C.; Warma, M. Asymptotic behavior of fractional order semilinear evolution equations. *Differ. Integral Equ.* **2013**, *26*, 757–780.

4. Trong, L.V. Decay mild solutions for two-term time fractional differential equations in Banach spaces. *J. Fixed Point Theory Appl.* **2016**, *18*, 417–432.
5. Singh, V.; Pandey, D.N. Controllability of multi-term time-fractional differential systems. *J. Control Decis.* **2020**, *7*, 109–125. [CrossRef]
6. Pardo, E.A.; Lizama, C. Mild solutions for multi-term time-fractional differential equations with nonlocal initial conditions. *Electron. J. Differ. Equ.* **2014**, *39*, 1–10.
7. Bharucha-Reid, A.T. *Random Integral Equations*; Academic Press: New York, NY, USA, 1972.
8. Tsokos, C.P.; Padgett, W.J. *Random Integral Equations with Applications in Life Sciences and Engineering*; Academic: New York, NY, USA, 1974.
9. Dhage, B.C.; Ntouyas, S.K. Existence and attractivity results for nonlinear first order random differential equations. *Opusc. Math.* **2010**, *30*, 411–429. [CrossRef]
10. Edsinger, R. Random Ordinary Differential Equations. Ph.D. Thesis, University of California, Berkeley, Berkeley, CA, USA, 1968.
11. Lungan, C.; Lupulescu, V. Random differential equations on time scales. *Electron. J. Differ. Equ.* **2012**, *86*, 1–14.
12. Yang, D.; Wang, J. Non-instantaneous impulsive fractional- order implicit differential equations with random effects. *Stoch. Anal. Appl.* **2017**, *35*, 719–741. [CrossRef]
13. Abbas, S.; Benchohra, M.; Henderson, J. Random Caputo-Fabrizio fractional differential inclusions. *Math. Model. Control* **2021**, *1*, 102–111. [CrossRef]
14. Singh, V.; Pandey, D.N. Mild Solutions for Multi-Term Time-Fractional Impulsive Differential Systems. *Nonlinear Dyn. Syst. Theory* **2018**, *18*, 307–318.
15. Chang, Y.K.; Ponce, R.V. Mild solutions for a multi-term fractional differential equation via resolvent operators. *AIMS Math.* **2021**, *6*, 2398–2417. [CrossRef]
16. Engl, H.W. A general stochastic fixed-point theorem for continuous random operators on stochastic domains. *J. Math. Anal. Appl.* **1978**, *66*, 220–231. [CrossRef]
17. Banaś, J.; Goebel, K. *Measures of Noncompactness in Banach Spaces*; Lect. Notes Pure Appl. Math. 60; Marcel Dekker: New York, NY, USA, 1980.
18. Deimling, K. *Nonlinear Functional Analysis*; Springer: New York, NY, USA, 1985.
19. O'Regan, D.; Precup, R. Existence criteria for integral equations in Banach spaces. *J. Inequalities Appl.* **2001**, *6*, 77–97. [CrossRef]
20. Kamenskii, M.; Obukhovskii, V.; Zecca, P. *Condensing Multivalued Maps and Semilinear Differential Inclusions in Banach Spaces*; De Gruyter: Berlin, Germany, 2001.
21. Mönch, H. Boundary value problems for nonlinear ordinary differential equations of second order in Banach spaces. *Nonlinear Anal.* **1980**, *4*, 985–999. [CrossRef]

Article

On Solvability Conditions for a Certain Conjugation Problem

Vladimir Vasilyev *[ ] and Nikolai Eberlein

Department of Applied Mathematics and Computer Modeling, Belgorod State National Research University, Pobedy Street 85, 308015 Belgorod, Russia; 649377@bsu.edu.ru
* Correspondence: vladimir.b.vasilyev@gmail.com; Tel.: +7-4722301300; Fax: +7-4722301012

Abstract: We study a certain conjugation problem for a pair of elliptic pseudo-differential equations with homogeneous symbols inside and outside of a plane sector. The solution is sought in corresponding Sobolev–Slobodetskii spaces. Using the wave factorization concept for elliptic symbols, we derive a general solution of the conjugation problem. Adding some complementary conditions, we obtain a system of linear integral equations. If the symbols are homogeneous, then we can apply the Mellin transform to such a system to reduce it to a system of linear algebraic equations with respect to unknown functions.

Keywords: pseudo-differential equation; conjugation problem; wave factorization; solvability condition

Citation: Vasilyev, V.; Eberlein, N. On Solvability Conditions for a Certain Conjugation Problem. *Axioms* **2021**, *10*, 234. https://doi.org/10.3390/axioms10030234

Academic Editor: Chris Goodrich

Received: 23 July 2021
Accepted: 17 September 2021
Published: 20 September 2021

Publisher's Note: MDPI stays neutral with regard to jurisdictional claims in published maps and institutional affiliations.

Copyright: © 2021 by the authors. Licensee MDPI, Basel, Switzerland. This article is an open access article distributed under the terms and conditions of the Creative Commons Attribution (CC BY) license (https://creativecommons.org/licenses/by/4.0/).

1. Introduction

The theory of pseudo-differential equations on manifolds with a smooth boundary was systematically developed, starting from the papers of M.I. Vishik and G.I. Eskin [1,2] in the middle of the last century. After this start, L. Boutet de Monvel [3] published a paper in which he suggested an algebraic variant of the theory, including the index theorem. These studies were continued and refined by S. Rempel and B.-W. Schulze [4], and then such results have became useful for situations of manifolds with non-smooth boundaries [5–7].

The first author has started to develop a new approach for non-smooth situations in the middle of the last century [8], and general concepts of the approach are presented in the book and latest papers [9–11]. This paper is related to this approach, and it is devoted to some generalizations of classical results for the Riemann boundary value problem [12,13] in which we consider model pseudo-differential equations in canonical non-smooth domains instead of the Cauchy–Riemann operator. These studies were indicated in [14], and here we develop these results, obtaining more exact and refined solvability conditions. We formulate the solvability conditions in terms of a system of linear algebraic equations similar to well-known Shapiro–Lopatinskii conditions [2]. The Mellin transform [15] is used to reduce the problem for homogeneous elliptic symbols to the mentioned algebraic system.

2. Auxiliaries

A pseudo-differential operator A in a domain $D \subset \mathbf{R}^m$ is defined by its symbol $A(\xi)$ in the following way

$$u(x) \longmapsto \int_D \int_{\mathbf{R}^m} A(\xi) e^{i(y-x)\cdot\xi} u(y) dy d\xi, \quad x \in D,$$

where the function u is defined in the domain D. The symbol $A(\xi)$ is a certain measurable function defined in $\mathbf{R}^m$. The space $H^s(D)$ consists of functions from Sobolev–Slobodetskii space $H^s(\mathbf{R}^m)$ with supports in $\overline{D}$. The norm in $H^s(D)$ is induced by the H^s-norm

$$||u||_s = \left(\int_{\mathbf{R}^m} |\tilde{u}(\xi)|^2 (1+|\xi|)^{2s} d\xi \right)^{1/2},$$

where $\tilde{u}$ is the Fourier transform of u:

$$\tilde{u}(\xi) = \int\limits_{\mathbf{R}^m} e^{ix\cdot\xi} u(x) dx.$$

We start our considerations from measurable symbols $A(\xi)$, satisfying the condition

$$c_1(1+|\xi|)^\alpha \leq |A(\xi)| \leq c_2(1+|\xi|)^\alpha$$

with positive constants c_1, c_2, and the number α, we call an order of the pseudo-differential operator A. Such operators are linear bounded operators $H^s(D) \to H^{s-\alpha}(D)$ [2].

In this paper, we consider plane case $m = 2$ and canonical plane domain $D = C_+^a = \{x \in \mathbf{R}^2 : x = (x_1, x_2), x_2 > a|x_1|, a > 0\}$. For such domains, the key role for the solvability description for the pseudo-differential equation

$$(Au)(x) = v(x), \quad x \in C_+^a,$$

takes the wave factorization concept for the symbol $A(\xi)$ [9].

Let us reiterate that the radial tube domain $T(C_+^a)$ over the cone C_+^a is called the following domain $\mathbf{R}^2 + iC_+^a$ of a two-dimensional complex space $\mathbf{C}^2$ [9].

Definition 1. *By wave factorization of $A(\xi)$ with respect to cone $C_+^a = \{x = (x_1, x_2) \in \mathbf{R}^2 : x_2 > a|x_1|, a > 0\}$, we mean its representation in the form*

$$A(\xi) = A_{\neq}(\xi) A_=(\xi),$$

where the factors $A_{\neq}(\xi), A_=(\xi)$ must satisfy the following conditions:
(1) $A_{\neq}(\xi)$ is defined, generally speaking, on the set $\{x \in \mathbf{R}^2 : a^2 x_2^2 \neq x_1^2\}$ only;
(2) $A_{\neq}(\xi)$ admits an analytical continuation into radial tube domain $T(\overset{*}{C}_+^a)$ over the cone $\overset{*}{C}_+^a = \{x \in \mathbf{R}^2 : ax_2 > |x_1|\}$, which satisfies the following estimate:

$$\left| A_{\neq}^{\pm 1}(\xi + i\tau) \right| \leq c(1 + |\xi| + |\tau|)^{\pm \mathfrak{x}}, \quad \forall \tau \in \overset{*}{C}_+^a.$$

The factor $A_=(\xi)$ has similar properties with $-\overset{*}{C}_+^a$ instead of $\overset{*}{C}_+^a$ and $\alpha - \mathfrak{x}$ instead of $\mathfrak{x}$. The number $\mathfrak{x}$ is called index of wave factorization of $A(\xi)$ with respect to cone C_+^a.

Let us note that if the factors $A_{\neq}(\xi), A_=(\xi)$ are homogeneous of order $\mathfrak{x}$ and $\alpha - \mathfrak{x}$, respectively, and then the symbol $A(\xi)$ is homogeneous of order α, then one can discusshomoge wave factorization. The corresponding definition is given in [9].

3. Statement of the Problem

Let us denote $\Gamma = \{x \in \mathbf{R}^2 : x_2 = a|x_1|, a > 0\}$. We study here the following conjugation problem. Finding a function $U(x)$ which consists of two components

$$U(x) = \begin{cases} U_+(x), & x \in C_+^a \\ U_-(x), & x \in \mathbf{R}^2 \setminus \overline{C_+^a} \end{cases}$$

in the space $H^s(\mathbf{R}^2 \setminus \Gamma)$, and the function should satisfy the following conditions

$$\begin{cases} (AU)(x) = 0, & x \in \mathbf{R}^2 \setminus \Gamma \\ \int\limits_{-\infty}^{+\infty} U_+(x_1, x_2) dx_2 = g_0(x_1), & x_1 \in \mathbf{R} \\ \int\limits_{-\infty}^{+\infty} U_-(x_1, x_2) dx_2 = g_1(x_1), & x_1 \in \mathbf{R} \\ u_+(x) - u_-(x) = g_2(x), & x \in \Gamma, \end{cases} \quad (1)$$

where u_+, u_- are boundary values of U from C_+^a and $\mathbf{R}^2 \setminus \overline{C_+^a}$, respectively, and the functions $g_0, g_1 \in H^{s+1/2}(\mathbf{R})$ and $g_2 \in H^{s-1/2}(\Gamma)$ are given. Since we seek a solution in the space H^s, then such spaces $H^{s\pm 1/2}$ are chosen according to the theorem on restriction on a hyper-plane [2].

If we consider the equation

$$(Au)(x) = 0, \quad x \in C_+^a, \tag{2}$$

separately, then we can use one of key results from the book [9], Theorem 8.1.2; more precisely, it is the following: if the symbol $A(\xi)$ admits the wave factorization with respect to the cone C_+^a with the index $\mathfrak{x}$ such that $\mathfrak{x} - s = n + \delta, n \in \mathbf{N}, |\text{ffi}| < 1/2$, then a general solution $u \in H^s(C_+^a)$ of Equation (2) has the following form

$$\tilde{u}(\xi) = A_{\neq}^{-1}(\xi) \sum_{k=0}^{n-1} \left(\tilde{a}_k(\xi_1 - a\xi_2)(\xi_1 + a\xi_2)^k + \tilde{b}_k(\xi_1 + a\xi_2)(\xi_1 - a\xi_2)^k \right),$$

where a_k, b_k are arbitrary functions from $H^{s_k}(\mathbf{R})$, $s_k = s - \mathfrak{x} + k + 1/2, k = 0, 1, \ldots, n-1$. Furthermore, we have a priori estimates

$$||u||_s \leq C \sum_{k=0}^{n-1} \left([a_k]_{s_k} + [b_k]_{s_k} \right),$$

where $[\cdot]_s$ denotes the $H^s(\mathbf{R})$-norm.

In this paper, we consider the case $n = 1$ so that we have the following formula for a general solution

$$\tilde{U}_+(\xi) = A_{\neq}^{-1}(\xi)(\tilde{a}_0(\xi_1 - a\xi_2) + \tilde{b}_0(\xi_1 + a\xi_2)).$$

For the second equation

$$(Au)(x) = 0, \quad x \in \mathbf{R}^2 \setminus \overline{C_+^a}. \tag{3}$$

we have an analogous formula for a general solution

$$\tilde{U}_-(\xi) = A_=^{-1}(\xi)(\tilde{c}_0(\xi_1 - a\xi_2) + \tilde{d}_0(\xi_1 + a\xi_2)),$$

where c_0, d_0 are a distinct pair of arbitrary functions.

Now, our main goal is to describe the procedure to uniquely determine four arbitrary functions in general solutions of the Equations (2) and (3) using boundary and integral conditions.

4. A System of Linear Integral Equations

Using properties of the Fourier transform [2], we write integral conditions in the form

$$\tilde{U}_+(\xi_1, 0) = A_{\neq}^{-1}(\xi_1, 0)(\tilde{a}_0(\xi_1) + \tilde{b}_0(\xi_1)),$$

$$\tilde{U}_-(\xi_1, 0) = A_=^{-1}(\xi_1, 0)(\tilde{c}_0(\xi_1) + \tilde{d}_0(\xi_1)).$$

It gives the first two relations

$$\begin{aligned} A_{\neq}^{-1}(\xi_1, 0)(\tilde{a}_0(\xi_1) + \tilde{b}_0(\xi_1)) &= \tilde{g}_0(\xi_1) \\ A_=^{-1}(\xi_1, 0)(\tilde{c}_0(\xi_1) + \tilde{d}_0(\xi_1)) &= \tilde{g}_1(\xi_1). \end{aligned} \tag{4}$$

We introduce new variables

$$\begin{cases} \xi_1 - a\xi_2 = t_1 \\ \xi_1 + a\xi_2 = t_2 \end{cases}$$

and re-denote
$$\tilde{u}_{\pm}\left(\frac{t_2+t_1}{2},\frac{t_2-t_1}{2a}\right) \equiv \tilde{V}_{\pm}(t_1,t_2),$$
$$A_{\neq}\left(\frac{t_2+t_1}{2},\frac{t_2-t_1}{2a}\right) \equiv a_{\neq}(t_1,t_2), \quad A_{=}\left(\frac{t_2+t_1}{2},\frac{t_2-t_1}{2a}\right) \equiv a_{=}(t_1,t_2),$$

so that the boundary values $u_{\pm}$ will be boundary values $v_{\pm}$ for new variables t_1, t_2. Thus, general solutions of the Equations (2) and (3) take the form

$$\tilde{V}_+(t_1,t_2) = a_{\neq}^{-1}(t_1,t_2)(\tilde{a}_0(t_1) + \tilde{b}_0(t_2)),$$

$$\tilde{V}_-(t_1,t_2) = a_{=}^{-1}(t_1,t_2)(\tilde{c}_0(t_1) + \tilde{d}_0(t_2)).$$

Therefore, using properties of the Fourier transform [2] we obtain

$$\int_{-\infty}^{+\infty} a_{\neq}^{-1}(t_1,t_2)(\tilde{a}_0(t_1) + \tilde{b}_0(t_2))dt_1 = \tilde{v}_+(0,t_2)$$

$$\int_{-\infty}^{+\infty} a_{\neq}^{-1}(t_1,t_2)(\tilde{a}_0(t_1) + \tilde{b}_0(t_2))dt_2 = \tilde{v}_+(t_1,0),$$

$$\int_{-\infty}^{+\infty} a_{=}^{-1}(t_1,t_2)(\tilde{c}_0(t_1) + \tilde{d}_0(t_2))dt_1 = \tilde{v}_-(0,t_2),$$

$$\int_{-\infty}^{+\infty} a_{=}^{-1}(t_1,t_2)(\tilde{c}_0(t_1) + \tilde{d}_0(t_2))dt_2 = \tilde{v}_-(t_1,0).$$

Let us introduce new notations

$$r_1(t_2) \equiv \int_{-\infty}^{+\infty} a_{\neq}^{-1}(t_1,t_2)dt_1, \quad r_2(t_1) \equiv \int_{-\infty}^{+\infty} a_{\neq}^{-1}(t_1,t_2)dt_2,$$

$$r_3(t_2) \equiv \int_{-\infty}^{+\infty} a_{=}^{-1}(t_1,t_2)dt_1, \quad r_4(t_1) \equiv \int_{-\infty}^{+\infty} a_{=}^{-1}(t_1,t_2)dt_2.$$

We rewrite integral relations by using the above notations.

$$\int_{-\infty}^{+\infty} a_{\neq}^{-1}(t_1,t_2)\tilde{a}_0(t_1)dt_1 + \tilde{b}_0(t_2)r_1(t_2)-$$

$$-\int_{-\infty}^{+\infty} a_{=}^{-1}(t_1,t_2)\tilde{c}_0(t_1)dt_1 - \tilde{d}_0(t_2)r_3(t_2) = \tilde{g}_{21}(t_2),$$

$$r_2(t_1)\tilde{a}_0(t_1) + \int_{-\infty}^{+\infty} a_{\neq}^{-1}(t_1,t_2)\tilde{b}_0(t_2)dt_2 - r_4(t_1)\tilde{c}_0(t_1)-$$

$$-\int_{-\infty}^{+\infty} a_{=}^{-1}(t_1,t_2)\tilde{d}_0(t_2))dt_2 = \tilde{g}_{22}(t_1),$$

where $\tilde{g}_{21}(t_2), \tilde{g}_{22}(t_1)$ are Fourier transforms of the function g_2, which is considered as two parts related to angle sides.

So, we have the following relations for determining the unknown functions $\tilde{a}_0, \tilde{b}_0, \tilde{c}_0, \tilde{d}_0$. Of course, according to the equalities (4), we can write

$$\tilde{b}_0(\xi_1) = A_{\neq}(\xi_1, 0)\tilde{g}_0(\xi_1) - \tilde{a}_0(\xi_1),$$
$$\tilde{d}_0(\xi_1) = A_{=}(\xi_1, 0)\tilde{g}_1(\xi_1) - \tilde{c}_0(\xi_1),$$

and can obtain the following integral system with respect to unknowns $\tilde{a}_0, \tilde{c}_0$:

$$\begin{cases} \int_{-\infty}^{+\infty} a_{\neq}^{-1}(t_1, t_2)\tilde{a}_0(t_1)dt_1 - \tilde{a}_0(t_2)r_1(t_2) - \\ \int_{-\infty}^{+\infty} a_{=}^{-1}(t_1, t_2)\tilde{c}_0(t_1)dt_1 + \tilde{c}_0(t_2)r_3(t_2) = \tilde{f}_1(t_2) \\ r_2(t_1)\tilde{a}_0(t_1) - \int_{-\infty}^{+\infty} a_{\neq}^{-1}(t_1, t_2)\tilde{a}_0(t_2)dt_2 - r_4(t_1)\tilde{c}_0(t_1) + \\ \int_{-\infty}^{+\infty} a_{=}^{-1}(t_1, t_2)\tilde{c}_0(t_2))dt_2 = \tilde{f}_2(t_1), \end{cases} \quad (5)$$

where we have denoted

$$\tilde{f}_1(t_2) = \tilde{g}_{21}(t_2) - A_{\neq}(t_2, 0)\tilde{g}_0(t_2)r_1(t_2) - A_{=}(t_2, 0)\tilde{g}_1(t_2)r_3(t_2)$$

$$\tilde{f}_2(t_1) = \tilde{g}_{22}(t_1) - \int_{-\infty}^{+\infty} a_{\neq}^{-1}(t_1, t_2)A_{\neq}(t_2, 0)\tilde{g}_0(t_2)dt_2 +$$

$$\int_{-\infty}^{+\infty} a_{=}^{-1}(t_1, t_2)A_{=}(t_2, 0)\tilde{g}_1(t_2)dt_2.$$

Finally, we obtain the following assertion.

Theorem 1. *If the symbol $A(\xi)$ admits wave factorization with respect to the cone C_+^a with the index $æ$ such that $æ - s = 1 + \delta, |\delta| < 1/2$, then unique solvability of the problem (1) is equivalent to unique solvability of the system (5).*

The next section is devoted to study the system (5).

5. Homogeneous Symbols and Applying the Mellin Transform

We consider here the case when the symbol $A(\xi)$ is positively homogeneous of order α and the factors $A_{\neq}(\xi)$ and $A_{=}(\xi)$ are positively homogeneous of order $æ$ and $\alpha - æ$, respectively.

Lemma 1. *The functions r_1, r_2 are positively homogeneous function of order $1 - æ$, and the functions r_3, r_4 are positively homogeneous functions of order $1 + æ - \alpha$.*

Proof. Let us verify. Indeed, for $\lambda > 0$, we have

$$r_1(\lambda t_2) = \int_{-\infty}^{+\infty} a_{\neq}^{-1}(t_1, \lambda t_2)dt_1,$$

and after the change of variable $t_1 = \lambda t$ we obtain

$$r_1(\lambda t_2) = \lambda \int_{-\infty}^{+\infty} a_{\neq}^{-1}(\lambda t, \lambda t_2)dt = \lambda \lambda^{-\varkappa} r_1(t_2) = \lambda^{1-\varkappa} r_1(t_2).$$

Analogously,

$$r_3(\lambda t_2) = \int_{-\infty}^{+\infty} a_{=}^{-1}(t_1, \lambda t_2)dt_1,$$

and after similar change we have

$$r_3(\lambda t_2) = \lambda \int_{-\infty}^{+\infty} a_{=}^{-1}(\lambda t, \lambda t_2)dt = \lambda \lambda^{-(\alpha-\varkappa)} r_3(t_2) = \lambda^{1+\varkappa-\alpha} r_3(t_2).$$

Similar conclusions are valid for r_2, r_4. □

Remark 1. *If $\varkappa = \alpha/2$, then all functions r_1, r_2, r_3, r_4 have the same order of homogeneity, which equals to $1 - \varkappa$.*

Lemma 2. *The functions $a_{\neq}^{-1}(t_1, t_2)r_1^{-1}(t_2), a_{\neq}^{-1}(t_1, t_2)r_2^{-1}(t_1)$ are homogeneous functions of order -1 with respect to variables t_1, t_2, and the functions $a_{=}^{-1}(t_1, t_2)r_3^{-1}(t_2), a_{=}^{-1}(t_1, t_2)r_4^{-1}(t_1)$ are homogeneous functions of order -1 too.*

Proof. According to Lemma 1, we have

$$a_{\neq}^{-1}(\lambda t_1, \lambda t_2)r_1^{-1}(\lambda t_2) = \lambda^{-\varkappa} a_{\neq}^{-1}(t_1, t_2)\lambda^{\varkappa-1} r_1^{-1}(t_2) =$$

$$\lambda^{-1} a_{\neq}^{-1}(t_1, t_2)r_1^{-1}(t_2).$$

Analogously,

$$a_{=}^{-1}(\lambda t_1, \lambda t_2)r_3^{-1}(\lambda t_2) = \lambda^{\varkappa-\alpha} a_{=}^{-1}(t_1, t_2)\lambda^{\alpha-\varkappa-1})r_3^{-1}(t_2) =$$

$$\lambda^{-1} a_{=}^{-1}(t_1, t_2)r_3^{-1}(t_2).$$

The same is valid for the left two functions. □

Let us note that Lemmas 1 and 2 are almost the same, as in [9].
Now, we divide by r_1 and r_2

$$\begin{cases} \int_{-\infty}^{+\infty} a_{\neq}^{-1}(t_1, t_2)r_1^{-1}(t_2)\tilde{a}_0(t_1)dt_1 - \tilde{a}_0(t_2) - \\ \int_{-\infty}^{+\infty} a_{=}^{-1}(t_1, t_2)r_1^{-1}(t_2)\tilde{c}_0(t_1)dt_1 + \tilde{c}_0(t_2)r_1^{-1}(t_2)r_3(t_2) = \tilde{f}_1(t_2)r_1^{-1}(t_2) \\ \tilde{a}_0(t_1) - \int_{-\infty}^{+\infty} a_{\neq}^{-1}(t_1, t_2)r_2^{-1}(t_1)\tilde{a}_0(t_2)dt_2 - r_4(t_1)r_2^{-1}(t_1)\tilde{c}_0(t_1) + \\ \int_{-\infty}^{+\infty} a_{=}^{-1}(t_1, t_2)r_2^{-1}(t_1)\tilde{c}_0(t_2))dt_2 = \tilde{f}_2(t_1)r_2^{-1}(t_1), \end{cases}$$

and we obtain the following system of two linear integral equations

$$\begin{cases} \int_{-\infty}^{+\infty} K_1(t_1,t_2)\tilde{a}_0(t_1)dt_1 - \tilde{a}_0(t_2) - \\ \int_{-\infty}^{+\infty} K_2(t_1,t_2)\tilde{c}_0(t_1)dt_1 + \tilde{c}_0(t_2)R(t_2) = F_1(t_2) \\ \tilde{a}_0(t_1) - \int_{-\infty}^{+\infty} K_3(t_1,t_2)\tilde{a}_0(t_2)dt_2 - Q(t_1)\tilde{c}_0(t_1) + \\ \int_{-\infty}^{+\infty} K_4(t_1,t_2)\tilde{c}_0(t_2))dt_2 = F_2(t_1), \end{cases} \quad (6)$$

after new notations with

$$K_1(t_1,t_2) = a_{\neq}^{-1}(t_1,t_2)r_1^{-1}(t_2), K_2(t_1,t_2) = a_{=}^{-1}(t_1,t_2)r_1^{-1}(t_2),$$

$$K_3(t_1,t_2) = a_{\neq}^{-1}(t_1,t_2)r_2^{-1}(t_1), K_4(t_1,t_2) = a_{=}^{-1}(t_1,t_2)r_2^{-1}(t_1),$$

$$R(t_2) = r_1^{-1}(t_2)r_3(t_2), F_1(t_2) = \tilde{f}_1(t_2)r_1^{-1}(t_2),$$

$$Q(t_1) = r_4(t_1)r_2^{-1}(t_1), F_2(t_1) = \tilde{f}_2(t_1)r_2^{-1}(t_1).$$

Lemma 3. *Let $æ = \alpha/2$. The kernels of integral operators K_1, K_2, K_3, K_4 are homogeneous of order -1, and the functions R, Q are homogeneous of order 0.*

Proof. Using Lemma 1 and Lemma 2, we obtain the required assertion. □

Now, we will rewrite the system (6) as a system of integral equations on the positive half-axis to apply the Mellin transform.

$$\begin{cases} \int_0^{+\infty} K_1(t_1,t_2)\tilde{a}_0(t_1)dt_1 + \int_{-\infty}^0 K_1(t_1,t_2)\tilde{a}_0(t_1)dt_1 - \tilde{a}_0(t_2) - \\ -\int_0^{+\infty} K_2(t_1,t_2)\tilde{c}_0(t_1)dt_1 + \int_{-\infty}^0 K_2(t_1,t_2)\tilde{c}_0(t_1)dt_1 + \tilde{c}_0(t_2)R(t_2) = F_1(t_2) \\ \tilde{a}_0(t_1) - \int_0^{+\infty} K_3(t_1,t_2)\tilde{a}_0(t_2)dt_2 - \int_{-\infty}^0 K_3(t_1,t_2)\tilde{a}_0(t_2)dt_2 - Q(t_1)\tilde{c}_0(t_1) + \\ + \int_0^{+\infty} K_4(t_1,t_2)\tilde{c}_0(t_2))dt_2 + \int_{-\infty}^0 K_4(t_1,t_2)\tilde{c}_0(t_2))dt_2 = F_2(t_1), \end{cases}$$

The next step is the following. We would like to transform the latter system to a 4×4 system on a positive half-axis. For this purpose, we introduce two additional unknown functions and new notations.

We denote for all $t_1 > 0$

$$M_1(t_1,t_2) = K_1(-t_1,t_2), \quad M_2(t_1,t_2) = K_2(-t_1,t_2),$$

and for all $t_2 > 0$

$$M_3(t_1,t_2) = K_3(t_1,-t_2), \quad M_4(t_1,t_2) = K_4(t_1,-t_2),$$

and we put also for $t > 0$

$$\tilde{a}_1(t) = \tilde{a}_0(-t), \quad \tilde{c}_1(t) = \tilde{c}_0(-t), \quad G_1(t) = F_1(-t), \quad G_2(t) = F_2(-t).$$

Thus, we have the following system of linear integral equations with respect to four unknown functions $\tilde{a}_0, \tilde{a}_1, \tilde{c}_0, \tilde{c}_1$ in which all kernel and functions are defined for positive t_1, t_2:

$$\begin{cases} \int_0^{+\infty} K_1(t_1,t_2)\tilde{a}_0(t_1)dt_1 + \int_0^{+\infty} M_1(t_1,t_2)\tilde{a}_1(t_1)dt_1 - \tilde{a}_0(t_2) - \\ - \int_0^{+\infty} K_2(t_1,t_2)\tilde{c}_0(t_1)dt_1 + \int_0^{+\infty} M_2(t_1,t_2)\tilde{c}_1(t_1)dt_1 + \tilde{c}_0(t_2)R(t_2) = F_1(t_2) \\ \int_0^{+\infty} K_1(t_1,-t_2)\tilde{a}_0(t_1)dt_1 + \int_0^{+\infty} M_1(t_1,-t_2)\tilde{a}_1(t_1)dt_1 - \tilde{a}_1(t_2) - \\ - \int_0^{+\infty} K_2(t_1,-t_2)\tilde{c}_0(t_1)dt_1 + \int_0^{+\infty} M_2(t_1,-t_2)\tilde{c}_1(t_1)dt_1 + \tilde{c}_1(t_2)R(-t_2) = G_1(t_2) \\ \tilde{a}_0(t_1) - \int_0^{+\infty} K_3(t_1,t_2)\tilde{a}_0(t_2)dt_2 - \int_0^{+\infty} M_3(t_1,t_2)\tilde{a}_1(t_2)dt_2 - Q(t_1)\tilde{c}_0(t_1) + \\ + \int_0^{+\infty} K_4(t_1,t_2)\tilde{c}_0(t_2))dt_2 + \int_0^{+\infty} M_4(t_1,t_2)\tilde{c}_1(t_2))dt_2 = F_2(t_1) \\ \tilde{a}_1(t_1) - \int_0^{+\infty} K_3(-t_1,t_2)\tilde{a}_0(t_2)dt_2 - \int_0^{+\infty} M_3(-t_1,t_2)\tilde{a}_1(t_2)dt_2 - Q(-t_1)\tilde{c}_1(t_1) + \\ + \int_0^{+\infty} K_4(-t_1,t_2)\tilde{c}_0(t_2))dt_2 + \int_0^{+\infty} M_4(-t_1,t_2)\tilde{c}_1(t_2))dt_2 = G_2(-t1). \end{cases}$$

Further, we introduce notation:

$$R(t_2) = \begin{cases} r_1, & t_2 > 0 \\ r_2, & t_2 < 0 \end{cases}, \quad Q(t_1) = \begin{cases} q_1, & t_1 > 0 \\ q_2, & t_1 < 0 \end{cases},$$

$K_i(t_1, -t_2) = k_i(t_1, t_2), M_i(t_1, -t_2) = m_i(t_1, t_2), i = 1, 2$, and $K_j(-t_1, t_2) = k_j(t_1, t_2)$, $M_j(-t_1, t_2) = m_j(t_1, t_2), j = 3, 4$.

Now we can rewrite our system as follows.

$$\begin{cases}
\int_0^{+\infty} K_1(t_1,t_2)\tilde{a}_0(t_1)dt_1 + \int_0^{+\infty} M_1(t_1,t_2)\tilde{a}_1(t_1)dt_1 - \tilde{a}_0(t_2) - \\
- \int_0^{+\infty} K_2(t_1,t_2)\tilde{c}_0(t_1)dt_1 + \int_0^{+\infty} M_2(t_1,t_2)\tilde{c}_1(t_1)dt_1 + r_1\tilde{c}_0(t_2) = F_1(t_2) \\
\int_0^{+\infty} k_1(t_1,t_2)\tilde{a}_0(t_1)dt_1 + \int_0^{+\infty} m_1(t_1,t_2)\tilde{a}_1(t_1)dt_1 - \tilde{a}_1(t_2) - \\
- \int_0^{+\infty} k_2(t_1,t_2)\tilde{c}_0(t_1)dt_1 + \int_0^{+\infty} m_2(t_1,t_2)\tilde{c}_1(t_1)dt_1 + r_2\tilde{c}_1(t_2) = G_1(t_2) \\
\tilde{a}_0(t_1) - \int_0^{+\infty} K_3(t_1,t_2)\tilde{a}_0(t_2)dt_2 - \int_0^{+\infty} M_3(t_1,t_2)\tilde{a}_1(t_2)dt_2 - q_1\tilde{c}_0(t_1) + \\
+ \int_0^{+\infty} K_4(t_1,t_2)\tilde{c}_0(t_2))dt_2 + \int_0^{+\infty} M_4(t_1,t_2)\tilde{c}_1(t_2))dt_2 = F_2(t_1) \\
\tilde{a}_1(t_1) - \int_0^{+\infty} k_3(t_1,t_2)\tilde{a}_0(t_2)dt_2 - \int_0^{+\infty} m_3(t_1,t_2)\tilde{a}_1(t_2)dt_2 - q_2\tilde{c}_1(t_1) + \\
+ \int_0^{+\infty} k_4(t_1,t_2)\tilde{c}_0(t_2))dt_2 + \int_0^{+\infty} m_4(t_1,t_2)\tilde{c}_1(t_2))dt_2 = G_2(t_1).
\end{cases} \quad (7)$$

Now, we can apply the Mellin transform to the system (7). Let us restate that the Mellin transform for the function f of one real variable is the following [15]

$$\hat{f}(\mu) = \int_0^{+\infty} x^{\mu-1} f(x) dx,$$

and the function $\hat{f}$ exists for a wide class of functions.

We will use the following notations for the Mellin transforms. For $K_i(t_1,t_2), k_i(t_1,t_2), M_i(t_1,t_2), m_i(t_1,t_2), i = 1,2$, the notation $\hat{K}_i(\mu), \hat{k}_i(\mu), \hat{M}_i(\mu), \hat{m}_i(\mu)$ denotes the Mellin transform of the functions $K_i(1,t), k_i(1,t), M_i(1,t), m_i(1,t)$, respectively. For $K_j(t_1,t_2), k_j(t_1,t_2), M_j(t_1,t_2), m_j(t_1,t_2), j = 3,4$, the notation $\hat{K}_j(\mu), \hat{k}_j(\mu), \hat{M}_j(\mu), \hat{m}_j(\mu)$ denotes the Mellin transform of the functions $K_j(t,1).k_j(t,1), M_j(t,1), m_j(t,1)$, respectively.

Applying the Mellin transform to the system (7), we obtain at least formally the following system of linear algebraic equations

$$\begin{cases}
(\hat{K}_1(\mu) - 1)\hat{a}_0(\mu) + \hat{M}_1(\mu)\hat{a}_1(\mu) + \\
(\hat{K}_2(\mu) + r_1)\hat{c}_0(\mu) + \hat{M}_2(\mu)\hat{c}_1(\mu) = \hat{F}_1(\mu) \\
\hat{k}_1(\mu)\hat{a}_0(\mu) + (\hat{m}_1(\mu) - 1)\hat{a}_1(\mu) - \\
\hat{k}_2(\mu)\hat{c}_0(\mu) + (\hat{m}_2(\mu) + r_2)\hat{c}_1(\mu) = \hat{G}_1(\mu) \\
(1 - \hat{K}_3(\mu))\hat{a}_0(\mu) - \hat{M}_3(\mu)\hat{a}_1(\mu) + \\
(\hat{K}_4(\mu) - q_1)\hat{c}_0(\mu) + \hat{M}_4(\mu)\hat{c}_1(\mu) = \hat{F}_2(\mu) \\
-\hat{k}_3(\mu)\hat{a}_0(\mu) + (1 - \hat{m}_3(\mu))\hat{a}_1(\mu) + \\
\hat{k}_4(\mu)\hat{c}_0(\mu) + (\hat{m}_4(\mu) - q_2)\hat{c}_1(\mu) = \hat{G}_2(\mu).
\end{cases} \quad (8)$$

A matrix of the (4×4)-system (8) is the following

$$\mathcal{A}(\mu) = \begin{pmatrix} \hat{K}_1(\mu) - 1 & \hat{M}_1(\mu) & \hat{K}_2(\mu) + r_1 & \hat{M}_2(\mu) \\ \hat{k}_1(\mu) & \hat{m}_1(\mu) - 1 & \hat{k}_2(\mu) & \hat{m}_2(\mu) + r_2 \\ 1 - \hat{K}_3(\mu) & -\hat{M}_3(\mu) & \hat{K}_4(\mu) - q_1 & \hat{M}_4(\mu) \\ -\hat{k}_3(\mu) & 1 - \hat{m}_3(\mu) & \hat{k}_4(\mu) & \hat{m}_4(\mu) - q_2 \end{pmatrix}.$$

6. Solvability Conditions

Here, we can formulate the following assertion on the solvability of system (5) for homogeneous kernels (see also [9]).

Theorem 2. *Let $A_{\neq}(\xi)$ and $A_{=}(\xi)$ be homogeneous non-vanishing functions of order $\alpha/2$ and $-\alpha/2$, respectively, and differentiable away from the origin, $r_1(t_2) \neq 0, \forall t_2 \neq 0, r_2(t_1) \neq 0, \forall t_1 \neq 0$. The system of linear integral Equation (5) is uniquely solvable if, and only if, the condition*

$$\inf |\det \mathcal{A}(\mu)| \neq 0, \quad \Re\mu = 1/2 \tag{9}$$

holds.

Proof. Basic elements of the proof were given in the above considerations and Lemmas 1–3. The condition (9) is related to properties of the Mellin transform [2,9,15].

Nevertheless, we will give some explanations. If we have the wave factorization, then we obtain the system (5). For homogeneous factors $A_{\neq}(\xi)$ and $A_{=}(\xi)$, the system (5) transforms into the system (7). The latter system of linear integral equations has kernels which are homogeneous of order -1. That is why we can apply the Mellin transform. If we have the expression

$$\int_0^{+\infty} K(t_1, t_2) u(t_1) dt_1,$$

in which the kernel $K(t_1, t_2)$ is a homogeneous function of order -1, then after applying the Mellin transform we obtain the following expression

$$\int_0^{+\infty} t_2^{\mu-1} \left(\int_0^{+\infty} K(t_1, t_2) u(t_1) dt_1 \right) dt_2.$$

The change of variable in the inner integral $t_2 = xt_1$ leads to the following integral

$$\int_0^{+\infty} t_1^{\mu-1} x^{\mu-1} \left(\int_0^{+\infty} t_1 K(t_1, xt_1) u(t_1) dt_1 \right) dx,$$

and after rearrangements of integrals we obtain the following product

$$\int_0^{+\infty} t_1^{\mu-1} u(t_1) dt_1 \int_0^{+\infty} x^{\mu-1} K(1, x) dx = \hat{u}(\mu) \hat{K}(\mu),$$

where $\hat{u}$ denotes the Mellin transform of u.

So, using the Mellin transform, we can obtain the system of linear algebraic Equation (8), which is equivalent to the system (7). Lemma 3 is needed for this purpose. The condition (9) is a necessary and sufficient condition for the unique solvability of such systems and the applicability of the inverse Mellin transform.

Since we suppose the factors $A_{\neq}, A_{=}$ are differentiable, then the Mellin transform is applicable for the kernels K_j. The functions under the integral can be assumed to be smooth enough, taking into account further approximation in H^s-spaces. □

Remark 2. *A priori estimates for a solution of the problem* (1) *can be obtained by the methods described in [9]. We will give these estimates in next papers.*

7. Conclusions

In this paper, we have shown that a certain conjugation problem can be reduced to a system of linear algebraic equations. One can consider other conjugation problems for homogeneous elliptic symbols using this approach. Perhaps it is reasonable to consider different boundary conditions which are local, such as Dirichlet and Neumann conditions.

Author Contributions: V.V. has suggested a general concept of the study, N.E. has proved main results. All authors have read and agreed to the published version of the manuscript.

Funding: This research received no external funding.

Conflicts of Interest: The authors declare no conflict of interest.

References

1. Vishik, M.I.; Eskin, G.I. Convolution equations in a bounded domain. *Russ. Math. Surv.* **1965**, *20*, 89–152. (In Russian) [CrossRef]
2. Eskin, G. *Boundary Value Problems for Elliptic Pseudodifferential Equations*; AMS: Providence, RI, USA, 1981.
3. Boutet de Monvel, L. Boundary problems for pseudodifferential operators. *Acta Math.* **1971**, *126*, 11–51. [CrossRef]
4. Rempel, S.; Schulze, B.-W. *Index Theory of Elliptic Boundary Problems*; Akademie: Berlin, Germany, 1982.
5. Nazarov, S.A.; Plamenevsky, B.A. *Elliptic Problems in Domains with Piecewise Smooth Boundaries*; Walter de Gruyter: Berlin, Germany; New York, NY, USA, 1994.
6. Schulze, B.-W. *Boundary Value Problems and Singular Pseudo-Differential Operators*; Wiley: Chichester, UK, 1998.
7. Schulze, B.-W.; Sternin, B.; Shatalov, V. *Differential Equations on Singular Manifolds; Semiclassical Theory and Operator Algebras*; Wiley-VCH: Berlin, Germany, 1998.
8. Vasilyev, V.B. Pseudodifferential equations on cones.r. *Differ. Equ.* **1995**, *31*, 1385–1395.
9. Vasil'ev, V.B. *Wave Factorization of Elliptic Symbols: Theory and Applications. Introduction to the Theory of Boundary Value Problems in Non-Smooth Domains*; Kluwer Academic Publishers: Dordrecht, The Netherlands; Boston, MA, USA; London, UK, 2000.
10. Vasilyev, V. Elliptic operators and their symbols. *Demonstr. Math.* **2019**, *52*, 361–369. [CrossRef]
11. Vasilyev, V. Operator Symbols and Operator Indices. *Symmetry* **2020**, *12*, 64. [CrossRef]
12. Gakhov, F.D. *Boundary Value Problems*; Dover Publications: Mineola, NY, USA, 1981.
13. Muskhelishvili, N.I. *Singular Integral Equations*; North Holland: Amsterdam, The Netherlands, 1976.
14. Vasilyev, V.B. On some transmission problems in a plane corner. *Tatra Mt. Math. Publ.* **2015**, *63*, 291–301.
15. Titchmarsh, E. *Introduction to Fourier Integrals Theory*; Moscow-Leningrad: Gostehteoretizdat, Russia, 1948. (In Russian)

Article

Closed-Form Solutions of Linear Ordinary Differential Equations with General Boundary Conditions

Efthimios Providas [1,*], Stefanos Zaoutsos [2] and Ioannis Faraslis [1]

[1] Department of Environmental Sciences, Gaiopolis Campus, University of Thessaly, 415 00 Larissa, Greece; faraslis@uth.gr
[2] Department of Energy Systems, Gaiopolis Campus, University of Thessaly, 415 00 Larissa, Greece; szaoutsos@uth.gr
* Correspondence: providas@uth.gr; Tel.: +30-2410-684-473

Abstract: This paper deals with the solution of boundary value problems for ordinary differential equations with general boundary conditions. We obtain closed-form solutions in a symbolic form of problems with the general n-th order differential operator, as well as the composition of linear operators. The method is based on the theory of the extensions of linear operators in Banach spaces.

Keywords: differential equations; differential operators; non-local boundary value problems; general conditions; integral conditions; multipoint conditions; composition of operators

1. Introduction

Differential equations model numerous phenomena and processes in sciences and engineering. Boundary value problems for elementary differential equations with classical boundary conditions have been studied exhaustively by many researchers and comprehensive material is now included in various standard texts. A more difficult and less investigated subject is the general or nonlocal boundary value problems. In many applications, the incorporation of general boundary conditions such as multipoint and integral conditions is inevitable. For example, in [1], the necessity of integral conditions in certain models of epidemics and population growth and the effects when neglecting these conditions are explained.

Ordinary differential equations with non-local boundary conditions were first studied at the beginning of the 20th century in [2–4], and later in [5]. Abstract non-local boundary value problems were considered in [6–8]. Operator methods for solving differential equations are analyzed in the books [9–11]. A description of the theory and the different directions of differential equations with non-local boundary conditions are given in the monograph [12]. An overview of non-local boundary value problems and their historical evolution can also be found in the survey papers [13–16]. Boundary value problems with integral constraints have been considered in [17–25], to mention but a few. Boundary value problems with multipoint and integral conditions have been studied in [26–32], and others. The present paper aims at providing a framework for symbolic computations for the solution of linear ordinary differential equations of order n with the most general multipoint and integral conditions, and boundary value problems for powers and products of differential operators.

In $C[a,b]$, let the general n-th order linear ordinary differential operator,

$$A = a_0(x)\frac{d^n}{dx^n} + a_1(x)\frac{d^{n-1}}{dx^{n-1}} + \cdots + a_{n-1}(x)\frac{d}{dx} + a_n(x), \qquad (1)$$

where the coefficients $a_i(x)$, $i = 0, \ldots, n$, are all continuous functions on the interval $[a,b]$, $a_0(x) \neq 0$, and $D(A) = C^n[a,b]$ and $R(A) = C[a,b]$ are its domain and range, respectively.

We are concerned with the solution in the closed-form of boundary value problem for the differential equation
$$Au(x) = f(x), \quad x \in (a,b), \tag{2}$$
and the boundary conditions
$$Y(u) = \mathbf{b}, \tag{3}$$
where $Y = \text{col}(Y_1, \ldots, Y_n)$ is a vector of linear bounded functionals of the general form
$$Y_i(u) = \sum_{j=0}^{m} \sum_{k=0}^{n-1} v_{ikj} u^{(k)}(x_j) + \sum_{k=0}^{n-1} \int_a^b h_{ik}(t) u^{(k)}(t) dt, \quad i = 1, \ldots, n, \tag{4}$$

where $Y_i \in [C^{n-1}[a,b]]^*$, $i = 1, \ldots, n$; the $m+1$ ordered points $a \leq x_0 < x_1 < \cdots < x_m \leq b$ are fixed boundary points, $h_{ik}(x)$, $i = 1, \ldots, n$, $k = 0, \ldots, n-1$, are continuous functions on $[a,b]$, v_{ikj}, $i = 1, \ldots, n$, $k = 0, \ldots, n-1$, $j = 0, \ldots, m$, are constants, and $u^{(k)}$ designates the k-th order derivative of u. The non-homogeneous term $f(x) \in C[a,b]$, while the non-homogeneous term $\mathbf{b} = \text{col}(\beta_1, \ldots, \beta_n)$ is a constant vector. The function $u(x) \in C^n[a,b]$ is the sought solution. We formulate the above problem in a convenient symbolic form and establish uniqueness solvability criteria and derive the solution in closed-form. Solution formulae to some special boundary value problems for composite differential operators are also obtained. The method is based on the theory of the extensions of linear operators in Banach spaces, see, for example, [33–36], and is an extension of the work [37] by the authors.

The paper is organized as follows. In Section 2, we give some results needed for the analysis in later sections. Sections 3 and 4 contain the main findings of our investigation. In Section 5, the implementation of the technique is explained by solving two example problems. Finally, some conclusions are quoted in Section 6.

2. Preliminaries

Let X, Y be Banach spaces and $P : X \to Y$ a linear operator. The operator P is *injective* or *one-to-one* if for every $u_1, u_2 \in D(P)$, $u_1 \neq u_2$ implies $Pu_1 \neq Pu_2$. The operator P is *surjective* or *onto* Y if $R(P) = Y$. If P is both injective and onto, then there exists the inverse operator $P^{-1} : Y \to X$ defined by $P^{-1}f = u$ if and only if $Pu = f$ for each $f \in Y$; in this case, $R(P^{-1}) = D(P)$.

The operator P is called *closed* if for every sequence u_m in $D(P)$ converging to u_0 with $Pu_m \to y_0$, $y_0 \in Y$, it follows that $u_0 \in D(P)$ and $Pu_0 = y_0$. A closed operator P is called *maximal* if $R(P) = Y$ and $\ker P \neq \{0\}$.

The operator P is *correct* if it is both injective and onto, and the inverse operator P^{-1} is bounded on Y. The problem $Pu = f$ is correct if the operator P is correct.

An operator $P_r : X \to Y$ is a *restriction* of P, or P is an *extension* of P_r, if $D(P_r) \subset D(P)$ and $P_r u = Pu$ for all $u \in D(P_r)$.

Let $\Psi = \text{col}(\Psi_1, \ldots, \Psi_n)$ be a column vector of functionals $\Psi_i \in X^*$, $i = 1, \ldots, n$, and $\mathbf{v} = (v_1, \ldots, v_n)$ a row vector of elements $v_i \in X$, $i = 1, \ldots, n$. By $\Psi(\mathbf{v})$, we symbolize the $n \times n$ matrix
$$\Psi(\mathbf{v}) = \begin{pmatrix} \Psi_1(v_1) & \cdots & \Psi_1(v_n) \\ \vdots & \ddots & \vdots \\ \Psi_n(v_1) & \cdots & \Psi_n(v_n) \end{pmatrix},$$
whose $\Psi_i(v_j)$ element is the value of the functional Ψ_i on the element v_j. It is easy to show that
$$\Psi(\mathbf{v}\mathbf{N}) = \Psi(\mathbf{v})\mathbf{N}. \tag{5}$$
where $\mathbf{N}$ is a $n \times m$ constant matrix.

Proposition 1. Let X, Y be real Banach spaces, $A : X \xrightarrow{on} Y$ a linear operator, $\mathbf{z} = (z_1, \ldots, z_n)$ a basis of $\ker A$, and $\widehat{A}$ the restriction of A defined by

$$\widehat{A} \subset A, \quad D(\widehat{A}) = \{u : u \in D(A), \Phi(u) = \mathbf{0}\},$$

where the components of the vector $\Phi = \mathrm{col}(\Phi_1, \ldots, \Phi_n)$ are linear bounded functionals on X. Then:

(i) The operator $\widehat{A}$ is injective if and only if

$$\det \Phi(\mathbf{z}) = \det \begin{pmatrix} \Phi_1(z_1) & \cdots & \Phi_1(z_n) \\ \vdots & \ddots & \vdots \\ \Phi_n(z_1) & \cdots & \Phi_n(z_n) \end{pmatrix} \neq 0.$$

(ii) If additionally to (i), the operator $\widehat{A}^{-1}$ is bounded on the whole Y, then the operator $\widehat{A}$ is correct.

Proof. (i) Let $\det \Phi(\mathbf{z}) \neq 0$. Take $u_0 \in \ker \widehat{A}$, then $\widehat{A} u_0 = 0$, $\Phi(u_0) = \mathbf{0}$ and

$$u_0 = \mathbf{za} = a_1 z_1 + \cdots + a_n z_n,$$

where $\mathbf{a} = \mathrm{col}(a_1, \ldots, a_n) \in \mathbb{R}_n$. Acting by the vector Φ on u_0, we get

$$\Phi(u_0) = \Phi(\mathbf{za}) = \Phi(\mathbf{z})\mathbf{a} = \mathbf{0}.$$

Since $\det \Phi(\mathbf{z}) \neq 0$ by hypothesis, it is implied that $\mathbf{a} = \mathbf{0}$ and so $u_0 = 0$. That is $\ker \widehat{A} = \{0\}$, and therefore $\widehat{A}$ is injective. Conversely, let $\det \Phi(\mathbf{z}) = 0$. Then, there exists a nonzero vector $\mathbf{c} = \mathrm{col}(c_1, \ldots, c_n) \in \mathbb{R}_n$ such that $\Phi(\mathbf{z})\mathbf{c} = \mathbf{0}$. Consider the element $u_0 = \mathbf{zc}$. Note that $u_0 \neq 0$, since the components of $\mathbf{z}$ are linearly independent, and that $\Phi(u_0) = \Phi(\mathbf{z})\mathbf{c} = \mathbf{0}$. This means that $u_0 \in D(\widehat{A})$. Furthermore, $\widehat{A} u_0 = A u_0 = (A\mathbf{z})\mathbf{c} = 0$. From the above, it follows that $u_0 \in \ker \widehat{A}$. Hence, $\widehat{A}$ is not injective.

(ii) Since $\widehat{A}$ is injective and the operator $\widehat{A}^{-1}$ is bounded on the whole Y by hypothesis, it follows that the operator $\widehat{A}$ is correct. □

Proposition 2. Let X, Y be real Banach spaces, $A : X \xrightarrow{on} Y$ a linear operator, and $\mathbf{z} = (z_1, \ldots, z_n)$ a basis of $\ker A$. If there exists a correct restriction $\widehat{A}$ of A defined by

$$\widehat{A} \subset A, \quad D(\widehat{A}) = \{u : u \in D(A), \Phi(u) = \mathbf{0}\},$$

where the components of the vector $\Phi = \mathrm{col}(\Phi_1, \ldots, \Phi_n)$ are linear bounded functionals on X, then A is closed and so A is a maximal operator.

Proof. Let $u_m \in D(A)$, $u_m \to u$ and $A u_m \to y$, $m \to \infty$. Denote $A u_m = y_m$. Since A is a linear operator and $\widehat{A}^{-1} y_m$ is a particular solution of $A u_m = y_m$, then every solution u_m to this equation can be represented as

$$u_m = \widehat{A}^{-1} y_m + \mathbf{za}_m = \widehat{A}^{-1} y_m + a_{m1} z_1 + \cdots + a_{mn} z_n, \tag{6}$$

where $\mathbf{a}_m = \mathrm{col}(a_{m1}, \ldots, a_{mn}) \in \mathbb{R}_n$. Acting by the vector Φ on (6), we obtain

$$\Phi(u_m) = \Phi(\mathbf{z})\mathbf{a}_m.$$

Since, by hypothesis, the operator $\widehat{A}$ is correct, it follows from Proposition 1 that $\det \Phi(\mathbf{z}) \neq 0$ and hence

$$\mathbf{a}_m = \Phi^{-1}(\mathbf{z})\Phi(u_m). \tag{7}$$

Substitution of (7) into (6) yields

$$u_m = \widehat{A}^{-1} y_m + \mathbf{z}\Phi^{-1}(\mathbf{z})\Phi(u_m).$$

Since $\widehat{A}^{-1}, \Phi_1, \ldots, \Phi_n$ are bounded and $u_m \to u$, $y_m \to y$ for $m \to \infty$, it follows that

$$u = \widehat{A}^{-1} y + \mathbf{z}\Phi^{-1}(\mathbf{z})\Phi(u).$$

Further, taking into account that $D(A) = D(\widehat{A}) \oplus \ker A$ [34], we conclude that $u \in D(A)$ and $Au = y$. So the operator A is closed and hence A is a maximal operator. □

3. General Boundary Conditions

In this Section, we study boundary value problems for ordinary differential equations with general homogeneous and nonhomogeneous boundary conditions.

Let now $X = Y = C[a,b]$ and $X^n = C^n[a,b]$. Let $A : X \to X$ be the n-th order linear operator in (1), $\widehat{A} : X \to X$ a correct restriction of A defined by

$$\begin{aligned} \widehat{A}u(x) &= Au(x), \quad x \in (a,b) \\ D(\widehat{A}) &= \{u : u \in D(A), \ \Phi(u) = \mathbf{0}\}, \end{aligned} \quad (8)$$

where $\Phi = \mathrm{col}(\Phi_1, \ldots, \Phi_n)$ is a vector of n linear bounded functionals $\Phi_i \in [X^{n-1}]^*$, $i = 1, \ldots, n$, and $\widehat{A}^{-1}$ the inverse of $\widehat{A}$.

For example, the operator

$$\begin{aligned} \widehat{A}u(x) &= Au(x), \quad x \in (a,b) \\ D(\widehat{A}) &= \{u : u \in X^n, \ u(x_0) = u'(x_0) = \cdots = u^{(n-1)}(x_0) = 0\}, \end{aligned}$$

where $\Phi_i(u) = u^{(i-1)}(x_0) = 0$, $i = 1, \ldots, n$, with x_0 being a fixed point in $[a,b]$, known as Cauchy boundary conditions, is correct. In the particular case where $A = \frac{d^n}{dx^n}$, the inverse $\widehat{A}$ and the unique solution of the correct problem $\widehat{A}u = f$, for any $f \in X$, is given explicitly by

$$u = \widehat{A}^{-1} f = \frac{1}{(n-1)!} \int_{x_0}^{x} (x-t)^{n-1} f(t) dt, \quad \forall f \in X.$$

From the above and Proposition 2, it is concluded that the n-th order linear operator A in (1) is closed and maximal.

3.1. Homogeneous Boundary Conditions

First, we consider the boundary value problem with homogeneous boundary conditions, namely

$$\begin{aligned} A_0 u &= Au = f, \quad x \in (a,b), \\ D(A_0) &= \{u : u \in D(A), \ Y(u) = \mathbf{0}\}, \end{aligned} \quad (9)$$

where the linear operator $A_0 : X \to X$ is a restriction of the n-th order linear operator A in (1), the components of the vector $Y = \mathrm{col}(Y_1, \ldots, Y_n) \in [X_n^{n-1}]^*$ are as in (3), (4), and $f \in X$.

Lemma 1. *The linear operator A_0 is a closed operator.*

Proof. Let u_r, $r = 1, 2, \ldots$, be a sequence in $D(A_0)$, $u_r \to u_0$ and $A_0 u_r \to f$. Then, $Au_r \to f$ and since A is a closed operator, we get that $u_0 \in D(A)$ and $Au_0 = f$. Moreover, since $Y_1, \ldots, Y_n$ are bounded functionals on X, we get

$$0 = Y_i(u_r) \to Y_i(u_0) = 0, \quad i = 1, \ldots, n, \quad r \to \infty.$$

This is that $u_0 \in D(A_0)$ and so $Au_0 = A_0 u_0 = f$. Hence, A_0 is a closed operator. □

Theorem 1. *Let A_0 be the linear operator defined by (9), $\mathbf{z} = (z_1, \ldots, z_n)$ a basis of $\ker A$, and $\widehat{A}^{-1}$ the inverse of the correct operator $\widehat{A}$ in (8). Then:*

(i) *The operator A_0 is injective if and only if $\det Y(\mathbf{z}) \neq 0$.*

(ii) *In addition, under (i), the operator A_0 is correct and the unique solution to the boundary value problem (9), for every $f \in X$, is given by*

$$u = A_0^{-1} f = \widehat{A}^{-1} f - \mathbf{z} Y^{-1}(\mathbf{z}) Y(\widehat{A}^{-1} f). \tag{10}$$

Proof. (i) Suppose $\det Y(\mathbf{z}) \neq 0$. Let $u \in \ker A_0$. Then, $A_0 u = Au = 0$ and $u = \mathbf{zc}$, where $\mathbf{c} = \mathrm{col}(c_1, \ldots, c_n)$ is a vector of arbitrary constants. Additionally, $Y(u) = Y(\mathbf{zc}) = Y(\mathbf{z})\mathbf{c} = 0$, which implies that $\mathbf{c} = 0$. That is $u = 0$ and consequently $\ker A_0 = \{0\}$. This proves that A_0 is injective. Conversely, let $\det Y(\mathbf{z}) = 0$. Then, there exists a nonzero vector of constants $\mathbf{c} = \mathrm{col}(c_1, \ldots, c_n)$ such that $Y(\mathbf{z})\mathbf{c} = 0$. Let the element $u_0 = \mathbf{zc}$ and notice that $u_0 \neq 0$ since the components of $\mathbf{z}$ are linearly independent. Then, $A_0 u_0 = Au_0 = A(\mathbf{zc}) = (A\mathbf{z})\mathbf{c} = 0$ and $Y(u_0) = Y(\mathbf{zc}) = Y(\mathbf{z})\mathbf{c} = 0$. That is $u_0 = \mathbf{zc} \in \ker A_0$ and as a consequence A_0 is not injective.

(ii) Let $\det Y(\mathbf{z}) \neq 0$. Then, from statement (i) follows that the opeartor A_0 is injective and hence there exists the inverse operator $A_0^{-1} : R(A_0) \subseteq X \to X$. Since by Lemma 1 the operator A_0 is closed, it is implied that A_0^{-1} is a closed operator. Furthermore, $\widehat{A}$ is a correct restriction of the linear operator A and therefore the general solution of the problem $Au = f$, for every $f \in X$, may be written as follows

$$u = \widehat{A}^{-1} f + \mathbf{zc}, \tag{11}$$

where $\mathbf{c} = \mathrm{col}(c_1, \ldots, c_n)$ is a vector of arbitrary constants. By requiring u to satisfy the boundary conditions in (9), we have

$$Y(u) = Y(\widehat{A}^{-1} f) + Y(\mathbf{z})\mathbf{c} = 0, \quad \text{and hence} \quad \mathbf{c} = -Y^{-1}(\mathbf{z}) Y(\widehat{A}^{-1} f).$$

Substitution of $\mathbf{c}$ into (11) yields (10), which is the unique solution of the boundary value problem $A_0 u = f$. In addition, it follows that $R(A_0) = X$ and since A_0^{-1} is a closed operator with $D(A_0^{-1}) = R(A_0) = X$, it is implied that A_0^{-1} is bounded on X. This proves that the operator A_0 is correct. □

3.2. Non-Homogeneous Boundary Conditions

Next, we consider the complete non-homogeneous boundary value problem

$$\begin{aligned} A_1 u &= Au = f, \quad x \in (a,b) \\ D(A_1) &= \{u : u \in D(A), \ Y(u) = \mathbf{b}\}, \end{aligned} \tag{12}$$

where $A_1 : X \to X$ is a restriction of the n-th order linear operator A in (1), the components of the vector $Y = \mathrm{col}(Y_1, \ldots, Y_n) \in [X_n^{n-1}]^*$ are as in (3), (4), $\mathbf{b} = \mathrm{col}(\beta_1, \ldots, \beta_n) \in \mathbb{R}_n$, and $f \in X$. It is noted that the operator A_1 is not linear, since its domain is a nonlinear set.

We state and prove the next theorem for the existence and construction of the unique solution of the boundary value problem (12).

Theorem 2. *Let A_1 be the operator defined by (12), $\mathbf{z} = (z_1, \ldots, z_n)$ a basis of $\ker A$, $\widehat{A}^{-1}$ the inverse of the correct operator $\widehat{A}$ in (8), and A_0 the operator defined in (9). Then, the operator A_1 is injective if and only if A_0 is injective. In this case, for every $f \in X$ and $\mathbf{b} \in \mathbb{R}_n$, the unique solution of (12) is given by*

$$u = A_1^{-1} f = \widehat{A}^{-1} f + \mathbf{z} Y^{-1}(\mathbf{z}) \Big[\mathbf{b} - Y(\widehat{A}^{-1} f) \Big]. \tag{13}$$

Proof. Suppose A_0 is injective. Then, $\ker A_0 = \{0\}$ and $\det Y(\mathbf{z}) \neq 0$ by Theorem 1. Let $u_1, u_2 \in D(A_1)$ and $A_1 u_1 = A_1 u_2 = f$. That is,

$$A_1 u_i = A u_i = f, \quad Y(u_i) = \mathbf{b}, \quad i = 1, 2,$$

from which we get

$$A_1 u_i = A u_i = A(u_i - \mathbf{z} Y^{-1}(\mathbf{z}) \mathbf{b}) = f, \quad Y\left(u_i - \mathbf{z} Y^{-1}(\mathbf{z}) \mathbf{b}\right) = 0, \quad i = 1, 2, \quad (14)$$

by taking into account that $A\mathbf{z} = 0$ and (5). From (14), it is implied that $u_i - \mathbf{z} Y^{-1}(\mathbf{z}) \mathbf{b} \in D(A_0)$ and

$$A_1 u_i = A_0 \left(u_i - \mathbf{z} Y^{-1}(\mathbf{z}) \mathbf{b} \right) = f, \quad i = 1, 2.$$

From Theorem 1, we have

$$u_i - \mathbf{z} Y^{-1}(\mathbf{z}) \mathbf{b} = A_0^{-1} f \quad \text{or} \quad u_i = A_0^{-1} f + \mathbf{z} Y^{-1}(\mathbf{z}) \mathbf{b}, \quad i = 1, 2.$$

Since A_0 is injective, it is concluded that $u_1 = u_2$ and therefore A_1 is an injective operator. Conversely, suppose A_1 is injective. Let $u \in \ker A_1$, which means

$$A_1 u = A u = 0, \quad Y(u) = \mathbf{b}, \quad \mathbf{b} \in \mathbb{R}_n. \quad (15)$$

It follows that $u = \mathbf{z} \mathbf{c}$, where $\mathbf{c}$ is a vector of constants, and

$$Y(u) = Y(\mathbf{z} \mathbf{c}) = Y(\mathbf{z}) \mathbf{c} = \mathbf{b}. \quad (16)$$

Since A_1 is injective, the system (16) has only one solution, that is $\det Y(\mathbf{z}) \neq 0$, and hence by Theorem 1 A_0 is injective.

Finally, under the hypothesis that A_1 is injective, for any $u \in D(A_1)$ that solves the completely nonhomogeneous problem $A_1 u = f$, we have

$$A_1 u = A u = A(u - \mathbf{z} Y^{-1}(\mathbf{z}) \mathbf{b}) = f, \quad Y\left(u - \mathbf{z} Y^{-1}(\mathbf{z}) \mathbf{b}\right) = 0. \quad (17)$$

This means that $u - \mathbf{z} Y^{-1}(\mathbf{z}) \mathbf{b} \in D(A_0)$ and

$$A_1 u = A_0 \left(u - \mathbf{z} Y^{-1}(\mathbf{z}) \mathbf{b} \right) = f.$$

The solution to this problem follows from Theorem 1, namely

$$\begin{aligned} u - \mathbf{z} Y^{-1}(\mathbf{z}) \mathbf{b} &= A_0^{-1} f \\ &= \widehat{A}^{-1} f - \mathbf{z} Y^{-1}(\mathbf{z}) Y(\widehat{A}^{-1} f), \end{aligned}$$

from where we get

$$\begin{aligned} u &= \widehat{A}^{-1} f - \mathbf{z} Y^{-1}(\mathbf{z}) Y(\widehat{A}^{-1} f) + \mathbf{z} Y^{-1}(\mathbf{z}) \mathbf{b} \\ &= \widehat{A}^{-1} f + \mathbf{z} Y^{-1}(\mathbf{z}) \left[\mathbf{b} - Y(\widehat{A}^{-1} f) \right]. \end{aligned}$$

□

4. Composition of Operators

In this Section, we investigate boundary value problems for special differential operators, specifically the k-th power of an operator and the product of two operators, with general homogeneous boundary conditions.

4.1. k-th Power of an Operator

The k-th power of an operator A^k is defined as the composition of the operator with itself, repeated k times, i.e.,

$$A^k = A(A^{k-1}) = \underbrace{AA\cdots A}_{k}, \quad k = 2, 3, \ldots.$$

If $A: X \to X$ is an n-th order linear differential operator with $D(A) = X^n$ then $A^k: X \to X$ is a kn-th order linear operator with $D(A^k) = X^{kn}$.

Let the boundary value problem

$$\begin{aligned} A_0^k u &= A^k u = f, \quad x \in (a, b), \\ D(A_0^k) &= \{u : u \in D(A^k), Y(u) = 0, Y(Au) = 0, \ldots, Y(A^{k-1}u) = 0\}, \end{aligned} \quad (18)$$

where the operator $A_0^k: X \to X$, the operators A and A_0 are defined as in (1) and (9), respectively, the components of the vector $Y = \mathrm{col}(Y_1, \ldots, Y_n)$ are as in (3), (4) where now $Y_i \in [X^{kn-1}]^*$, $i = 1, \ldots, n$, and $f \in X$. We state the following theorem.

Theorem 3. *Let A_0^k be the linear operator defined in (18) and $\mathbf{z} = (z_1, \ldots, z_n)$ be a basis of $\ker A$. Then:*

(i) *The operator A_0^k is injective if and only if $\det Y(\mathbf{z}) \neq 0$.*

(ii) *Moreover, under (i), the operator A_0^k is correct and the unique solution to the boundary value problem (18), for any $f \in X$, is given by*

$$u = (A_0^k)^{-1} f = A_0^{-k} f. \quad (19)$$

Proof. (i) Let $\det Y(\mathbf{z}) \neq 0$. Then, by Theorem 1, the operator A_0 is injective. Further, the operator A_0^k is injective as a composition of injective operators. Conversely, let A_0^k be injective. Then, $\ker A_0^k = \{0\}$, and from the well known relation, which holds for any linear operator A_0,

$$\ker A_0 \subseteq \ker A_0^2 \subseteq \ldots \subseteq \ker A_0^k, \quad k \in \mathbb{N},$$

follows that $\ker A_0 = \{0\}$, i.e., A_0 is injective. Then, by Theorem 1, we have $\det Y(\mathbf{z}) \neq 0$.

(ii) Let $\det Y(\mathbf{z}) \neq 0$. Then, by Theorem 1, the operator A_0 is correct. Observe that the problem (18), for any $f \in X$, by setting $Au = v_1$, $Av_1 = v_2$, $\ldots$, $Av_{k-2} = v_{k-1}$, $Av_{k-1} = f$ can be decomposed into the k boundary value problems:

$$\begin{aligned} Av_{k-1} &= f, & Y(v_{k-1}) &= 0 & \text{or} && A_0 v_{k-1} &= f, \\ Av_{k-2} &= v_{k-1}, & Y(v_{k-2}) &= 0 & \text{or} && A_0 v_{k-2} &= v_{k-1}, \\ & & &\cdots \\ Av_1 &= v_2, & Y(v_1) &= 0 & \text{or} && A_0 v_1 &= v_2, \\ Au &= v_1, & Y(u) &= 0 & \text{or} && A_0 u &= v_1. \end{aligned}$$

By applying Theorem 1 successively, we get

$$\begin{aligned} v_{k-1} &= A_0^{-1} f = \hat{A}^{-1} f - \mathbf{z} Y^{-1}(\mathbf{z}) Y(\hat{A}^{-1} f), \\ v_{k-2} &= A_0^{-1} v_{k-1} = A_0^{-1}\left(A_0^{-1} f\right) = A_0^{-2} f, \\ &\cdots \\ v_1 &= A_0^{-1} v_2 = A_0^{-1}\left(A_0^{-(k-2)} f\right) = A_0^{-(k-1)} f, \\ u &= A_0^{-1} v_1 = A_0^{-1}\left(A_0^{-(k-1)} f\right) = A_0^{-k} f, \end{aligned}$$

which is (19) with
$$A_0^{-k} = \underbrace{A_0^{-1} A_0^{-1} \cdots A_0^{-1}}_{k}.$$

Finally, since $R(A_0^{-k}) = X$ and A_0^{-k} is bounded as a composition of bounded operators, it is concluded that the operator A_0^k is correct. □

For the important category of boundary value problems for $k = 2$, we state the following corollary, which follows immediately from Theorem 3.

Corollary 1. *The boundary value problem*
$$\begin{aligned} A_0^2 u &= A^2 u = f, \\ D(A_0^2) &= \{u : u \in D(A^2), Y(u) = \mathbf{0}, Y(Au) = \mathbf{0}\}, \end{aligned} \quad (20)$$

is correct if and only if $\det Y(\mathbf{z}) \neq 0$ and the unique solution, for every $f \in X$, is given by
$$\begin{aligned} u &= A_0^{-2} f \\ &= \widehat{A}^{-2} f - \mathbf{z} Y^{-1}(\mathbf{z}) Y(\widehat{A}^{-2} f) \\ &\quad - \left[\widehat{A}^{-1}\mathbf{z} - \mathbf{z} Y^{-1}(\mathbf{z}) Y(\widehat{A}^{-1}\mathbf{z})\right] Y^{-1}(\mathbf{z}) Y(\widehat{A}^{-1} f). \end{aligned} \quad (21)$$

4.2. Product of Two Operators

Here, we are looking at yet another special boundary value problem, which is the generalization of Corollary 1. In particular, we consider the boundary value problem
$$\begin{aligned} A_0 \tilde{A}_0 u &= A^2 u = f, \quad x \in (a,b), \\ D(A_0 \tilde{A}_0) &= \{u : u \in D(A^2), \tilde{Y}(u) = \mathbf{0}, Y(Au) = \mathbf{0}\}, \end{aligned} \quad (22)$$

where A, A_0 are defined as in (1) and (9), respectively, and $\tilde{A}_0 : X \to X$ is a restriction of A defined by
$$\begin{aligned} \tilde{A}_0 u &= Au, \quad x \in (a,b), \\ D(\tilde{A}_0) &= \{u : u \in D(A), \tilde{Y}(u) = \mathbf{0}\}, \end{aligned}$$

wherein $\tilde{Y} = \mathrm{col}(\tilde{Y}_1, \ldots, \tilde{Y}_n)$ with $\tilde{Y}_i \in [X^{n-1}]^*$ being defined by
$$\tilde{Y}_i(u) = \sum_{j=0}^{m} \sum_{k=0}^{n-1} \tilde{v}_{ikj} u^{(k)}(x_j) + \sum_{k=0}^{n-1} \int_a^b \tilde{h}_{ik}(t) u^{(k)}(t) dt, \quad i = 1, \ldots, n, \quad (23)$$

where $\tilde{h}_{ik}(x)$, $i = 1, \ldots, n$, $k = 0, \ldots, n-1$, are continuous functions on $[a,b]$, $\tilde{v}_{ikj}$, $i = 1, \ldots, n$, $k = 0, \ldots, n-1$, $j = 0, \ldots, m$, are constants.

Theorem 4. *Let A_0 and $\tilde{A}_0$ be the linear operators defined in (22), $\mathbf{z} = (z_1, \ldots, z_n)$ a basis of $\ker A$, and $\widehat{A}^{-1}$ the inverse of the correct operator $\widehat{A}$ in (8). Then:*

(i) *The operator $A_0 \tilde{A}_0$ is injective if and only if*
$$\det Y(\mathbf{z}) \neq 0, \quad \det \tilde{Y}(\mathbf{z}) \neq 0. \quad (24)$$

(ii) *Furthermore, under (24), the operator $A_0 \tilde{A}_0$ is correct and the unique solution to the boundary value problem (22), for any $f \in X$, is given by*
$$\begin{aligned} u &= \widehat{A}^{-2} f - \mathbf{z} \tilde{Y}^{-1}(\mathbf{z}) \tilde{Y}(\widehat{A}^{-2} f) \\ &\quad - \left[\widehat{A}^{-1}\mathbf{z} - \mathbf{z} \tilde{Y}^{-1}(\mathbf{z}) \tilde{Y}(\widehat{A}^{-1}\mathbf{z})\right] Y^{-1}(\mathbf{z}) Y(\widehat{A}^{-1} f). \end{aligned} \quad (25)$$

Proof. (i)–(ii) By setting $\tilde{A}_0 u = v$, the problem (22) may be decomposed into the following two boundary value problems:

$$Av = f, \quad Y(v) = \mathbf{0} \quad \text{or} \quad A_0 v = f, \tag{26}$$
$$Au = v, \quad \tilde{Y}(u) = \mathbf{0} \quad \text{or} \quad \tilde{A}_0 u = v. \tag{27}$$

By Theorem 1, the boundary value problem (26) is correct if and only if $\det Y(\mathbf{z}) \neq 0$ and its unique solution is given by

$$v = A_0^{-1} f = \hat{A}^{-1} f - \mathbf{z} Y^{-1}(\mathbf{z}) Y(\hat{A}^{-1} f). \tag{28}$$

Similarly, the boundary value problem (27) is correct if and only if $\det \tilde{Y}(\mathbf{z}) \neq 0$ and its solution is

$$u = \tilde{A}_0^{-1} v = \hat{A}^{-1} v - \mathbf{z} \tilde{Y}^{-1}(\mathbf{z}) \tilde{Y}(\hat{A}^{-1} v). \tag{29}$$

Substitution of (28) into (29) yields

$$\begin{aligned}
u &= \hat{A}^{-1}\left[\hat{A}^{-1} f - \mathbf{z} Y^{-1}(\mathbf{z}) Y(\hat{A}^{-1} f)\right] \\
&\quad - \mathbf{z} \tilde{Y}^{-1}(\mathbf{z}) \tilde{Y}\left(\hat{A}^{-1}\left[\hat{A}^{-1} f - \mathbf{z} Y^{-1}(\mathbf{z}) Y(\hat{A}^{-1} f)\right]\right) \\
&= \hat{A}^{-2} f - \hat{A}^{-1} \mathbf{z} Y^{-1}(\mathbf{z}) Y(\hat{A}^{-1} f) \\
&\quad - \mathbf{z} \tilde{Y}^{-1}(\mathbf{z}) \tilde{Y}(\hat{A}^{-2} f) + \mathbf{z} \tilde{Y}^{-1}(\mathbf{z}) \tilde{Y}(\hat{A}^{-1} \mathbf{z}) Y^{-1}(\mathbf{z}) Y(\hat{A}^{-1} f) \\
&= \hat{A}^{-2} f - \mathbf{z} \tilde{Y}^{-1}(\mathbf{z}) \tilde{Y}(\hat{A}^{-2} f) \\
&\quad - \left[\hat{A}^{-1} \mathbf{z} - \mathbf{z} \tilde{Y}^{-1}(\mathbf{z}) \tilde{Y}(\hat{A}^{-1} \mathbf{z})\right] Y^{-1}(\mathbf{z}) Y(\hat{A}^{-1} f),
\end{aligned}$$

which is (25). Thus, the operator $A_0 \tilde{A}_0$ is correct if and only if $\det Y(\mathbf{z}) \neq 0$ and $\det \tilde{Y}(\mathbf{z}) \neq 0$, and the unique solution of $A_0 \tilde{A}_0 u = f$ is given explicitly by (25). □

5. Examples

To explain the implementation of the results presented in the previous section and to show the efficiency of the proposed solution routine, we solve two example problems.

Example 1. *Consider the differential Equation [29]*

$$u''(x) + u(x) = 0, \quad 0 < x < \frac{\pi}{2},$$

with the constraints

$$\int_0^{\pi/2} u(x) dx = 1, \quad \int_0^{\pi/2} u(x) \sin x \, dx = 1.$$

Comparing with (12), it is natural to take

$$Au(x) = u''(x) + u(x), \quad D(A) = C^2[0, \frac{\pi}{2}], \quad f(x) = 0,$$

$$Y_1(u(x)) = \int_0^{\pi/2} u(x) dx, \quad Y_2(u(x)) = \int_0^{\pi/2} u(x) \sin x \, dx, \quad \beta_1 = \beta_2 = 1,$$

and $\hat{A}$ and A_1 defined on

$$D(\hat{A}) = \{u : u \in D(A), u(0) = u'(0) = 0\},$$
$$D(A_1) = \{u : u \in D(A), Y(u) = \mathbf{b}\},$$

respectively. Observe that the only solution of $\hat{A} u = 0$ is $u \equiv 0$, $z_1 = \cos x$ and $z_2 = \sin x$ are two linearly independent solutions of $Au = 0$, and the matrix

$$Y(z) = \begin{pmatrix} Y_1(z_1) & Y_1(z_2) \\ Y_2(z_1) & Y_2(z_2) \end{pmatrix}$$

is non-singular. Thus, by applying Theorem 2, we get the unique solution

$$u(x) = zY^{-1}(z)\mathbf{b} = \frac{2\sin x + (\pi - 4)\cos x}{\pi - 2}.$$

Example 2. *Let the second order ordinary differential equation*

$$u''(x) - \frac{6}{x}u'(x) + \frac{12}{x^2}u(x) = 2x^4, \quad x \in (1,2), \tag{30}$$

subjected to non-local boundary conditions

$$\begin{aligned}
8u\left(\frac{3}{2}\right) - \frac{3}{4}u(2) &= \frac{177}{22}\int_1^2 \frac{1}{t^3}u(t)dt, \\
u'\left(\frac{5}{4}\right) - \frac{12}{5}u\left(\frac{5}{4}\right) &= \frac{625}{768}\int_1^2 \left(\frac{1}{t^2}u'(t) - \frac{3}{t^3}u(t)\right)dt.
\end{aligned} \tag{31}$$

Observe that this problem can be written as follows:

$$\left[\frac{d}{dx} - \frac{3}{x}\right]^2 u(x) = 2x^4, \quad x \in (1,2),$$

$$8u\left(\frac{3}{2}\right) - \frac{3}{4}u(2) - \frac{177}{22}\int_1^2 \frac{1}{t^3}u(t)dt = 0,$$

$$\left[\frac{d}{dx} - \frac{3}{x}\right]u(x)\bigg|_{x=\frac{5}{4}} - \frac{625}{768}\int_1^2 \frac{1}{t^2}\left(\left[\frac{d}{dt} - \frac{3}{t}\right]u(t)\right)dt = 0.$$

Comparing now with (22), we take

$$\begin{aligned}
Au(x) &= \left[\frac{d}{dx} - \frac{3}{x}\right]u(x), \quad D(A) = C^1[1,2], \\
f(x) &= 2x^4, \\
\tilde{Y}(u) &= 8u\left(\frac{3}{2}\right) - \frac{3}{4}u(2) - \frac{177}{22}\int_1^2 \frac{1}{t^3}u(t)dt, \\
Y(Au) &= Au(x)|_{x=\frac{5}{4}} - \frac{625}{768}\int_1^2 \frac{1}{t^2}Au(t)dt,
\end{aligned}$$

and we set

$$\begin{aligned}
\tilde{A}_0 u &= Au, \quad D(\tilde{A}_0) = \{u : u \in D(A), \tilde{Y}(u) = 0\}, \\
A_0 v &= Av, \quad D(A_0) = \{v : v \in D(A), Y(v) = 0\}.
\end{aligned}$$

Further, let the correct operator $\widehat{A}$ defined by

$$\widehat{A}u = Au = f, \quad D(\widehat{A}) = \{u : u \in D(A), u(1) = 0\},$$

and its inverse given by

$$\widehat{A}^{-1}f(x) = e^{\int_1^x \frac{3}{s}ds}\int_1^x e^{-\int_1^t \frac{3}{s}ds}f(t)dt = x^3\int_1^x \frac{1}{t^3}f(t)dt.$$

Finally, note that $z(x) = x^3$ is a fundamental solution of the homogeneous equation $Au = 0$ and that

$$Y(z) \neq 0, \quad \tilde{Y}(z) \neq 0.$$

Hence, from Theorem 4, it follows that the non-local boundary value problem (30), (31) has a unique solution, which after substituting into the formula (25), is

$$u(x) = \frac{x^3(x^3-1)}{3}.$$

6. Discussion

A method for constructing closed-form solutions to boundary value problems for ordinary differential equations with general multipoint and integral boundary conditions has been presented. Ready to use solution formulae in a symbolic form have been derived for some classes of boundary value problems. Specifically, we considered the following boundary value problems:

$$\begin{aligned}
A_0 u &= f, & D(A_0) &= \{u : u \in D(A), Y(u) = \mathbf{0}\}, \\
A_1 u &= f, & D(A_1) &= \{u : u \in D(A), Y(u) = \mathbf{b}\}, \\
A_0^k u &= f, & D(A_0^k) &= \{u : u \in D(A^k), Y(u) = \mathbf{0}, Y(Au) = \mathbf{0}, \ldots, Y(A^{k-1}u) = \mathbf{0}\}, \\
A_0 \tilde{A}_0 u &= f, & D(A_0 \tilde{A}_0) &= \{u : u \in D(A^2), \tilde{Y}(u) = \mathbf{0}, Y(Au) = \mathbf{0}\},
\end{aligned}$$

where the operators $A_0, A_1, \tilde{A}_0$ are restrictions of the n-th order linear differential operator A in (1) and $Y, \tilde{Y}$ are vectors of linear bounded functionals as in (4) and (23), respectively, describing general non-local boundary conditions.

The proposed methodology can be specialized to other categories of boundary value problems and extended to some classes of partial differential equations.

Author Contributions: Conceptualization, E.P., S.Z. and I.F.; validation, S.Z. and I.F.; formal analysis, E.P. All authors have read and agreed to the published version of the manuscript.

Funding: This research received no external funding.

Acknowledgments: The authors would like to thank the anonymous reviewers for their valuable suggestions and comments.

Conflicts of Interest: The authors declare no conflict of interest.

References

1. Busenberg, S.; Cooke, K.L. The effect of integral conditions in certain equations modelling epidemics and population growth. *J. Math. Biol.* **1980**, *10*, 13–32. [CrossRef] [PubMed]
2. Picone, M. Sui valori eccezionali di un parametro da cui dipende un'equazione dif- ferenziale lineare del secondo ordine. *Annali della Scuola Normale Superiore di Pisa-Classe di Scienze* **1910**, *11*, 1–141. (In Italian)
3. Sommerfeld, A. Ein Beitrag zur hydrodynamischen Erklärung der turbulenten Flüssigkeitsbewegung. *Proc. Int. Math. Congr.* **1909**, *3*, 116–124.
4. Tamarkin, J. Some general problems of the theory of ordinary linear differential equations and expansion of an arbitrary function in series of fundamental functions. *Math. Z.* **1928**, *27*, 1–54. [CrossRef]
5. Il'in, V.A.; Moiseev, E.I. An a priori estimate for the solution of a problem associated with a nonlocal boundary value problem of the first kind. *Differ. Equ.* **1988**, *24*, 519–526.
6. Dezin, A.A. On the general theory of boundary value problems. *Math. USSR-Sb.* **1976**, *100*, 171–180. [CrossRef]
7. Grubb, G. A characterization of the nonlocal boundary value problems associated with an elliptic operator. *Annali della Scuola Normale Superiore di Pisa-Classe di Scienze* **1968**, *22*, 425–513.
8. Vishik, M.I. On general boundary problems for elliptic differential equations. *Am. Math. Soc. Transl.* **1952**, *1*, 187–246.
9. Gorbachuk, M.L. *Boundary Value Problems for Operator Differential Equations*; Springer: Berlin, Germany, 1991. [CrossRef]
10. Krein, S.G. *Linear Equations in Banach Spaces*; Birkhäuser: Basel, Switzerland, 1982. [CrossRef]
11. Lions, J.L.; Magenes, E. *Non-Homogeneous Boundary Value Problems and Applications*; Springer: Berlin/Heidelberg, Germany, 1972; Volume 1. [CrossRef]
12. Skubachevskii, A.L. Nonclassical boundary-value problems. *Int. J. Math. Sci.* **2008** *155*, 199–334. [CrossRef]
13. Krall, A.M. The development of general differential and general differential-boundary systems. *Rocky Mt. J. Math.* **1975**, *5*, 493–542. [CrossRef]
14. Ma, R. A survey on nonlocal boundary value problems. *Appl. Math. E-Notes* **2007**, *7*, 257–279.

15. Štikonas, A. A survey on stationary problems, Green's functions and spectrum of Sturm–Liouville problem with nonlocal boundary conditions. *Nonlinear Anal.-Model. Control* **2014**, *19*, 301–334. [CrossRef]
16. Whyburn, W.M. Differential equations with general boundary conditions. *Bull. Am. Math. Soc.* **1942**, *48*, 692–704. [CrossRef]
17. Denche, M.; Kourta, A. Boundary value problem for second-order differential operators with nonregular integral boundary conditions. *Rocky Mt. J. Math.* **2006**, *36*, 893–913. [CrossRef]
18. Gallardo, J.M. Second-Order differential operators with integral boundary conditions and generation of analytic semigroups. *Rocky Mt. J. Math.* **2000**, *30*, 1265–1291. [CrossRef]
19. Jankowski, T. Differential equations with integral boundary conditions. *J. Comput. Appl. Math.* **2002**, *147*, 1–8. [CrossRef]
20. Jones, W.R. Differential systems with integral boundary conditions. *J. Differ. Equ.* **1967**, *3*, 191–202. [CrossRef]
21. Kalenyuk, P.I.; Kuduk, G.; Kohut, I.V.; Nytrebych, Z.M. Problem with integral conditions for differential-operator equation. *J. Math. Sci.* **2015**, *208*, 267–276. [CrossRef]
22. Liu, L.; Hao, X. ; Wu Y. Positive solutions for singular second order differential equations with integral boundary conditions. *Math. Comput. Model.* **2013**, *57*, 836–847. [CrossRef]
23. Zhang, L.; Xuan, Z. Multiple positive solutions for a second-order boundary value problem with integral boundary conditions. *Bound. Value Probl.* **2016**, *60*. [CrossRef]
24. Zhang, X.; Ge, W. Positive solutions for a class of boundary-value problems with integral boundary conditions. *Comput. Math. Appl.* **2009**, *58*, 203–215. [CrossRef]
25. Adomian, G. Integral Boundary Conditions. In *Solving Frontier Problems of Physics: The Decomposition Method. Fundamental Theories of Physics (An International Book Series on the Fundamental Theories of Physics: Their Clarification, Development and Application)*; Springer: Dordrecht, The Netherlands, 1994; Volume 60, pp. 196–210. [CrossRef]
26. Chen, S.; Ni, W.; Wang, C. Positive solution of fourth order ordinary differential equation with four-point boundary conditions. *Appl. Math. Lett.* **2006**, *19*, 161–168. [CrossRef]
27. Dovletov, D.M. On a nonlocal boundary value problem of the second kind for the Sturm- Liouville operator in the differential and difference statements. *e-J. Anal. Appl. Math.* **2018**, *1*, 37–55. [CrossRef]
28. Krall, A.M. Differential operators and their adjoints under integral and multiple point boundary conditions. *J. Differ. Equ.* **1968**, *4*, 327–336. [CrossRef]
29. Ojika, T.; Welsh, W. A numerical method for the solution of multi-point problems for ordinary differential equations with integral constraints. *J. Math. Anal. Appl.* **1979**, *72*, 500–511. [CrossRef]
30. Parasidis, I.N.; Providas, E. Exact solutions to problems with perturbed differential and boundary operators. In *Analysis and Operator Theory*; Springer Optimization and Its Applications; Rassias, T., Zagrebnov, V., Eds.; Springer: Cham, Switzerland, 2019; Volume 146. pp. 301–317. [CrossRef]
31. Sadybekov, M.A.; Imanbaev, N.S. A regular differential operator with perturbed boundary condition. *Math. Notes* **2017**, *101*, 878–887. [CrossRef]
32. Aida-Zade, K.R.; Abdullaev, V.M. On the solution of boundary value problems with nonseparated multipoint and integral conditions. *Differ. Equ.* **2013**, *49*, 1114–1125. [CrossRef]
33. Biyarov, B.N. Normal extensions of linear operators. *Eurasian Math. J.* **2016**, *7* 17–32.
34. Kokebaev, B.K.; Otelbaev, M.; Shynibekov, A.N. About Restrictions and Extensions of operators. *DAN SSSR* **1983**, *271*, 1307–1310. (In Russian)
35. Oinarov, R.O.; Parasidis, I.N. Correct extensions of operators with finite defect in Banach spaces. *Izv. Akad. Kaz. SSR* **1988** *5*, 42–46. (In Russian)
36. Parasidis, I.N.; Tsekrekos, P. Some quadratic correct extensions of minimal operators in Banach spaces. *Oper. Matrices* **2010**, *4*, 225–243. [CrossRef]
37. Parasidis, I.N.; Providas, E.; Zaoutsos, S. On the Solution of Boundary Value Problems for Ordinary Differential Equations of Order n and 2n with General Boundary Conditions. In *Computational Mathematics and Variational Analysis*; Springer Optimization and Its Applications; Daras, N., Rassias, T., Eds.; Springer: Cham, Switzerland, 2020; Volume 159, pp. 299–314. [CrossRef]

Article

A Dynamic Model of Multiple Time-Delay Interactions between the Virus-Infected Cells and Body's Immune System with Autoimmune Diseases

Hoang Pham

Department of Industrial and Systems Engineering, Rutgers University, Piscataway, NJ 08854, USA; hopham@soe.rutgers.edu

Citation: Pham, H. A Dynamic Model of Multiple Time-Delay Interactions between the Virus-Infected Cells and Body's Immune System with Autoimmune Diseases. *Axioms* **2021**, *10*, 216. https://doi.org/10.3390/axioms10030216

Academic Editor: Davron Aslonqulovich Juraev

Received: 17 July 2021
Accepted: 23 August 2021
Published: 7 September 2021

Publisher's Note: MDPI stays neutral with regard to jurisdictional claims in published maps and institutional affiliations.

Copyright: © 2021 by the author. Licensee MDPI, Basel, Switzerland. This article is an open access article distributed under the terms and conditions of the Creative Commons Attribution (CC BY) license (https:// creativecommons.org/licenses/by/ 4.0/).

Abstract: The immune system is a complex interconnected network consisting of many parts including organs, tissues, cells, molecules and proteins that work together to protect the body from illness when germs enter the body. An autoimmune disease is a disease in which the body's immune system attacks healthy cells. It is known that when the immune system is working properly, it can clearly recognize and kill the abnormal cells and virus-infected cells. But when it doesn't work properly, the human body will not be able to recognize the virus-infected cells and, therefore, it can attack the body's healthy cells when there is no invader or does not stop an attack after the invader has been killed, resulting in autoimmune disease.; This paper presents a mathematical modeling of the virus-infected development in the body's immune system considering the multiple time-delay interactions between the immune cells and virus-infected cells with autoimmune disease. The proposed model aims to determine the dynamic progression of virus-infected cell growth in the immune system. The patterns of how the virus-infected cells spread and the development of the body's immune cells with respect to time delays will be derived in the form of a system of delay partial differential equations. The model can be used to determine whether the virus-infected free state can be reached or not as time progresses. It also can be used to predict the number of the body's immune cells at any given time. Several numerical examples are discussed to illustrate the proposed model. The model can provide a real understanding of the transmission dynamics and other significant factors of the virus-infected disease and the body's immune system subject to the time delay, including approaches to reduce the growth rate of virus-infected cell and the autoimmune disease as well as to enhance the immune effector cells.

Keywords: immune system; virus-infected cell; effector cell; autoimmune disease; time-delay virus-immune model

1. Introduction

Human beings are constantly exposed to germs such as bacteria, viruses and toxins (chemicals produced by microbes) that enter into the human body that make-up the infections and diseases that will eventually make people sick. The body is made up of many types of cells. Usually, cells grow and divide to produce new cells. A body's well-working immune system can prevent germs from entering the body and destroys any infectious microorganisms that do invade the body [1–3]. As long as our immune system is working smoothly, we often do not pay much attention to it or even do not know that it is there. However, if it stops working properly because it is weak or cannot fight particularly germs or the diseases, then we become sick. The germs that our body has never encountered before are also likely to make us sick [4]. Some germs will only make you ill the first time you come into contact with them. When the body senses danger from a virus or infection, the immune system will respond and attack it.

The human immune system is complex and it is the body's defense system. It is a complex network consisting of many parts including cells, tissues, molecules and organs working together to defend the body against invaders as well as to fight the infections and diseases when germs enter our body [1,2,5]. The skin is also a part of the immune system that prevents germs from entering the body [4]. Our immune system, believe it or not, works very hard to keep us healthy. The main tasks of the body's immune system are to attack and destroy substances that are foreign to our body, such as bacteria and viruses, or limit the extent of their harm if they get in [5].

When our immune system is working properly, it can recognize which cells are ours and which substances are foreign to our body. It then activates, mobilizes, attacks and kills foreign invader germs that can cause us harm. In fact, our immune system learns about any germs after we have been exposed to them. Our body develops antibodies to protect us from those specific germs [1,6]. When we are given a vaccine for example, our immune system builds up antibodies to the foreign cells in the vaccine and will quickly remember these foreign cells and destroy them if we are exposed to them in the future. However, when our immune system is not working properly, the body attacks normal and healthy cells when there is no invader or does not stop an attack after the invader has been killed, resulting in autoimmune disease [1–3,5].

Developing mathematical models to predict the growth of tumors, virus-infected cells and immune cells have been of interest in the area of cancer epidemiology research [7–10] and infectious disease epidemiology [11,12] in the past few decades. Many models [9,10,13–16] have been proposed using the ordinary differential equations and partial differential equations in the past several decades and using the delay partial differential equations in recent years for characterizing tumor-immune dynamic growth, but there is still no consensus on the modeling due to the complexity of virus-infected and tumor cancer growth in the body's immune system and the growth patterns of the tumors and virus-infected cells [16]. Many researchers [7,17–24] have used the existing prey–predator modeling concept [25,26] to study and model the tumor–immune interactions [7,27,28] and the effects of tumor growth [17,29,30]. To simplify an understanding of the interaction between tumor and immune cells, several researchers used the concept of the prey–predator system [24,29]. Here, the immune cells play the role of the predator, while tumor or virus-infected cells of the prey. In other words, the predator is the immune system that kills the tumor cells (prey) [24].

The modeling studies of prey–predator systems and its related applications have beentremendously interesting in recent years to various disciplines including population disease [31,32], life expectancy [33,34], biomathematics [35,36], cancel growth [29,37–42] and engineering science [23,35,43]. Many researchers have studied various dynamic prey–predator models including a two-dimension predator–prey model [44–48], multi-predator models [20,23,49], multi-prey models [50–52] and time-delay prey–predator models [23,36,43,50] with various applications in biomathematics [53], population disease [22,23,29,30,54,54–59] and recent COVID-19 disease analysis [1,12,60–65].

Haque et al. [54] analyzed a predator–prey model using standard disease incidence. Naji and Mustafa [56] studied a dynamic model of eco-epidemiology considering nonlinear disease incidence rates with an infective type of disease in prey. Mukhopadhyaya and Bhattacharyya [36] studied the effect of delay on a prey–predator model with disease in the prey considering a Holling type II functional response. Wang et al. [43] studied a predator–prey model with distributed delays. Huang et al. [22] recently studied a stochastic predator–prey model with a Holling II increasing function in the predator and discussed the analytic results of the dynamics of the stochastic predator–prey model.

Jana and Kar [23] studied a three-dimensional epidemiological dynamic model incorporating time delay in the model for considering it as the time taken by a susceptible prey to become infected. Lestari et al. [29] discussed an epidemic model of cancer with chemotherapy in the form of a system of non-linear differential equations with three subpopulations. They presented the point of equilibrium and numerically determined the reproduction number and the growth rate of cancer cells. Pham [63] studied a model to estimate the number of deaths related to COVID-19 based on the US data and recently,

Pham [64] studied a mathematical model that considers the time-dependent effects of various pandemic restrictions and changes related to COVID-19 such as reopening states, social distancing, reopening schools and face mask mandates in communities.

In this paper, we develop a new mathematical model considering the multiple time-delay interactions between the immune cells and virus-infected cells with an autoimmune disease in the form of delay partial differential equations. The model can be used to determine the dynamic progression of the virus-infected cell growth and observe the patterns of how the virus-infected cells spread in the body's immune system with respect to time delays. In Section 2, we discuss all the model assumptions and the mathematical time-delay virus-immune model development of the body's immune system considering the multiple time-delay interactions between the immune cells (or effector cells) and virus-infected cells with an autoimmune disease. The model aims to predict the dynamic progression of virus-infected cell growth in the immune system. Section 3 discusses several numerical examples to illustrate the proposed model and shows numerical results with various cases whether a virus-infected free state can be reached or not as the time progresses. Section 4 discusses a brief conclusion and future research problems.

2. A Mathematical Model with Multiple Time-Delay Interactions between Infected-Virus and Immune Effector Cells

As mentioned earlier, many researchers [7,17,22,28] have developed various prey–predator models and recently developed mathematical models to investigate the interactions between the tumor cells and immune systems, and tumor-immune cells with consideration of an interaction between the tumor and immune cells with a time delay. In this section, we discuss a new virus-immune time-delay model of the body's immune system with considerations of the multiple interactions between the virus-infected cells and body's immune cells with an autoimmune disease. With the same concept of the prey–predator models in the literature, here, in this new model, the immune effector cells play the role of the predator while the virus-infected cells play prey. The effector cell, usually used to describe cells in the immune system, is a cell that performs a specific function in response to a stimulus or defends the body in an immune response. We first describe a list of our modeling assumptions, also based on a recent study by Lestari et al. [29], and then present a derivation of the mathematical modeling results as follows.

Notation: We use the following notation throughout the paper:

a = the intrinsic growth rate per unit time

b = the elimination rate of the virus-infected cells by the healthy immune system (effector cells) per cells and unit time

c = the death rate of the healthy immune system per unit time

d = degree of recruitment of maximum immune-effector cells in relation with virus-infected cells per unit of time

e = capacity of the virus-infected cells per unit of time

f = the rate that the immune system attacks the body's own healthy (effector) cells, resulting in autoimmune disease per cells and unit of time

g = constant factor of growth rate per unit of time

h = the half saturation constant (cells)

k = the half saturation for virus-infected cleanup (cells)

m = the degree of inactivation of effector cells by virus-infected cells per cells and unit of time

p = parameter of virus-infected cleanup by immune-effector cells per unit of time

s = growth rate of immune-effector cells per unit of time

$I(t)$ = the number of healthy immune-effector cells at time t

$V(t)$ = the number of virus-infected cells at time t

2.1. Immune Cell Model Formulation

In apopulation of healthy immune-cell or effector cells (in this case as the predator), we assume the following:

1. The effector cell has a constant growth rate, s, of effector cells [29].
2. The effector cell has a natural death rate, c, of effector cells [29]. There is an increase in the number of effector cells by the growthrate d with a maximum degree of recruitment of immune-effector cells in response to the shift toward virus-infected cells [29] with a τ_3 time delay.
3. There is a constant rate f of the immune system attacking the body's own healthy (effector) cells, resulting in an autoimmune disease. The constant f, in general, will be very small compared to c, so that when I is not too large, then the term $f\,I^2$ will be negligible compared to cI.
4. There will be a reduction in the number of effector cells due to their interaction with the virus-infected cells witha constant rate m [29].

We can derive a mathematical equation based on the assumptions (1–3) and the result is as follows:

$$\frac{\partial I(t)}{\partial t} = s - cI + \frac{dV(t-\tau_3)I(t-\tau_3)}{h+V(t-\tau_3)} \tag{1}$$

We can derive a mathematical equation based on the assumptions (4–5) and the result is as follows:

$$\frac{\partial I(t)}{\partial t} = -fI^2 - mIV \tag{2}$$

From Equations (1) and (2), a model of the rate of the immune-effector cells governing the interactions between the virus-infected and virus-infected cells over time can be presented as follows:

$$\frac{\partial I(t)}{\partial t} = s - cI + \frac{dV(t-\tau_3)I(t-\tau_3)}{h+V(t-\tau_3)} - fI^2 - mIV. \tag{3}$$

2.2. Virus-Infected Cell Model Formulation

In a population of virus-infected cells (in this case as prey), which is when a virus infects a host, a virus invades the healthy immune cells of its host and also can infect other cells, we assume the following:

5. The virus-infected cell has a constant growth rate, a, ref. [29] with consideration of a constant factor of growth rate, g, and a τ_1 time delay before the virus is to be infected.
6. There will be a constant elimination rate of the virus-infected cells by the healthy immune system (effector cells), b, by a τ_2 time delay. In other word, b measures how efficiently the effector cells kill the virus-infected cells.
7. The number of virus-infected cells will decline by a constant parameter of the virus-infected cleanup of effector cells, p, ref. [29] with a τ_3 time delay.
8. There will be a reduction in the number of virus-infected cells by a constant rate e that encounters of the two virus-infected cells per unit of time in competing with each other due to the limited number of host cells. The constant rate e here can be considered to be very small.

We can derive a mathematical equation based on the assumptions (6–7) and the result is as follows:

$$\frac{\partial V(t)}{\partial t} = aV(t-\tau_1)e^{-gV(t-\tau_1)} - bVI(t-\tau_2) \tag{4}$$

Here, the constant parameter b measures how efficiency effector cells kill virus-infected cells. From assumptions (8–9), we can derive a mathematical equation and the result is as follows:

$$\frac{\partial V(t)}{\partial t} = -eV^2 - p\frac{V(t-\tau_3)I(t-\tau_3)}{k+V(t-\tau_3)}. \tag{5}$$

From Equations (4) and (5), a model of the rate of the virus-infected cells overtime can be presented as follows:

$$\frac{\partial V(t)}{\partial t} = aV(t-\tau_1)e^{-gV(t-\tau_1)} - bVI(t-\tau_2) - eV^2 - p\frac{V(t-\tau_3)I(t-\tau_3)}{k+V(t-\tau_3)}. \qquad (6)$$

Thus, from Equations (3) and (6), a new virus-immune time-delay model for the body's immune system with considerations of multiple interactions between the virus infected cells and body's immune cells with autoimmune disease is given as follows:

$$\begin{array}{l}\frac{\partial I(t)}{\partial t} = s - cI + \frac{dV(t-\tau_3)I(t-\tau_3)}{h+V(t-\tau_3)} - fI^2 - mIV \\ \frac{\partial V(t)}{\partial t} = aV(t-\tau_1)e^{-gV(t-\tau_1)} - bVI(t-\tau_2) - eV^2 - p\frac{V(t-\tau_3)I(t-\tau_3)}{k+V(t-\tau_3)}.\end{array} \qquad (7)$$

If we do not consider the effect of the chemotherapy drug from the model studied by Lestari et al. [29], then their model [29] can be slightly considered as a special case of our model, as given in Equation (7), where $f = 0$, $e = 0$, $g = 0$, $\tau_1 = 0$, $\tau_2 = 0$ and $\tau_3 = 0$.

We now wish to determine the number of immune-infector cells $I(t)$ and virus-infected cells $V(t)$ at any given time. We developed a program using R software to calculate and plot the two functions $I(t)$ and $V(t)$ with respect to time t, as will be discussed in the next section.

3. Model Analysis

In this section, we present an analysis of the proposed model. Table 1 shows the parameter values that we use in our analysis based on some existing studies [29,39–41] for the illustration of our model. Any other sets of parameter values can be easily applied from the model.

Table 1. Model parameter values.

$a = 0.43$/day	$b = 43 \times 10^{-7}$/cells/day	$c = 4.12 \times 10^{-2}$/day
$d = 15 \times 10^{-5}$/day	$e = 4 \times 10^{-8}$/day	$f = 4 \times 10^{-7}$/day
$g = 3 \times 10^{-6}$/day	$h = 20.2$ (cells)	$k = 10^5$/cells
$m = 2 \times 10^{-11}$ cells/day	$p = 341 \times 10^{-12}$/day	$s = 7000$ cells/day

In this study, we consider various initial numbers of virus-infected cells and numbers of immune-effect cells from 15,000 to 30,000 and from 50,000 to 75,000, respectively, to explore if the results depend on those initial numbers of cells. We discuss below several cases based on various parameter values of the virus-infected growth rates, a, the elimination rate of the virus-infected cells by the immune-effector cells, b and the growth rate of the immune-effector cells, s, as follows:

Case 1: When $a = 0.43$, $b = 43 \times 10^{-7}$, $s = 7000$.

We first assume that the initial number of virus-infected cells is $V_0 = 30,000$ and the initial number of immune-effector cells is $I_0 = 50,000$. From Figure 1a,b, we can observe that the initial number of virus-infected cells and immune-effector cells are 30,000 and 50,000, respectively, as expected. The virus-infected counts begins to increase and it reaches the highest point at around the 14th day as $(V,I) = (72,248, 81,228)$ and starts to decrease slowly, where $(V,I) = (31,905, 90,578)$, at the 300th day. As seen in the graphs in Figure 1a, on the one hand, the number of immune-effector cells keeps increasing but starts to slowly stabilize after the 100th day at the level of 90,578. On the other hand, the number of virus-infected cells first begins to increase until it reaches the maximum number of infected cells at 72,304 (see Figure 1b) then startsto decrease and slowly stabilize after around the 280th day and stays at just above the level of the initial number of virus-infected cells, at 31,900 cells. It seems that in this case, with a given growth rate of effector cells $s = 7000$ cells per day and avirus-infected growth rate $a = 0.43$, it will not be able to reach the virus free state. Figure 1c,d show the relationship between the immune-effector cells and the virus-infected cells. Figure 1e,f show the 3D relationships of the effector cells, the immune-effector cells and time.

We observe the same results above even when the initial number of immune effector cells is $I_0 = 75{,}000$ (see Figure 1g,h) as well as the same results when the initial number of virus-infected cells is reduced to $V_0 = 15{,}000$ (see Figure 1i,j), respectively. It is worth noting that the initial number of virus-infected cells and immune-effector cells do not influence the end result of whether the body is of virus free stage or not. This shows that our model can be used to obtain the results without needing to know the exact initial number of virus-infected cells or the number of immune effector cells in the body.

Let us consider when $I_0 = 75{,}000$ and $V_0 = 15{,}000$: From Figure 1k,l, we observe that the virus-infected counts keep increasing significantly from the beginning until around the 50th day as $(V,I) = (31{,}291, 90{,}521)$ and slowly stabilizes around the 100th day at the level of 31,770, while the number of immune-effector cells also keeps increasing but starts to slowly stabilize after the 60th day at the level of 90,560. In this case ($I_0 = 75{,}000$ and $V_0 = 15{,}000$), the result is the same as all the cases above that, the body will not be able to reach the virus free state. This concludes that the initial number of virus-infected cells and immune-effector cells do not influence the end results.

Model comparison: We now use the model studied by Lestari et al. [29], as we mentioned earlier, to compare their modeling result (i.e., without the effect of the chemotherapy drug and when $f = 0$, $e = 0$, $g = 0$, $\tau_1 = 0$, $\tau_2 = 0$ and $\tau_3 = 0$) to our model from Equation (7). From Figure 1m, we observe that the number of immune-effector cells, I, keeps increasing but starts to slowly stabilize after the 150th day at the level of 169,669 cells. The virus-infected count, V, (see Figure 1n) begins to increase and it reaches the highest point at around the 14th day with $(V,I) = (107{,}189, 102{,}683)$ but starts to decrease sharply until it reaches the virus free state as $(V,I) = (0, 168{,}064)$ after the 100th day with the number of immune-effector cells around 168,064 cells.

In this example, when those values, e, f and g, are not equal to zero, our proposed model shows that one cannot reach the virus free state because we consider the autoimmune disease factor in our model, where one can reach the virus free state at the 100th day by using the model developed by Lestari et al. [29], since they did not consider the autoimmune disease factor in their study.

Case 2: This is the same as Case 1, except $s = 10{,}000$ (instead of $s = 7000$).

From Figure 2a,b, we can observe that the initial number of virus-infected cells and immune-effector cells are 30,000 and 50,000, respectively, as expected. It should be noted that the number of immune-effector cells (see Figure 2a) keeps increasing but starts to slowly stabilize after the 250th day at the level of 114,790. The virus-infected counts (see Figure 2b) begins to increase and it reaches the highest point at around the 8th day with $(V,I) = (55{,}468, 89{,}222)$ and starts to decrease significantly until it reaches the virus free state at $(V,I) = (0, 114{,}790)$ after the 298th day, where the number of immune-effector cells is 114,790. Figure 2c,d show the relationship between the immune-effector cells and the virus-infected cells. Figure 2e,f show the 3D relationships of the effector cells, the immune-effector cells and time.

The result is about the same even when the initial number of immune effector cells to be as $I_0 = 75{,}000$, see Figure 2g,h. The result is also about the same, even when the initial number of virus-infected cells is reduced to $V_0 = 15{,}000$ cells (see Figure 2i,j). It is worth noting that the initial number of virus-infected cells and immune-effector cells do not influence the end result.

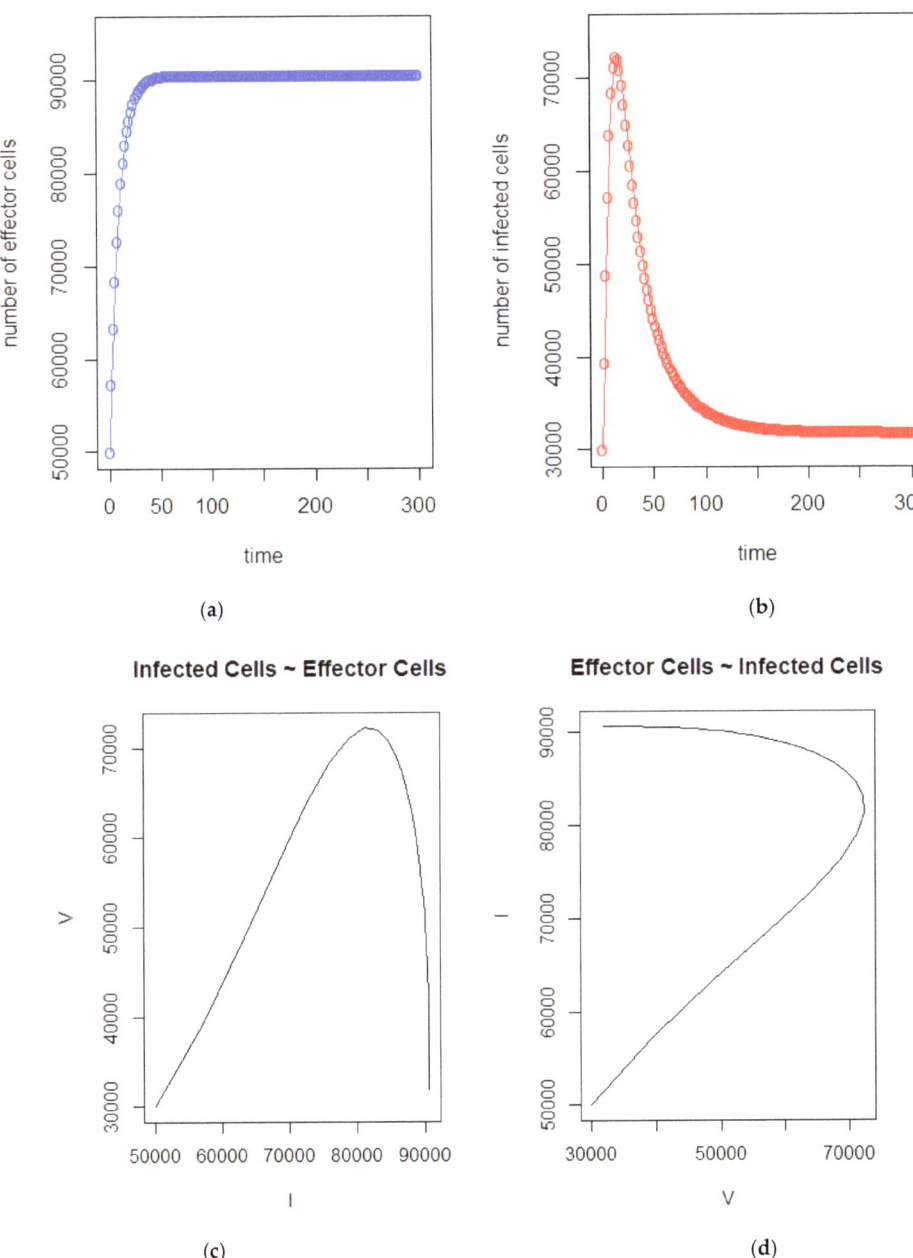

Figure 1. *Cont.*

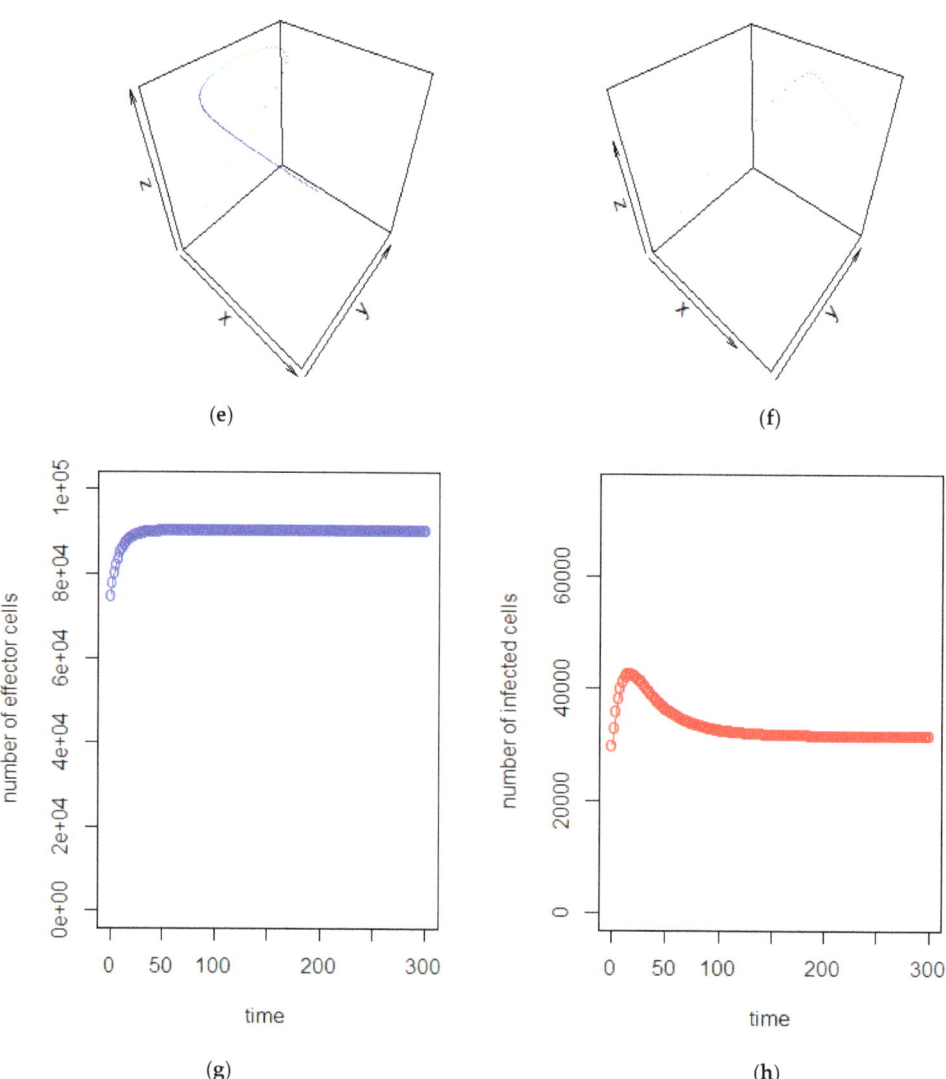

Figure 1. *Cont.*

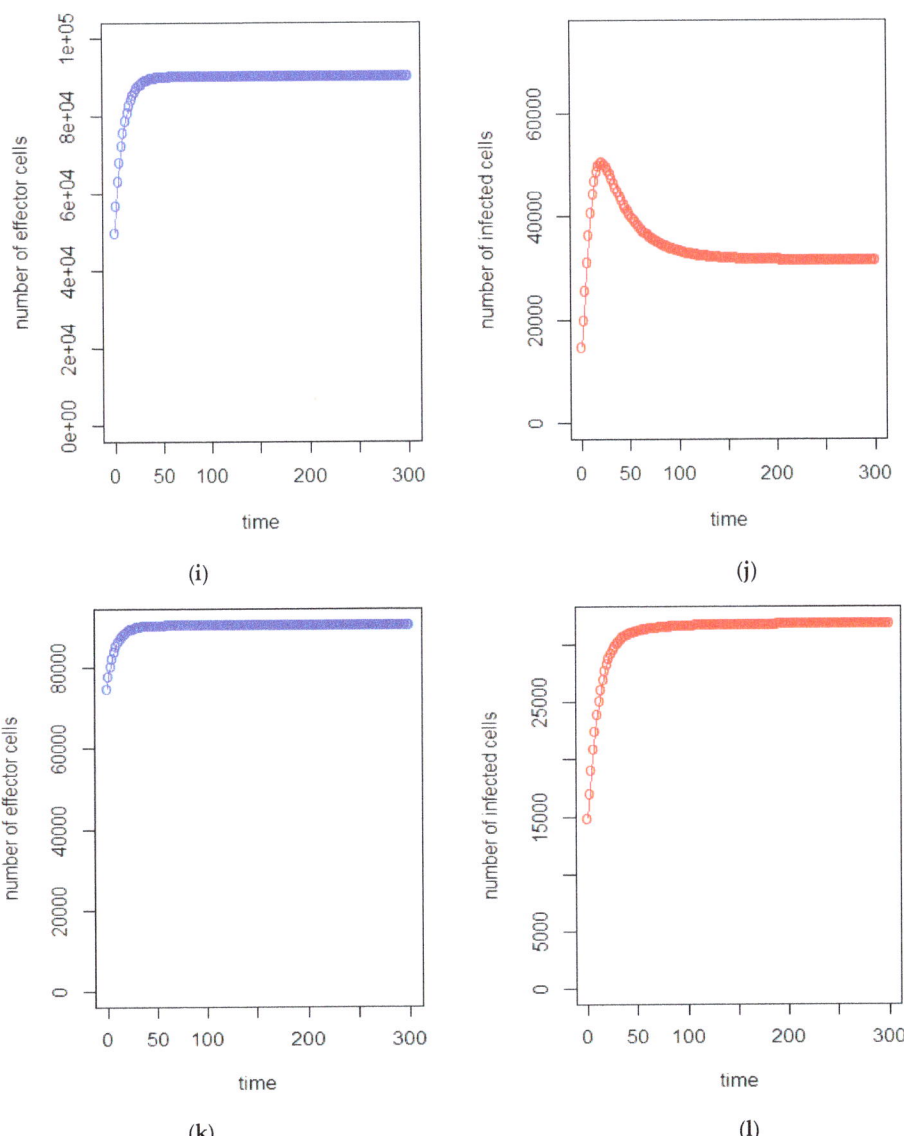

Figure 1. *Cont.*

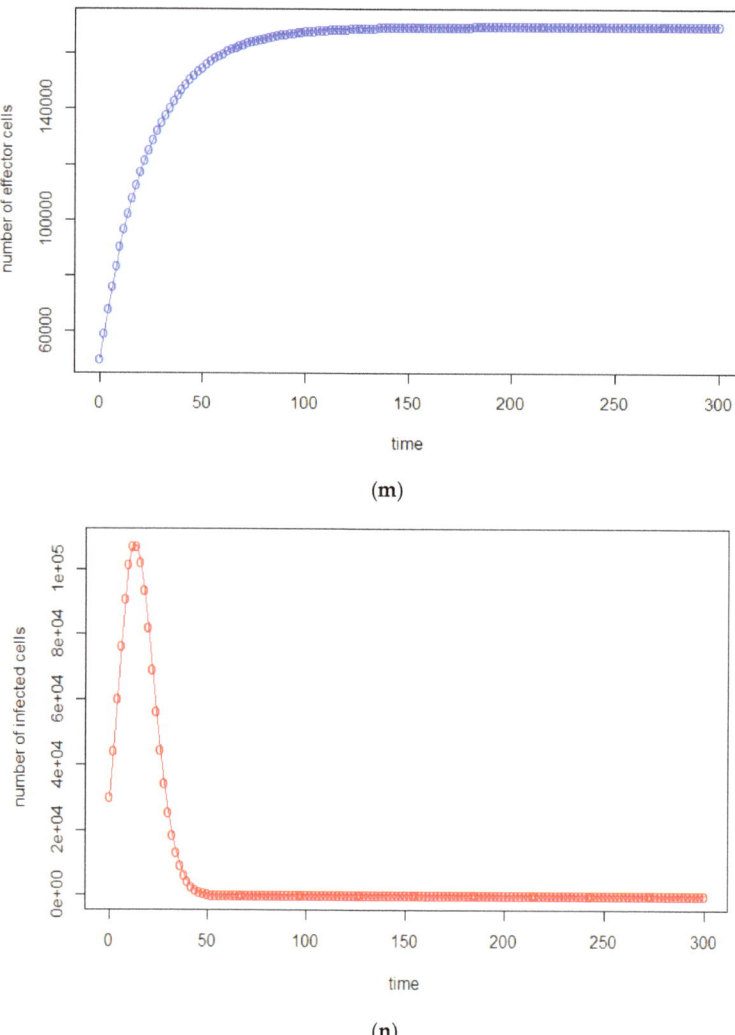

Figure 1. (**a**,**b**): The immune-effector cells (**a**) and virus-infected cells (**b**) vs. time (days), (note: (**a**) is on the left; (**b**) is on the right); (**c**,**d**): The relationship between the immune-effector cells and virus-infected cells.; (**e**,**f**): 3-D relationships of the effector cells, immune-effector cells and unit of time (days); (**g**,**h**): The immune-effector cells (**g**) and virus-infected cells (**h**) vs. time (days). The same as Case 1, except I_0 = 75,000 cellsvs. time (days); (**i**,**j**): The immune-effector cells (**i**) and virus-infected cells (**j**) vs. time (days). The same as Case 1, except V_0 = 15,000 cells; (**k**,**l**): The immune-effector cells (**k**) and virus-infected cells (**l**) vs. time (days). The same as Case 1, except V_0 = 15,000 cells and I_0 = 75,000 cells; (**m**): The immune-effector cells vs. time (days); (**n**): The virus-infected cells vs. time (days).

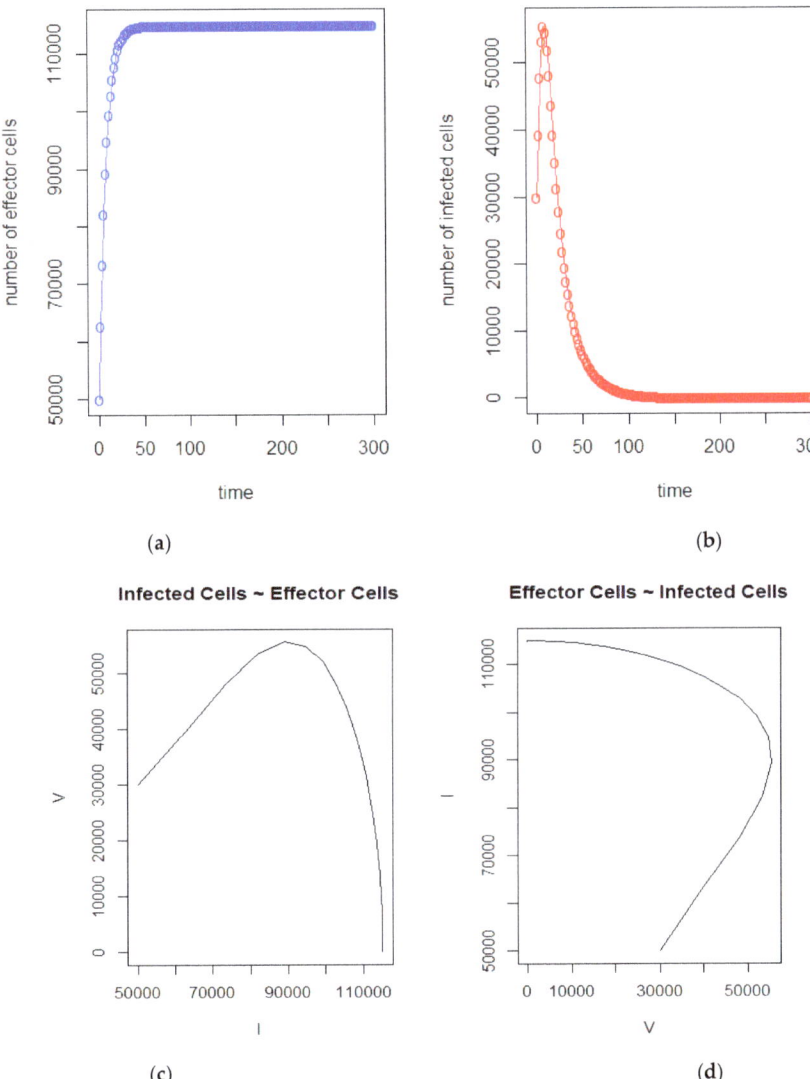

Figure 2. *Cont.*

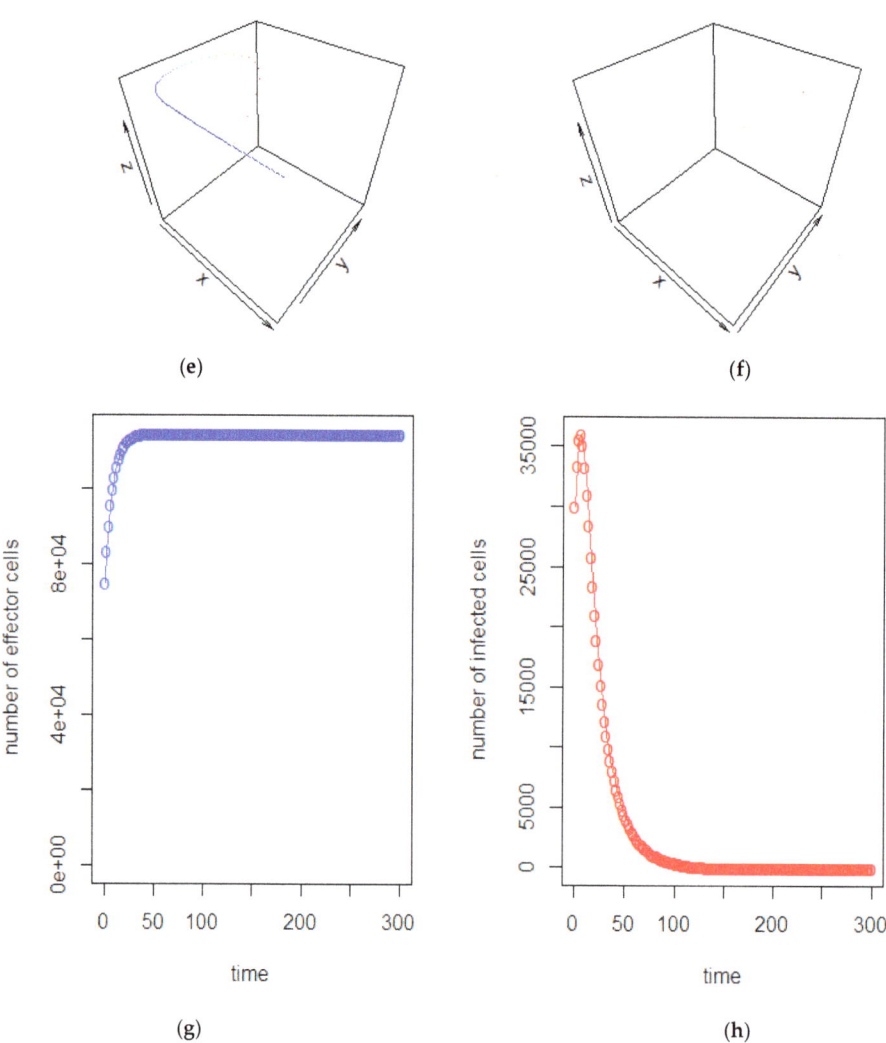

Figure 2. *Cont.*

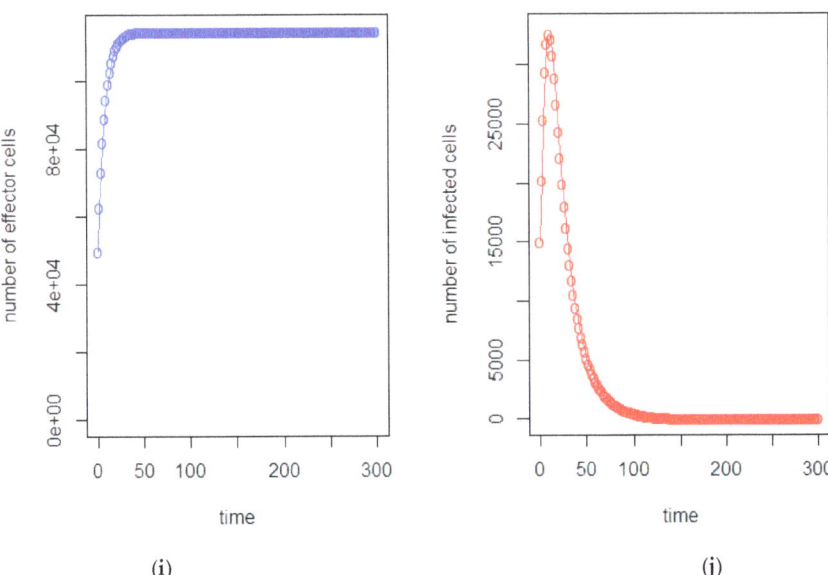

(i) (j)

Figure 2. (**a**,**b**): The immune-effector cells (**a**) and virus-infected cells (**b**) vs. time (days); (**c**,**d**): The relationship between the immune-effector cells and virus-infected cells; (**e**,**f**): 3D relationships of the effector cells, immune-effector cells and unit of time (days); (**g**,**h**): The immune-effector cells (**g**) and virus-infected cells (**h**) vs. time (days). The same as Case 2, except I_0 = 75,000 cells; (**i**,**j**): The immune-effector cells (**i**) and virus-infected cells (**j**) vs. time (days). The same as Case 2, except V_0 = 15,000 cells.

Case 3: This is the same as Case 1 (i.e., $b = 43 \times 10^{-7}$, $s = 7000$), except $a = 0.043$.

From Figure 3a,b, we can observe that the initial number of virus-infected cells and immune-effector cells are 30,000 and 50,000, respectively, as expected. It should be noted that the number of immune-effector cells (see Figure 3a) keeps increasing but starts to slowly stabilize after the 50th day at the level of 90,310, where the virus-infected count (see Figure 3b) starts to decrease sharply until it reaches the virus free stateat $(V,I) = (0, 90,310)$ after the 50th day.

The result is about the same, even when the initial number of immune effector cells is I_0 = 75,000, see Figure 3c,d. The result is also about the same, even when the initial number of virus-infected cells is reduced to V_0 = 15,000 (see Figure 3e,f). It is worth noting that the initial number of virus-infected cells and immune-effector cells do not influence the end result.

Case 4: The same as Case 3 (i.e., $b = 43 \times 10^{-7}$, $a = 0.043$), except s = 10,000 (instead of s = 7000).

From Figure 4a,b, we can observe that the initial number of virus-infected cells and immune-effector cells are 30,000 and 50,000, respectively. It should be noted that the number of immune-effector cells (see Figure 4a) keeps increasing but starts to slowly stabilize after the 40th day at the level of 114,416 where the virus-infected count (see Figure 4b) starts to decrease significantly until it reaches the virus free state at $(V,I) = (0, 114,416)$ after the 40th day. The result is about the same, even when the initial number of immune effector cells is I_0 = 75,000, see Figure 4c,d. The result is also about the same, even when the initial number of virus-infected cells is reduced to V_0 = 15,000 (see Figure 4e,f). It is worth noting that the initial number of virus-infected cells and immune-effector cells do not influence the end result.

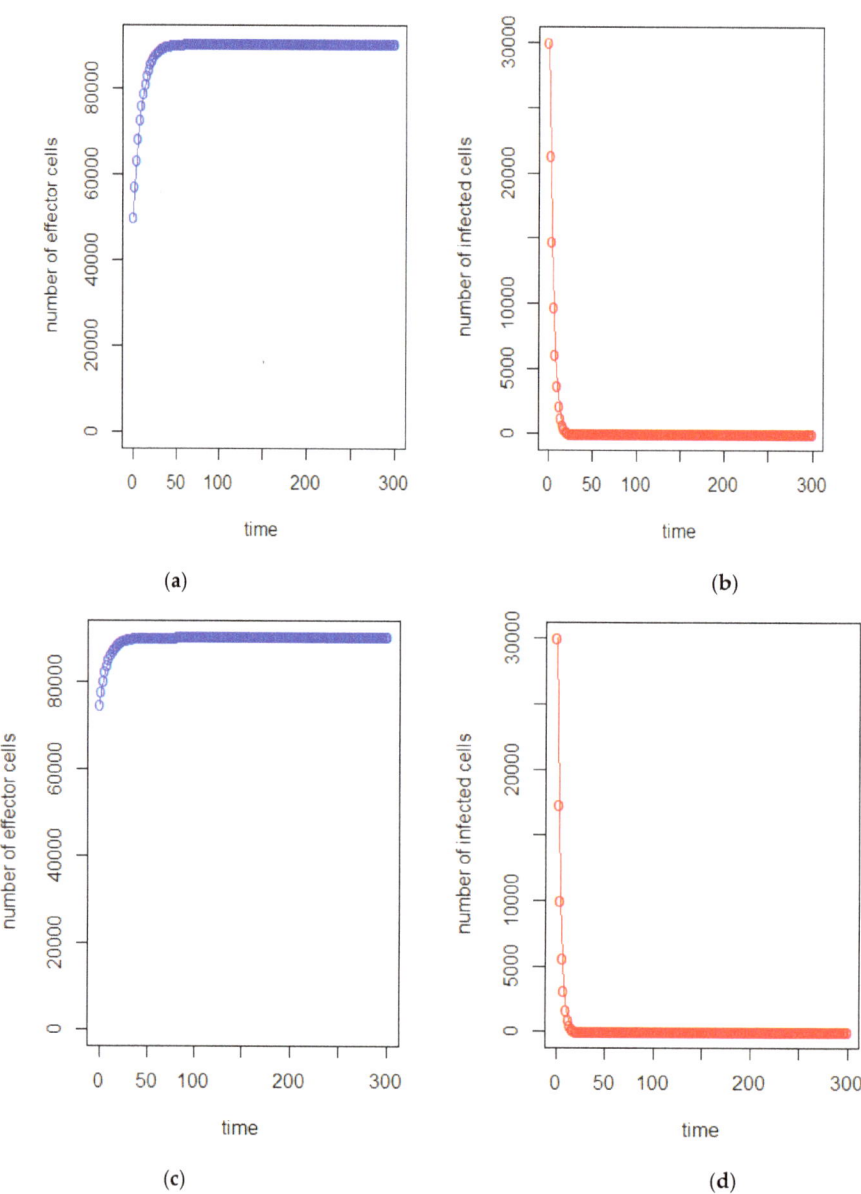

Figure 3. *Cont.*

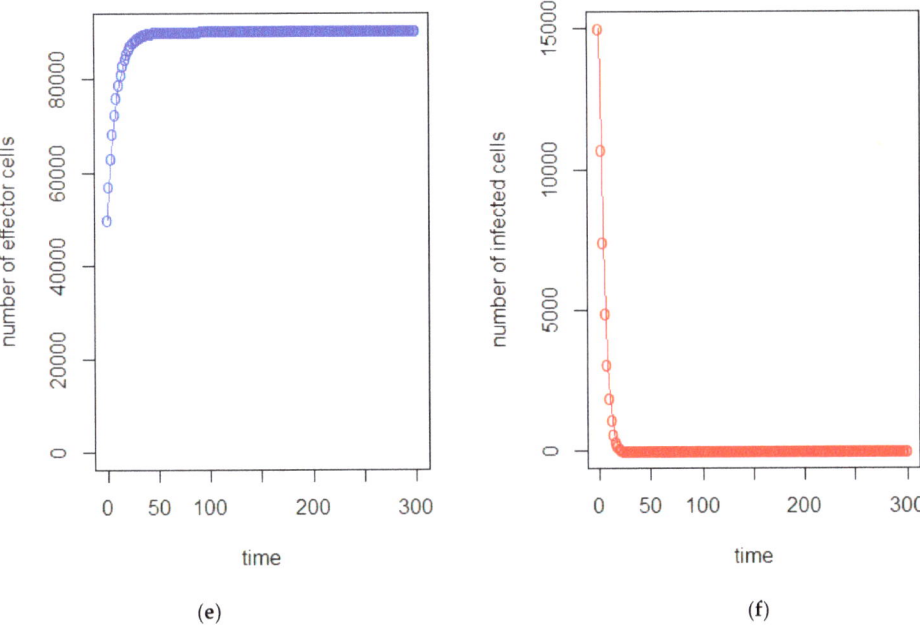

(e)   (f)

Figure 3. (**a**,**b**): The immune-effector cells (**a**) and virus-infected cells (**b**) vs. time (days); (**c**,**d**): The immune-effector cells (**c**) and virus-infected cells (**d**) vs. time (days). The same as Case 3, except I_0 = 75,000 cells; (**e**,**f**): The immune-effector cells (**e**) and virus-infected cells (**f**) vs. time (days). The same as Case 3, except V_0 = 15,000 cells.

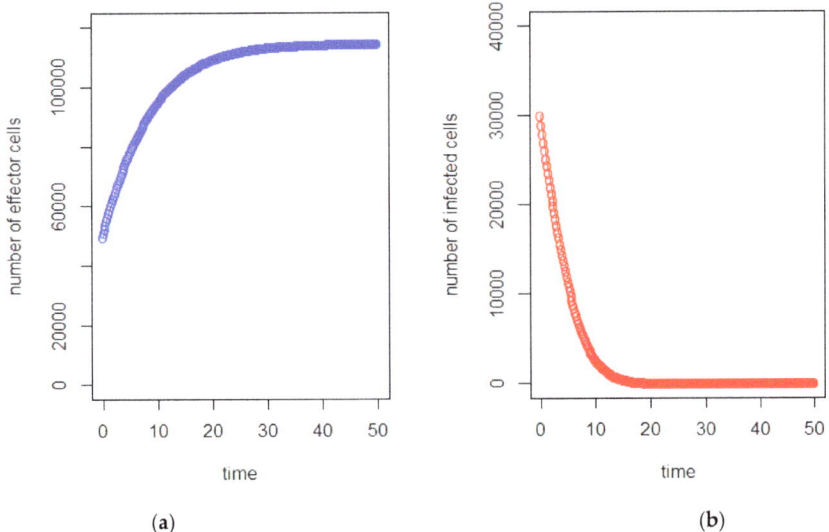

(a)   (b)

Figure 4. *Cont.*

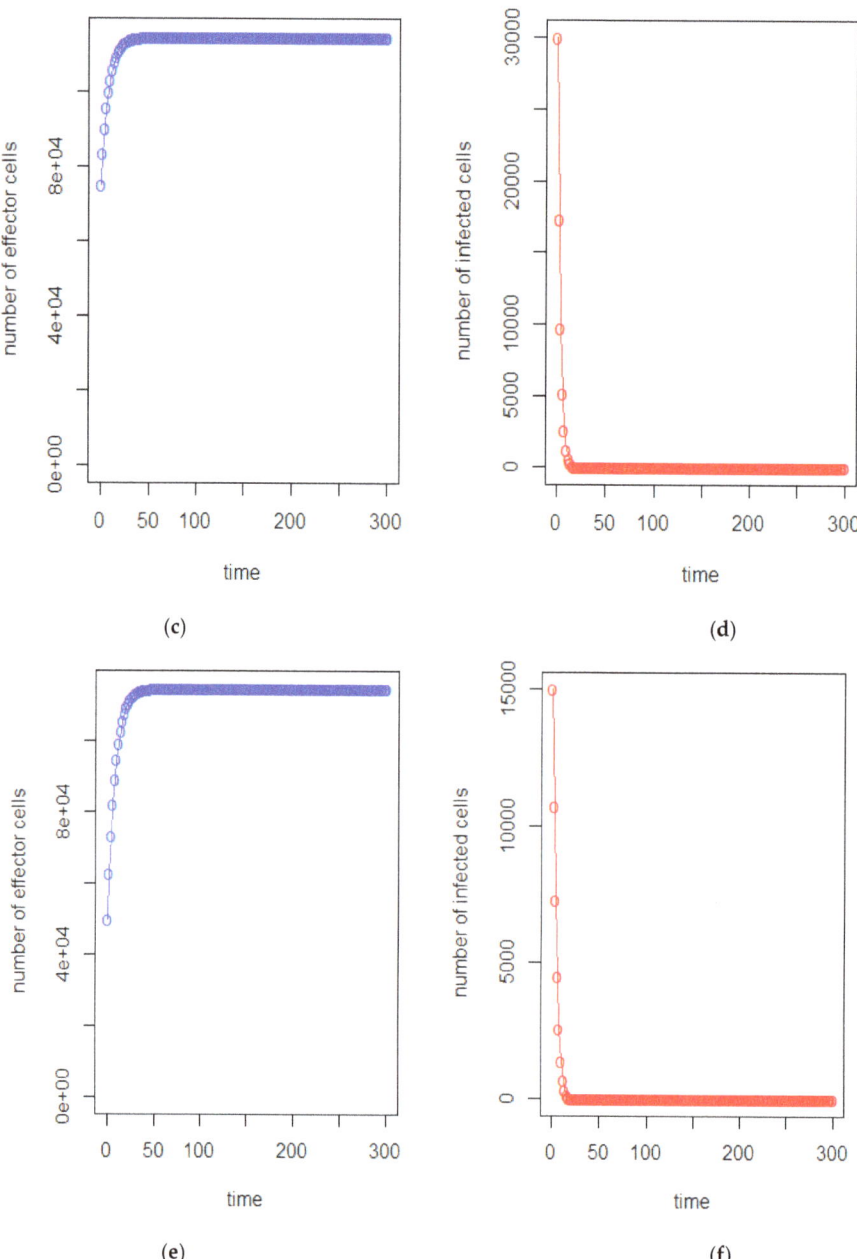

Figure 4. (**a**,**b**): The immune-effector cells (**a**) and virus-infected cells (**b**) vs. time (days); (**c**,**d**): The immune-effector cells (**c**) and virus-infected cells (**d**) vs. time (days). The same as Case 4, except I_0 = 75,000 cells; (**e**,**f**): The immune-effector cells (**e**) and virus-infected cells (**f**) vs. time (days). The same as Case 4, except V_0 = 15,000 cells.

Case 5: Same as Case 1 (i.e., $s = 7000$), except $b = 4.3 \times 10^{-4}$.

Table 2 below shows the parameter values that we will use to analyze here, the same as Case 1 except the value b is 4.3×10^{-4}. Any other sets of parameter values can be easily applied from the model.

Table 2. Model parameter values.

$a = 0.43$/day	$b = 4.3 \times 10^{-4}$/(cells·day)	$c = 4.12 \times 10^{-2}$ 0.0412/day
$d = 15 \times 10^{-5}$/day	$e = 4 \times 10^{-8}$/day	$f = 0.0000004$/day
$g = 3 \times 10^{-6}$/day	$h = 20.2$ (cells)	$k = 10^5$/cells
$m = 2 \times 10^{-11}$/day	$p = 341 \times 10^{-12}$/cell	$s = 7000$ cells/day

Note that b = the elimination rate of the virus-infected cells by the healthy immune system (effector cells).

From Figure 5a,b, we can observe that the initial number of virus-infected cells and immune-effector cells are 30,000 and 50,000, respectively. It should be noted that the number of immune-effector cells (see Figure 5a) keeps increasing but starts to slowly stabilize after the 30th day at the level of 89,920, where the virus-infected count (see Figure 5b) starts to decrease significantly right after the first day and it quickly reaches the virus free state at $(V,I) = (0, 60,467)$ after the third day.

The result is about the same, even when the initial number of immune effector cells is $I_0 = 75,000$, see Figure 5c,d. The result is also about the same, even when the initial number of virus-infected cells is reduced to $V_0 = 15,000$ (see Figure 5e,f). It is worth noting that the initial number of virus-infected cells and immune-effector cells do not influence the end result.

Case 6: Same as Case 5, except $s = 10,000$ cells/day.

From Figure 6a,b, we can observe that the initial number of virus-infected cells and immune-effector cells are 30,000 and 50,000, respectively. It should be noted that the number of immune-effector cells (see Figure 6a) keeps increasing but starts to slowly stabilize after the 30th day at the level of 113,310, where the virus-infected counts (see Figure 6b) starts to decrease significantly right after the firstday and it quickly reaches the virus free state at $(V,I) = (0, 68,680)$ after the thirdday.

The result is about the same, even when the initial number of immune effector cells is as $I_0 = 75,000$, see Figure 6c,d. The result is also about the same, even when the initial number of virus-infected cells is reduced to $V_0 = 15,000$ (see Figure 6e,f). It is worth noting that the initial number of virus-infected cells and immune-effector cells do not influence the end result.

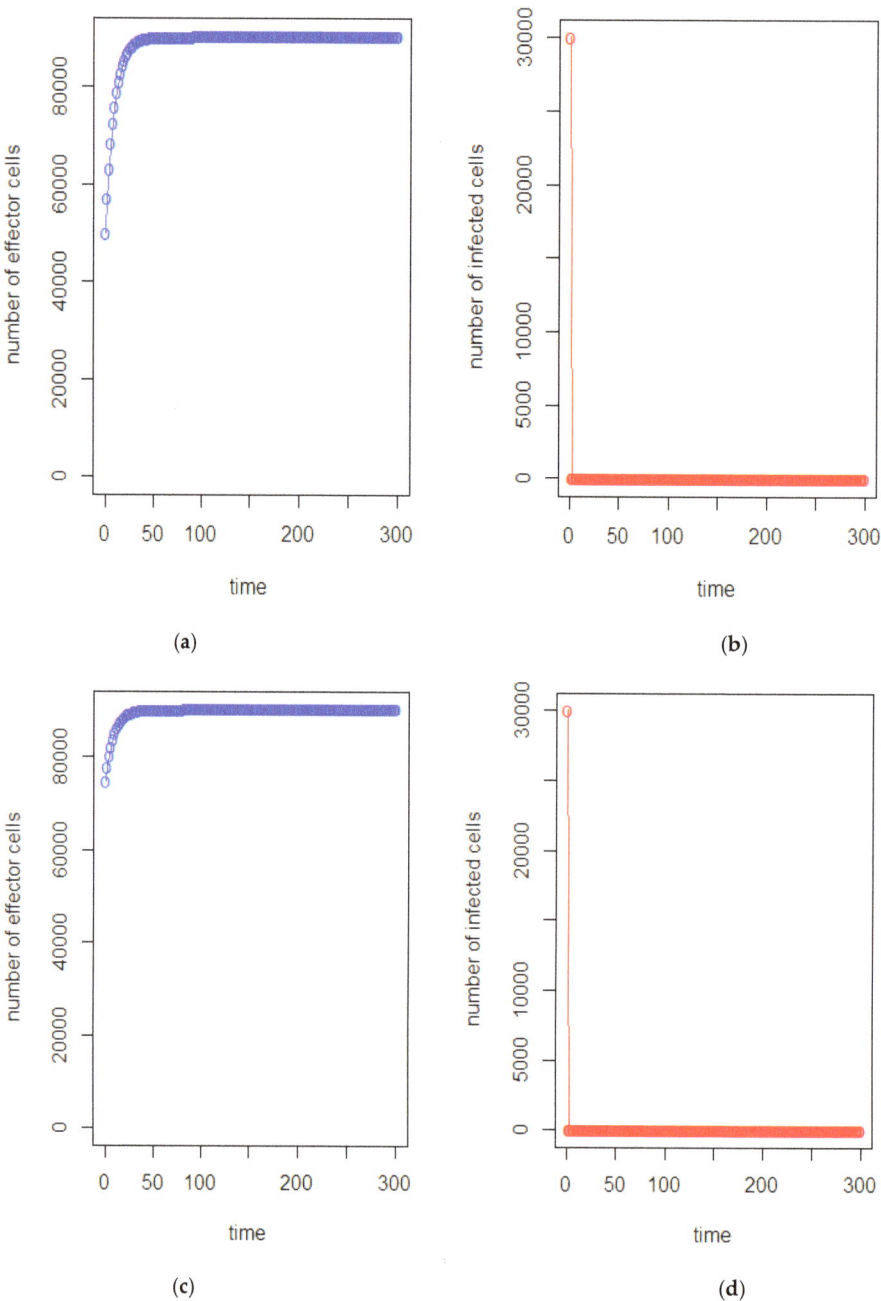

Figure 5. *Cont.*

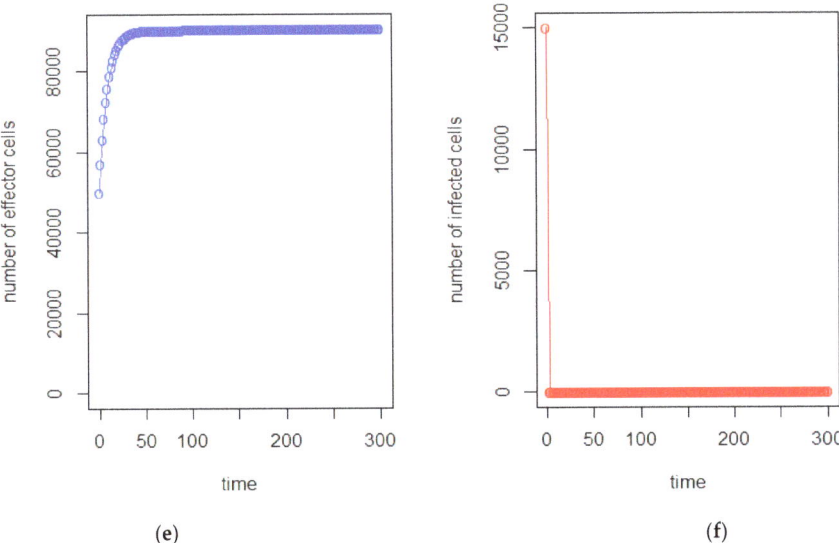

(e) (f)

Figure 5. (**a**,**b**): The immune-effector cells (**a**) and virus-infected cells (**b**) vs. time (days); (**c**,**d**):When the initial number of immune-effector cells I_0 = 75,000 cells. The immune-effector cells (**c**) and virus-infected cells (**d**) vs. time (days); (**e**,**f**): When the initial number of virus-infected cells V_0 = 15,000 cells. The immune-effector cells (**e**) and virus-infected cells (**f**) vs. time (days).

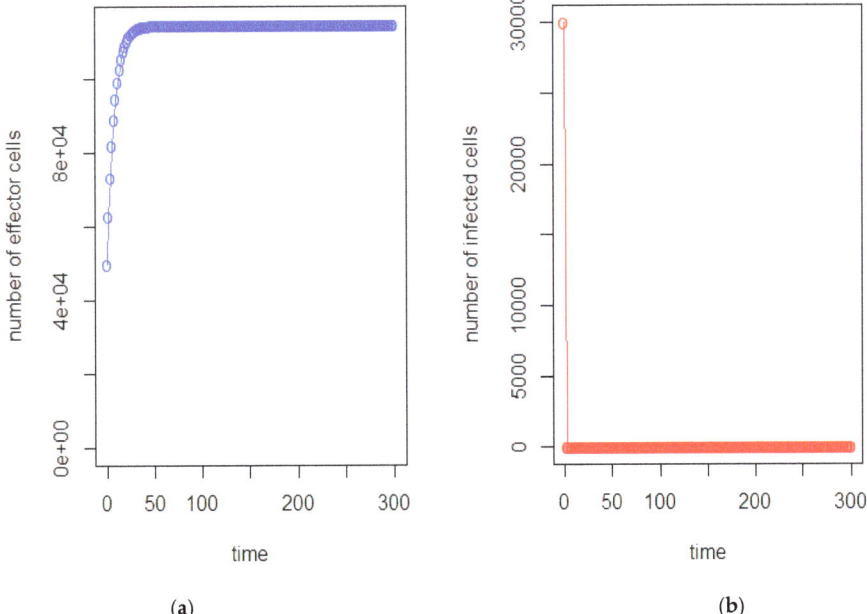

(a) (b)

Figure 6. *Cont.*

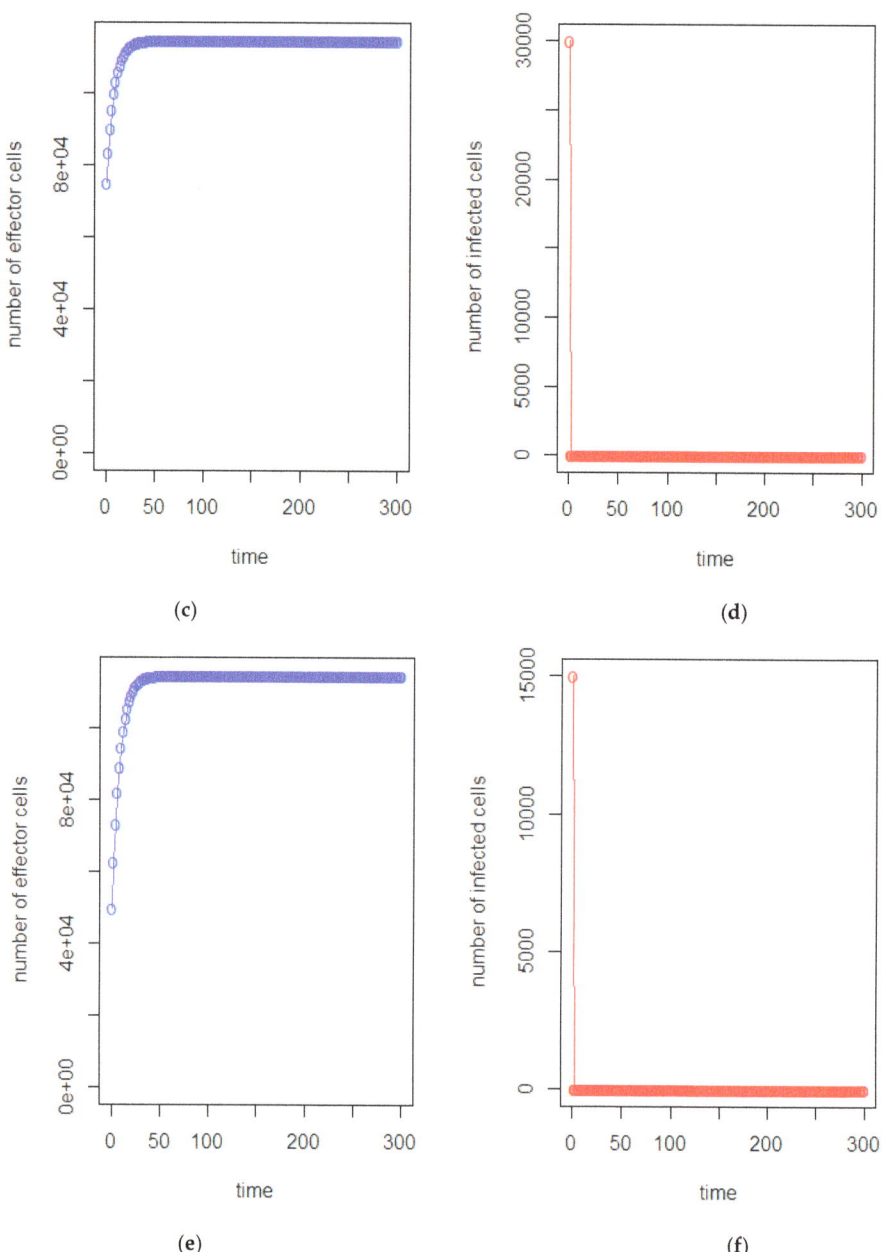

Figure 6. (**a**,**b**): The immune-effector cells (**a**) and virus-infected cells (**b**) vs. time (days); (**c**,**d**): When I_0 = 75,000 cells. The immune-effector cells (**c**) and virus-infected cells (**d**) vs. time (days); (**e**,**f**): When V_0 = 15,000 cells. The immune-effector cells (**e**) and virus-infected cells (**f**) vs. time (days).

4. Conclusions

This paper discusses a mathematical model of the body's immune system, considering the multiple time-delay interactions between the immune cells and virus-infected cells with an autoimmune disease using the delay partial differential equations. The model can be used to determine the dynamic progression of virus-infected cell growth and observe the patterns of how the virus-infected cells spread in the body's immune system with respect to time delays. The model can be used to predict when the virus-infected free state can be reached as the time progresses as well as the number of body's immune cells as any given time. From the numerical examples, we observe that the initial number of virus-infected cells and immune-effector cells that are needed to obtain the solutions of the delay partial differential equations do not influence the end results. We plan to broaden our model in a near future by considering the chemotherapy drug treatment subject to the time delays.

Funding: This research received no external funding.

Conflicts of Interest: The author declares no conflict of interest.

References

1. Thompson, E. The immune system. *J. Am. Med. Assoc. JAMA* **2015**, *313*, 16. [CrossRef]
2. CRI Staff, How Does the Immune System Work? 30 April 2019. Available online: https://www.cancerresearch.org/blog/april-2019/how-does-the-immune-system-work-cancer?gclid=CjwKCAjwsNiIBhBdEiwAJK4khgG7w-9Ugv3HMGc1VWZbRuhFBOfIPMW2Qo3Dv1-VH1HGvmJruZjwxxoC3HYQAvD_BwE (accessed on 15 May 2021).
3. Guide to Your Immune System. 2020. Available online: https://www.webmd.com/cold-and-flu/ss/slideshow-immune-system (accessed on 20 May 2021).
4. Immune System, Cleveland Clinic. 2021. Available online: https://my.clevelandclinic.org/health/articles/21196-immune-system (accessed on 10 May 2021).
5. Newman, T. How the Immune System Works. *Medical News Today*. 11 January 2018. Available online: https://www.medicalnewstoday.com/articles/320101 (accessed on 20 May 2021).
6. How Does the Immune System Work? 2020. Available online: https://www.ncbi.nlm.nih.gov/books/NBK279364/ (accessed on 20 May 2021).
7. Ng, J.; Stovezky, Y.R.; Brenner, D.J.; Formenti, S.C.; Shuryak, I. Development of a model to estimate the association between delay in cancer treatment and local tumor control and risk of metastates. *JAMA Netw. Open* **2021**, *4*, 1–10. [CrossRef]
8. Talkington, A.; Durrett, R. Estimating tumor growth rates in vivo. *Bull. Math. Biol.* **2015**, *77*, 1934–1954. [CrossRef]
9. Vaghi, C.; Rodallec, A.; Fanciullino, R.; Ciccolini, J.; Mochel, J.P.; Mastri, M. Population modeling of tumor growth curves and the reduced Gompertz model improve prediction of the age of experimental tumors. *PLoS Comput. Biol.* **2020**, *16*, e1007178. Available online: https://journals.plos.org/ploscompbiol/article?id=10.1371/journal.pcbi.1007178 (accessed on 25 April 2021). [CrossRef]
10. Yin, D.J.; Moes, A.R.; van Hasselt, J.G.C.; Swen, J.J.; Guchelaar, H.-J. A Review of Mathematical Models for Tumor Dynamics and Treatment Resistance Evolution of Solid Tumors. *CPT Pharmacomet. Syst. Pharmacol.* **2019**, *8*, 720–737. [CrossRef] [PubMed]
11. Bekiros, S.; Kouloumpou, D. SBDiEM: A new mathematical model of infectious-disease dynamics. *Chaos Solitons Fractals* **2020**, *136*, 109828. [CrossRef]
12. Rothan, H.A.; Byrareddy, S.N. The epidemiology and pathogenesis of coronavirus disease (COVID-19) outbreak. *J. Autoimmun.* **2020**, *2020*, 102433. [CrossRef] [PubMed]
13. Jackson, T.L.; Byrne, H.M. A mathematical model to study the effects of drug resistance and vasculature on the response of solid tumors to chemotherapy. *Math. Biosci.* **2000**, *164*, 17–38. [CrossRef]
14. Meacham, C.E.; Morrison, S.J. Tumour heterogeneity and cancer cell plasticity. *Nature* **2013**, *501*, 328–337. [CrossRef] [PubMed]
15. Sun, X.; Bao, J.; Shaob, Y. Mathematical modeling of therapy-induced cancer drug resistance: Connecting cancer mechanisms to population survival rates. *Sci. Rep.* **2016**, *6*, 22498. [CrossRef]
16. Taniguchi, K.; Okami, J.; Kodama, K.; Higashiyama, M.; Kato, K. Intratumor heterogeneity of epidermal growth factor receptor mutations in lung cancer and its correlation to the response to gefitinib. *Cancer Sci.* **2008**, *99*, 929–9354. [CrossRef] [PubMed]
17. Al-Huniti, N.; Feng, Y.; Yu, J.; Lu, Z.; Nagase, M.; Zhou, D.; Sheng, J. Tumor growth dynamic modeling in oncology drug development and regulatory approval: Past, present, and future Opportunities. *CPT Pharmacomet. Syst. Pharmacol.* **2020**, *9*, 419–427. [CrossRef] [PubMed]
18. Cai, Y.; Kang, Y.; Wang, W. A stochastic SIRS epidemic model with nonlinear incidence rate. *Appl. Math. Comput.* **2017**, *305*, 221–240. [CrossRef]
19. Cai, Y.; Jiao, J.; Gui, Z.; Liu, Y.; Wang, W. Environmental variability in a stochastic epidemic model. *Appl. Math. Comput.* **2018**, *329*, 210–226. [CrossRef]
20. Farajzadeh, M.; Doust, F.; Haghighifar, D.; Baleanu, D. The stability of gauss model having one-prey and two-predators. *Abstr. Appl. Anal.* **2012**, *2012*, 219640. [CrossRef]
21. He, X.; Zheng, S. Protection zone in a diffusive predator-prey model with Beddington DeAngelis functional response. *J. Math. Biol.* **2016**, *75*, 239–257. [CrossRef]
22. Huang, Y.; Shi, W.; Wei, C.; Zhang, S. A stochastic predator–prey model with Holling II increasing function in the predator. *J. Biol. Dyn.* **2021**, *15*, 1–18. [CrossRef]

23. Jana, S.; Kar, T. Modeling and analysis of a prey-predator system with disease in the prey. *Chaos Solitons Fractals* **2013**, *47*, 42–53. [CrossRef]
24. Kaur, G.; Ahmad, N. On study of immune response to tumor cells in prey-predator system. *Int. Sch. Res. Not.* **2014**, *2014*, 346597. [CrossRef]
25. Bandyopadhyay, M.; Chattopadhyay, J. Ratio-dependent predator-prey model: Effect of environmental fluctuation and stability. *Nonlinearity* **2005**, *18*, 913–936. [CrossRef]
26. Liu, Q.; Jiang, D.; Hayat, T.; Alsaedi, A. Dynamics of a stochastic predator–prey model with stage structure for predator and Holling type II functional response. *J. Nonlinear Sci.* **2018**, *28*, 1151–1187. [CrossRef]
27. Bonate, P.L. Comprehensive overview of tumor growth modeling. In Proceedings of the 9th American Conference of Pharmacometrics (ACOP), San Diego, CA, USA, 7–11 October 2018.
28. Singh, M.; Poonam, K. Qualitative analysis of a predator-prey model in the presence of additional food to predator and constant-yield predator harvesting. *Univ. J. Appl. Math. Comput.* **2019**, *7*, 20–34.
29. Lestari, E.R.; Arifah, H. Dynamics of a mathematical model of cancer cells with chemotherapy. *J. Phys. Conf. Ser.* **2019**, *1320*, 1–8. [CrossRef]
30. Li, S.; Wang, X. Analysis of a stochastic predator–prey model with disease in the predator and Beddington–Deangelis functional response. *Adv. Differ. Equ.* **2015**, *2015*, 224. [CrossRef]
31. Liming, X.C.; Li, M.; Guo, B. Stability analysis of an HIV/AIDS epidemics model with treatment. *J. Comput. Appl.* **2009**, *229*, 313–323.
32. Wahyuda, C.N.; Lestari, D. Local stability of AIDS epidemic model through treatment and vertical transmission with time delay. *J. Phys. Conf. Ser.* **2016**, *693*, 1–10.
33. Alho, J.M. Forecasting life expectancy: A statistical look at model choice and use of auxiliary series. In *Old and New Perspectives on Mortality Forecasting. Demographic Research Monographs*; Bengtsson, T., Keilman, N., Eds.; Springer: Berlin/Heidelberg, Germany, 2019.
34. Pham, H. Modeling U.S. mortality and risk-cost optimization on life expectancy. *IEEE Trans. Reliab.* **2011**, *60*, 125–133. [CrossRef]
35. Derouich, M.; Boutayeb, A. An Avian influenzam mathematical model. *Appl. Math. Sci.* **2008**, *2*, 1749–1760.
36. Mukhopadhyaya, B.; Bhattacharyya, R. Dynamics of a delay-diffusion prey-predator Model with disease in the prey. *J. Appl. Math. Comput.* **2005**, *17*, 361–377. [CrossRef]
37. Aparico, J.P.; Chavez, C.C. Mathematical modelling of tuberculosis epidemics. *Math. Biosci. Eng.* **2009**, *6*, 209–237.
38. Chavez, C.C.; Sony, B.J. Dynamical models of tuberculosis and their applications. *Math. Biosci. Eng.* **2004**, *1*, 361–404. [CrossRef]
39. De Pillis, L.G.; Gu, W.; Fister, K.R.; Head, T.; Maples, K.; Murugan, A.; Neal, T.; Yoshida, K. Chemotherapy for tumors: An analysis of the dynamics and a study of quadratic and linear optimal control. *Math. Biosci.* **2006**, *209*, 292–315. [CrossRef]
40. Kuznetsov, V.; Makalkin, I.; Taylor, M.; Perelson, A. Nonlinear dynamics of immunogenic tumors: Parameter estimation and global bifurcation analysis. *Bull. Math. Biol.* **1994**, *56*, 295–321. [CrossRef]
41. Tsygvintsev, A.; Marino, S.; Kirschner, D.E. A Mathematical Model of Gene Therapy for The Treatment of Cancer. In *Mathematical Models and Methods in Biomedicine*; Springer: New York, NY, USA, 2012; pp. 357–373.
42. Waziri, A.S.; Estomih, S.; Oluwole, D. Mathematical modelling of HIV/AIDS dynamic with treatment and vertical transmission. *Appl. Math.* **2012**, *3*, 77–89. [CrossRef]
43. Wang, Y.; Kuang, C.; Ding, S.; Zhang, S. Stability and bifurcation of a stage-structured predator-prey model with both discete and distributed delays. *Chaos Solitons Fractals* **2013**, *46*, 19–27. [CrossRef]
44. Kumar, S.; Chattopadh, S. A bioeconomic model of two equally dominated prey and one predator system. *Mod. Appl. Sci.* **2010**, *4*, 84–96. [CrossRef]
45. Liu, M. Dynamics of a stochastic regime-switching predator-prey model with modified Leslie–Gower Holling-type II schemes and prey harvesting. *Nonlinear Dyn.* **2019**, *96*, 417–442. [CrossRef]
46. Liu, M.; Bai, C. Dynamics of a stochastic one-prey two-predator model with Lévy jumps. *J. Comput. Appl. Math.* **2016**, *284*, 308–321. [CrossRef]
47. Liu, M.; Bai, C.; Deng, M.; Du, B. Analysis of stochastic two-prey one-predator model with Lévy jumps. *Physica A* **2016**, *445*, 176–188. [CrossRef]
48. Tripathi, J.; Abbas, S.; Thakur, M. Local and global stability analysis of a two prey one predator model with help. *Commun. Nonlinear Sci.* **2014**, *19*, 3284–3297. [CrossRef]
49. Pang, P.; Wang, M. Strategy and stationary pattern in a three-species predator–prey model. *J. Differ. Equ.* **2004**, *200*, 245–273. [CrossRef]
50. Jiao, J.; Wang, R.; Chang, H.; Liu, X. Codimension bifurcation analysis of a modified Leslie–Gower predator-prey model with two delays. *Int. J. Bifurcat. Chaos* **2018**, *28*, 1850060. [CrossRef]
51. Nie, L.; Peng, J.; Teng, Z.; Hu, L. Existence and stability of periodic solution of a Lotka–Volterra predator-prey model with state dependent impulsive effects. *J. Comput. Appl. Math.* **2009**, *224*, 544–555. [CrossRef]
52. Tian, Y.; Sun, K.; Chen, L. Comment on existence and stability of periodic solution of a Lotka–Volterra predator–prey model with state dependent impulsive effects. *J. Comput. Appl. Math.* **2010**, *234*, 2916–2923. [CrossRef]
53. Law, G.R. What do epidemiologists mean by 'population mixing'? *Pediatric Blood Cancer* **2008**, *51*, 155–160. [CrossRef]

54. Haque, M.; Zhen, J.; Ventrino, E. An eco-epidemiological predator-prey model with standard disease incidence. *Math. Methods Appl. Sci.* **2008**, *35*, 875–898.
55. Jiang, D. Analysis of a predator-prey model with disease in the prey. *Int. J. Biomath.* **2013**, *6*, 1350012.
56. Naji, R.K.; Mustafa, A.N. The dynamics of an eco-epidemiological model with nonlinear incidence rate. *J. Appl. Math.* **2012**, *11*, 853–862. [CrossRef]
57. Pal, P.; Haque, M.; Mandal, P. Dynamics of a predator–prey model with disease in the predator. *Math. Methods Appl. Sci.* **2013**, *37*, 2429–2450. [CrossRef]
58. Xiao, Y.; Chen, L. Modeling and analysis of a predator–prey model with disease in the prey. *Math. Biosci.* **2001**, *171*, 59–82. [CrossRef]
59. Xu, R.; Zhang, S. Modelling and analysis of a delayed predator–prey model with disease in the predator. *Appl. Math. Comput.* **2013**, *224*, 372–386. [CrossRef]
60. Battegay, M. 2019-novel Coronavirus (2019-nCoV): Estimating the case fatality rate—A word of caution. *Swiss Med. Wkly.* **2020**, *150*, 506. [CrossRef]
61. Chin, W.H.; Chu, J.T.S.; Perara, M.R.A.; Hui, K.P.Y.; Yen, M.C.W.; Chan, M.; Paris, M.C.W.; Poon, L.L.M. Stability of SARS-CoV-2 in different environmental conditions. *Lancet Mircobe* **2020**, *20*, e145. [CrossRef]
62. Kucharski, J. Early dynamics of transmission and control of COVID-19: A mathematical modelling study. *Lancet Infect. Dis.* **2020**, *20*, 553–558. [CrossRef]
63. Pham, H. On estimating the number of deaths related to Covid-19. *Mathematics* **2000**, *8*, 655. [CrossRef]
64. Pham, H. Estimating the COVID-19 death toll by considering the time-dependent effects of various pandemic restrictions. *Mathematics* **2020**, *8*, 1628. [CrossRef]
65. Pham, H.; Pham, D.H. A novel generalized logistic dependent model to predict the presence of breast cancer based on biomarkers. *Concurr. Comput. Pract. Exp.* **2020**, *32*, 1. [CrossRef]

Article

Reinitializing Sea Surface Temperature in the Ensemble Intermediate Coupled Model for Improved Forecasts

Sittisak Injan [1], Angkool Wangwongchai [1], Usa Humphries [1,*], Amir Khan [2] and Abdullahi Yusuf [3,4]

[1] Department of Mathematics, Faculty of Science, King Mongkut's University of Technology Thonburi, Bangkok 10140, Thailand; sittisak.inja@mail.kmutt.ac.th (S.I.); angkool.wan@kmutt.ac.th (A.W.)
[2] Department of Mathematics and Statistics, University of Swat, Charbagh 25000, Khyber Pakhtunkhwa, Pakistan; amirkhan@uswat.edu.pk
[3] Department of Computer Engineering, Biruni University, Istanbul 34010, Turkey; yusufabdullahi@fud.edu.ng
[4] Department of Mathematics, Federal University Dutse, Jigawa Gida Sitin, Dutse 720101, Nigeria
* Correspondence: usa.wan@kmutt.ac.th; Tel.: +66-2470-8722

Citation: Injan, S.; Wangwongchai, A.; Humphries, U.; Khan, A.; Yusuf, A. Reinitializing Sea Surface Temperature in the Ensemble Intermediate Coupled Model for Improved Forecasts. *Axioms* **2021**, *10*, 189. https://doi.org/10.3390/axioms10030189

Academic Editors: Davron Aslonqulovich Juraev and Samad Noeiaghdam

Received: 21 April 2021
Accepted: 9 August 2021
Published: 17 August 2021

Publisher's Note: MDPI stays neutral with regard to jurisdictional claims in published maps and institutional affiliations.

Copyright: © 2021 by the authors. Licensee MDPI, Basel, Switzerland. This article is an open access article distributed under the terms and conditions of the Creative Commons Attribution (CC BY) license (https://creativecommons.org/licenses/by/4.0/).

Abstract: The Ensemble Intermediate Coupled Model (EICM) is a model used for studying the El Niño-Southern Oscillation (ENSO) phenomenon in the Pacific Ocean, which is anomalies in the Sea Surface Temperature (SST) are observed. This research aims to implement Cressman to improve SST forecasts. The simulation considers two cases in this work: the control case and the Cressman initialized case. These cases are simulations using different inputs where the two inputs differ in terms of their resolution and data source. The Cressman method is used to initialize the model with an analysis product based on satellite data and in situ data such as ships, buoys, and Argo floats, with a resolution of 0.25 × 0.25 degrees. The results of this inclusion are the Cressman Initialized Ensemble Intermediate Coupled Model (CIEICM). Forecasting of the sea surface temperature anomalies was conducted using both the EICM and the CIEICM. The results show that the calculation of SST field from the CIEICM was more accurate than that from the EICM. The forecast using the CIEICM initialization with the higher-resolution satellite-based analysis at a 6-month lead time improved the root mean square deviation to 0.794 from 0.808 and the correlation coefficient to 0.630 from 0.611, compared the control model that was directly initialized with the low-resolution in-situ-based analysis.

Keywords: cressman method; EICM; ENSO; SSTA

MSC: 37M05; 37N10; 65K05

1. Introduction

The El Niño–Southern Oscillation (ENSO) is used to describe the Sea Surface Temperature Anomaly (SSTA) in the equatorial Pacific Ocean and ocean–atmosphere system fluctuations in the Southern Hemisphere. Scientists now use the term ENSO warm event to describe the phenomenon where the SST in the eastern and central parts of the Pacific region is warmer than normal, while the term ENSO cold event is now used to describe the phenomenon where the SST in the central and eastern parts of the Pacific region is colder than normal. Many countries in the world are affected by these two phenomena, especially countries in the equatorial parts of the Pacific Ocean. The ENSO is also associated with abnormal climatic conditions, leading to droughts in southern Africa and other areas of the Southern Hemisphere, such as Australia; for example, the Australian continent experienced a drought in 1997 as a result of the ENSO phenomenon. At present, the hot weather in Australia is believed to be the cause of forest fires in Victoria and New South Wales. Southeast Asia, comprising Indonesia, the Philippines, Malaysia, Singapore, Brunei, and Papua New Guinea, experienced the greatest incidence of forest fires in 1997–1998. Moreover, other countries in the region, such as Thailand, Laos, Cambodia, and Vietnam, suffered from drought conditions at this time. The ENSO has been identified as the dominant cause of climate variability around the equatorial Pacific Ocean. It connects the air circulation in the atmosphere with the temperature of water flowing into the Pacific Ocean. International research has shown that the ENSO phenomenon affects more than 70% of the global temperature, although it occurs in the Pacific Ocean.

The modelling of ENSO phenomena has improved, in terms of prediction skills, to within a range of 12 months in advance, based on analyses of the relationships between the atmosphere and ocean. Several studies have been conducted to predict ENSO phenomena using different methods [1–6]. Studies have reviewed the efficacy of many models, in an attempt to rule out changes related to ENSO phenomena [7]. The Hybrid Oceanic and Atmospheric System Model (HCM) has been studied to explain climate variability in the tropical Pacific Ocean system [8]. An intermediate coupled model (ICM) has been studied and developed with a variety of methods, in order to improve ENSO forecasting results [9]. Scientists in the Institute of Oceanology, Chinese Academy of Sciences, have studied the evolution of the SST in the tropical Pacific Ocean, as predicted using the IOCAS ICM model. A unique feature is how the temperature of the sub-surface water, entrained into the mixed layer, is parameterized [10]. SST data have been used to predict ENSO phenomena, as an essential geophysical variable that can act as a predictor of atmospheric conditions [1]. The simplest model that can be used to predict the ENSO phenomenon is the EICM. The EICM is constructed from an Intermediate Ocean Model (IOM), which seeks to couple the ocean with entrainment temperature, SST, and wind stress in the tropical Pacific Ocean; however, the observation of oceanic data is very difficult, for various reasons [4], and the resulting inaccuracies in the input data result in incorrect ENSO, leading to incorrect assessment of the model status and its predictions [11]. Therefore, it is necessary to find a procedure that can lead to predictions of the model which are in agreement with the observed data. The unstable data problem may not occur if one uses satellite data, as the model grid resolution is lower than that of the satellite data [12]. For the above reason, discovering an optimal method is necessary for improving initial data, to make them consistent with observation data. Hence, the Cressman initialization method may serve as a potential means to provide the initial data in the EICM.

The data assimilation method is a technique of statistical combination that combines the forecasted result with the initial observation data. This technique is used to correct the initial data that are to be fed into the EICM [13,14]. The process of data assimilation between oceanic and atmospheric improved the El Nino forecasts compared to the forecasting result without data assimilation [15]. The Cressman method has been used to correct the SST data when there are difficulties in measuring the temperatures at exact locations and exact times over vast areas, with satellite-measured observations of sea surface temperature from the MODIS Aqua spectroradiometer [16]. The Cressman method may improve results slightly compared to other methods but is suitable for SST, as shown by [16,17]. Artur et al. (2015) uses Cressman, but applies it to satellite-measured sea surface temperature from the MODIS Aqua spectroradiometer, using a coupled ecosystem model [16]. This procedure provides more correct input data, which may lead to more accurate forecasts and more reliable predictions [18]. Therefore, this work aims to improve the SSTA prediction of the ENSO phenomenon with EICM, using the Cressman initialization method. The Optimum Interpolation Sea Surface Temperature (OISST) data from the Advanced Very High-Resolution Radiometer (AVHRR) was analyzed through the data assimilation process to be used as an input of EICM.

2. Materials and Methods

2.1. Ensemble Intermediate Coupled Model

The EICM was developed from the ICM, in order to improve ENSO phenomenon forecasting results [13], using a different method to generate initial ensemble members with the Markov stochastic random model. Furthermore, the studies of Evensen [19,20] provided a set of initial conditions for an ICM with 100 members. The EICM consists of three main parts: the IOM, the anomaly model for T_e and the wind stress [21]. This model has been used to predict ENSO phenomena in tropical regions of the Pacific Ocean [6,22,23]. The anomaly model for T_e has been implemented using the prediction of Hybrid Coupled Model (HCM) simulations [24,25] in ENSO. The EICM framework is shown in Figure 1. Keenlyside and Kleeman [26] developed an IOM model to predict the upper ocean currents near the equator, where the model was based on the Baroclinic Euclidean model [27]. These ocean models are able to simulate the variance in SST over the year in several ways. The role of SST in the ENSO has been widely accepted, especially in the eastern Pacific. The SST variance is regulated by zonal and meridional advection and entrainment processes [28]. Zonal currents play important roles in the calculation of the SST in the central Pacific.

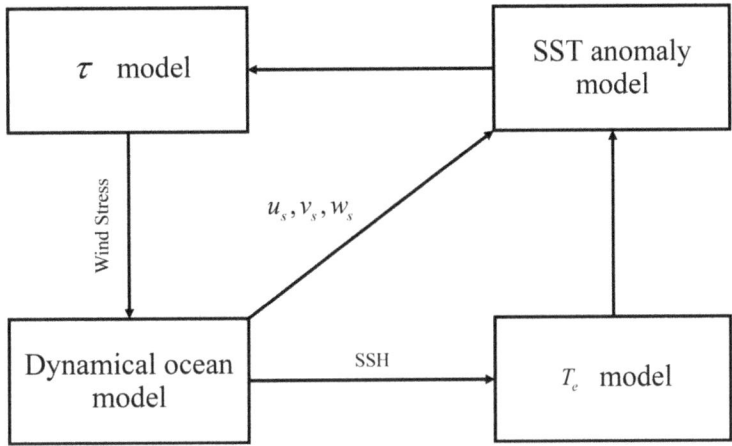

Figure 1. General simple structure of the EICM [29].

The SST component of the model can be formulated using Equation (1). The features of the model that are different from the traditional EICMs are given special attention, which includes simulating the anomalies in the thermocline depth and defining the sub-surface temperature parameters. This model consists of different horizontally blended layers, which serve as the starting point for the various modifications to the traditional EICM. The vertical diffusivity temperature treatment is analogous to the vertical diffusion of momentum in the non-linear component of the model. The equation for SSTA that is implemented in the SST component is written as follows:

$$\begin{aligned}
\frac{\partial T'}{\partial t} = & -u'\frac{\partial \overline{T}}{\partial x} - (\overline{u}+u')\frac{\partial \overline{T}}{\partial x} - v'\frac{\partial \overline{T}}{\partial y} - (\overline{v}+v')\frac{\partial \overline{T}}{\partial y} \\
& -\{(\overline{w}+w')M(-\overline{w}-w') - \overline{w}M(-\overline{w})\}\frac{(\overline{T}_e - \overline{T})}{H} \\
& -(\overline{w}+w')M(-\overline{w}-w')\frac{(\overline{T}_e - \overline{T})}{H} - \alpha T' \\
& +\frac{\kappa_h}{H}\nabla_h \cdot (H\nabla_h T') + \frac{2\kappa_v}{H(H+H_2)}(T'_e - T'),
\end{aligned} \quad (1)$$

where T' and T'_e are the SSTA and water temperature below the mixed layer, respectively; the mean values of the SSTA and the water temperature below the mixed layer are represented by $\overline{T}$ and $\overline{T}_e$, respectively; the parameters u', v', and w' are the corresponding anomaly fields; $\overline{u}$ and $\overline{v}$ are the prescribed seasonally varying mean zonal and meridional currents in the mixed layer, respectively, and $\overline{w}$ is the prescribed seasonally varying mean entrainment velocity at the base of the mixed layer, which are all obtained from the dynamical ocean model; $M(x)$ is the Heaviside step function; $-\alpha T'$ is the surface heat flux term, which is parameterized as being negatively proportional to the local SST anomalies with the thermal damping coefficient; H is the depth of the mixed layer; H_2 is the depth of the second layer [21]; κ_h is the coefficient for horizontal diffusivity; κ_v is the coefficient for vertical diffusivity; and $\nabla_h = \left(\frac{\partial}{\partial x}, \frac{\partial}{\partial y}\right)$ is the horizontal divergence operator.

2.2. Cressman Scheme

The Cressman technique was developed by George Cressman in 1959, and is the process of modifying the background table point values (derived from the forecast model) by a linear combination of residual values between the predicted and observed values. This technique involves continuously inserting station data into a user-defined latitude–longitude grid, through a grid at a smaller radius of influence, for increased accuracy. The scheme starts with a background field from a numerical forecast, with the background values at each grid point being continuously adjusted based on nearby observations. The advantages of the Cressman method include the associated ease and speed of calculation, combining forecast data into the background field, and offering generally satisfactory results. These multi-faceted advantages of the Cressman method have made it very popular. One disadvantage of the model is that large deviations are often observed around the edges [30]. The Cressman method is not suitable for multiple observations, as it does not take the observational error into account, which is another disadvantage of this method. In this work, the simulation is divided into two cases: the "control" case and the "Cressman initialized" case. Both of these cases are simulations using different inputs where the two inputs differ in terms of their resolution and data source. The control initialized model uses the coarse 2 × 2 degree ICOADS SST data, which provides an analysis that is based solely on in situ data. The Cressman method is used to initialize the model with an analysis product that is based on satellite data AND in situ data, such as ships, buoys, and Argo floats, with a resolution of 0.25 × 0.25 degrees. A schematic of the grid points of the Cressman method is shown in Figure 2.

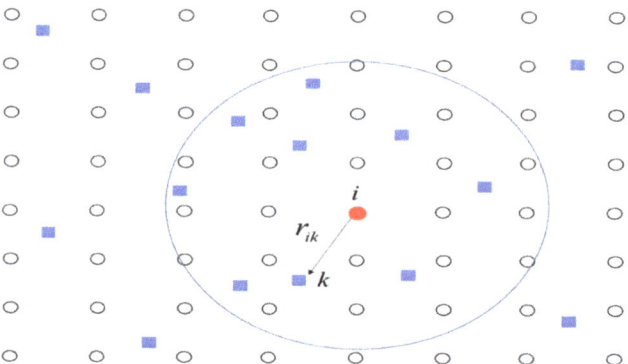

Figure 2. Model data coordinates (circle) and observation data (square) [31].

The model state is assumed to be univariate and is represented as a grid point value. Previous estimates of the model state (background), provided by previous forecasts, are represented by f_i^n, while observations of the same parameter are represented by f_k^o, where $k = 1, ..., K$. The Cressman analytical equation is defined by the model state f_i^{n+1} at each grid point i, according to Equation (2):

$$f_i^{n+1} = f_i^n + \frac{\sum_{k=1}^{K} w_{ik}^n (f_k^o - f_k^n)}{\sum_{k=1}^{K} w_{ik}^n + \varepsilon^2}, \qquad (2)$$

where f_i^n is grid point i of the model at the nth iteration evaluation, f_k^o is observation point k, f_k^n is the estimated value at point k of the nth iteration evaluation, ε^2 is the estimation of the error between the model and observed data, and K is the number of observation points. The following equations determine the weights w_{ik}^n:

$$w_{ik}^n = \begin{cases} \frac{R_n^2 - r_{ik}^2}{R_n^2 + r_{ik}^2} & \text{where } r_{ik}^2 \leq R_n^2 \\ 0 & \text{where } r_{ik}^2 > R_n^2, \end{cases} \qquad (3)$$

where r_{ik} is the distance between observation point k and grid point i and R_n is the radius of the nth iteration evaluation.

2.3. Optimum Interpolation Sea Surface Temperature

NOAA 1/4° daily Optimum Interpolation Sea Surface Temperature (or daily OISST) is an analysis created by combining observations from various platforms, including satellites, boats, buoys, and Argo floats, on a regular global grid. This method includes adjusting the biases of satellite and ship observations (referring to buoys), in order to compensate for platform differences and sensor bias, with a spatially complete SST map being created with corrections to fill the gaps. Satellite data from the Advanced Very High Resolution Radiometer (AVHRR) are the primary input, which have allowed for high temporal and spatial coverage, from late 1981 to the present. The AVHRR is a broadband sensor, featuring three bands in the visible and near-infrared, and three bands in the infrared spectral domain. It is used to represent various world phenomena, in terms of meteorology, soil analysis, and ocean analysis; for example, in the calculation of vegetation indices, cloud properties, dust, snow, ice, fire detection, sea ice concentration, and SST. The spatial resolution of AVHRR is 1 km, at the lowest point reached by a celestial body during its apparent orbit around a given point of observation. These instruments are operated on satellites such as the National Oceanic and Atmospheric Administration (NOAA-11) and European Meteorological Operational (MetOp) satellites. The SST is generated in real-time, using the AVHRR infrared transponder from High-Resolution Picture Transmission (HRPT) using Seaspace's Tera Scan software and NOAA's multi-channel regression algorithm [32,33]. The individual SST scenes are daily combined day–night mean grids for the U.S. East coast, where overpasses occur at roughly 1:30 a.m., 9:30 a.m., 1:30 p.m., and 9:30 p.m. (local time) each day. In this work, we considered combining day- and night-time temperatures, in order to represent the daily averaged sea-surface temperature. Daily grids are combined into 3-day, 7-day, monthly, seasonal, and yearly average grids. As this work required monthly SSTA data, NOAA's daily day–night data were analyzed as monthly data.

The optimum interpolation (OI) sea surface temperature (SST) analysis was performed on a regular grid using irregularly spaced data. The analysis is based on the weighted sum of the data using linear weights OI, which is determined by regression [34].

$$r_k = \sum_{i=1}^{N} w_{ik} q_i, \qquad (4)$$

where r_k is the analyzed SST, w_{ik} is linear weights, q_i are the SST, k is grid points, index i is for data and N is the number of data. The values of q and r are defined as differences from the analysis in the previous time step, and q and r are usually different from the first guess reference system. The average of the analysis correlation error $\langle \pi_i \pi_j \rangle$ is assumed to be representative and is proportional of a Gaussian distribution, as follows:

$$\langle \pi_i \pi_j \rangle = exp\left[\frac{-(x_i - x_j)^2}{\lambda_x^2} + \frac{-(y_i - y_j)^2}{\lambda_y^2}\right] \tag{5}$$

where $\langle \pi_i \pi_j \rangle$ is the average of the analysis correlation error, x and y are the zonal and meridional data and analysis locations, and λ_x and λ_y are the zonal and meridional spatial scales. The weight was then determined according to Reynolds and Smith 1994 by

$$\sum_{i=1}^{N}\left(\langle \pi_i \pi_j \rangle + \epsilon_i^2 \delta_{ij}\right) w_{ik} = \langle \pi_i \pi_j \rangle, \tag{6}$$

where ϵ_i is the noise-to-signal standard deviation ratio, δ_{ij} is data correlation error and w_{ik} is a linear weighting coefficient. The averages of the data errors is assumed to be uncorrelated between the different observations. Therefore, the data correlation error is $\delta_{ij} = 1$ for $i = j$, and 0 otherwise. Data types currently include ships, buoys, and day and night satellite data for each instrument. Spatial functions are defined for each of these quantities with different ϵ_i fields for each data type.

2.4. Extended Reconstructed Sea Surface Temperature

The SSTA data of the National Oceanic and Atmospheric Administration (NOAA, Washington, DC, USA) monthly SST analysis dataset, provided by the Extended Reconstructed Sea Surface Temperature (ERSST) project, were used as input data for the EICM [35,36]. The ERSST dataset is a global monthly SST dataset that derived entirely from in situ observational data and it is provided on a coarse 2×2 degree grid, [37,38]. These data were obtained from the International comprehensive Ocean–Atmosphere Dataset (ICOADS) from 1854 until present, as derived from Argo floats above 5 m. The ERSST is suitable for long-term and basin-wide global studies, and uses smooth, specific, and short-term models in the dataset. ERSST version 5 is the newest version of the ERSST, which has improved SST spatial and temporal variability. The newest version of ERSST improves absolute SST, shifting from using the Nighttime Marine Air Temperature (NMAT) to the use of SST buoys as a reference to correct ship SST biases. The ERSST data were used as input to the EICM, as shown in Table 1.

Table 1. Details of the data in this research.

Detail	ERSST	OISST	IN SITU
Source	National Oceanic and Atmospheric Administration (NOAA)		
Name of data	Sea Surface Temperature Anomaly (SSTA)		
Longitude	0° E to 360° E	0.125° E to 359.875° E	137° E to 265° E
Latitude	−90° N to 90° N	−89.875° N to 89.875° N	−8° N to 9° N
Time	1854-01 to Present	1981-09 to Present	1977-01 to Present
Resolution	2°×2°	0.25°×0.25°	Point
Search	https://www.ncdc.noaa.gov/data-access/marineocean-data/extended-reconstructed-sea-surface-temperature-ersst [accessed on 5 August 2020]	https://www.ncdc.noaa.gov/oisst [accessed on 22 December 2020]	https://www.pmel.noaa.gov/tao/drupal/disdel/index.html [accessed on 27 February 2021]

The Cressman analysis procedure was used to improve the accuracy of the model from AVHRR. Subsequently, the system was switched back to the EICM format. Figure 3 shows an outline of how the CIEICM works automatically, using the inputs from the NOAA.

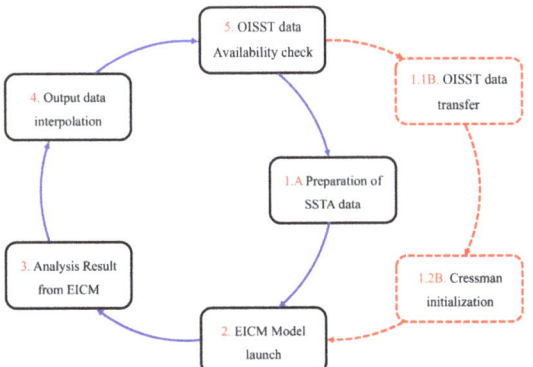

Figure 3. CIEICM with the operational assimilation working scheme.

3. Results and Discussion

Simulation was performed for the SSTA data of OISST in the EICM and CIEICM, using historical data spanning from 1995 to 2019. The results of EICM and CIEICM simulations were compared with the observed data from NOAA/PMEL TAO buoy network, in order to determine whether the Cressman method had worked properly. A comparison of the SSTA result samples from each simulation with the observations was carried out, in order to assess the accuracy of the preliminary models. The absorption validation with EICM consists of three parts: First, the results of both simulations are compared with the OISST data to determine if the initial method is working correctly. Second, the results of both simulations are compared with source data, to verify the accuracy of the two simulations. Finally, the results of the model are computed, in SSTA index format, in the Niño 3.4 area and compared with data from Hadley Center's Sea Ice and Sea Surface Temperature (HadISST) [39]. Figure 4 shows the Pacific region SSTA data for January 1995, including SSTA data from ERSST (which was the EICM import data, shown in Figure 4a). The SSTA data of EICM from ICOADS are shown in Figure 4b. The SSTA data, obtained from OISST, are shown in Figure 4c, while SSTA data of CIEICM that Cressman initialized from OISST are shown in Figure 4d.

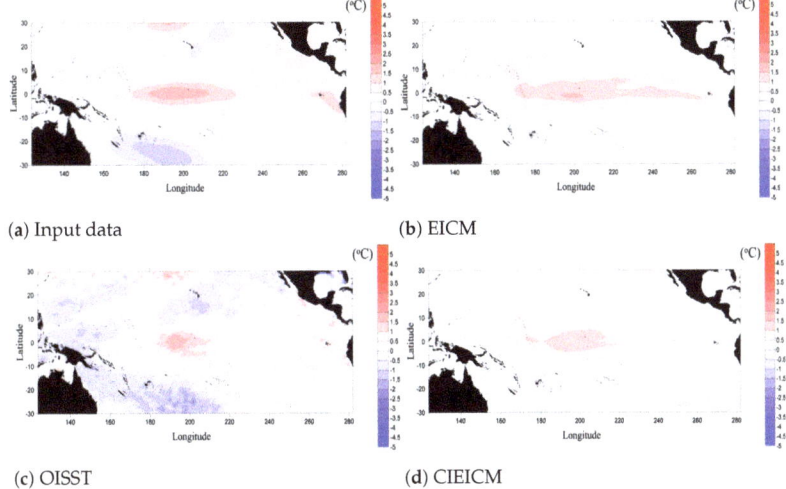

Figure 4. Comparison of the sample results from the EICM and CIEICM.

From Figure 4, the OISST data yielded SSTAs lower than those from ERSST and EICM. CIEICM found that most SSTAs were lower than EICM, and similar to those from ERSST and OISST, as expected. Figure 5 shows a comparison of SSTA data from EICM and CIEICM in January 1995, where the blue color means that Cressman Initialized resulted in lower SSTA values. White and red colours indicate that Cressman Initialized increased the SSTA value of the model. The simulation results showed that SSTAs were mostly reduced from EICM, which was consistent with Figure 4.

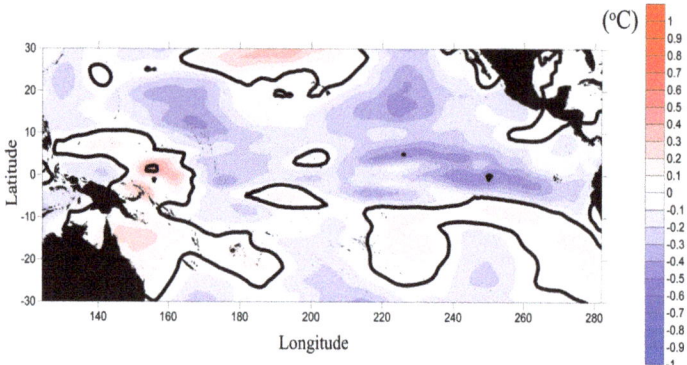

Figure 5. Difference between EICM and CIEICM SSTA data in January 1995.

Visual comparison may not be sufficient and, therefore, a statistical method must be utilised to assess the validity of the model. The statistics used were the Root Mean Square Deviation, shown as Equation (7), Correlation Coefficient, shown as Equation (8), and Standard Deviation of Error, shown in Equation (9).

Root Mean Square Deviation:

$$RMSD = \sqrt{\frac{\sum_{i=1}^{n}(m_i - o_i)^2}{n}}. \tag{7}$$

Correlation Coefficient:

$$R = \frac{\sum_{i=1}^{n}((m_i - \overline{m}) \times (o_i - \overline{o}))}{\sqrt{\sum_{i=1}^{n}(m_i - \overline{m})^2 \times \sum_{i=1}^{n}(o_i - \overline{o})^2}}. \tag{8}$$

Standard Deviation:

$$SD = \sqrt{\frac{\sum_{i=1}^{n}(e_i - \overline{e})^2}{n}}. \tag{9}$$

In the above equations, m_i denotes the SSTA data from the EICM and CIEICM, o_i denotes the SSTA data from OISST, $\overline{m}$ denotes the mean SSTA data from EICM and CIEICM, $\overline{o}$ denotes the mean SSTA data from OISST, n is the number of grids of data, e_i is the Error information between the model and OISST $(m_i - o_i)$, and $\overline{e}$ is the mean of the Error. Comparing SSTA data from EICM and CIEICM with the OISST from 1995 to 2019 (i.e., over 25 years), it was found that, when using Cressman Initialized, there was less discrepancy, as the Root Mean Square Deviation value decreased from 0.616 to 0.605. The correlation between the simulation and OISST increased from 0.535 to 0.548, and there was less variation in the error (which decreased from 0.896 to 0.869), as shown in Table 2.

Table 2. Comparison of accuracy between EICM and CIEICM with OISST data.

Model Type	Root Mean Square Deviation (°C)	Correlation Coefficient (−)	Standard Deviation (°C)	Significance of the Correlation Coefficient (p-Value)
EICM	0.616	0.535	0.896	0
CIEICM	0.605	0.548	0.869	0

When comparing the CIEICM with OISST, the absorption algorithm worked correctly. A comparison of the precision of both models and OISST with in situ data from 1995 to 2019 was carried out, in order to validate the model and to correlate the model with situational data, where the in situ data were obtained from the NOAA/PMEL TAO buoy network [40,41], for which the buoy locations in the central Pacific Ocean to predict ENSO-related climate variations are shown in Figure 6.

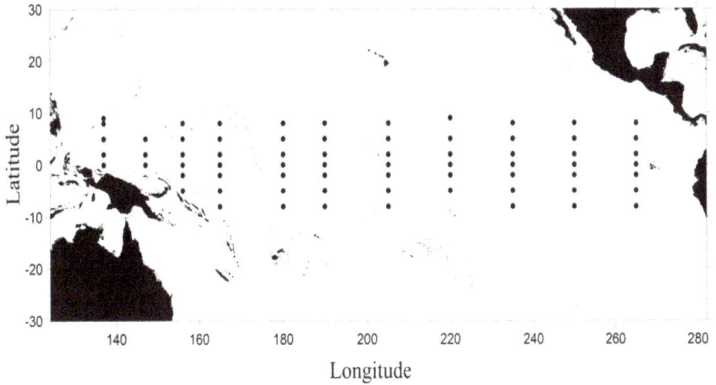

Figure 6. Location of measurement data from NOAA/PMEL TAO buoy network.

Figure 7 shows the correlation of SSTA data between OISST and source data, with black lines showing that the OISST data and in situ data were highly correlated. If the SSTA value is below the black line, the OISST data were higher than the source data. If the SSTA value is higher than the black line, the OISST data were lower than the source data. From the figure, the data from OISST had a correlation coefficient of 0.968.

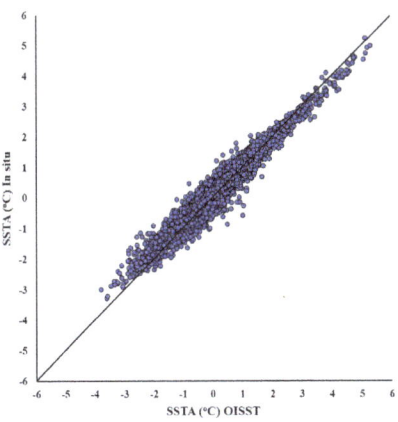

Figure 7. Correlation of the SSTA from the OISST with in situ data.

Figure 8a shows the SSTA relationship between EICM or CIEICM and OISST, where blue indicates the relationship between EICM and OISST, and red shows the relationship between CIEICM and OISST. It was found that the EICM was less correlated than that of the CIEICM (at 0.624 and 0.637, respectively). Figure 8b shows the the SSTA relationship between EICM and CIEICM and in situ data, where the blue colour shows the relationship between EICM and in situ data, and red shows the relationship between CIEICM and in situ data. It was found that the EICM was less correlated than that of the CIEICM (at 0.614 and 0.632, respectively).

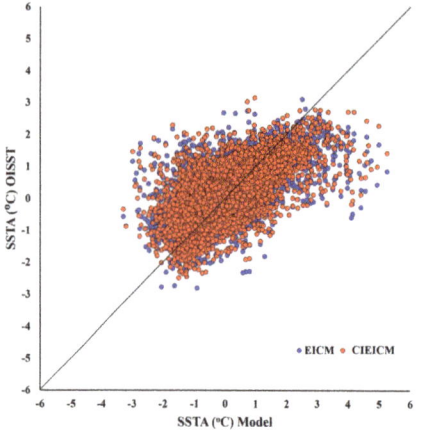
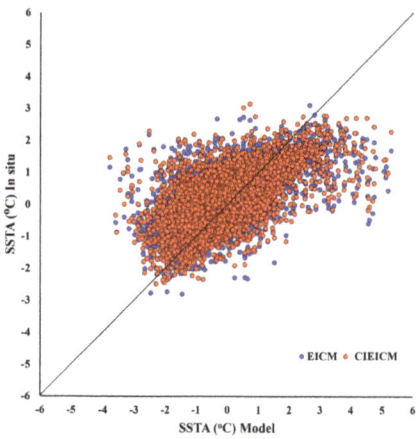

(a) Correlation of the SSTA from the EICM EICM and CIEICM with OISST data.

(b) Correlation of the SSTA from the and CIEICM with in-situ data.

Figure 8. Correlation of the SSTA from the EICM and CIEICM with OISST and in-situ data.

Comparison of the accuracy of the two simulations with in situ data for each month from 1995 to 2019 found that the CIEICM was able to reduce the model error. During the period from June to August, the model tolerance was better than in other months, where the $RMSD$ from CIEICM was approximately 0.024 less than EICM. When comparing the model relationships, it was found that the CIEICM was more correlated than that of the model; it was also found that the EICM had a correlation of approximately 0.16, as shown in Table 3.

The EICM error was compared with the in situ data, and the CIEICM error was compared with 300 in situ data by the Mann–Whitney U statistical method. Considering the statistical value of the Mann–Whitney U Test, it was found that the value was 41,416.5, and the Asymptotic Significance (1-tailed) value was 0.455, which was compared with the statistical significance level to conclude. As a result of the analysis, the Asymptotic Significance (1-tailed) value was less than the significance level of 0.05. The error from the EICM is greater than the error from the CIEICM, as shown in Table 4.

Table 3. Comparison of accuracy between EICM and CIEICM with in situ data.

Month Lead	Measurements	Root Mean Square Deviation		Correlation Coefficients	
		EICM	CIEICM	EICM	CIEICM
Jan	1413	0.691	0.669	0.777	0.789
Feb	1402	0.559	0.535	0.766	0.778
Mar	1402	0.533	0.513	0.691	0.718
Apr	1391	0.55	0.521	0.606	0.629
May	1392	0.668	0.646	0.405	0.399
Jun	1369	0.772	0.743	0.318	0.336
Jul	1383	0.774	0.744	0.436	0.434
Aug	1389	0.767	0.733	0.419	0.476
Sep	1394	0.737	0.71	0.516	0.556
Oct	1384	0.691	0.652	0.723	0.746
Nov	1394	0.684	0.663	0.758	0.763
Dec	1396	0.710	0.665	0.770	0.788
Mean		0.676	0.652	0.600	0.616

Table 4. Statistical data analysis results by the Mann–Whitney U Test method.

	Mann-Whitney U	Wilcoxon W	Z	Asymp. Sig (1-Tailed)
$RMSD$	41,416.5	86,566.5	−1.688	0.0455

The Taylor diagram shows a comparison of the SSTA Index in Niño 3.4 (180° E–240° E and 5° S–5° N). The red dot is the SSTA data from the EICM for each ensemble of 100 ensemble and HadISST data. The blue dot is the SSTA data from the CIEICM for each ensemble of 100 ensembles and data from HadISST. The red cross is the SSTA mean from all EICM 100 ensembles and data from HadISST, and the blue cross is the SSTA mean from all CIEICM 100 ensembles and data from HadISST. Figure 9 shows the prediction results for each month. It was found that, in the forecast using imported data from December 1994 to November 2019, the forecast from January 1995 to December 2019 had a forecast period of 1 month. The $RMSD$ of each ensemble was 0.53 to 0.55 and the $RMSD$ of the Mean Ensemble was 0.49 over a 12-month forecast period. Forecasts for December 1995 through December 2019 had an $RMSD$ of each ensemble from 1.2 to 1.4 and a Mean Ensemble $RMSD$ of 0.8. When forecasting SSTA in the near-term, there was little discrepancy, while the longer period SSTA forecast increased the error.

Figure 10 shows the SSTA Index at Niño 3.4, in order to determine the ENSO phenomenon, where the black line is the SSTA Index data from HadISST, the red line is the SSTA Index data from EICM, and the blue line is the SSTA Index data. If the Niño 3.4 index is greater than 0.5, then the El Niño is defined to occur. If the value of the index is between 0.5 and 1, a weak El Niño is defined to occur. If it is between 1 and 1.5, it is defined to be a moderate El Niño. If it is between 1.5 and 2, it is defined to be strong. Finally, if it is greater than 2, it is defined to be very strong. On the other hand, if the Oceanic Niño index is negative and lower than −0.5, then the La Niña is defined to occur. The level of the phenomenon is defined to be divided into weak, moderate, and strong, similar to that for El Niño. The shorter the forecast is, the less accurate the forecast; the longer the forecast, the greater the error.

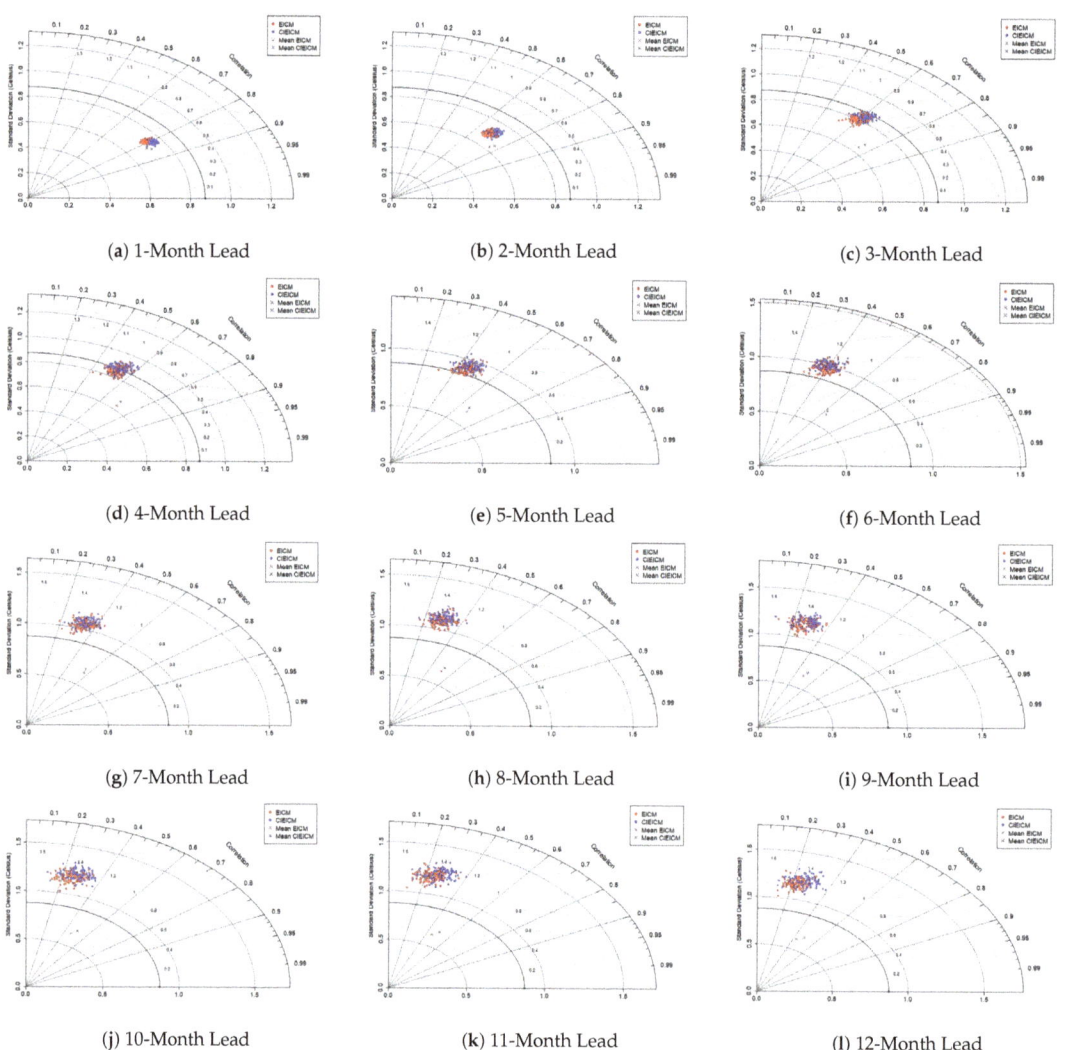

Figure 9. Taylor diagram comparing the SSTA index at Niño 3.4 (180° E–240° E and 5° S–5° N).

Comparing the accuracy of the two simulations with HadISST for each month's lead time from 1995 to 2019, the CIEICM was able to reduce the model error. With 1–8 months lead time, the CIEICM yielded better forecasts than the EICM. With 9–12 months lead time, the two simulations showed no significant differences in terms of $RMSD$, with the mean $RMSD$ from CIEICM being approximately 0.017 less than EICM. When comparing the model relationships, it was found that the CIEICM was more correlated than the EICM, with an increase of approximately 0.11. In the ranges 1, 2–6, and 7–12 months lead, the relationship was at the levels of very strong, strong, and moderate, respectively, as shown in Table 5.

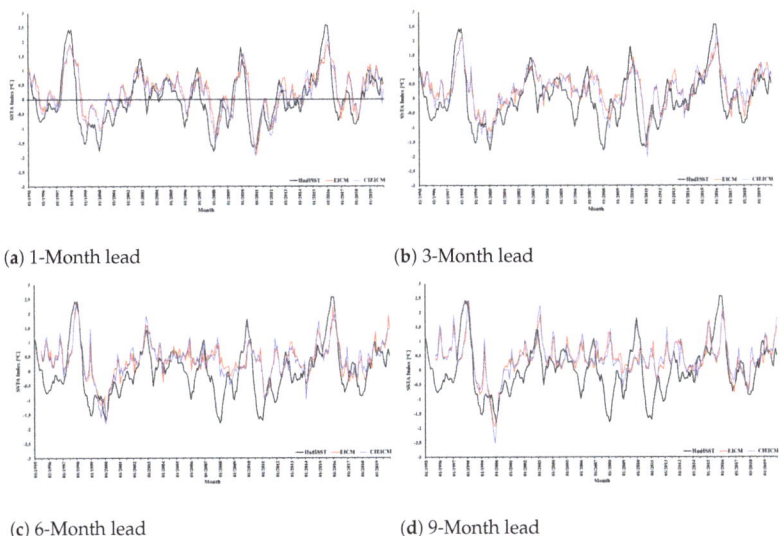

(a) 1-Month lead (b) 3-Month lead

(c) 6-Month lead (d) 9-Month lead

Figure 10. ENSO Phenomenon from 1995 to 2019.

Table 5. Accuracy comparison between EICM, CIEICM, and HadISST data.

Month Lead	Measurements	Root Mean Square Deviation		Correlation Coefficients	
		EICM	CIEICM	EICM	CIEICM
1-Month Lead	300	0.555	0.51	0.827	0.842
2-Month Lead	299	0.654	0.624	0.759	0.77
3-Month Lead	298	0.673	0.651	0.744	0.749
4-Month Lead	297	0.721	0.716	0.706	0.698
5-Month Lead	296	0.76	0.754	0.659	0.676
6-Month Lead	295	0.808	0.794	0.611	0.63
7-Month Lead	294	0.861	0.843	0.55	0.563
8-Month Lead	293	0.897	0.877	0.506	0.509
9-Month Lead	292	0.904	0.89	0.494	0.481
10-Month Lead	291	0.889	0.889	0.499	0.464
11-Month Lead	290	0.881	0.871	0.45	0.499
12-Month Lead	289	0.871	0.855	0.434	0.489
Mean		0.790	0.773	0.603	0.614

Figure 11 shows the Pacific SSTA data in November 2015, which includes SSTA data from EICM and SSTA data from CIEICM. It was found that, during the El Niño phenomenon, CIEICM simulations yielded higher SSTAs than EICMs in the Niño region, as shown in Figure 11a,b, respectively. Figure 11c shows the difference between EICM and CIEICM, where blue means that carrying out Cressman Initialized resulted in a decrease in the model's SSTA, while white and red meant that Cressman Initialization increased the model's SSTA. Most SSTA values were reduced from EICM.

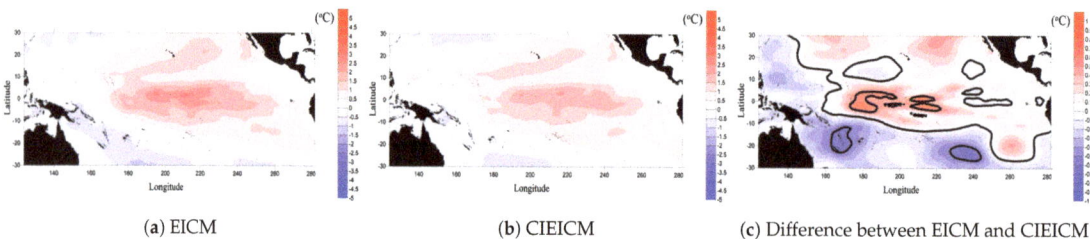

(a) EICM (b) CIEICM (c) Difference between EICM and CIEICM

Figure 11. Comparison of the El Niño range.

Figure 12 shows the Pacific SSTA data in January 2000, which includes SSTA data from EICM and SSTA data from CIEICM. It was found that, during the La Niña phenomenon, CIEICM simulation led to lower SSTA than EICM during Niño region, as shown in Figure 12a,b. Figure 12c shows the difference between EICM and CIEICM, where blue means Cressman Initialized decreased the model SSTA, while white and red indicate that Cressman Initialization increased the model SSTA. Most SSTA values were reduced from EICM.

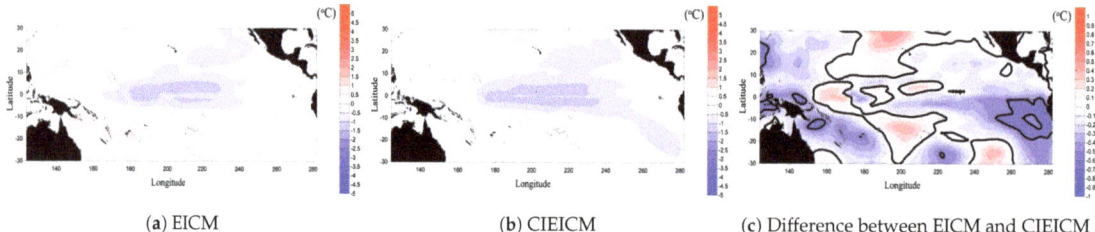

(a) EICM (b) CIEICM (c) Difference between EICM and CIEICM

Figure 12. Comparison of the El Niño range.

4. Conclusions

Accurately predicting the occurrence of ENSO phenomena several months in advance is a major goal of climate research. It is necessary to find a procedure that can yield model predictions which agree with the observed data. This study, therefore, focused on improving the predictions of ENSO phenomena in the Pacific using the Cressman method to improve SST forecasts with the EICM and CIEICM from 1995 to 2019. The work consisted of three main parts. Part 1, comparing the results from EICM and CIEICM with OISST data, it was found that the Cressman method works properly because the Cressman Initialization method minimizes the error. From the statistical analysis, Root Mean Square Deviation decreased from 0.616 to 0.605, and the relationship between simulation and OISST increased from 0.535 to 0.548. Part 2 compares SST data between EICM that initializing with ICOADS and SST data from NOAA/PMEL TAO buoy network and compares SST data between CIEICM that initializing with Cressman using OISST and SST data from NOAA/PMEL TAO buoy network. It was found that the CIEICM was able to reduce the error each month. The $RMSD$ value decreased from 0.676 to 0.652, and the correlation coefficient increased from 0.600 to 0.616. The results showed that the Asymptotic Significance (1-tailed) value was less than the significance level of 0.05. This means that the test has a greater EICM error than the CIEICM error is shown in Table 4. The last part compared the SSTA index at Niño 3.4 of the two simulations using HadISST data. It was found that the CIEICM was more accurate than the EICM. Moreover, the CIEICM simulation was more correlated with data from HadISST than EICM. Both simulations were able to predict the occurrence of ENSO phenomena well in the first six months, and had a strong correlation. This confirmed the reliability of the algorithm using OISST data in conjunction with the EICM to obtain the CIEICM, which yielded better prediction results.

Several studies have attempted to solve problems associated with the input data used in the model to increase ENSO prediction accuracy. The results of the study of Ji and Leetmaa indicated that adequate physical parameters of data assimilation can improve forecasting results and, thus, improve predictive skills [42]. Chen demonstrated the effects and necessity of data assimilation on ICM and pointed out that assimilation is key in improving the prediction skills of the current ENSO model [43]. Many assimilation techniques have been used for the initial simulation of the ocean atmosphere [44]. A four-dimensional variational (4D-Var) data assimilation method has been implemented in an improved model of the tropical Pacific, in order to improve the accuracy of the model. The results showed that 4D-Var can effectively reduce the error in ENSO analysis [45–47]. However, improving forecasting skills by assimilation is not the only method that can be achieved. The need for assimilation may create an imbalance between early ocean conditions and models. The results of the study of Zavala-Garay suggest that there is little room for improvement in predictive skills, as a result of the highly limited data assimilation method [48]. These imbalances and errors in the model can be a significant limiting factor in forecasting skills, especially for predictions that occur in the northern winter. Most studies to date have focused on improving the model through data assimilation, while future studies are expected to focus on optimal error study efforts and actual forecast limit estimates [49]. Some limiting factors cannot be avoided using data assimilation; they have to be addressed through modifications to the model.

Author Contributions: Conceptualization, S.I. and A.K.; methodology, S.I.; software, S.I.; validation, S.I., A.W., U.H. and A.Y.; formal analysis, S.I.; investigation, S.I.; resources, A.W. and A.Y.; data curation, S.I.; writing original draft preparation, S.I.; writing review and editing, S.I.; visualization, A.W. and A.K.; supervision, U.H.; project administration, U.H.; funding acquisition, S.I. All authors have read and agreed to the published version of the manuscript.

Funding: The authors would like to acknowledge the Petchra Pra Jom Klao Ph.D. Research Scholarship from King Mongkut's University of Technology Thonburi and the authors would like to thank Office of National Higher Education Science Research and Innovation Policy Council [number B16F630087].

Institutional Review Board Statement: Not applicable.

Informed Consent Statement: Not applicable.

Data Availability Statement: Not applicable.

Acknowledgments: The authors would like to thank Office of National Higher Education Science Research and Innovation Policy Council [number B16F630087], Thai Research Fund (TRF) [RDG6230004], and ARDA [PRP6405031190]. The authors would like to acknowledge the Petchra Pra Jom Klao Ph.D. Research Scholarship from King Mongkut's University of Technology Thonburi.

Conflicts of Interest: The authors declare no conflict of interest.

Abbreviations

The following abbreviations are used in this manuscript:

AVHRR	Advanced Very High-Resolution Radiometer
EICM	Ensemble Intermediate Coupled Model
CIEICM	Cressman Initialized Ensemble Intermediate Coupled Model
ENSO	El Niño–Southern Oscillation
ERSST	Extended Reconstructed Sea Surface Temperature
HadISST	Hadley Center's Sea Ice and Sea Surface Temperature
HCM	Hybrid Coupled Model
HRPT	High-Resolution Picture Transmission
ICM	Intermediate Coupled Model

IOM	Intermediate Ocean Model
MetOp	Meteorological Operational
NMAT	Night-time Marine Air Temperature
NOAA	National Oceanic and Atmospheric Administration
OISST	Optimum Interpolation Sea Surface Temperature
R	Correlation Coefficient
$RMSD$	Root Mean Square Deviation
SD	Standard Deviation
SST	Sea Surface Temperature
SSTA	Sea Surface Temperature Anomaly

References

1. Merchant, C.J.; Embury, O.; Roberts-Jones, J.; Fiedler, E.; Bulgin, C.; Corlett, G.K.; Good, S.; Mclaren, A.J.; Rayner, N.; Morak-Bozzo, S.; et al. Sea surface temperature datasets for climate applications from Phase 1 of the European Space Agency Climate Change Initiative (SST CCI). *Geosci. Data J.* **2014**, *1*, 179–191. [CrossRef]
2. Bjerknes, J. Atmospheric teleconnections from the equatorial Pacific. *Mon. Weather Rev.* **1969**, *97*, 163–172. [CrossRef]
3. McCreary, J.P., Jr. A model of tropical ocean-atmosphere interaction. *Mon. Weather Rev.* **1969**, *111*, 370–387. [CrossRef]
4. Wyrtki, K. El Niño-the dynamic response of the equatorial Pacific Ocean to atmospheric forcing. *J. Phys. Oceanogr.* **1975**, *5*, 572–584. [CrossRef]
5. Cane, M.A.; Zebiak, S.E.; Dolan, S.C. Experimental forecasts of El Niño. *Nature* **1986**, *321*, 827–832. [CrossRef]
6. Zhang, R.-H.; Zheng, F.; Zhu, J.; Wang, Z. A successful real-time forecast of the 2010–11 La Niña event. *Sci. Rep.* **2013**, *3*, 1–7. [CrossRef]
7. Barnston, A.G.; Tippett, M.K.; L'Heureux, M.L.; Li, S.; DeWitt, D.G. Skill of Real-Time Seasonal ENSO Model Predictions during 2002–11: Is Our Capability Increasing? *Bull. Am. Meteorol. Soc.* **2012**, *93*, 631–651. [CrossRef]
8. Barnett, T.P.; Latif, M.; Graham, N.; Flugel, M.; Pazan, S; White, W. ENSO and ENSO-related predictability. Part I: Prediction of equatorial Pacific sea surface temperature with a hybrid coupled ocean-atmosphere model. *J. Clim.* **1993**, *6*, 1545–1566. [CrossRef]
9. Kang, I.S.; Kug, J.S. An El Nino prediction system using an intermediate ocean and a statistical atmosphere. *Geophys. Res. Lett.* **2000**, *27*, 1167–1170. [CrossRef]
10. Zhang, R.H.; Gao, C. The IOCAS intermediate coupled model (IOCAS ICM) and its real-time predictions of the 2015-16 El Niño event. *Sci. Bull.* **2016**, *66*, 1061–1070. [CrossRef]
11. Zhang, S.; Harrison, M.J.; Wittenberg, A.T.; Rosati, A.; Anderson, J.L.; Balaji, V. Initialization of an ENSO forecast system using a parallelized ensemble filter. *Mon. Weather Rev.* **2005**, *133*, 3176–3201. [CrossRef]
12. Webb, D.J.; Cowaed, A.C.; Snaith, H.M. A comparison of ocean model data and satellite observations of features affecting the growth of the North Equatorial Counter Current during the strong 1997–1998 El Niño. *Ocean. Sci.* **2020**, *16*, 565–574. [CrossRef]
13. Zheng, F.; Zhang, R.-H. Ensemble hindcasts of SST anomalies in the tropical Pacific using an intermediate coupled model. *Geophys. Res. Lett.* **2006**, *33*, 1–5. [CrossRef]
14. Zheng, F.; Zhu, J.; Wang, H.; Zhang, R.-H. Ensemble Hindcasts of ENSO Events over the Past 120 Years Using a Large Number of Ensembles. *Adv. Atmos. Sci.* **2009**, *26*, 359–372. [CrossRef]
15. Chen, D.; Zebiak, S.E.; Busalacchi, A.J.; Cane, M.A. An improved procedure for El Niño forecasting: Implications for predictability. *Science* **1995**, *269*, 1699–1702. [CrossRef] [PubMed]
16. Artur, N.; Lidia, D.G.; Macigj, J.; Macigi, K. Assimilation of the satellite SST data in the 3D CEMBS model. *Oceanologia* **2015**, *57*, 17–24.
17. Abbasi, M.R.; Chegini, V.; Sadrinasab, M.; Siadatmousavi, S.M. Capabilities of data assimilation in correcting sea surface temperature in the Persian Gulf. *Pollution* **2017**, *3*, 273–283.
18. Zheng, F.; Zhu, J. Improved ensemble mean forecasting of ENSO events by a zero mean stochastic error model of an intermediate coupled model. *Clim. Dyn.* **2016**, *47*, 3901–3915. [CrossRef]
19. Evensen, G. The ensemble Kalman filter: Theoretical formulation and practical implementation. *Ocean. Dyn.* **2003**, *53*, 343–367. [CrossRef]
20. Evensen, G. Sampling strategies and square root analysis schemes for the EnKF. *Ocean. Dyn.* **2003**, *54*, 539–560. [CrossRef]
21. Zebiak, S.E.; Cane, M.A. A model El Niño-Southern Oscillation. *Mon. Weather Rev.* **1987**, *115*, 2262–2278. [CrossRef]
22. Zhang, R.-H.; Zebiak, S.E.; Kleeman, R. A new intermediate coupled model for El Niño simulation and prediction. *Geophys. Res. Lett.* **2003**, *30*, 1–4. [CrossRef]
23. Zhang, R.-H.; Kleeman, R.,; Zebiak, S.E.; Keenlyside N.; Raynaud, S. An empirical parameterization of subsurface entrainment temperature for improved SST simulations in an intermediate ocean model. *J. Clim.* **2005**, *18*, 350–371. [CrossRef]
24. Zhu, J.; Zhou, G.; Zhang, R.-H.; Sun, Z. On the role of ocean entrainment temperature (Te) in decadal changes of El Niño/Southern Oscillation. *Ann. Geophys.* **2011**, *29*, 529–540. [CrossRef]
25. Zhu, J.; Zhou, G.; Zhang, R.-H.; Sun, Z. Improving ENSO prediction in a hybrid coupled model with an embedded entrainment temperature parameterization. *Int. J. Climatol.* **2013**, *33*, 343–355. [CrossRef]

26. Keenlyside, N.; Kleeman, R. Annual cycle of equatorial zonal currents in the Pacific. *Geophys. Res. Lett.* **2002**, *107*, 1–13. [CrossRef]
27. McCreary, J.P. A linear stratified ocean model of the equatorial undercurrent. *Philos. Trans. R. Soc. Lond.* **1981**, *298*, 603–635.
28. Gill, A.E. An estimation of sea-level and surface-currents during the 1972 El Niño and consequenct thermal effects. *J. Phys. Oceanogr.* **1983**, *13*, 486–606. [CrossRef]
29. Zhang, R.-H.; Busalacchi, A.J. The Roles of Atmospheric Stochastic Forcing (SF) and Oceanic Entrainment Temperature (Te) in Decadal Modulation of ENSO. *J. Clim.* **2007**, *21*, 674–704. [CrossRef]
30. Cressman, G.P. An operational objective analysis system. *Mon. Weather Rev.* **1959**, *87*, 367–374. [CrossRef]
31. Kalnay, E. *Atmospheric Modeling, Data Assimilation and Predictability*; Cambridge University Press: Cambridge, UK, 2003; Volume 87, pp. 136–204.
32. Li, X.; Pichel, W.; Maturi, E.; Clemente-Colon, P.; Sapper, J. Deriving the operational nonlinear multichannel sea surface temperature algorithm coefficients for NOAA-15 AVHRR/3. *Int. J. Remote Sens.* **2001**, *22*, 699–704. [CrossRef]
33. Li, X.; Pichel, W.; Clemente-Colon, P.; Krasnopolsky, V.; Sapper, J. Validation of coastal sea and lake surface temperature measurements derived from NOAA/AVHRR data. *Int. J. Remote. Sens.* **2001**, *22*, 1285–1303. [CrossRef]
34. Reynolds, R.W.; Smith, T.M. Improved global sea surface temperature analyses. *J. Clim.* **1994**, *7*, 929–948. [CrossRef]
35. Smith, T.M.; Reynolds, R.W.; Peterson, T.C.; Lawrimore, J. Improvements NOAAs Historical Merged Land-Ocean Temp Analysis (1880–2006). *J. Clim.* **2008**, *21*, 2283–2296. [CrossRef]
36. Xue, Y.; Smith, T.M.; Reynolds, R.W. Interdecadal Changes of 30-Yr SST Normals during 1871–2000. *J. Clim.* **2003**, *16*, 1601–1612. [CrossRef]
37. Eric, F.; Elizabeth, C.K.; Philip, B.; Thomas, C.; Lydia, G.; Boyin, H.; Chunying, L.; Shawn, R.S.; Steven, J.W.; Zhang, H.-M. The International Comprehensive Ocean-Atmosphere Data Set— Meeting Users Needs and Future Priorities. *Front. Mar. Sci.* **2019**, *23*, 1–8.
38. Liu, W.; Huang, B.; Thorne, P.W.; Banzon, V.F.; Zhang, H.M.; Freeman, E.; Lawrimore, J.; Peterson, T.C.; Smith, T.M.; Woodruff, S.D. Extended Reconstructed Sea Surface Temperature Version 4 (ERSST.v4): Part II. Parametric and Structural Uncertainty Estimations. *J. Clim.* **2015**, *28*, 931–951. [CrossRef]
39. Rayner, N.A.; Parker, D.E.; Horton, E.B.; Folland, C.K.; Alexander, L.V.; Rowell, D.P.; Kent, E.C.; Kaplan, A. Global analyses of sea surface temperature, sea ice, and night marine air temperature since the late nineteenth century. *J. Geophys. Res.* **2003**, *108*, 4407. [CrossRef]
40. McPhaden, M.J.; Busalacchi, A.J.; Anderson, D.L.T. A TOGA retrospective. *Oceanography* **2010**, *23*, 86–103. [CrossRef]
41. McPhaden, M.J.; Busalacchi, A.J.; Cheney, R.; Donguy, J.R.; Gage, K.S.; Halpern, D.; Ji, M.; Julian, P.; Meyers, G.; Mitchum, G.T.; et al. The Tropical Ocean-Global Atmosphere (TOGA) observing system: A decade of progress. *J. Geophys. Res.* **2020**, *103*, 14169–14240.
42. Ji, M.; Leetmaa, A. Impact of Data Assimilation on Ocean Initialization and El Niño Prediction. *Mon. Weather Rev.* **1997**, *125*, 742–753. [CrossRef]
43. Chen, D. Coupled data assimilation for enso prediction. *Adv. Geosci.* **2010**, *18*, 45–62.
44. Oberhuber, J.M. Predicting the El Niño event with a global climate model. *Geophys. Res. Lett.* **1998**, *25*, 2273–2276. [CrossRef]
45. Gao, C.; Wu, X.; Zhang, R.H. Testing a Four-Dimensional Variational Data Assimilation Method Using an Improved Intermediate Coupled Model for ENSO Analysis and Prediction. *Adv. Atmos. Sci.* **2016**, *33*, 875–888. [CrossRef]
46. Gao, C.; Zhang, R.H.; Wu, X.; Sun, J. Idealized experiments for optimizing model parameters using a 4D-Variational method in an intermediate coupled model of ENSO. *Adv. Atmos. Sci.* **2018**, *35*, 410–422. [CrossRef]
47. Injan, S.; Wangwongchai, A.; Humphries, U. Application of Data Assimilation and the Relationship between ENSO and Precipitation. *Math. Comput. Appl.* **2021**, *26*, 1–16.
48. Zavala-Garay, J.; Moore, A.M.; Kleeman, R. Influence of stochastic forcing on ENSO prediction. *J. Geophys. Res. Ocean.* **2004**, *109*, 1–10. [CrossRef]
49. Tang, Y.; Zhang, R.H.; Liu, T.; Duan, W.; Yang, D.; Zheng, F.; Ren, H.; Lian, T.; Gao, C.; Chen, D.; et al. Progress in ENSO prediction and predictability study. *Natl. Sci. Rev.* **2018**, *5*, 826–839. [CrossRef]

Article

Application of a Generalized Secant Method to Nonlinear Equations with Complex Roots

Avram Sidi

Computer Science Department, Technion—Israel Institute of Technology, Haifa 32000, Israel; asidi@cs.technion.ac.il

Abstract: The secant method is a very effective numerical procedure used for solving nonlinear equations of the form $f(x) = 0$. In a recent work (A. Sidi, Generalization of the secant method for nonlinear equations. *Appl. Math. E-Notes*, 8:115–123, 2008), we presented a generalization of the secant method that uses only one evaluation of $f(x)$ per iteration, and we provided a local convergence theory for it that concerns real roots. For each integer k, this method generates a sequence $\{x_n\}$ of approximations to a real root of $f(x)$, where, for $n \geq k$, $x_{n+1} = x_n - f(x_n)/p'_{n,k}(x_n)$, $p_{n,k}(x)$ being the polynomial of degree k that interpolates $f(x)$ at $x_n, x_{n-1}, \ldots, x_{n-k}$, the order s_k of this method satisfying $1 < s_k < 2$. Clearly, when $k = 1$, this method reduces to the secant method with $s_1 = (1 + \sqrt{5})/2$. In addition, $s_1 < s_2 < s_3 < \cdots$, such that $\lim_{k \to \infty} s_k = 2$. In this note, we study the application of this method to simple complex roots of a function $f(z)$. We show that the local convergence theory developed for real roots can be extended almost as is to complex roots, provided suitable assumptions and justifications are made. We illustrate the theory with two numerical examples.

Keywords: secant method; generalized secant method; complex roots

MSC: 65H05

Citation: Sidi, A. Application of a Generalized Secant Method to Nonlinear Equations with Complex Roots. *Axioms* **2021**, *10*, 169. https://doi.org/10.3390/axioms10030169

Academic Editors: Davron Aslonqulovich Juraev and Samad Noeiaghdam

Received: 7 June 2021
Accepted: 22 July 2021
Published: 29 July 2021

Publisher's Note: MDPI stays neutral with regard to jurisdictional claims in published maps and institutional affiliations.

Copyright: © 2021 by the author. Licensee MDPI, Basel, Switzerland. This article is an open access article distributed under the terms and conditions of the Creative Commons Attribution (CC BY) license (https://creativecommons.org/licenses/by/4.0/).

1. Introduction

Let α be the solution to the equation $f(x) = 0$. An effective iterative method used for solving this equation that makes direct use of $f(x)$ (but no derivatives of $f(x)$) is the *secant method* that is discussed in many books on numerical analysis. See, for example, Atkinson [1], Dahlquist and Björck [2], Henrici [3], Ralston and Rabinowitz [4], and Stoer and Bulirsch [5]. See also the recent note [6] by the author, in which the treatment of the secant method and those of the Newton–Raphson, regula falsi, and Steffensen methods are presented in a unified manner.

Recently, this method was generalized by the author in [7] as follows: Starting with $x_0, x_1, \ldots, x_k$, $k + 1$ initial approximations to α, we generate a sequence of approximations $\{x_n\}$, via the recursion

$$x_{n+1} = x_n - \frac{f(x_n)}{p'_{n,k}(x_n)}, \quad n = k, k+1, \ldots, \tag{1}$$

$p'_{n,k}(x)$ being the derivative of the polynomial $p_{n,k}(x)$ that interpolates $f(x)$ at the points $x_n, x_{n-1}, \ldots, x_{n-k}$. (Thus, $p_{n,k}(x)$ is of degree k.) Clearly, the case $k = 1$ is simply the secant method. In [7], we also showed that, provided $x_0, x_1, \ldots, x_k$ are sufficiently close to α, the method converges with order s_k, that is, $\lim_{n \to \infty} \frac{|x_{n+1} - \alpha|}{|x_n - \alpha|^{s_k}} = C \neq 0$ for some constant C, and that $1 < s_k < 2$. (We call s_k the *order of convergence* of the method or the *order of the method* for short.) Here s_k is the only positive root of the polynomial $s^{k+1} - \sum_{i=0}^{k} s^i$. We also have that

$$\frac{1+\sqrt{5}}{2} = s_1 < s_2 < s_3 < \cdots < 2; \quad \lim_{k \to \infty} s_k = 2.$$

Actually, rounded to four significant figures,

$s_1 \doteq 1.618$, $s_2 \doteq 1.839$, $s_3 \doteq 1.928$, $s_4 \doteq 1.966$, $s_5 \doteq 1.984$, $s_6 \doteq 1.992$, $s_7 \doteq 1.996$, etc.

Note that to compute x_{n+1} we need knowledge of only $f(x_n), f(x_{n-1}), \ldots, f(x_{n-k})$, and because $f(x_{n-1}), \ldots, f(x_{n-k})$ have already been computed, $f(x_n)$ is the only new quantity to be computed. Thus, each step of the method requires $f(x)$ to be computed only once. From this, it follows that the efficiency index of this method is simply s_k and that this index approaches 2 by increasing k even moderately, as can be concluded from the values of $s_1, \ldots, s_7$ given above.

In this work, we consider the application of this method to simple complex roots of a function $f(z)$, where z is the complex variable. Let us denote a real or complex root of $f(z)$ by α again; that is, $f(\alpha) = 0$ and $f'(\alpha) \neq 0$. Thus, starting with $z_0, z_1, \ldots, z_k$, $k+1$ initial approximations to α, we generate a sequence of approximations $\{z_n\}$ via the recursion

$$z_{n+1} = z_n - \frac{f(z_n)}{p'_{n,k}(z_n)}, \quad n = k, k+1, \ldots, \tag{2}$$

$p'_{n,k}(z)$ being the derivative of the polynomial $p_{n,k}(z)$ that interpolates $f(z)$ at the points $z_n, z_{n-1}, \ldots, z_{n-k}$. As in [7], we can use Newton's interpolation formula to generate $p_{n,k}(z)$ and $p'_{n,k}(z)$. Thus

$$p_{n,k}(z) = f(z_n) + \sum_{i=1}^{k} f[z_n, z_{n-1}, \ldots, z_{n-i}] \prod_{j=0}^{i-1} (z - z_{n-j}) \tag{3}$$

and

$$p'_{n,k}(z_n) = f[z_n, z_{n-1}] + \sum_{i=2}^{k} f[z_n, z_{n-1}, \ldots, z_{n-i}] \prod_{j=1}^{i-1} (z_n - z_{n-j}). \tag{4}$$

Here, $g[\zeta_0, \zeta_1, \ldots, \zeta_m]$ is the divided difference of order m of the function $g(z)$ over the set of points $\{\zeta_0, \zeta_1, \ldots, \zeta_m\}$ and is a symmetric function of these points. For details, we refer the reader to [7].

As proposed in [7], we generate the $k+1$ initial approximations as follows: We choose the approximations z_0, z_1 first. We then generate z_2 by applying our method with $k = 1$ (that is, with the secant method). Next, we apply our method to z_0, z_1, z_2 with $k = 2$ and obtain z_3, and so on, until we have generated all $k+1$ initial approximations, via

$$z_{n+1} = z_n - \frac{f(z_n)}{p'_{n,n}(z_n)}, \quad n = 1, 2, \ldots, k-1. \tag{5}$$

Remark 1.

1. Instead of choosing z_1 arbitrarily, we can generate it as $z_1 = z_0 + f(z_0)$ as suggested in Brin [8], which is quite sensible since $f(z)$ is small near the root α. We can also use the method of Steffensen—which uses only $f(z)$ and no derivatives of $f(z)$—to generate z_1 from z_0; thus,

$$z_1 = z_0 - \frac{[f(z_0)]^2}{f(z_0 + f(z_0)) - f(z_0)}.$$

2. It is clear that, in case $f(z)$ takes on only real values along the Re z axis and we are looking for nonreal roots of $f(z)$, at least one of the initial approximations must be chosen to be nonreal.
3. We would like to mention that Kogan, Sapir, and Sapir [9] have proposed another generalization of the secant method for simple real roots of nonlinear equations $f(x) = 0$ that

resembles our method described in (1). In the notation of (1), this method produces a sequence of approximations $\{x_n\}$ via

$$x_{n+1} = x_n - \frac{f(x_n)}{p'_{n,n}(x_n)}, \quad n = 1, 2, \ldots, \tag{6}$$

starting with arbitrary x_0 and x_1, and it is of order 2. Note that, in (6), $p_{n,n}(x)$ interpolates $f(x)$ at the points $x_0, x_1, \ldots, x_n$, hence is of degree n, which is tending to infinity. In (1), $p_{n,k}(x)$ is of degree k, which is fixed.

4. Yet another generalization of the secant method for finding simple real roots of $f(x)$ was recently given by Nijmeijer [10]. This method too requires no derivative information, requires one evaluation of $f(x)$ per iteration, and has the same order of convergence as our method. It follows an idea of applying a convergence acceleration method, such as Aitken's Δ^2-process, to approximations obtained from the secant method, as proposed by Han and Potra [11]. Because Nijmeijer's method is not based on polynomial interpolation, it is completely different from our method, however. For Aitken's Δ^2-process, see [1–5]. See also [12] (Chapter 15) by the author.

In the next section, we analyze the local convergence properties of the method as it is applied to complex roots. We show that the analysis of [7] can be extended to the complex case following some clever manipulation. We prove that the order s_k of the method is the same as that we discovered in the real case. In Section 3, we provide two numerical examples to confirm the results of our convergence analysis.

2. Local Convergence Analysis

We now turn to the analysis of the sequence $\{z_n\}_{n=0}^{\infty}$ that is generated via (2). Our treatment covers all $k \geq 1$.

In our analysis, we will make use of the Hermite-Genocchi formula that provides an integral representation for divided differences (For a proof of this formula, see Atkinson [1], for example). Even though this formula is usually stated for functions defined on real intervals, it is easy to verify (see Filipsson [13], for example) that it also applies to functions defined in the complex plane under proper assumptions. Thus, provided $g(z)$ is analytic on E, a bounded closed convex set in the complex plane, and provided $\zeta_0, \zeta_1, \ldots, \zeta_m$ are in E, there holds

$$g[\zeta_0, \zeta_1, \ldots, \zeta_m] = \int \cdots \int_{S_m} g^{(m)}(t_0\zeta_0 + t_1\zeta_1 + \cdots + t_m\zeta_m)\, dt_1 \cdots dt_m, \quad t_0 = 1 - \sum_{i=1}^{m} t_i. \tag{7}$$

Here S_m is the m-dimensional simplex defined as

$$S_m = \left\{ (t_1, \ldots, t_m) \in \mathbb{R}^m : t_i \geq 0,\ i = 1, \ldots, m,\ \sum_{i=1}^{m} t_i \leq 1 \right\}. \tag{8}$$

We note that (7) holds whether the ζ_i are distinct or not. We also note that $g[\zeta_0, \zeta_1, \ldots, \zeta_m]$ is a symmetric and continuous function of its arguments.

By the conditions we have imposed on $g(z)$, it is easy to see that the integrand $g^{(m)}(\sum_{i=0}^{m} t_i\zeta_i)$ in (7) is always defined because $\sum_{i=0}^{m} t_i\zeta_i$ is in the set E and $g(z)$ is analytic on E. This is so because, by (7) and (8),

$$(t_1, \ldots, t_m) \in S_m \Rightarrow t_i \geq 0,\ i = 0, 1, \ldots, m,\ \text{and}\ \sum_{i=0}^{m} t_i = 1,$$

which implies that $\sum_{i=0}^{m} t_i\zeta_i$ is a convex combination of $\zeta_0, \zeta_1, \ldots, \zeta_m$ hence is in the set $C = \operatorname{conv}\{\zeta_0, \zeta_1, \ldots, \zeta_m\}$, the convex hull of the points $\zeta_0, \zeta_1, \ldots, \zeta_m$, and $C \subseteq E$. Consequently, taking moduli on both sides of (7), we obtain, for all ζ_i in E,

$$|g[\zeta_0, \zeta_1, \ldots, \zeta_m]| \leq \int \cdots \int_{S_m} \left| g^{(m)} \left(\sum_{i=0}^{m} t_i \zeta_i \right) \right| dt_1 \cdots dt_m$$

$$\leq \frac{\|g^{(m)}\|}{m!}, \quad \|g^{(m)}\| = \max_{z \in E} |g^{(m)}(z)|. \tag{9}$$

In addition, since $\sum_{i=0}^{m} t_i = 1$ in (7), as $\zeta_i \to \hat{\zeta}$ for all $i = 0, 1, \ldots, m$, there hold $\sum_{i=0}^{m} t_i \zeta_i \to \hat{\zeta}$ and $g^{(m)}(\sum_{i=0}^{m} t_i \zeta_i) \to g^{(m)}(\hat{\zeta})$, and hence

$$\lim_{\substack{\zeta_i \to \hat{\zeta} \\ i=0,1,\ldots,m}} g[\zeta_0, \zeta_1, \ldots, \zeta_m] = g[\underbrace{\hat{\zeta}, \hat{\zeta}, \ldots, \hat{\zeta}}_{m+1 \text{ times}}] = \frac{g^{(m)}(\hat{\zeta})}{m!}. \tag{10}$$

In (9) and (10), we have also invoked the fact that (see [14] (p. 346), for example)

$$\int \cdots \int_{S_m} dt_1 \cdots dt_m = \frac{1}{m!}.$$

We will make use of these in the proof of our main theorem that follows. This theorem and its proof are almost identical to that given in [7] once we take into account, where and when needed, the fact that we are now working in the complex plane. For convenience, we provide all the details of the proof.

Theorem 1. *Let α be a simple root of $f(z)$, that is, $f(\alpha) = 0$, but $f'(\alpha) \neq 0$. Let B_r be the closed disk of radius r containing α as its center, that is,*

$$B_r = \{z \in \mathbb{C} : |z - \alpha| \leq r\}. \tag{11}$$

Let $f(z)$ be analytic on B_r. Choose a positive integer k and let $z_0, z_1, \ldots, z_k$ be distinct initial approximations to α. Generate $z_{k+1}, z_{k+2}, \ldots$ via

$$z_{n+1} = z_n - \frac{f(z_n)}{p'_{n,k}(z_n)}, \quad n = k, k+1, \ldots, \tag{12}$$

where $p_{n,k}(z)$ is the polynomial of interpolation to $f(z)$ at the points $z_n, z_{n-1}, \ldots, z_{n-k}$. Then, provided $z_0, z_1, \ldots, z_k$ are in B_r and sufficiently close to α, we have the following cases:

1. *If $f^{(k+1)}(\alpha) \neq 0$, the sequence $\{z_n\}$ converges to α, and*

$$\lim_{n \to \infty} \frac{\epsilon_{n+1}}{\prod_{i=0}^{k} \epsilon_{n-i}} = \frac{(-1)^{k+1}}{(k+1)!} \frac{f^{(k+1)}(\alpha)}{f'(\alpha)} \equiv L; \quad \epsilon_n = z_n - \alpha \quad \forall n. \tag{13}$$

The order of convergence is s_k, $1 < s_k < 2$, where s_k is the only positive root of the polynomial $g_k(s) = s^{k+1} - \sum_{i=0}^{k} s^i$ and satisfies

$$2 - 2^{-k-1}e < s_k < 2 - 2^{-k-1} \quad \text{for } k \geq 2; \quad s_k < s_{k+1}; \quad \lim_{k \to \infty} s_k = 2, \tag{14}$$

e being the base of natural logarithms, and

$$\lim_{n \to \infty} \frac{|\epsilon_{n+1}|}{|\epsilon_n|^{s_k}} = |L|^{(s_k - 1)/k}, \tag{15}$$

which also implies that

$$s_k = \lim_{n \to \infty} \frac{\log |\epsilon_{n+1}/\epsilon_n|}{\log |\epsilon_n/\epsilon_{n-1}|}. \tag{16}$$

2. If $f(z)$ is a polynomial of degree at most k, the sequence $\{z_n\}$ converges to α, and

$$\lim_{n\to\infty} \frac{\epsilon_{n+1}}{\epsilon_n^2} = \frac{f''(\alpha)}{2f'(\alpha)}; \quad \epsilon_n = z_n - \alpha \quad \forall\, n. \tag{17}$$

Thus, $\{z_n\}$ converges with order 2 if $f''(\alpha) \neq 0$, and with order greater than 2 if $f''(\alpha) = 0$.

Proof. We start by deriving a closed-form expression for the error in z_{n+1}. Subtracting α from both sides of (12), and noting that

$$f(z_n) = f(z_n) - f(\alpha) = f[z_n, \alpha](z_n - \alpha),$$

we have

$$z_{n+1} - \alpha = \left(1 - \frac{f[z_n,\alpha]}{p'_{n,k}(z_n)}\right)(z_n - \alpha) = \frac{p'_{n,k}(z_n) - f[z_n,\alpha]}{p'_{n,k}(z_n)}(z_n - \alpha). \tag{18}$$

We now note that

$$p'_{n,k}(z_n) - f[z_n,\alpha] = \{p'_{n,k}(z_n) - f'(z_n)\} + \{f'(z_n) - f[z_n,\alpha]\}, \tag{19}$$

and that

$$f'(z_n) - p'_{n,k}(z_n) = f[z_n, z_n, z_{n-1}, \ldots, z_{n-k}] \prod_{i=1}^{k}(z_n - z_{n-i}) \tag{20}$$

and

$$f'(z_n) - f[z_n,\alpha] = f[z_n, z_n] - f[z_n, \alpha] = f[z_n, z_n, \alpha](z_n - \alpha). \tag{21}$$

Note that (20) can be obtained by starting with the divided difference representation of $f(z) - p_{n,k}(z)$, namely, $f(z) - p_{n,k}(z) = f[z, z_n, z_{n-1}, \ldots, z_{n-k}]\prod_{i=0}^{k}(z - z_{n-i})$, and by computing $\lim_{z\to z_n}[f(z) - p_{n,k}(z)]/\prod_{i=0}^{k}(z - z_{n-i})$ via L'Hôpital's rule.

For simplicity of notation, let

$$-f[z_n, z_n, z_{n-1}, \ldots, z_{n-k}] = \widehat{D}_n \quad \text{and} \quad f[z_n, z_n, \alpha] = \widehat{E}_n, \tag{22}$$

and rewrite (19) and (20) as

$$p'_{n,k}(z_n) - f[z_n,\alpha] = \widehat{D}_n \prod_{i=1}^{k}(\epsilon_n - \epsilon_{n-i}) + \widehat{E}_n \epsilon_n, \tag{23}$$

$$p'_{n,k}(z_n) = f'(z_n) + \widehat{D}_n \prod_{i=1}^{k}(\epsilon_n - \epsilon_{n-i}). \tag{24}$$

Substituting these into (18), we finally obtain

$$\epsilon_{n+1} = C_n \epsilon_n; \quad C_n \equiv \frac{p'_{n,k}(z_n) - f[z_n,\alpha]}{p'_{n,k}(z_n)} = \frac{\widehat{D}_n \prod_{i=1}^{k}(\epsilon_n - \epsilon_{n-i}) + \widehat{E}_n \epsilon_n}{f'(z_n) + \widehat{D}_n \prod_{i=1}^{k}(\epsilon_n - \epsilon_{n-i})}. \tag{25}$$

We now prove that convergence takes place. First, let us assume without loss of generality that $f'(z) \neq 0$ for all $z \in B_r$, and set $m_1 = \min_{z\in B_r}|f'(z)| > 0$. (This is possible since $\alpha \in B_r$ and $f'(\alpha) \neq 0$, and we can choose r as small as we wish to also guarantee $m_1 > 0$.) Next, let $M_s = \max_{z\in B_r}|f^{(s)}(z)|/s!$, $s = 1, 2, \ldots$. Thus, assuming that $\{z_n, z_{n-1}, \ldots, z_{n-k}\} \subset B_r$ and noting that B_r is a convex set, we have by (9) that

$$|\widehat{D}_n| \leq M_{k+1}, \quad |\widehat{E}_n| \leq M_2, \quad \text{because } \{\alpha, z_n, z_{n-1}, \ldots, z_{n-k}\} \subset B_r.$$

Next, choose the ball $B_{t/2}$ sufficiently small (with $t/2 \leq r$) to ensure that $m_1 > 2M_{k+1}t^k + M_2 t/2$. It can now be verified that, provided $z_n, z_{n-1}, \ldots, z_{n-k}$ are all in $B_{t/2}$, there holds

$$|C_n| \leq \frac{M_{k+1} \prod_{i=1}^{k} |\epsilon_n - \epsilon_{n-i}| + M_2 |\epsilon_n|}{m_1 - M_{k+1} \prod_{i=1}^{k} |\epsilon_n - \epsilon_{n-i}|}$$

$$\leq \frac{M_{k+1} \prod_{i=1}^{k} (|\epsilon_n| + |\epsilon_{n-i}|) + M_2 |\epsilon_n|}{m_1 - M_{k+1} \prod_{i=1}^{k} (|\epsilon_n| + |\epsilon_{n-i}|)} \leq \overline{C},$$

where

$$\overline{C} \equiv \frac{M_{k+1} t^k + M_2 t/2}{m_1 - M_{k+1} t^k} < 1.$$

Consequently, by (25), $|\epsilon_{n+1}| \leq \overline{C}|\epsilon_n| < |\epsilon_n|$, which implies that $z_{n+1} \in B_{t/2}$, just like $z_n, z_{n-1}, \ldots, z_{n-k}$. Therefore, if $z_0, z_1, \ldots, z_k$ are chosen in $B_{t/2}$, then $|C_n| \leq \overline{C} < 1$ for all $n \geq k$, hence $\{z_n\} \subset B_{t/2}$ and $\lim_{n \to \infty} z_n = \alpha$.

As for (13) when $f^{(k+1)}(\alpha) \neq 0$, we proceed as follows: By the fact that $\lim_{n \to \infty} z_n = \alpha$, we first note that, by (20) and (21),

$$\lim_{n \to \infty} p'_{n,k}(z_n) = f'(\alpha) = \lim_{n \to \infty} f[z_n, \alpha], \qquad (26)$$

and thus $\lim_{n \to \infty} C_n = 0$. This means that $\lim_{n \to \infty} (\epsilon_{n+1}/\epsilon_n) = 0$ and, equivalently, that $\{z_n\}$ converges with order greater than 1. As a result,

$$\lim_{n \to \infty} (\epsilon_n / \epsilon_{n-i}) = 0 \quad \text{for all } i \geq 1,$$

and

$$\epsilon_n / \epsilon_{n-i} = o(\epsilon_n / \epsilon_{n-j}) \quad \text{as } n \to \infty, \text{ for } j < i.$$

Consequently, expanding in (25) the product $\prod_{i=1}^{k}(\epsilon_n - \epsilon_{n-i})$, we have

$$\prod_{i=1}^{k}(\epsilon_n - \epsilon_{n-i}) = \prod_{i=1}^{k}\left(-\epsilon_{n-i}[1 - \epsilon_n/\epsilon_{n-i}]\right)$$

$$= (-1)^k \left(\prod_{i=1}^{k} \epsilon_{n-i}\right)[1 + O(\epsilon_n/\epsilon_{n-1})] \quad \text{as } n \to \infty. \qquad (27)$$

Substituting (27) into (25), and defining

$$D_n = \frac{\widehat{D}_n}{p'_{n,k}(z_n)}, \quad E_n = \frac{\widehat{E}_n}{p'_{n,k}(z_n)}, \qquad (28)$$

we obtain

$$\epsilon_{n+1} = (-1)^k D_n \left(\prod_{i=0}^{k} \epsilon_{n-i}\right)[1 + O(\epsilon_n/\epsilon_{n-1})] + E_n \epsilon_n^2 \quad \text{as } n \to \infty. \qquad (29)$$

Dividing both sides of (29) by $\prod_{i=0}^{k} \epsilon_{n-i}$, and defining

$$\sigma_n = \frac{\epsilon_{n+1}}{\prod_{i=0}^{k} \epsilon_{n-i}}, \qquad (30)$$

we have

$$\sigma_n = (-1)^k D_n [1 + O(\epsilon_n/\epsilon_{n-1})] + E_n \sigma_{n-1} \epsilon_{n-k-1} \quad \text{as } n \to \infty. \qquad (31)$$

Now, by (10), (22), and (26),

$$\lim_{n\to\infty} D_n = -\frac{1}{(k+1)!}\frac{f^{(k+1)}(\alpha)}{f'(\alpha)}, \quad \lim_{n\to\infty} E_n = \frac{f^{(2)}(\alpha)}{2f'(\alpha)}. \tag{32}$$

Because $\lim_{n\to\infty} D_n$ and $\lim_{n\to\infty} E_n$ are finite, and because $\lim_{n\to\infty}(\epsilon_n/\epsilon_{n-1}) = 0$ and $\lim_{n\to\infty} \epsilon_{n-k-1} = 0$, it follows that there exist a positive integer N and positive constants $\beta < 1$ and D, with $|E_n \epsilon_{n-k-1}| \leq \beta$ when $n > N$, for which (31) gives

$$|\sigma_n| \leq D + \beta|\sigma_{n-1}| \quad \text{for all } n > N. \tag{33}$$

Using (33), it is easy to show that

$$|\sigma_{N+s}| \leq D\frac{1-\beta^s}{1-\beta} + \beta^s|\sigma_N|, \quad s = 1, 2, \ldots,$$

which, by the fact that $\beta < 1$, implies that $\{\sigma_n\}$ is a bounded sequence. Making use of this fact, we have $\lim_{n\to\infty} E_n \sigma_{n-1}\epsilon_{n-k-1} = 0$. Substituting this into (31), and invoking (32), we next obtain $\lim_{n\to\infty} \sigma_n = (-1)^k \lim_{n\to\infty} D_n = L$, which is precisely (13).

That s_k, the order of the method, as defined in the statement of the theorem, satisfies (14) and (15) follows from Traub [15] (Chapter 3). We provide a simplified treatment of this topic in Appendix A.

This completes the proof of part 1 of the theorem.

When $f(z)$ is a polynomial of degree at most k, we first observe that $f^{(k+1)}(z) = 0$ for all z, which implies that $p_{n,k}(z) = f(z)$ for all z, hence also $p'_{n,k}(z) = f'(z)$ for all z. Therefore, we have that $p'_{n,k}(z_n) = f'(z_n)$ in the recursion of (12). Consequently, (12) becomes

$$z_{n+1} = z_n - \frac{f(z_n)}{f'(z_n)}, \quad n = k, k+1, \ldots,$$

which is the recursion for the Newton–Raphson method. Thus, (17) follows. This completes the proof of part 2 of the theorem. □

3. Numerical Examples

In this section, we present two numerical examples that we treated with our method. Our computations were done in quadruple-precision arithmetic (approximately 35-decimal-digit accuracy). Note that in order to verify the theoretical results concerning iterative methods with order greater than unity, we need to use computer arithmetic of high precision (preferably, of variable precision, if available) because the number of correct significant decimal digits in the z_n increases dramatically from one iteration to the next as we are approaching the solution.

In both examples below, we take $k = 2$. We choose z_0 and z_1 and compute z_2 using one step of the secant method, namely,

$$z_2 = z_1 - \frac{f(z_1)}{f[z_0, z_1]}. \tag{34}$$

Following that, we compute $z_3, z_4, \ldots$, via

$$z_{n+1} = z_n - \frac{f(z_n)}{f[z_n, z_{n-1}] + f[z_n, z_{n-1}, z_{n-2}](z_n - z_{n-1})}, \quad n = 2, 3, \ldots. \tag{35}$$

In our examples, we have carried out our computations for several sets of z_0, z_1, and we have observed essentially the same behavior that we observe in Tables 1 and 2.

Example 1. Consider $f(z) = 0$, where $f(z) = z^3 - 8$, whose solutions are $\alpha_r = 2e^{i2\pi r/3}$, $r = 0, 1, 2$. We would like to obtain the root $\alpha_1 = 2e^{i2\pi/3} = -1 + i\sqrt{3}$. We chose $z_0 = 2i$ and $z_1 = -2 + 2i$. The results of our computations are given in Table 1.

Table 1. Results obtained by applying the generalized secant method with $k = 2$, as shown in (34) and (35), to the equation $z^3 - 8 = 0$, to compute the root $\alpha_1 = -1 + i\sqrt{3}$. The entries denoted "**" mean that the limit of the extended-precision arithmetic has been reached.

n	$\|\epsilon_n\|$	$\dfrac{\epsilon_{n+1}}{\epsilon_n \epsilon_{n-1} \epsilon_{n-2}}$	$\dfrac{\log\|\epsilon_{n+1}/\epsilon_n\|}{\log\|\epsilon_n/\epsilon_{n-1}\|}$
0	$1.035D + 00$	-	-
1	$1.035D + 00$	-	-
2	$4.808D - 01$	$-8.972D - 02 + i\,1.015D - 01$	2.516
3	$6.979D - 02$	$1.224D - 01 - i\,2.727D - 02$	1.437
4	$4.355D - 03$	$1.009D - 01 - i\,4.079D - 02$	2.023
5	$1.591D - 05$	$4.561D - 02 - i\,9.794D - 02$	1.839
6	$5.223D - 10$	$3.793D - 02 - i\,7.268D - 02$	1.839
7	$2.967D - 18$	$3.741D - 02 - i\,7.579D - 02$	1.838
8	$2.083D - 33$	**	**
9	$0.000D + 00$	**	**

From (13) and (16) in Theorem 1, we should have

$$\lim_{n\to\infty} \frac{\epsilon_{n+1}}{\epsilon_n \epsilon_{n-1} \epsilon_{n-2}} = \frac{(-1)^3}{3!} \frac{f'''(\alpha_1)}{f'(\alpha_1)} = \frac{1}{24}(1 - i\sqrt{3}) = 0.04166\cdots - i\,0.07216\cdots$$

and

$$\lim_{n\to\infty} \frac{\log|\epsilon_{n+1}/\epsilon_n|}{\log|\epsilon_n/\epsilon_{n-1}|} = s_2 = 1.83928\cdots,$$

and these seem to be confirmed in Table 1. Furthermore, in infinite-precision arithmetic, z_9 should have close to 60 correct significant figures; we do not see this in Table 1 due to the fact that the arithmetic we have used to generate Table 1 can provide an accuracy of at most 35 digits.

Example 2. Consider $f(z) = 0$, where $f(z) = \sin(iz) - \cos z$. $f(z)$ has infinitely many roots $\alpha_r = (1 - i)(\pi/4 + r\pi)$, $r = 0, \pm 1, \pm 2, \ldots$. We would like to obtain the root $\alpha_0 = (1 - i)\pi/4$. We chose $z_0 = 1.5 - i1.3$ and $z_1 = 0.6 - i0.5$. The results of our computations are given in Table 2.

Table 2. Results obtained by applying the generalized secant method with $k = 2$, as shown in (34) and (35), to the equation $\sin(iz) - \cos z = 0$, to compute the root $\alpha_0 = (1 - i)\pi/4$. The entries denoted "**" mean that the limit of the extended-precision arithmetic has been reached.

n	$\|\epsilon_n\|$	$\dfrac{\epsilon_{n+1}}{\epsilon_n \epsilon_{n-1} \epsilon_{n-2}}$	$\dfrac{\log\|\epsilon_{n+1}/\epsilon_n\|}{\log\|\epsilon_n/\epsilon_{n-1}\|}$
0	$6.608D - 01$	-	-
1	$3.403D - 01$	-	-
2	$1.341D - 01$	$3.163D - 01 + i\,1.397D - 01$	2.743
3	$1.043D - 02$	$1.466D - 01 - i\,1.846D - 01$	1.774
4	$1.122D - 04$	$-2.943D - 03 - i\,1.117D - 01$	1.934
5	$1.755D - 08$	$9.223D - 03 - i\,1.614D - 01$	1.766
6	$3.320D - 15$	$-7.686D - 04 - i\,1.658D - 01$	1.857
7	$1.084D - 27$	**	**
8	$9.630D - 35$	**	**

From (13) and (16) in Theorem 1, we should have

$$\lim_{n\to\infty} \frac{\epsilon_{n+1}}{\epsilon_n \epsilon_{n-1} \epsilon_{n-2}} = \frac{(-1)^3}{3!} \frac{f'''(\alpha_1)}{f'(\alpha_1)} = -\frac{i}{6} = -i\,0.16666\cdots$$

and
$$\lim_{n\to\infty} \frac{\log|\epsilon_{n+1}/\epsilon_n|}{\log|\epsilon_n/\epsilon_{n-1}|} = s_2 = 1.83928\cdots,$$

and these seem to be confirmed in Table 2. Furthermore, in infinite-precision arithmetic, z_8 should have close to 50 correct significant figures; we do not see this in Table 2 due to the fact that the arithmetic we have used to generate Table 2 can provide an accuracy of at most 35 digits.

Remark 2. *In relation to the examples we have just presented, we would like to discuss the issue of estimating the relative errors $|\epsilon_n/\alpha|$ in the z_n. This should help the reader when studying the numerical results included in Tables 1 and 2. Starting with (13) and (15), we first note that, for all large n,*
$$|\epsilon_{n+1}| \approx |L|^{(s_k-1)/k} |\epsilon_n|^{s_k}.$$

Therefore, assuming also that $\alpha \neq 0$, we have
$$|\epsilon_{n+1}/\alpha| \approx D|\epsilon_n/\alpha|^{s_k}, \quad D = \left(|L|^{1/k}|\alpha|\right)^{s_k-1}.$$

Now, if z_n has $q > 0$ correct significant figures, we have $|\epsilon_n/\alpha| = O(10^{-q})$. If, in addition, $D = O(10^r)$ for some r, then we will have
$$|\epsilon_{n+1}/\alpha| \approx O(10^{r-qs_k}).$$

For simplicity, let us consider the case $r = 0$, which is practically what we have in the two examples we have treated. Then z_{n+1} has approximately qs_k correct significant decimal digits. That is, if z_n has q correct significant decimal digits, then, due to the fact that $s_k > 1$, z_{n+1} will have s_k times as many correct significant decimal digits as z_n.

Funding: This research received no external funding.

Acknowledgments: The author would like to thank Tamara Kogan for drawing his attention to the paper [9] mentioned in the Introduction.

Conflicts of Interest: The author declares no conflict of interest.

Appendix A

Before ending, we would like to provide a brief treatment of the order of convergence of our method stated in (14) and (15) by considering
$$\frac{\epsilon_{n+1}}{\prod_{i=0}^{k}\epsilon_{n-i}} = L \quad \forall n \quad \Leftrightarrow \quad \epsilon_{n+1} = L\prod_{i=0}^{k}\epsilon_{n-i} \quad \forall n,$$

instead of (13). We will show that $|\epsilon_{n+1}| = Q|\epsilon_n|^{s_k}$ is possible if s_k is a solution to the polynomial equation $s^{k+1} = \sum_{i=0}^{k} s^i$ and $Q = |L|^{(s_k-1)/k}$. (For a more detailed treatment, we refer the reader to [15] (Section 3.3)).

We start by expressing all $|\epsilon_{n-i}|$ in terms of $|\epsilon_n|$. We have
$$|\epsilon_{n-i}| = \frac{|\epsilon_n|^{1/s_k^i}}{Q^{m_i}}, \quad m_i = \sum_{j=1}^{i}\frac{1}{s_k^j}, \quad i = 1, 2, \ldots.$$

Substituting this into $|\epsilon_{n+1}| = |L|\prod_{i=0}^{k}|\epsilon_{n-i}|$, we obtain
$$Q|\epsilon_n|^{s_k} = |L|\,|\epsilon_n|\prod_{i=1}^{k}\frac{|\epsilon_n|^{1/s_k^i}}{Q^{m_i}} = \frac{|L|}{Q^M}|\epsilon_n|^{\rho}; \quad \rho = \sum_{i=0}^{k}\frac{1}{s_k^i}, \quad M = \sum_{i=1}^{k}m_i.$$

Of course, this is possible when $s_k = \rho$ and $Q^{M+1} = |L|$.

Now, the requirement that $s_k = \rho$ is the same as $s_k^{k+1} = \sum_{i=0}^{k} s_k^i$, which implies that the order s_k should be a root of the polynomial

$$g_k(s) = s^{k+1} - \sum_{i=0}^{k} s^i = \frac{s^{k+2} - 2s^{k+1} + 1}{s-1}.$$

By Descartes' rule of signs, $g_k(s)$ has only one positive root, which we denote by $\tilde{s}$. Since $g_k(1) = -k < 0$ and $g_k(2) = 1 > 0$, we have that $1 < \tilde{s} < 2$. The remaining k roots of $g_k(s)$ are the zeroes of the polynomial $\tilde{g}(s) = g_k(s)/(s - \tilde{s}) = \sum_{j=0}^{k} c_j s^j$, the c_j satisfying $\tilde{s} c_0 = 1$ and $\tilde{s} c_j - c_{j-1} = 1, j = 1, \ldots, k$, hence

$$c_j = \frac{1}{\tilde{s}} \sum_{i=0}^{j} \frac{1}{\tilde{s}^i}, \quad j = 0, 1, \ldots, k \quad \Rightarrow \quad 0 < c_0 < c_1 \cdots < c_{k-1} < c_k = 1.$$

Therefore, by the Eneström–Kakeya theorem, all k roots of $\tilde{g}(s)$ are in the unit disk. We thus conclude that $\tilde{s} = s_k$ since we already know that the order of our method is greater than 1. (For Descartes' rule of signs and the Eneström–Kakeya theorem, see, for example, Henrici [16] (pp. 442, 462)).

Next, we note that $g_k(s) = s g_{k-1}(s) - 1$. Therefore, $g_{k-1}(s_{k-1}) = 0$ implies $g_k(s_{k-1}) = -1 < 0$, which, along with $g_k(2) = 1 > 0$, implies that $s_{k-1} < s_k < 2$. Therefore, the sequence $\{s_k\}_{k=1}^{\infty}$ is monotonically increasing and is bounded from above by 2. Consequently, $\lim_{k \to \infty} s_k = \hat{s}$ exists and $\hat{s} \leq 2$. Now,

$$g_k(s_k) = 0 \quad \Rightarrow \quad s_k^{k+2} - 2s_k^{k+1} + 1 = 0 \quad \Rightarrow \quad s_k^2 - 2s_k = -\frac{1}{s_k^k}.$$

Upon letting $k \to \infty$ on both sides, we obtain $\hat{s}^2 - 2\hat{s} = 0$, which gives $\hat{s} = 2$.

The expression given for M can be simplified considerably as we show next. First, it is easy to verify that

$$M = \sum_{i=1}^{k} \frac{k-i+1}{s_k^i} = \frac{1}{s_k^k} \sum_{i=1}^{k} i s_k^{i-1}.$$

Next,

$$s_k^k M = \left(\frac{d}{ds} \sum_{i=0}^{k} s^i \right)\bigg|_{s=s_k} = \left(\frac{d}{ds} \frac{s^{k+1} - 1}{s - 1} \right)\bigg|_{s=s_k} = \frac{(k+1)s^k(s-1) - (s^{k+1} - 1)}{(s-1)^2}\bigg|_{s=s_k}.$$

By $s^{k+1} - 1 = (s-1) \sum_{i=0}^{k} s^i$, this becomes

$$s_k^k M = \frac{(k+1)s^k - \sum_{i=0}^{k} s^i}{s-1}\bigg|_{s=s_k} = \frac{k s_k^k - \sum_{i=0}^{k-1} s_k^i}{s_k - 1}.$$

Now, by the fact that $g_k(s_k) = 0$, we have $\sum_{i=0}^{k-1} s_k^i = s_k^{k+1} - s_k^k$. Consequently,

$$M = \frac{k - (s_k - 1)}{s_k - 1} \quad \Rightarrow \quad M + 1 = \frac{k}{s_k - 1},$$

which is the required result.

References

1. Atkinson, K.E. *An Introduction to Numerical Analysis*, 2nd ed.; John Wiley & Sons Inc.: New York, NY, USA, 1989.
2. Dahlquist, G.; Björck, Å. *Numerical Methods in Scientific Computing: Volume I*; SIAM: Philadelphia, PA, USA, 2008.
3. Henrici, P. *Elements of Numerical Analysis*; Wiley: New York, NY, USA, 1964.
4. Ralston, A.; Rabinowitz, P. *A First Course in Numerical Analysis*, 2nd ed.; McGraw-Hill: New York, NY, USA, 1978.
5. Stoer, J.; Bulirsch, R. *Introduction to Numerical Analysis*, 3rd ed.; Springer: New York, NY, USA, 2002.

6. Sidi, A. Unified treatment of regula falsi, Newton–Raphson, secant, and Steffensen methods for nonlinear equations. *J. Online Math. Appl.* **2006**, *6*, 1–13.
7. Sidi, A. Generalization of the secant method for nonlinear equations. *Appl. Math. E-Notes* **2008**, *8*, 115–123.
8. Brin, L.Q. *Tea Time Numerical Analysis*; Southern Connecticut State University: New Haven, CT, USA, 2016.
9. Kogan, T.; Sapir, L.; Sapir, A. A nonstationary iterative second-order method for solving nonlinear equations. *Appl. Math. Comput.* **2007**, *188*, 75–82. [CrossRef]
10. Nijmeijer, M.J.P. A method to accelerate the convergence of the secant algorithm. *Adv. Numer. Anal.* **2014**, *2014*, 321592. [CrossRef]
11. Han, W.; Potra, F.A. Convergence acceleration for some root finding methods. *Comput. Suppl.* **1993**, *9*, 67–78.
12. Sidi, A. *Practical Extrapolation Methods: Theory and Applications*; Number 10 in Cambridge Monographs on Applied and Computational Mathematics; Cambridge University Press: Cambridge, UK, 2003.
13. Filipsson, L. Complex mean-value interpolation and approximation of holomorphic functions. *J. Approx. Theory* **1997**, *91*, 244–278. [CrossRef]
14. Davis, P.J.; Rabinowitz, P. *Methods of Numerical Integration*, 2nd ed.; Academic Press: New York, NY, USA, 1984.
15. Traub, J.F. *Iterative Methods for the Solution of Equations*; Prentice Hall: Englewood Cliffs, NJ, USA, 1964.
16. Henrici, P. *Applied and Computational Complex Analysis*; Wiley: New York, NY, USA, 1974; Volume 1.

Article

Critical Indices and Self-Similar Power Transform

Simon Gluzman

Materialica and Research Group, Bathurst St. 3000, Apt. 606, Toronto, ON M6B 3B4, Canada; simongluzmannew@gmail.com

Abstract: "Odd" factor approximants of the special form suggested by Gluzman and Yukalov (J. Math. Chem. 2006, 39, 47) are amenable to optimization by power transformation and can be successfully applied to critical phenomena. The approach is based on the idea that the critical index by itself should be optimized through the parameters of power transform to be calculated from the minimal sensitivity (derivative) optimization condition. The critical index is a product of the algebraic self-similar renormalization which contributes to the expressions the set of control parameters typical to the algebraic self-similar renormalization, and of the power transform which corrects them even further. The parameter of power transformation is, in a nutshell, the multiplier connecting the critical exponent and the correction-to-scaling exponent. We mostly study the minimal model of critical phenomena based on expansions with only two coefficients and critical points. The optimization appears to bring quite accurate, uniquely defined results given by simple formulas. Many important cases of critical phenomena are covered by the simple formula. For the longer series, the optimization condition possesses multiple solutions, and additional constraints should be applied. In particular, we constrain the sought solution by requiring it to be the best in prediction of the coefficients not employed in its construction. In principle, the error/measure of such prediction can be optimized by itself, with respect to the parameter of power transform. Methods of calculation based on optimized power-transformed factors are applied and results presented for critical indices of several key models of conductivity and viscosity of random media, swelling of polymers, permeability in two-dimensional channels. Several quantum mechanical problems are discussed as well.

Keywords: minimal sensitivity; optimization; power transform; critical index

Citation: Gluzman, S. Critical Indices and Self-Similar Power Transform. *Axioms* **2021**, *10*, 162. https://doi.org/10.3390/axioms10030162

Academic Editor: Davron Aslonqulovich Juraev

Received: 16 June 2021
Accepted: 23 July 2021
Published: 26 July 2021

Publisher's Note: MDPI stays neutral with regard to jurisdictional claims in published maps and institutional affiliations.

Copyright: © 2021 by the author. Licensee MDPI, Basel, Switzerland. This article is an open access article distributed under the terms and conditions of the Creative Commons Attribution (CC BY) license (https://creativecommons.org/licenses/by/4.0/).

1. Introduction

Often, the limit-problems are characterized by power laws. Accurate analytical formulae for deterministic and random systems, such as composites, suspensions, and porous media, can be derived by employing approximants, when the low-concentration series are supplemented with information on the high-concentration regime near divergence points signifying the physical percolation effects [1,2].

Sometimes, the results can be improved by simple transformation to the divergent quantities. It is often assumed that even a naive inverse transformation of the original truncated series could be accomplished. Generally, one can introduce the so-called power transform defined in [3]. Below, we consider application of power transform to calculation of the critical properties and of the critical index in particular. It is possible to find the critical index as an explicit function of the parameters defining the power transformation. Instead of optimization of critical amplitudes as in [3–7], we express and optimize the critical index directly. The critical index is obtained then by finding some optimal power transformation, demanding the critical index to be minimally sensitive to the parameters of power transform.

Indeed, as we are interested in the critical index, it seems more natural to optimize the index directly, rather than to try to find from optimizing some other quantities, even the critical amplitudes. Furthermore, as the critical indices are not sensitive to the fine details of interactions, or to the parameters imposed in the course of optimization, we apply rather generic optimization conditions to rather short expansions.

Various power transformations are known in statistics. Most notable are the Tukey's ladder of powers [8] and Box-Cox transformation [9]. The Box-Cox transformation is nothing else but the celebrated replica-trick. It is intimately related to the concept of highly optimized tolerance (HOT) [10]. HOT is viewed as the mechanism behind the power laws.

Formally, we study the behaviour of a real function $\phi(x)$ of a real variable $x \in [0, \infty)$. This function is typically defined by a complicated problem that does not allow for an explicit knowledge of the functional form. However, one can attempt to develop some kind of perturbation theory. The perturbation theory yields truncated asymptotic expansions representing the function

$$\phi(x) \simeq \phi_k(x) \qquad (x \to 0) \tag{1}$$

at a small variable $x \to 0$, with $k = 0, 1, \ldots$ being perturbation order. The perturbative series of kth order can be written as an truncated expansion in powers of x as

$$\phi_k(x) = a_0 + \sum_{n=1}^{k} a_n x^n . \tag{2}$$

Unless otherwise stated, $a_0 = 1$ is chosen. The series (2) will be subject to renormalization. The most difficult region for approximating is that of the large variable. Our main interest here will be to find the large-variable behaviour of the function, where its asymptotic form is expected to be

$$\phi(x) \simeq Bx^\beta \qquad (x \to \infty) . \tag{3}$$

The constant B will be called the critical amplitude and the power β is the critical exponent.

We have to transform the truncated series into convergent expressions with the asymptotic behavior (3). We are going to calculate β first, and B second, by employing the technique of power transforms applied to factor approximants [3].

One can consider the index k as the discrete time. Then, the truncated series (2) corresponds to the points of the trajectory of the dynamical system. The velocity which governs the passage from one point to another is approximated from the available truncated series. Finding the stable fixed points for such a dynamical system means to be able to find the sum (1) as $k \to \infty$.

In the vicinity of the fixed point, the functional self-similarity relation between the consecutive functional approximations can be employed constructively. Moreover, due to the self-similarity relation, one can look for the fixed point representing the sought sum in the analytical form, e.g., the power-transformed factor approximant can be considered as such a representation [3]. For such approximants, one can analytically express the critical index and amplitude, and optimize the critical index with respect to the parameters of power transform. To define the power transform in full, we have to apply additional, optimization conditions of rather general nature. More specifically, the program could be accomplished by means of the power-transformed "odd" factor approximants [3]. Furthermore, the conditions on fixed points using power transform parameters as control parameters take particularly simple form and optimization leads to some polynomial-type equations, relatively easy to solve numerically.

The general idea of application of various optimal conditions in the space of approximations is due to V.I. Yukalov [11,12]. For the particular example, minimal difference condition was put forward in [13]. Minimal sensitivity condition was exploited in [14]. Kleinert's variational-perturbation method also employs minimal derivative conditions [15]. However, the method is very difficult to adapt to the purpose of finding an analytical expression for the sought quantity.

The method of coherent anomaly was suggested and developed by M. Suzuki [16,17]. He applied the ideas of the renormalization group, but specifically considered it in the space of approximations. The physical quantities are considered with respect to the approximation parameter, discrete in most cases, just as in the general approach briefly discussed above. The method allows to estimate critical indices but does not bring the concrete approximant whatsoever. Absence of control parameters makes it necessary to consider high-order approximations to the sought quantity in order to achieve better results.

Various self-similar techniques were applied to calculations with short series [5,6,18], but they all appear to be designed for providing lower and upper bounds for the index. Such methods combine at least two separate methods and are applied through optimizing the expressions for critical amplitudes [2,5]. Such an indirect approach to the critical indices computations can be remedied by reformulating it for the direct optimization of the expressions for critical indices. However, to accomplish the task, it turned out to be necessary to extend the class of approximations in order to be able to select possible stable solutions within the bounds found by other methodologies.

Because of the limited information supplied by the short series, the problem of critical properties should be attacked by various methods, under various assumptions [2]. The method considered in the current paper does have some unique qualities, such as incorporating both critical and sub-critical indices simultaneously, while the optimization procedure is applied directly to the critical index. It appears to give more stable performance than all other methods based on finding the index as a control parameter from the critical amplitude optimization.

2. Power Transform and Minimal Sensitivity Condition for the Index

It is possible to improve the quality of approximants by employing power transforms [3]. To this end, we defined the power transform of the reduced expansion (2) as

$$P_k(x, m) \equiv \phi_k^m(x) \, . \tag{4}$$

In turn, the power transform (4) can be expanded in powers of x giving

$$P_k(x, m) \simeq \sum_{n=0}^{k} b_n(m) x^n \, . \tag{5}$$

After the self-similar renormalization of the expansion (5) is accomplished by means of, say, factor approximants, we arrive at an approximant $\mathcal{P}_k^*(x, m)$. Then, we accomplish the inverse transformation

$$F_k(x, m) = [\mathcal{P}_k^*(x, m)]^{1/m} \, . \tag{6}$$

Thus, in distinction with the statistics, we rigidly demand the existence of a direct and inverse transformation, eradicating an uncertainty in the matter of interpretation existing in the statistical applications [8,9].

The powers $m_k = m_k(x)$ are defined by the variational condition

$$\frac{\partial F_k(x, m)}{\partial m} = 0 \, . \tag{7}$$

Finally, the corresponding approximation for the sought function is given by

$$\phi_k^*(x) = F_k(x, m_k) \, . \tag{8}$$

Suppose that after all transformations performed explicitly one can find an explicit expression for the critical index $\beta_k(m)$, in kth order. When we are interested in the power law appearing in the large-variable limit and the critical index in particular, the condition (7) reduces to the requirement that the critical index should not depend on the parameter of transformation expressed in the following form:

$$\frac{\partial \beta_k(m)}{\partial m} = 0 \, . \tag{9}$$

The differentiation of the critical index leads to the minimal sensitivity condition on the parameter of transformation m.

Although equation (9) is quite intuitive, it can be explained following the idea put forward in [7]. Assume that we found the expression for the critical index

$$\beta(m) \approx \beta_k(m),$$

in kth approximation. Correspondingly, we can find the critical amplitude for the kth approximation $B_k(m) \approx B(m)$. Here, we should also recognize that $B(m) = B(\beta(m))$. As at large x

$$\phi_k^*(x) \simeq Bx^\beta,$$

the differentiation (7) can be performed, and the following expression obtained for large x,

$$\frac{\partial \phi_k^*(x)}{\partial m} \simeq x^\beta \left(\frac{\partial B}{\partial \beta} + B \ln x \right) \frac{\partial \beta}{\partial m}. \tag{10}$$

Thus, as long as we are interested in large x critical properties, in order to make the approximation $F_k(x, m)$ minimally sensitive to the transformation parameters it is sufficient to satisfy the equation (9).

To put the meaning into the formulas one is bound to select the self-similar approximants. Following the work in [3], we consider the kth order "odd" self-similar factor approximant

$$\mathcal{P}_k^*(x) = 1 + \mathcal{A}x \prod_{i=1}^{N_k-1} (1 + \mathcal{A}_i x)^{s_i}, \quad N_k = \frac{1+k}{2}, \quad k = 3, 5, \ldots. \tag{11}$$

The parameters $\mathcal{A}$, $\mathcal{A}_i$ and s_i have to be defined from an odd number of k conditions. The necessary number of conditions typically is extracted from the conditions of asymptotic equivalence with the truncated series.

In our study, we insist only that the total number of conditions employed should be equal to k. The approximant can be considered as "odd" [3], while factor approximants of "even" type, recently considered as subject of optimization [6], are simply not amenable to the optimization by power transform. They should be approached by some other optimization techniques developed previously in [6,7].

With $k = 3$ we simply have

$$\mathcal{P}_3^*(x) = 1 + \mathcal{A}x(1 + \mathcal{A}_1 x)^{s_1}. \tag{12}$$

As $x \to \infty$

$$F_3^*(x, m) \simeq \left(\mathcal{A}(m)(\mathcal{A}_1(m))^{s_1(m)} \right)^{1/m} x^{(1+s_1(m))/m}.$$

Minding that $\mathcal{A}$, $\mathcal{A}_1$ and s_1 are functions of m, the critical index can be found

$$\beta \approx \beta_3(m) = \frac{1 + s_1(m)}{m}, \tag{13}$$

but only as the function of parameter m. It has to be found from the simplified minimal sensitivity condition

$$\frac{\partial \beta_3(m)}{\partial m} = 0. \tag{14}$$

After the particular value of $m_3 \equiv m^*$ is found, one can find the the critical index from the formula (13), and the critical amplitude

$$B \approx B_3(m^*) = \left(\mathcal{A}(m^*)\mathcal{A}_1(m^*)^{s_1(m^*)} \right)^{1/m^*}.$$

The form of approximation expressed by the formula (12) is common for all self-similar approximants known to us, such as roots, factors, additive approximants, continued roots [3,19,20], i.e., it is not sensitive at all to the assumptions made while constructing the approximations in higher-orders, and includes the basic, single step of the algebraic self-similar transformation, as explained in [20,21].

For $k = 5$ we have

$$P_5^*(x) = 1 + \mathcal{A}x(1 + \mathcal{A}_1 x)^{s_1}(1 + \mathcal{A}_2 x)^{s_2}.$$

As $x \to \infty$

$$F_5^*(x,m) \simeq \left(\mathcal{A}(m)(\mathcal{A}_1(m))^{s_1(m^*)}(\mathcal{A}_2(m))^{s_2(m)}\right)^{1/m} x^{(1+s_1(m)+s_2(m))/m}.$$

With

$$\beta \approx \beta_5(m) = \frac{1 + s_1(m) + s_2(m)}{m}, \tag{15}$$

the parameter $m_5 \equiv m^*$ has to be found from the minimal sensitivity condition

$$\frac{\partial \beta_5(m)}{\partial m} = 0, \tag{16}$$

One can also find the critical amplitude

$$B \approx B_5(m^*) = \left(\mathcal{A}(m^*)(\mathcal{A}_1(m^*))^{s_1(m^*)}(\mathcal{A}_2(m^*))^{s_2(m^*)}\right)^{1/m^*}.$$

In order to simplify formulas without modifying the expressions for critical index and minimal sensitivity equations, we are going to set $\mathcal{A}_2(m) = 1$ (see, e.g., in [7]). However, imposing such conditions in lower-order would not bring a solution to the optimization problem.

The form of the expressions for critical induces (13) and (15), suggests that in special case when the optimal value is found as $m^* = 0$ (or close), one can start, or include into consideration at par with other approximations, some logarithmic-type approximations as discussed, e.g., in [22]. As $m = 1$, the formulas (13) and (15) will correspond to the results for the critical index originating only from the factor approximants (11) per se.

In the general case of arbitrary m, the critical index is a product of the algebraic transformation which contributes the set of control parameters s_i, and of the power transform which corrects them in two ways. First, s_i are getting "dressed" becoming dependent on the parameter of power transform m and, second, they are divided by m.

The parameter of power transformation m has a simple meaning of the multiplier connecting the critical exponent with the correction-to-scaling exponent, to be explained below in the Section 3.1. Both the power transform and minimal sensitivity condition are of a non-perturbative origin, based on generic ideas on improving the quality of perturbative calculations. The expression (15) can be with ease generalized to the kth order,

$$\beta \approx \beta_k(m) = \frac{1 + s_1(m) + s_2(m) + \ldots s_k(m)}{m}.$$

3. Critical Point at Infinity

As long as we are concerned with the low-order calculations, it makes sense to simplify the problem even further. In order to analytically solve the minimal problem, with only two non-trivial coefficients, a_1 and a_2, let us impose an additional condition on the amplitudes, namely,

$$\mathcal{A}_1 = \mathcal{A}.$$

This condition does not directly involve the shape of the minimization condition imposed exclusively on the critical index. Imposing the constraint also allows to greatly simplify calculations and obtain a unique solution for the optimization problem in lowest non-trivial order. Besides, it also guarantees the existence of the (approximate) solution by guaranteeing positive amplitudes.

From the asymptotic equivalence with the truncated series, we find

$$\mathcal{A}_1(m) = \mathcal{A}(m) = a_1 m, \quad s_1(m) = \frac{a_1^2 m - a_1^2 + 2a_2}{2a_1^2 m}. \tag{17}$$

From the optimization condition, we find the optimal control parameter

$$m^* = \frac{2(a_1^2 - 2a_2)}{3a_1^2}. \tag{18}$$

It leads to the critical index

$$\beta \approx \beta_3 = \frac{9a_1^2}{8(a_1^2 - 2a_2)}, \tag{19}$$

and critical amplitude

$$B \approx B_3 = \left(\frac{2}{3}\right)^{\frac{9a_1^2}{8(a_1^2 - 2a_2)}} \left(a_1 - \frac{2a_2}{a_1}\right)^{\frac{9a_1^2}{8(a_1^2 - 2a_2)}}. \tag{20}$$

See Section 3.4 for more general formulas (17) in the case of $\mathcal{A}_1(m) \neq \mathcal{A}(m)$.

For longer series, the optimization condition possesses multiple solutions and additional constraints should be applied. In particular, we require that the chosen solution should be best in prediction of the coefficients a_k not employed in its construction.

The optimization condition leads to the equation on high-order polynomials in the parameter m, so that an exact solution cannot be found. On the other hand, the class of solutions is significantly broadened because the complex-conjugate roots are now allowed. In some important examples presented below they appear to give the best results when the constraints are imposed. In principle, the error/measure of such prediction can be optimized by itself, with respect to the parameter of power transform.

3.1. Swelling of Polymers

An important characteristic of polymer chains is their swelling factor. It is just the ratio of the mean-square end-to-end distance of the chain, with interactions between its segments, to the value of the mean-square end-to-end distance of the chain, without such interactions.

Two-dimensional polymers are often met in chemistry and biology. For such polymers, perturbation theory with respect to weak interactions can be developed [23,24]. It can be reduced to a truncated series in a single dimensionless interaction parameter g. For the the expansion factor $Y(g)$, it gives

$$Y(g) \simeq 1 + \frac{1}{2}g - 0.12154525g^2 + 0.02663136g^3 - 0.13223603g^4, \tag{21}$$

as $g \to 0$.

In the strong-interaction limit [25,26], one expects the power law behavior

$$Y(g) \simeq Bg^\beta \quad (g \to \infty), \tag{22}$$

with the critical exponent $\beta = 1/2$. One also considers the critical index $\nu \equiv \frac{1}{2}(1+\beta)$, which gives $\nu = 0.75$.

Application of the formulas presented above gives $m^* = 1.31491$, $\beta = \beta_3 = 0.570382$, $B = B_3 = 0.787252$. Correspondingly, $\nu \approx 0.785191$, well within the bounds $\nu = 0.77525 \pm 0.021747$ calculated in [2].

In the case of a three-dimensional polymer coil, perturbation theory [23] for the swelling factor leads to series of the same type as (21), but with the coefficients

$$a_1 = \frac{4}{3}, \quad a_2 = -2.075385396, \quad a_3 = 6.296879676,$$

$$a_4 = -25.05725072, \quad a_5 = 116.134785, \quad a_6 = -594.71663.$$

The strong-coupling limit of the swelling factor $Y(g)$ can be found from [24], and

$$Y(g) \simeq 1.531 g^{0.3544} \quad (g \to \infty), \tag{23}$$

By applying our approach, we obtain $m^* = 2.22321$, $\beta = \beta_3 = 0.337351$, $B = B_3 = 1.44279$ and the critical index $\nu \approx 0.584338$. Complete swelling factor looks rather simple,

$$Y(g) = \phi_3^*(g) = \left(1 + \frac{2.96427g}{(2.96427g + 1)^{0.25}}\right)^{0.449801}.$$

The result for the critical index is located within the bounds $\nu = 0.5814 \pm 0.006$ found earlier in [5], based on the same input form the truncated series.

One may expect that the higher-order terms will have some effect on the results. There are multiple solutions to the minimal sensitivity equation. The best solution to the minimal sensitivity problem corresponds to

$$m_1^* = 2.31467.$$

The corresponding approximant reads as follows,

$$Y(g) = \triangleleft_5^*(g) = \left(1 + \frac{3.08622g}{(g+1)^{0.0869825}(6.21143g + 1)^{0.0954871}}\right)^{0.432028}. \tag{24}$$

Asymptotically, at large g

$$\phi_5^*(g) \simeq B g^\beta,$$

and the critical indices $\beta = 0.353196$, $\nu = 0.588299$, and the critical amplitude $B = 1.50912$. Compared to all other solutions to the minimal sensitivity condition, formula (24) gives the smallest average error in estimation of the 5th and 6th order coefficients in the expansion. The results fit within the bounds marked by the numerical result $\nu = 0.5886$ [24], and by the numerical result $\nu = 0.5877$ of [27], or even by a slightly lower value of 0.5876 obtained in [28].

Intriguingly, the "odd" factor approximants contain also the correction-to-scaling exponent Δ defined in [2,24,29], as follows:

$$Y(g) \simeq B g^\beta \left(1 + \frac{B_1}{B} g^{-\Delta}\right) \quad (g \to \infty). \tag{25}$$

After extraction of the critical behavior, and by rewriting the transformed factor approximants for $k = 3, 5$, one can see that in kth order

$$\Delta_k = \beta_k(m) \, m.$$

Note that the technique of optimization from [6], which also dwells on the factor approximants, always brings $\Delta \equiv 1$. The case of $k = 3$ is also qualitative, always bringing $\Delta = 0.75$.

However, for $k = 5$, we simply calculate $\Delta_5 \approx 0.82$, in a good agreement with the numerical estimate $\Delta = 0.93$ from [24]. Thus the parameter m of the power transformation is the multiplier connecting the two exponents. The connection is of general nature and not limited to the case of polymers. Analysis of available literature does confirm that in all cases considered in this section, such simultaneous estimation of the two indices works reasonably well.

3.2. Schwinger Model

The Schwinger model [30,31] represents the Euclidean quantum electrodynamics interacting with a Dirac fermion field, defined on a lattice in $(1+1)$ dimensions. The model reflects confinement, chiral symmetry breaking, and charge shielding, sharing therefore the key properties with quantum chromodynamics. The ground state of the model, given as a function of the dimensionless variable $x = m/g$. Here, m stands for electron mass and g is the coupling parameter. It also has the dimension of mass. The energy $E = M - 2m$, corresponds to a vector boson of mass $M(x)$.

The expansion at small-x for the ground-state energy [32–35] is known in the following form:

$$E(x) \simeq 0.5642 - 0.219x + 0.1907x^2 \quad (x \to 0). \tag{26}$$

In the complementary, large-x limit [35–38], there is a power law,

$$E(x) \simeq Bx^\beta + O\left(x^{-1}\right) \quad (x \to \infty), \tag{27}$$

with $B = 0.6418$, $\beta = -1/3$.

To standardize calculations, let us first normalize the expansion (26) to unity at $x = 0$. Elementary calculations according to the formulas presented above, give $m^* = -2.32446$, $\beta \approx \beta_3 = -0.322656$, $B_3 = 1.03374$. Restoring the original units, we find the critical amplitude $B = 0.583237$. The result for the index fits within the bounds $\beta = -0.311 \pm 0.2$ found in [5].

3.3. Harmonium

An N-electron harmonium atom is described by the Hamiltonian

$$\hat{H} = \frac{1}{2}\sum_{i=1}^{N}\left(-\nabla_i^2 + \omega^2 r_i^2\right) + \frac{1}{2}\sum_{i \neq j}^{N}\frac{1}{r_{ij}}, \tag{28}$$

where dimensionless variables are used, $\frac{\omega^2}{2}$ stands for the harmonic oscillator force constant [39], and $r_i \equiv |\mathbf{r}_i|$, $r_{ij} \equiv |\mathbf{r}_i - \mathbf{r}_j|$.

Following Cioslowski [39], we consider a two-electron harmonium atom with $N = 2$. The ground-state energy for a rigid potential diverges as the power law [39] at large ω,

$$E(\omega) = 3\omega + O\left(\omega^{1/2}\right) \quad (\omega \to \infty). \tag{29}$$

At a shallow harmonic potential, the energy can be expanded [39] in powers of ω giving in low orders the following truncation:

$$E(\omega) \simeq 1.19055\,\omega^{2/3} + 2.36603\,\omega + 0.122492\,\omega^{4/3} \quad (\omega \to 0). \tag{30}$$

By introducing the new variable $x \equiv \omega^{1/3}$, Equation (30) could be reduced to

$$E(x^3) \simeq 1.19055 x^2 (1 + 1.98734 x + 0.102887 x^2) \quad (x \to 0). \tag{31}$$

Applying our method to the expression within brackets, we find $m^* = 0.631933$, $\beta_3 = 1.18684$, $B_3 = 1.31048$, and reconstruct the large ω behaviour

$$E(\omega) \simeq B\,\omega^\beta \quad (\omega \to \infty), \tag{32}$$

with $B = 3.10063$, $\beta = 1.06228$. The error of 6% for the critical index should be considered as quite satisfactory for calculations with such short truncation as (30). The critical index is well within the bounds $\beta = 1.049 \pm 0.031$ found in [5].

3.4. Nonlinear Schrödinger Equation

The following nonlinear Hamiltonian

$$\hat{H}_{NLS} = -\frac{1}{2}\frac{d^2}{dx^2} + \frac{1}{2}x^2 + g|\psi|^2, \tag{33}$$

defines the one-dimensional stationary nonlinear Schrödinger equation for the wave function ψ of the Bose-condensed atoms in a harmonic trap. Here, g is a dimensionless coupling parameter.

The energy levels $E(g)$ for the Hamiltonian (33) can be represented in the form $E(g) = \left(n + \frac{1}{2}\right)e(g)$, where $n = 0, 1, 2, \ldots$ labels the eigenvalues. The following expansion for function $e(g)$ in powers of the effective coupling:

$$e_5(g) = 1 + g - \frac{1}{8}g^2 + \frac{1}{32}g^3 - \frac{1}{128}g^4 + \frac{3}{2048}g^5, \tag{34}$$

can be found in [20,40]. Moreover, for the strong-coupling limit we have

$$e(g) = \frac{3}{2}g^{2/3} + O\left(g^{-2/3}\right) \quad (g \to \infty). \tag{35}$$

There are multiple solutions to the minimal sensitivity equation. The best solution to the minimal sensitivity problem corresponds to the complex-conjugate pair

$$m_1^* = 1.27731 + 0.320816i, \quad m_2^* = 1.27731 - 0.320816i.$$

The real part of the corresponding approximants gives the following approximant,

$$\phi_5^*(g) = $$
$$0.5(1 + (1.27731 + 0.320816i)g(1 + (0.437208 - 0.179573i)g)^{-0.253536+0.0807663i} \times$$
$$(g+1)^{0.110002+0.0795681i})^{0.736436-0.184967i} + \tag{36}$$
$$0.5(1 + (1.27731 - 0.320816i)g(1 + (0.437208 + 0.179573i)g)^{-0.253536-0.0807663i} \times$$
$$(g+1)^{0.110002-0.0795681i})^{0.736436+0.184967i}.$$

Asymptotically, at large g

$$\phi_5^*(g) \simeq Bg^\beta,$$

and gives a very good critical index $\beta = 0.660389$, and critical amplitude $B = 1.50916$.

Compared to all other solutions to the minimal sensitivity condition, Formula (36) gives the smallest (by order of magnitude) error in estimation of the 5th order coefficient in the expansion.

One can also model Bose-condensate within spherically-symmetrical traps by the following effective Hamiltonian:

$$\hat{H}_r = \frac{1}{2}\left(-\frac{d^2}{dr^2} + r^2\right) + \frac{g}{4\pi r^2}\chi^2, \tag{37}$$

for the radial part of the condensate wave function $\chi(r)$ [41].

The ground state energy can be approximated by the expansions

$$E(c) \simeq \frac{3}{2} + \frac{1}{2}c - \frac{3}{16}c^2 + \frac{9}{64}c^3 - \frac{35}{256}c^4 \quad (c \to 0), \tag{38}$$

and
$$E(c) = \frac{5}{4}c^{2/5} + O\left(c^{-2/5}\right) \quad (c \to \infty), \tag{39}$$

where $c = \frac{g}{(2\pi)^{3/2}}$.

Applying our method, we find $m^* = 13/6 \approx 2.17$, $\beta_3 \approx 0.35$, $B_3 = 1.34$, and reconstruct the large c behaviour
$$E(c) \simeq Bc^\beta,$$

with $B = 1.34, \beta = 0.35$. The third-order coefficient (divided by a_0) could be estimated as well, and is equal to 0.112, in a reasonable agreement with the $a_3/a_0 \approx 0.14$ from the expansion (38).

Avoiding restrictions on the amplitudes by keeping $\mathcal{A}_1(m) \neq \mathcal{A}(m)$, we reconstruct the unrestricted approximation

$$F_3^*(x,m) = \left(1 + \mathcal{A}(m)\,x\,(1 + \mathcal{A}_1(m)\,x)^{s_1(m)}\right)^{1/m}.$$

and find
$$\mathcal{A}(m) = a_1 m,$$
$$\mathcal{A}_1(m) = \frac{a_1{}^4(-m^2) + 6a_1{}^4 m - 5a_1{}^4 - 12a_1{}^2 a_2 m + 12a_1{}^2 a_2 - 24a_1 a_3 + 12a_2{}^2}{6a_1(a_1{}^2 m - a_1{}^2 + 2a_2)}, \tag{40}$$
$$s(m) = -\frac{3\left(a_1{}^2 m - a_1{}^2 + 2a_2\right)^2}{a_1{}^4 m^2 - 6a_1{}^4 m + 5a_1{}^4 + 12a_1{}^2 a_2 m - 12a_1{}^2 a_2 + 24a_1 a_3 - 12a_2{}^2}.$$

Optimization condition remains of the same form and brings optimal $m^* = 6.0512$, with excellent critical index $\beta_3 \approx 0.39$, and good critical amplitude $B_3 \approx 1.325$, while the resulting approximant is given as follows,

$$\phi_3^*(c) = \frac{3}{2}\left(2.01707 c (0.343481 c + 1)^{1.35922} + 1\right)^{0.165257}.$$

The fourth-order coefficient (divided by a_0) could be estimated as well, and is equal to 0.088, in reasonable agreement with the $a_4/a_0 \approx 0.091$ from the expansion (38). It also reconstructs the large c behaviour
$$E(c) = Bc^\beta + O(c^{-1}),$$

with the results $B = 1.325, \beta = 0.39$.

3.5. Quartic Oscillator

To conclude our variations on the theme of perturbed harmonic oscillators, let us discuss the very popular quantum model of quartic anharmonic oscillator [42], with the Hamiltonian
$$\hat{H} = -\frac{1}{2}\frac{d^2}{dx^2} + \frac{1}{2}x^2 + gx^4,$$

with the anharmonicity parameter $g \in [0, \infty)$.

One can construct the perturbation theory for the ground-state energy in the parameter $g \to 0$,

$$e(g) \simeq \frac{1}{2} + \frac{3}{4}g - \frac{21}{8}g^2 + \frac{333}{16}g^3 - \frac{30885}{128}g^4 + \frac{916731}{256}g^5 - \frac{65518401}{1024}g^6. \tag{41}$$

The expansion in the parameter $g \to 0$ is divergent also in high orders.

For large g the series for $e(g)$ have fractional powers, with the leading term from the strong-coupling expansion of the energy given as follows,
$$e(g) \simeq Bg^\beta \quad (g \to \infty),$$

with $\beta = 1/3, B \approx 0.667986$.

However, setting $\mathcal{A}_2(m) = 1$ does not lead to a completely satisfactory solution, although some very reasonable estimates for the critical index and amplitude could be found following the standard calculations, i.e., optimal complex-conjugate pair

$$m_1^* = 0.0810036 - 5.35446i, \quad m_2^* = 0.0810036 + 5.35446i,$$

with excellent critical index $\beta_5 \approx 0.3213$, and good critical amplitude $B_5 \approx 0.6979$. However, an overall picture described by a complete approximant $\phi_5^*(g)$ is not so good, as we were not able to construct an accurate expression for all coupling constants.

However, lifting the restriction on amplitudes, and finding $\mathcal{A}_2(m)$ at par with other parameters from the asymptotic equivalence with the truncated weak-coupling expansion, allows to get a consistent expression for all g. Using five non-trivial conditions from the weak-coupling expansion and expressing the approximant parameters as functions of m, one can find m from the condition of minimal error in predicting the 6th order coefficient, not yet employed. Along this pass we find the following approximant:

$$\phi_5^*(g) = \frac{1}{2}\left(1 + \frac{16.7519g(1+1.8618g)^{2.42764}}{(1+25.4873g)^{0.0154605}}\right)^{0.0895423}. \tag{42}$$

with quite reasonable estimates for the critical index $\beta \approx 0.3055$ with error of 8.3%, and the critical amplitude $B \approx 0.7333$. To achieve high accuracy, we still need considerably more terms [22]. The approximant (42) gives good results in pure extrapolation, predicting the ground state energy with the error remaining less than 3% up to quite large $g = 50$, compared with the numerical results [43]. The self-similar approximants from [4] are also rather accurate, giving the critical index close to 0.3.

The example below shows the value of studying such a model system as the quartic oscillator, as the series to be studied does resemble very much the series for the quartic oscillator.

Let us consider also the energy gap $\delta(z)$ between the lowest and first excited states of the vector boson for the massive Schwinger model in Hamiltonian lattice theory [30,36], which can be represented for strong coupling constants g as follows:

$$\delta(z) \simeq 1 + 2z - 10z^2 + \\ 78.66667z^3 - 736.2222z^4 + 7572.929z^5 - 82736.69z^6 + 942803.4z^7, \tag{43}$$

where $z = x^2$, $x = \frac{1}{g^2 a^2}$, and a is the lattice spacing. The strong increase of the coefficients makes the series in powers of z widely divergent. They do resemble the truncated expression for the quartic oscillator. The transition from the lattice formulation to the continuous limit requires taking the limit $a \to 0$, implying $z \to \infty$. In this limit, the gap behaves as

$$\delta(z) \sim z^\beta \quad (z \to \infty),$$

with $\beta = 1/4$.

There are two good solutions to the optimization problem. The complex-conjugate pair

$$m_1^* = 7.13701 - 1.35426i, \quad m_2^* = 7.13701 + 1.35426,$$

with $\beta_5 = 0.22433$, (10% error), and real solution $m^* = 10.8324$, with $\beta_5 = 0.237578$ (5% error). The former solution incurs slightly smaller average error in predicting unexploited three higher-order terms from the expansion (43). Yet, the latter solution is still preferable as it brings a smaller maximal error of 2.7% in prediction of the remaining three coefficients from the truncation (43). The real solution corresponds to the approximant

$$\phi_5^*(z) = \left(1 + 21.6649z(1+z)^{0.964953}(1+6.35479z)^{0.608593}\right)^{0.0923153}.$$

4. Finite Critical Point

Consider another ubiquitous situation when the function $\phi(x)$ of a real variable x exhibits critical behavior,

$$\phi(x) \simeq B(x_c - x)^\beta, \text{ as } x \to x_c - 0, \tag{44}$$

with a positive or negative critical index β, and critical amplitude B at a finite critical point x_c. The approach developed for the case of a critical point located at infinity can be applied with minor modifications when the critical point x_c is finite and its position is known, in conjunction with transformation

$$z = \frac{x}{x_c - x}, \tag{45}$$

while $x = \frac{zx_c}{z+1}$. In the low-order case of $k = 3$, just like in the case of critical point located at infinity, we obtain explicit and uniquely defined expressions,

$$\mathcal{A}_1(m) = \mathcal{A}(m) = a_1 m x_c, \quad s_1(m) = \frac{a_1^2 m x_c - a_1^2 x_c - 2a_1 + 2a_2 x_c}{2a_1^2 m x_c}. \tag{46}$$

The optimization problem remains of the same form, and by solving it we find optimal control parameter

$$m^* = \frac{2}{3}\left(1 + \frac{2(a_1 - a_2 x_c)}{a_1^2 x_c}\right). \tag{47}$$

It leads to the critical index

$$\beta \approx \beta_3 = -\frac{9a_1^2 x_c}{8(a_1^2 x_c + 2a_1 - 2a_2 x_c)}, \tag{48}$$

and critical amplitude

$$B \approx B_3 = \left(\frac{2}{3}\right)^{\frac{9a_1^2 x_c}{8(a_1^2 x_c + 2a_1 - 2a_2 x_c)}} \left(x_c\left(-\frac{2a_2 x_c}{a_1} + a_1 x_c + 2\right)\right)^{\frac{9a_1^2 x_c}{8(a_1^2 x_c + 2a_1 - 2a_2 x_c)}}. \tag{49}$$

A number of examples considered below cover the most interesting for us physical cases. The problems of conductivity, viscosity, and permeability arising in different contexts can be solved based on seemingly sparse information.

4.1. 2D Ising Model

Consider spin-1/2 Ising model characterized by the Hamiltonian

$$\hat{H} = -\frac{J}{2}\sum_{\langle ij \rangle} s_i^z s_j^z,$$

with $s_j^z \equiv \frac{S_j^z}{S}$, on a square lattice, with the ferromagnetic interaction J of nearest neighbors, for spins $S_j^z = \pm 1/2$ [44]. The dimensionless interaction parameter is defined as $g \equiv \frac{J}{k_B T}$, where k_B stands for the Boltzmann constant and T is temperature.

On the square lattice, a high-temperature expansion of the susceptibility χ in powers of dimensionless inverse temperature g could be obtained in rather high orders [44]. The starting terms of the expansion are given as follows,

$$\chi(g) \simeq 1 + 4g + 12g^2. \tag{50}$$

It is expected [45] that in the vicinity of the threshold $g_c = (\frac{2}{\log(1+\sqrt{2})})^{-1} \approx 0.440687$, the 2D susceptibility diverges as

$$\chi(g) \sim (g_c - g)^{-\gamma},$$

with exact $\gamma = \frac{7}{4}$.

Application of the formulas presented above gives $m^* = 0.423062$, $\gamma = \beta_3 = -1.77279$, $B_3 = 0.139076$. The critical index is well within the bounds $\gamma = 1.76 \pm 0.15$, found in [46]. It is significantly better than our previous result $\gamma = 1.923 \pm 0.077$, obtained by applying the technique of root approximants [5].

4.2. Conductivity of Percolating Systems

The conductivity for site percolation model is explained as a transport of classical particles through a random medium [47,48]. The minimal model of such phenomena is the Lorenz 2D gas, which is realized on a square lattice with a fraction of sites being excluded at random. If f stands for the concentration of conducting or not excluded sites in the Lorenz model, then $x = 1 - f$ stands for the concentration of excluded sites. Through the diffusion coefficient for random walkers on such a lattice one can express the macroscopic conductivity [47]. The transport ceases to exist at the critical density of the excluded sites x_c corresponding to the site percolation threshold [49], and the conductivity behaves as

$$\sigma(x) \propto (x_c - x)^t \quad (x \to x_c - 0), \tag{51}$$

with $x_c = 0.4073$, $t = 1.310$. Perturbation theory in powers of the variable $x = 1 - f$ [47], gives for the two-dimensional square lattice the expansion

$$\sigma(x) \simeq 1 - \pi x + 1.28588 x^2 \quad (x \to 0). \tag{52}$$

In our approach, we obtain $m^* = -0.549066$, $t = \beta_3 = 1.36596$, $B_3 = 5.52404$. The estimate for the critical index is well within the bounds $t = 1.291 \pm 0.1$ found in [46].

The effective conductivity for the three-dimensional site percolation is studied similarly to the two-dimensional one. The conductivity also exhibits the critical behaviour [50–54], described just as in the equation (51), but with $x_c = 0.688$, $t = 1.9$. Perturbation theory gives

$$\sigma(x) \simeq 1 - 2.52x + 1.52x^2 \quad (x \to 0). \tag{53}$$

Using our method, we get $m^* = -0.421068$, $t = \beta_3 = 1.78119$, $B_3 = 3.40252$. The estimate for the critical index is well within the bounds $t = 1.82 \pm 0.09$, calculated in [46].

4.3. Permeability of the Two-Dimensional Channels

Let us consider the case of a Darcy flow in the two-dimensional channel bounded by the surfaces $z = \pm b(1 + \epsilon \cos x)$, where ϵ is termed waviness. The permeability $K(\epsilon)$ behaves critically [2,5,55]. Precisely, it tends to zero as

$$K(\epsilon) \sim (\epsilon_c - \epsilon)^\kappa, \text{ as } \epsilon \to \epsilon_c - 0, \tag{54}$$

with $\epsilon_c = 1$, $\kappa = \frac{5}{2}$. An expression for permeability can be derived by iterative perturbation method. It gives a truncated expansion in powers of the waviness [2,55–57].

The permeability, for $b = 0.5$, has the expansion

$$K(\epsilon) \simeq 1 - 3.14963\, \epsilon^2 + 4.08109\, \epsilon^4, \text{ as } \epsilon \to 0, \tag{55}$$

Application of the formulas presented above gives $m^* = -0.305188$, $\kappa = \beta_3 = 2.4575$, $B_3 = 1.10206$. The critical index is within the bounds $\kappa = 2.372 \pm 0.19$ found in [5], but is much closer to the upper bound and to the expected value of $5/2$ than to the center.

The permeability, for $b = 0.25$ [56], has the expansion

$$K(\varepsilon) \simeq 1 - 3.03748\,\varepsilon^2 + 3.54570\,\varepsilon^4, \text{ as } \varepsilon \to 0. \tag{56}$$

Application of the formulas presented above gives $m^* = -0.284699$, $\kappa = \beta_3 = 2.63436$, $B_3 = 0.670918$. The critical index is located well within the bounds $\kappa = 2.543 \pm 0.2$ found in [5].

4.4. Effective Viscosity

The elasticity problem of perfectly rigid spherical inclusions randomly embedded into an incompressible matrix is analogous to the problem of high-frequency effective viscosity of a hard-sphere suspension [58–60]. The viscosity, considered as a function of the variable $f \equiv \frac{4\pi}{3} r_s^3 \rho$, ($\rho \equiv \frac{N}{V}$), in which r_s is the sphere radius and ρ is average density, exhibits the critical behaviour

$$\eta(f) \propto (f_c - f)^{-S} \quad (f \to f - 0), \tag{57}$$

where $f_c = 0.637$, $S = 1.75$ [61]. The small f-expansion form [60], reads as

$$\eta(f) \simeq 1 + \frac{5}{2} f + 5.0022 f^2 \quad (f \to 0). \tag{58}$$

Using our method, we find $m^* = 0.436789$, $S = -\beta_3 = 1.71708$, $B_3 = 0.24717$. The estimate for the critical index appears to fit within the bounds $S = 1.726 \pm 0.06$, obtained in [5]. In what follows, one should distinguish the critical index for viscosity (elasticity) S from the critical index for superconductivity denoted as s.

4.5. Critical Index for Superconductivity in Random 3D Case

In the limiting case of a three-dimensional randomly distributed perfectly conducting inclusions, the effective conductivity σ_e is expected to tend to infinity as a power law, with critical index s, as the concentration of inclusions f tends to $f_c \approx 0.637$, the maximal value in 3D, and

$$\sigma_e(f) \sim (f_c - f)^{-s}.$$

The superconductivity critical index s is expected to have the value of 0.73 ± 0.01 [54]. There is also a slightly larger estimate, $s \approx 0.76$ [62].

For sample generation, the Random Sequential Adsorption protocol was employed [1]. The consecutive objects were placed randomly in the cell, rejecting those that overlap with the previously absorbed one. For macroscopically isotropic composites, the expansion for scalar effective conductivity was found to be

$$\sigma_e = 1 + 3f + 3f^2 + 4.80654 f^3 + O(f^{\frac{10}{3}}).$$

Application of the formulas presented above gives $m^* = 0.919937$, $s = -\beta_3 = 0.815273$, $B_3 = 1.09667$.

For the effective conductivity we find the following compact expression

$$\sigma_e(f) = \phi_3^*(f) = \left(1 + \frac{1.758 f}{\left(\frac{0.758 f + 0.637}{0.637 - f}\right)^{0.25} (0.637 - f)}\right)^{1.08703},$$

which can be expanded in powers of f. It does exceptionally well in estimating the coefficient $a_3 \approx 4.7963$, deviating from the expected value only 0.21%. The 4th-order coefficient can be estimated as well, $a_4 \approx 6.68$.

The low-order estimate appears to be self-consistent, because the estimate for the critical index is supported by a simultaneous excellent result for a_3. Of course, one can attempt to incorporate the third-order coefficient explicitly, i.e., to devise the approximant

$\mathcal{P}_3^*(f)$ with $\mathcal{A} \neq \mathcal{A}_1$. Computations are still possible to perform in a symbolic form. In such case the estimates become very close to the numerical results of [54], and to the results of calculations with various approximants [2], based on the same number of terms,

$$\sigma_e(f) = \phi_3^*(f) =$$
$$\frac{1}{2}((1 - \frac{(2.55836+0.117254i)f(1+\frac{(1.28492-0.744437i)f}{0.637-f})^{-0.042703+0.0208865i}}{f-0.637})^{0.745398-0.034163i} +$$
$$(1 + \frac{(2.55836-0.117254i)f(1-\frac{(1.28492+0.744437i)f}{f-0.637})^{-0.042703-0.0208865i}}{0.637-f})^{0.745398+0.034163i}), \quad (59)$$

leading to the critical index s = 0.714, and the critical amplitude B = 1.46. The 4th order coefficient can be estimated as $a_4 \approx 6.01$. But to confirm or reject the results one will have to rely on some novel, additional information, which is not available at the moment.

4.6. Critical Index for Superconductivity of Honeycomb Array

Finally, consider a regular honeycomb array of perfectly conducting (superconducting) disks. As their volume fraction $f \to f_c$, the effective conductivity of the array goes to infinity as a power law similar to previous example, with critical index s, as the concentration of disks f tends to $f_c = \frac{\pi}{3\sqrt{3}}$.

It is always instructive to study the regular case by yet different methods, other than employed in [63], and estimate the critical index s. The exact value for the index is expected to be 1/2. The small-f polynomial has the following form,

$$\sigma_e \simeq 1 + 2f + 2f^2 + 2f^3 + 4.14933f^4 + 6.29865f^5 + 8.44798f^6.$$

There are multiple solutions to the optimization problem understood as solving the minimal sensitivity equation. Note that we have to apply first the transformation $z(f) = \frac{f}{f_c - f}$, as in the formula (45), and then construct the critical index $\beta_5(m)$ (mind the sign!) and approximant $\phi_5^*(z(f))$, as shown above in the Section 2.

The best solution to the minimal sensitivity problem corresponds to the complex–conjugate pair

$$m_1^* = 2.46436 - 0.462709i, \qquad m_2^* = 2.46436 + 0.462709i.$$

The real part of the corresponding approximants gives

$$\sigma_e(f) = \phi_5^*(f) =$$
$$\frac{1}{2}((1 + (3.2629 + 0.850614i)f(1 + \frac{(1.56155-0.331863i)f}{0.6046-f})^{0.375221+0.347239i} \times$$
$$(\frac{1}{0.6046-f})^{0.788789-0.137954i})^{0.391967-0.0735958i} +$$
$$(1 + (3.2629 - 0.850614i)f(\frac{1}{0.6046-f})^{0.788789+0.137954i} \times$$
$$(1 + \frac{(1.56155+0.331863i)f}{0.6046-f})^{0.375221-0.347239i})^{0.391967+0.0735958i}). \quad (60)$$

Asymptotically, as $f \to f_c$,

$$\phi_5^*(g) \simeq B(f_c - f)^\beta,$$

with reasonable estimates for critical index $-\beta = s = 0.471655$, and critical amplitude B = 0.862639. The expected value for the index is 1/2 [63], the same as for other regular arrays of inclusions [1,64].

Consider for reassurance also the toy model represented by the function

$$\phi(x) = \frac{\sqrt{\frac{1+x}{1-x}} - 1}{x},$$

with transition at the point $x_c = 1$, index $\beta = 1/2$ and amplitude $B = \sqrt{2}$.

As $x \to 0$
$$\phi(x) \simeq 1 + \frac{x}{2} + \frac{x^2}{2} + \frac{3x^3}{8} + \frac{3x^4}{8} + \frac{5x^5}{16} + \frac{5x^6}{16}.$$

We find that the same conclusions as were reached above for realistic problems, apply here. Namely, there are multiple solutions of the higher-order optimization condition and the best result is achieved for the complex–conjugate pair

$$m_1^* = 2.89498 - 1.37644i, \quad m_2^* = 2.89498 + 1.37644i.$$

The real part of the corresponding approximants gives

$$\phi_5^*(f) = \frac{1}{2} \times$$
$$\left((1 - \frac{(1.44749 + 0.688222i)x(\frac{1}{1-x})^{0.720503 + 0.570413i}(1 - \frac{(1.3549 + 1.04464i)x}{x-1})^{-0.194988 - 0.0166872i}}{x-1}\right)^{0.281736 - 0.133954i} + \quad (61)$$
$$(1 - \frac{(1.44749 - 0.688222i)x(\frac{1}{1-x})^{0.720503 - 0.570413i} \times (1 - \frac{(1.3549 - 1.04464i)x}{x-1})^{-0.194988 + 0.0166872i}}{x-1})^{0.281736 + 0.133954i}),$$

while the critical index $\beta = -0.503966$. As was already suggested above, the good result for the index does not necessarily preclude correspondingly accurate results for the critical amplitude $B \approx 0.451$, as the optimization procedure involves the critical index, but not the critical amplitude. The formula (61) for the effective conductivity appears to be smart, meaning that it does more accurately (by orders of magnitude) than other solutions to the optimization problem, predict the average error for the remaining 2 coefficients from the expansion for conductivity.

4.7. Compressibility Factor of Hard-Disks fluids

The state of hard-disks fluids is described by the compressibility factor

$$Z = \frac{P}{\rho k_B T} = Z(f) \quad \left(f \equiv \frac{\pi \rho}{4} a_s^2\right), \quad (62)$$

in which P is pressure, ρ is density, T is temperature, a_s is the disk diameter, k_B stands for Boltzmann constant, and f is called packing fraction. The compressibility factor exhibits critical behaviour at a finite critical point. This behavior has been found from phenomenological equations as

$$Z(f) \simeq B(f_c - f)^\beta \quad (f \to f_c - 0), \quad (63)$$

with the fitted parameters $f_c = 1$ and $\beta = -2$ [65,66], although these are not asymptotically exact values. For low packing fractions the compressibility factor is represented by the virial expansion

$$Z(f) \simeq 1 + 2f + 3.12802 f^2 + 4.25785 f^3 + 5.3369 f^4 + 6.36296 f^5 + 7.35186 f^6$$
$$+ 8.3191 f^7 + 9.27215 f^8 + 10.2163 f^9, \quad (64)$$

from in [67,68]. Using the same optimization methods as above, we find $m^* = 0.808791$, $\beta_5 = -1.80539$, $B_5 = 1.84906$ and the approximant

$$\phi_5^*(f) = \left(1 + \frac{1.61758 \left(\frac{1}{1-f}\right)^{0.831449} f^{1.23641}}{(1-f)\left(1 + \frac{1.23536 f}{1-f}\right)^{0.371267}}\right), \quad (65)$$

which employs the terms up to the 4th order in f and predicts the rest with maximal error of just 0.52%. Simply setting $m^* = 1$ leads to a simpler expression

$$\phi_5^*(f, 1) = 1 + \frac{2f \left(\frac{1 - 0.122783 f}{1-f}\right)^{2.29987}}{\left(\frac{1}{1-f}\right)^{0.453479}},$$

which works with almost the same accuracy in predicting the coefficients from the truncated series (64). The latter formula may yet have advantage in predicting the higher-order coefficients. Such coefficients were only estimated up to 15th order (see, e.g., the book [2] and references therein), but are not exact unlike the others.

5. Concluding Remarks

We conclude that "odd" factor approximants of the special form represented by the Formula (11) are amenable to optimization by power transformation and can be successfully applied to the critical phenomena of various physical nature.

The novelty of the current approach amounts to the idea that the critical index by itself should be optimized through the parameters of power transform calculated from the optimization condition. The critical index is a product of the algebraic transformation which contributes to the expressions the set of control parameters s_i, and of the power transform which corrects them.

The parameter of power transformation m has the simple meaning of the multiplier connecting the critical exponent with the correction-to-scaling exponent. Both, the power transform and minimal sensitivity conditions are of a non-perturbative nature. As we optimize the critical index directly the results for the indices appear to be more accurate than obtained from optimization of critical amplitudes with a subsequent determination of the indices with optimal conditions calculated for the amplitudes.

We study mostly the minimal model of critical phenomena based on expansions with only two coefficients and critical points. The power-transformed factor approximants of the form (11) are asymptotically equivalent to such series. The approximants are optimized by complementing them with natural optimization conditions. The minimal sensitivity condition imposed on the critical index appears to bring quite accurate, uniquely defined results given by simple formulas. Surprisingly, many important cases of critical phenomena are being covered by the simple formulae. The knowledge of higher-order coefficients appears to be excessive as critical indices could be estimated from only two low-order coefficients. The multitude of the unknown higher-order coefficients can be mimicked by optimized power-transformed factor approximations.

For the longer series, the optimization condition possesses multiple solutions and additional constraints should be applied. In particular we require that the chosen solution is to be best in prediction of the coefficients a_k not employed in its construction. In principle, the error/measure of such prediction can be optimized by itself, with respect to the parameter of power transform. The latter idea dwells on the requirement of independence of the critical indices from the higher-order coefficients a_n. Some other approaches to long series were discussed in the preceding work [2,22].

Methods of calculation based on optimized power-transformed factors are applied and results presented for critical indices of several key models of conductivity and viscosity of random media, swelling of polymers, and permeability in two-dimensional channels. Several quantum mechanical problems based on the strongly perturbed harmonic oscillator are discussed as well.

Accurate calculations with short truncated series are possible and accurate in quite a few important cases, because the higher-order coefficients appear to be redundant close to the critical point, and critical indices could be estimated just from two low-order coefficients by imposing some universal conditions of non-perturbative nature. Power transform extends the class of approximations and brings some unique quality, such as incorporating both critical and sub-critical indices simultaneously. The optimization procedure is developed to estimate the indices together.

Convergence is not always as fast as in the examples presented above. To improve results one would try adding more terms from the truncation when possible, and apply different optimized approximations, or even introducing control functions instead of control parameters [1,22]. On the other hand, the methodology is limited to a short series because the optimization procedure developed above is based on analytical expressions for the parameters of approximants. Such limitation could be overcome by resorting to different self-similar approximants of the type described in [20].

One can think (see, e.g., in [18]) that for an approach based on optimization to be successful, Nature by itself should be organized in such a way that certain quantities called critical indices play a special role of stabilizing various physical phenomena in the vicinity of their respective critical points by making them minimally sensitive to the parameters of power transform.

Critical indices are introduced into consideration by assuming a trial power law. They could be found from special optimization conditions of general nature selecting a unique fixed point/sum of the asymptotic series associated with a given truncation. Such fixed points also incorporate the critical behavior which appears to be optimal.

Funding: This research received no external funding

Informed Consent Statement: Not applicable

Conflicts of Interest: The author declares no conflict of interest.

References

1. Gluzman, S.; Mityushev, V.; Nawalaniec, W. *Computational Analysis of Structured Media*; Academic Press: Cambridge, MA, USA, 2017.
2. Drygaś, P.; Gluzman, S.; Mityushev, V.; Nawalaniec, W. *Applied Analysis of Composite Media*; Woodhead Publishing: Sawston, UK, 2020.
3. Gluzman, S.; Yukalov, V.I. Self-Similar Power Transforms in Extrapolation Problems. *J. Math. Chem.* **2006**, *39*, 47–56. [CrossRef]
4. Gluzman, S.; Yukalov, V.I. Extrapolation of perturbation theory expansions by self-similar approximants. *Eur. J. Appl. Math.* **2014**, *25*, 595–628. [CrossRef]
5. Gluzman, S.; Yukalov, V.I. Critical indices from self-similar root approximants. *Eur. Phys. J. Plus* **2017**, *132*, 535. [CrossRef]
6. Gluzman, S. Optimized Factor Approximants and Critical Index. *Symmetry* **2021**, *13*, 903. [CrossRef]
7. Yukalov, V.I.; Gluzman, S. Optimization of Self-Similar Factor Approximants. *Mol. Phys.* **2009**, *107*, 2237–2244. [CrossRef]
8. Tukey, J.W. *Exploratory Data Analysis*; Addison-Wesley: Reading, MA, USA, 1977.
9. Box, G.E.; Cox, D.R. An analysis of transformations. *J. R. Stat. Soc. Ser. B* **1964**, *26*, 211–252. [CrossRef]
10. Carlson, J.M.; Doyle, J. HOT: A mechanism for power laws in designed systems. *Phys. Rev. E* **1999**, *60*, 1412–1427. [CrossRef] [PubMed]
11. Yukalov, V.I. Theory of perturbations with a strong interaction. *Mosc. Univ. Phys. Bull.* **1976**, *51*, 10–15.
12. Yukalov, V.I. Model of a hybrid crystal. *Theor. Math. Phys.* **1976**, *28*, 652–660. [CrossRef]
13. Kadanoff, L.P.; Houghton, A. Numerical evaluations of the critical properties of the two-dimensional Ising model. *Phys. Rev. B* **1975**, *11*, 377–386. [CrossRef]
14. Stevenson, P.M. The effective exponent $\gamma(Q)$ and the slope of the β-function. *Phys. Lett. B* **2016**, *761*, 428–430. [CrossRef]
15. Kleinert, H. *Path Integrals in Quantum Mechanics, Statistics, Polymer Physics and Financial Markets*; World Scientific: Singapore, 2006.
16. Suzuki, M. Statistical Mechanical Theory of Cooperative Phenomena.I. General Theory of Fluctuations, Coherent Anomalies and Scaling Exponents with Simple Applications to Critical Phenomena. *J. Phys. Soc. Jpn.* **1986**, *55*, 4205–4230. [CrossRef]
17. Suzuki, M. Continued-Fraction CAM Theory. *J. Phys. Soc. Jpn.* **1988**, *57*, 1–4. [CrossRef]
18. Yukalov, V.I.; Gluzman, S. Critical Indices as Limits of Control Functions. *Phys. Rev. Lett.* **1997**, *79*, 333–336. [CrossRef]
19. Gluzman, S.; Yukalov, V.I. Additive self-similar approximants. *J. Math. Chem.* **2017**, *55*, 607–622. [CrossRef]
20. Gluzman, S.; Yukalov, V.I. Self-similar continued root approximants. *Phys. Lett.* **2012**, *377*, 124–128. [CrossRef]
21. Gluzman, S.; Yukalov, V.I. Algebraic self-similar renormalization in theory of critical phenomena. *Phys. Rev. E* **1997**, *55*, 3983–3999. [CrossRef]
22. Gluzman, S. Padé and post-Padé approximations for critical phenomena. *Symmetry* **2020**, *12*, 1600. [CrossRef]
23. Muthukumar, M.; Nickel, B.G. Perturbation theory for a polymer chain with excluded volume interaction. *J. Chem. Phys.* **1984**, *80*, 5839–5850. [CrossRef]
24. Muthukumar, M.; Nickel, B.G. Expansion of a polymer chain with excluded volume interaction. *J. Chem. Phys.* **1987**, *86*, 460–476. [CrossRef]
25. Grosberg, A.Y.; Khokhlov, A.R. *Statistical Physics of Macromolecules*; AIP Press: Woodbury, NY, USA, 1994.
26. Pelissetto, A.; Vicari, E. Critical phenomena and renormalization-group theory. *Phys. Rep.* **2002**, *368*, 549–727. [CrossRef]
27. Li, B.; Madras, N.; Sokal, A.D. Critical exponents, hyperscaling, and universal amplitude ratios for two- and three-dimensional self-avoiding walks. *J. Stat. Phys.* **1995**, *80*, 661–754. [CrossRef]
28. Clisby, N. Accurate estimate of the critical exponent for self-avoiding walks via a fast implementation of the pivot algorithm. *Phys. Rev. Lett.* **2010**, *104*, 055702. [CrossRef]

29. Caracciolo, S.; Guttmann, A.J.; Jensen, I.; Pelissetto, A.; Rogers, A.N.; Sokal, A.D. Correction-to-scaling exponents for two-dimensional self-avoiding walks. *J. Stat. Phys.* **2005**, *120*, 1037–1100. [CrossRef]
30. Schwinger, J. Gauge invariance and mass. *Phys. Rev.* **1962**, *128*, 2425–2428. [CrossRef]
31. Banks, T.; Susskind, L.; Kogut, J. Strong-coupling calculations of lattice gauge theories: (1 + 1)-dimensional exercises. *Phys. Rev. D* **1976**, *13*, 1043–1053. [CrossRef]
32. Carrol, A.; Kogut, J.; Sinclair, D.K.; Susskind, L. Lattice gauge theory calculations in 1 + 1 dimensions and the approach to the continuum limit. *Phys. Rev. D* **1976**, *13*, 2270–2277. [CrossRef]
33. Vary, J.P.; Fields, T.J.; Pirner, H.J. Chiral perturbation theory in the Schwinger model. *Phys. Rev. D* **1996**, *53*, 7231–7238. [CrossRef] [PubMed]
34. Adam, C. The Schwinger mass in the massive Schwinger model. *Phys. Lett. B* **1996**, *382*, 383–388. [CrossRef]
35. Striganesh, P.; Hamer, C.J.; Bursill, R.J. A new finite-lattice study of the massive Schwinger model. *Phys. Rev. D* **2000**, *62*, 034508. [CrossRef]
36. Hamer, C.J.; Weihong, Z.; Oitmaa, J. Series expansions for the massive Schwinger model in Hamiltonian lattice theory. *Phys. Rev. D* **1997**, *56*, 55–67. [CrossRef]
37. Coleman, S. More about the massive Schwinger model. *Ann. Phys.* **1987**, *101*, 239–267. [CrossRef]
38. Hamer, C.J. Lattice model calculations for $SU(2)$ Yang-Mills theory in 1 + 1 dimensions. *Nucl. Phys. B* **1977**, *121*, 159–175. [CrossRef]
39. Cioslowski, J. Robust interpolation between weak-and strong-correlation regimes of quantum systems. *J. Chem. Phys.* **2012**, *136*, 044109. [CrossRef]
40. Yukalov, V.I.; Yukalova, E.P.; Gluzman, S. Self-similar interpolation in quantum mechanics. *Phys. Rev. A* **1998**, *58*, 96–115. [CrossRef]
41. Courteille, P.W.; Bagnato, V.S.; Yukalov, V.I. Bose-Einstein Condensation of Trapped Atomic Gases. *Laser Phys.* **2001**, *11*, 659–800.
42. Bender, C.M.; Wu, T.T. Anharmonic oscillator. *Phys. Rev.* **1969**, *184*, 1231–1260. [CrossRef]
43. Hioe, F.T.; McMillen, D.; Montroll, E.W. Quantum theory of anharmonic oscillators: energy levels of a single and a pair of coupled oscillators with quartic coupling. *Phys. Rep.* **1978**, *43*, 305–335. [CrossRef]
44. Butera, P.; Comi, M. A library of extended high-temperature expansions of basic observables for the spin-S Ising models on two- and three-dimensional lattices. *J. Stat. Phys.* **2002**, *109*, 311–315. [CrossRef]
45. Baxter, R.J. *Exactly Solved Models in Statistical Mechanics*; Academic: London, UK, 1982.
46. Gluzman, S. Critical Index for Conductivity, Elasticity, Superconductivity. Results and Methods. In *Mechanics and Physics of Structured Media*; Andrianov, I., Gluzman, S., Mityushev, V., Eds.; Elsevier: Amsterdam, The Netherlands, 2021.
47. Nieuwenhuizen, T.M.; van Velthoven, P.F.J.; Ernst, M.H. Diffusion and long-time tails in a two-dimensional site-percolation model. *Phys. Rev. Lett.* **1986**, *57*, 2477–2480. [CrossRef]
48. Frenkel, D. Velocity auto-correlation functions in a 2d lattice Lorentz gas: Comparison of theory and computer simulation. *Phys. Lett.* **1987**, *121*, 385–389. [CrossRef]
49. Grassberger, P. Conductivity exponent and backbone dimension in 2d percolation. *Phys. A* **1999**, *262*, 251–263. [CrossRef]
50. Ziff, R.M.; Torquato, S. Percolation of disordered jammed sphere packings. *J. Phys. A Math. Theor.* **2017**, *50*, 085001. [CrossRef]
51. Kirkpatrick, S. Percolation and Conduction. *Rev. Mod. Phys.* **1973**, *45*, 574–588. [CrossRef]
52. Hofling, F.; Franosch, T.; Frey, E. Localization transition of the three-dimensional Lorenz model and continuum percolation. *Phys. Rev. Lett.* **2006**, *96*, 165901. [CrossRef]
53. Bauer, T.; Hofling, F.; Munk, T.; Frey, E.; Franosch, T. The localization transition of the two-dimensional Lorenz model. *Eur. Phys. J. Spec. Top.* **2010**, *189*, 103–118. [CrossRef]
54. Clerc, J.P.; Giraud, G.; Laugie, J.M.; Luck, J.M. The electrical conductivity of binary disordered systems, percolation clusters, fractals and related models. *Adv. Phys.* **1990**, *39*, 191–309. [CrossRef]
55. Adler, P.M. *Porous Media. Geometry and Transport*; Butterworth-Heinemann: New York, NY, USA, 1992.
56. Malevich, A.E.; Mityushev, V.V.; Adler, P.M. Stokes flow through a channel with wavy walls. *Acta Mech.* **2006**, *182*, 151–182. [CrossRef]
57. Gluzman, S. Nonlinear approximations to critical and relaxation processes. *Axioms* **2020**, *9*, 126. [CrossRef]
58. Batchelor, G.K.; Green, J.T. The determination of the bulk stress in a suspension of spherical to order c^2. *J. Fluid Mech.* **1972**, *56*, 401–427. [CrossRef]
59. Brady, J.F. The rheological behavior of concentrated colloidal dispersions. *J. Chem. Phys.* **1993**, *99*, 567–581. [CrossRef]
60. Wajnryb, E.; Dahler, J.S. The Newtonian viscosity of a moderately dense suspensions. *Adv. Chem. Phys.* **1997**, *102*, 193–314.
61. Losert, W.; Bocquet, L.; Lubensky, T.C.; Gollub, J.P. Particle dynamics in sheared granular matter. *Phys. Rev. Lett.* **2000**, *85*, 1428–1431. [CrossRef]
62. Bergman, D.J.; Stroud, D. Physical properties of macroscopically inhomogeneous media. *Solid State Phys.* **1992**, *46*, 148–270.
63. Drygaś, P.; Filshtinski, L.A.; Gluzman, S.; Mityushev, V. Conductivity and elasticity of graphene-type composites. In *2D and Quasi-2D Composite and Nano Composite Materials, Properties and Photonic Applications*; McPhedran, R., Gluzman, S., Mityushev, V., Rylko, N., Eds.; Elsevier: Amsterdam, The Netherlands, 2020; Chapter 8, pp. 193–231.
64. Perrins, W.T.; McKenzie, D.R.; McPhedran, R.C. Transport properties of regular array of cylinders. *Proc. R. Soc. A* **1979**, *369*, 207–225.
65. Mulero, A.; Cachadina, I.; Solana, J.R. The equation of state of the hard-disc fluid revisited. *Mol. Phys.* **2009**, *107*, 1457–1465. [CrossRef]

66. Santos, A.; Lopez de Haro, M.; Bravo Yuste, S. An accurate and simple equation of state for hard disks. *J. Chem. Phys.* **1995**, *103*, 4622–4625. [CrossRef]
67. Clisby, N.; McCoy, B.M. Ninth and tenth order virial coefficients for hard spheres in D dimensions. *J. Stat. Phys.* **2006**, *122*, 15–57. [CrossRef]
68. Maestre, M.A.G.; Santos, A.; Robles, M.; Lopez de Haro, M. On the relation between virial coefficients and the close-packing of hard disks and hard spheres. *J. Chem. Phys.* **2011**, *134*, 084502. [CrossRef] [PubMed]

Article

Classical Partition Function for Non-Relativistic Gravity

Mir Hameeda [1,2,3]**, Angelo Plastino** [4,5,*]**, Mario Carlos Rocca** [4,5,6] **and Javier Zamora** [3,5]

1. Department of Physics, Government Degree College, Tangmarg, Kashmir 193402, India; hme123eda@gmail.com
2. School of Physics, Damghan University, Damghan P.O. Box 3671641167, Iran
3. Inter University Centre for Astronomy and Astrophysics, Pune 194409, India; javierzamora055@gmail.com
4. Departamento de Física, Universidad Nacional de La Plata, La Plata 1900, Argentina; rocca@fisica.unlp.edu.ar
5. Consejo Nacional de Investigaciones Cientcas y Tecnologicas, (IFLP-CCT-CONICET), C. C. 727, La Plata 1900, Argentina
6. Departamento de Matemática, Universidad Nacional de La Plata, La Plata 1900, Argentina
* Correspondence: plastino@fisica.unlp.edu.ar

Abstract: We considered the canonical gravitational partition function Z associated to the classical Boltzmann–Gibbs (BG) distribution $\frac{e^{-\beta H}}{\mathcal{Z}}$. It is popularly thought that it cannot be built up because the integral involved in constructing Z diverges at the origin. Contrariwise, it was shown in (Physica A 497 (2018) 310), by appeal to sophisticated mathematics developed in the second half of the last century, that this is not so. Z can indeed be computed by recourse to (A) the analytical extension treatments of Gradshteyn and Rizhik and Guelfand and Shilov, that permit tackling some divergent integrals and (B) the dimensional regularization approach. Only one special instance was discussed in the above reference. In this work, we obtain the classical partition function for Newton's gravity in the **four** cases that immediately come to mind.

Keywords: partition functions; analytical extensions; guelfand's and gradshteyn's; classical gravity

MSC: 32A70; 46N55; 82B03; 82B05

1. Introduction

This paper is a continuation and generalization of [1]. It involves mathematical ideas that were fully explored there (and references therein), in which a canonical ensemble at the temperature T was concocted for a two-particle gravitation system and fully analyzed. For brevity's sake, the accompanying (involved) math will not be repeated here. It is advisable to have [1] at hand in trying to follow our discussion below. A very important concept is that of generalized dimensionally regularized partition function Z, which we abbreviate as $GDR[\mathcal{Z}$. Dimensional regularization works in ν dimensions. We call ν_0 the actual physical dimension. We use the common notation $\beta = 1/(k_B T)$, with T as the temperature. To simplify numerical computations, we set $k_B = 1$, and also the gravitational constant G equal to unity. The domain of integration is called D-.

An important point that will emerge below is that of the behavior of the partition function $\mathcal{Z}$ as a function of the inverse temperature β. $\mathcal{Z}$ is a sum or integral of terms of the form $\exp[-\beta E_M]$, with E_m some energy. Thus, the second derivative $d\mathcal{Z}/d\beta^2$ is positive. Thus, its curvature, when plotted against β, *cannot change*.

Self-gravitating systems exhibit peculiarities that might perhaps defy ordinary common sense. We highlight here the following: [2]

(1) As gravitational binding gets tighter, the kinetic energy augments and so does, as a consequence, the temperature;
(2) As gravitational binding gets weaker, the kinetic energy decreases, and as a consequence, the temperature diminishes;
(3) The specific heat becomes negative if the system can freely expand or contract.

2. Basic Instance: Partition Function for the Distance-Case $0 \leq r < \infty$

The first case is the most important one. Conceptually, it is the most complex of the four to be addressed. Algebraically, however, it is the simplest instance. We deal with the gravitational interaction between two masses m and M. The partition function is then:

$$\mathcal{Z} = \mathcal{Z}_{\nu_0} = (-1)^{\nu_0} GDR[\mathcal{Z}_\nu]_{\nu=\nu_0}, \qquad (1)$$

and $\mathcal{Z}_\nu$ is given by classical phase space (ν dimensions):

$$\mathcal{Z}_\nu = \int_D e^{-\beta\left(\frac{p^2}{2m} - \frac{GmM}{r}\right)} d^\nu x \, d^\nu p. \qquad (2)$$

This quantity was already used for three dimensions in [1]. We then have:

$$\mathcal{Z}_\nu = \left[\frac{2\pi^{\frac{\nu}{2}}}{\Gamma(\frac{\nu}{2})}\right]^2 \int\int_0^\infty (rp)^{\nu-1} e^{-\beta\left(\frac{p^2}{2m} - \frac{GmM}{r}\right)} dr \, dp. \qquad (3)$$

By appeal to the relation below, employed in [1]:

$$\int_0^\infty x^{\nu-1} e^{\frac{\beta}{x}} dx = \cos(\pi\nu)\beta^\nu \Gamma(-\nu), \qquad (4)$$

we obtain:

$$\mathcal{Z}_\nu = \left[\frac{2\pi^{\frac{\nu}{2}}}{\Gamma(\frac{\nu}{2})}\right]^2 \frac{\cos(\pi\nu)}{2} (\beta GmM)^\nu \left(\frac{2m}{\beta}\right)^{\frac{\nu}{2}} \Gamma\left(\frac{\nu}{2}\right) \Gamma(-\nu). \qquad (5)$$

To apply GDR, we introduce $f(\nu)$:

$$f(\nu) = \frac{2\pi^\nu}{\Gamma(\frac{\nu}{2})} \frac{(\beta GmM)^\nu u}{\nu(\nu-1)(\nu-2)} \left(\frac{2m}{\beta}\right)^{\frac{\nu}{2}}, \qquad (6)$$

and in three dimensions:

$$\mathcal{Z}_\nu = f(\nu)\Gamma(3-\nu). \qquad (7)$$

Following [1], we Laurent expand around $\nu = 3$ to obtain:

$$\mathcal{Z}_\nu = -\frac{2}{3\sqrt{\pi}} \frac{(2\pi^2 \beta G^2 m^3 M^2)^{\frac{3}{2}}}{(\nu-3)} - \frac{1}{3\sqrt{\pi}} (2\pi^2 \beta G^2 m^3 M^2)^{\frac{3}{2}} \times$$

$$\left[\ln\left(8\pi^2 \beta G^2 m^3 M^2\right) + 3C - \frac{17}{3}\right] + \sum_{n=1}^\infty a_n (\nu-3)^n. \qquad (8)$$

Therefore:

$$\mathcal{Z} = \frac{1}{3\sqrt{\pi}} (2\pi^2 \beta G^2 m^3 M^2)^{\frac{3}{2}} \left[\ln\left(8\pi^2 \beta G^2 m^3 M^2\right) + 3C - \frac{17}{3}\right] \qquad (9)$$

Remind that the mean energy $<\mathcal{U}> = -\frac{1}{\mathcal{Z}}\frac{\partial \mathcal{Z}}{\partial \beta}$. Accordingly, we obtain:

$$<\mathcal{U}> = -\frac{\sqrt{2}\pi^{5/2} (G^2 m^3 M^2 \beta)^{3/2} (3C + \log(8\pi^2 G^2 m^3 M^2 \beta) - 5)}{\beta \mathcal{Z}}. \qquad (10)$$

For the specific heat C, we use the definition $C = \frac{\partial <\mathcal{U}>}{\partial T}$, and we have:

$$C = \frac{\pi^{5/2} G^4 m^6 M^4 \beta (3C + \log(8\pi^2 G^2 m^3 M^2 \beta) - 3)}{\sqrt{2}\sqrt{G^2 m^3 M^2 \beta} T \mathcal{Z}}$$

$$-\frac{\beta}{T}\left[\frac{\sqrt{2}\pi^{5/2}(G^2m^3M^2x)^{3/2}(3C+\log(8\pi^2G^2m^3M^2x)-5)}{\beta\mathcal{Z}}\right]^2. \quad (11)$$

We express a word of caution regarding the plotting the above relations. They involve big quantities like m and M and also very small ones like G and k_B. In order to make sense of the associated computational mess, we are forced to appeal to a simple scenario with $m = M = G = k_B = 1$ and set the horizontal coordinate to $x = 1/T$. Notice that we will be dealing with gigantic temperatures in the order of 10^{22} Kelvin. See Figures 1–3.

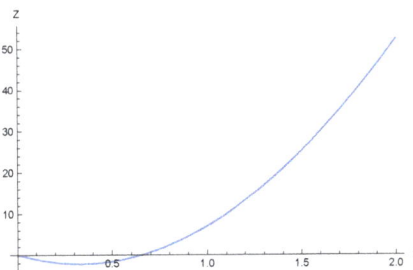

Figure 1. $m = M = G = k_B = 1$, and $x = 1/T$. Plot of $\mathcal{Z}(x)$. Here, $x = \beta$, $G = m = M = 1$. For small x, we see that Z is negative. These x-values are, of course, unphysical. Z grows as T diminishes.

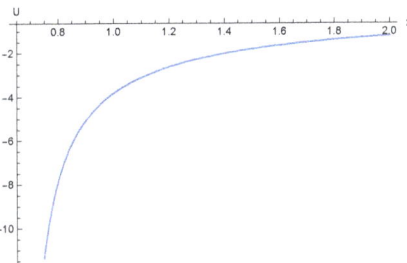

Figure 2. $m = M = G = k_B = 1$, and $x = 1/T$. Plot of $<\mathcal{U}(x)>$. At large temperatures (near the origin), the binding and kinetic energies, as well as the temperature are large, as prescribed by item (1) of the introduction. As T decreases, so does the binding and kinetic energies, as indicated by item (2) of the Introduction.

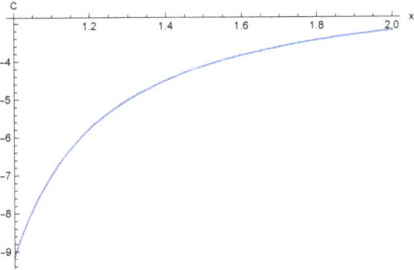

Figure 3. Plot of the specific heat $C(x)$. Here, $x = \beta$, $G = m = M = 1$. It is negative, as prescribed by item (3) of the Introduction. The specific heat tends to vanish as the temperature drops, as expected because of the third law of thermodynamics, that, remarkably enough, is obeyed classically here.

We see that our dimensionally regularized partition function predicts the expected gravitational behavior.

3. Partition Function for the Bound System $R_0 \leq r \leq R_1$

Now, suppose that we deal with two masses: one has mass M and radius R_0 and the second is a point mass m. Both are contained in a spherical box of radius R_1. Of course, in this paper, tidal forces are ignored. The system is bound by construction.

The partition function now turns out to be:

$$\mathcal{Z} = \left[\frac{2\pi^{\frac{3}{2}}}{\Gamma(\frac{3}{2})}\right]^2 \int_0^\infty dp \int_{R_0}^{R_1} dr (rp)^2 e^{-\beta\left(\frac{p^2}{2m} - \frac{GmM}{r}\right)}. \tag{12}$$

We appeal to the well-known result:

$$\int x^2 e^{\frac{a}{x}} dx = \frac{1}{6}\left[e^{\frac{a}{x}} x\left(a^2 + ax + 2x\right) - a^3 E_i\left(\frac{a}{x}\right)\right] + b. \tag{13}$$

Here, E_i is the integral exponential function [3] and b is an arbitrary constant. We obtain thus, in three dimensions:

$$\mathcal{Z} = \frac{\pi^{\frac{3}{2}} V_1}{2}\left\{e^{\frac{\beta GmM}{R_1}}\left[\left(\frac{\beta GmM}{R_1}\right)^2 + \frac{\beta GmM}{R_1} + 2\right] - \left(\frac{\beta GmM}{R_1}\right)^3 E_i\left(\frac{\beta GmM}{R_1}\right)\right\}$$

$$-\frac{\pi^{\frac{3}{2}} V_0}{2}\left\{e^{\frac{\beta GmM}{R_0}}\left[\left(\frac{\beta GmM}{R_0}\right)^2 + \frac{\beta GmM}{R_0} + 2\right] - \left(\frac{\beta GmM}{R_0}\right)^3 E_i\left(\frac{\beta GmM}{R_0}\right)\right\}, \tag{14}$$

where V_1 is the volume of a sphere of radius R_1 and V_0 is the volume of a sphere of radius R_0. For the mean energy, we have:

$$<\mathcal{U}> = -(1/(2R_1^3 R_0^3 \mathcal{Z})) \times$$

$$3GmM\pi^{3/2}\left(-V_1 R_1 R_0^3 e^{\frac{GmM\beta}{R_1}}(R_1 + GmM\beta) + R_1^3 V_0 R_0 e^{\frac{GmM\beta}{R_0}}(R_0 + GmM\beta) + V_1 R_0^3 GmM^2\right.$$

$$\left.\beta^2 Ei\left(\frac{GmM\beta}{R_1}\right) - R_1^3 V_0 GmM^2 \beta^2 Ei\left(\frac{GmM\beta}{R_0}\right)\right), \tag{15}$$

and for the specific heat:

$$C = -\frac{3\pi^{3/2} GmM^2 \beta}{\mathcal{Z} T R_1^3 R_0^3} \times$$

$$\left(-V_1 R_1 R_0^3 e^{\frac{GmM\beta}{b}} + R_1^3 V_0 R_0 e^{\frac{GmM\beta}{R_0}} + V_1 R_0^3 GmM\beta Ei\left(\frac{GmM\beta}{R_1}\right) - \right.$$

$$\left. R_1^3 V_0 GmM\beta Ei\left(\frac{GmM\beta}{R_0}\right)\right)$$

$$-\frac{\beta}{T}\left[(1/(2R_1^3 R_0^3 \mathcal{Z}))\right.$$

$$3GmM\pi^{3/2}\left(-V_1 R_1 R_0^3 e^{\frac{GmM\beta}{R_1}}(R_1 + GmM\beta) + R_1^3 V_0 R_0 e^{\frac{GmM\beta}{R_0}}(R_0 + GmM\beta) + V_1 R_0^3 GmM^2\right.$$

$$\left.\left.\beta^2 Ei\left(\frac{GmM\beta}{R_1}\right) - R_1^3 V_0 GmM^2 \beta^2 Ei\left(\frac{GmM\beta}{R_0}\right)\right)\right]^2. \tag{16}$$

See Figures 4–6.

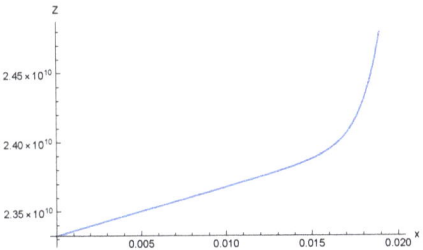

Figure 4. Plot of $\mathcal{Z}(x)$. Here, $x = \beta/R_1$, $R_0 = 1$, $R_1 = 1000$, $G = m = M = 1$.

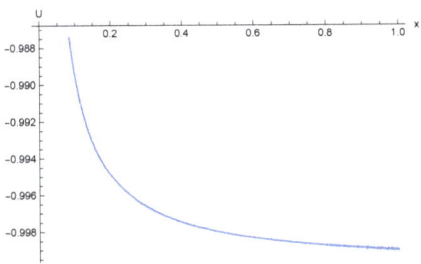

Figure 5. Plot of $<\mathcal{U}(x)>$. Here, $x = \beta/R_1$, $R_0 = 1$, $G = m = M = 1$. Note that the size of the container diminishes as x grows, so that the binding energy has to grow with x.

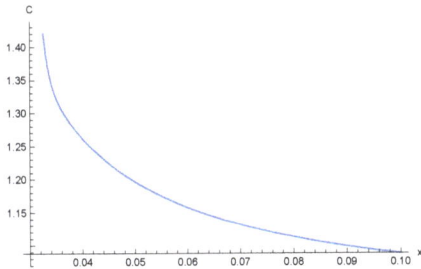

Figure 6. Plot of $C(x)$ versus x. Here, $x = \beta/R_1$, $R_0 = 1$, $G = m = M = 1$. The binding energy increases as β grows and the mean energy U is negative. Thus, the specific heat $= -\beta^2 \frac{dU}{d\beta}$ is positive. Notably enough, it tends to obey the third law in a classical scenario, as bound by construction.

4. Partition Function for $0 \leq r \leq R_1$

We now consider the case of two point masses m and M are enclosed in a container of radius R_1. The three-dimensional partition function is now:

$$\mathcal{Z} = -\left[\frac{2\pi^{\frac{3}{2}}}{\Gamma(\frac{3}{2})}\right]^2 \int_0^\infty dp \int_0^{R_1} dr (rp)^2 e^{-\beta\left(\frac{p^2}{2m} - \frac{GmM}{r}\right)} \qquad (17)$$

Evaluating the integral corresponding to the momenta, we arrive at:

$$\mathcal{Z} = -4\pi^{\frac{5}{2}}\left(\frac{2m}{\beta}\right)^{\frac{3}{2}} \int_0^{R_1} dr\, r^2 e^{\frac{\beta GmM}{r}}. \qquad (18)$$

We see that the resulting integral displays a singularity at the origin. We evaluate this integral using the Guelfand [4] method for calculating integrals of powers of x, expanding the exponential in power series around the origin. We are led to:

$$\mathcal{Z} = -4\pi^{\frac{5}{2}} \left(\frac{2m}{\beta}\right)^{\frac{3}{2}} \sum_{n=0}^{\infty} \frac{(\beta GmM)^n}{n!} \int_0^{R_1} r^{2-n} dr. \qquad (19)$$

We need here the result (see [4]):

$$\int_0^{R_1} r^{2-s} dr = \frac{R_1^{3-s}}{3-s} \quad ; \quad s \neq 3$$

$$\int_0^{R_1} r^{-1} dr = \ln R_1 \quad ; \quad s = 3. \qquad (20)$$

Using it, we obtain:

$$\mathcal{Z} = -4\pi^{\frac{5}{2}} \left(\frac{2m}{\beta}\right)^{\frac{3}{2}} \left[(\beta GmM)^3 \frac{\ln R_0}{3!} - 3 \sum_{n=0; n\neq 3}^{\infty} \left(\frac{\beta GmM}{R_1}\right)^n \frac{R_1^3}{n!(n-3)}\right]. \qquad (21)$$

Remember that we usually call C the Euler–Mascheroni constant. We now need a further result that is given in reference [5] and reads:

$$\sum_{s=0; s\neq 3} \frac{y^s}{s!(s-3)} = \frac{y^3}{3!}[\psi(4) - \ln|y| + E_1(y)] - \frac{e^y}{3!}[y^2 + y + 2], \qquad (22)$$

so that we finally obtain:

$$\mathcal{Z} = -4\pi^{\frac{5}{2}} \left(\frac{2m}{\beta}\right)^{\frac{3}{2}} \left\{ \frac{R_1^3 e^{\frac{\beta GmM}{R_1}}}{3!} \left[\left(\frac{\beta GmM}{R_1}\right)^2 + \frac{\beta GmM}{R_1} + 2\right] + \right.$$

$$\left. \frac{(\beta GmM)^3}{3!} \left[\ln(\beta GmM) - \psi(4) - E_i\left(\frac{\beta GmM}{R_1}\right)\right] \right\} \qquad (23)$$

where ψ is the poly-gamma function. For the mean energy, one has:

$$<\mathcal{U}> = \frac{GmM\pi^{5/2}}{\mathcal{Z}} \left(2R_1^2 e^{\frac{GmM\beta}{R_1}} + 2R_1 GmM\beta e^{\frac{GmM\beta}{R_1}} - 3GmM^2\beta^2 + 2CGmM^2\beta^2 - \right.$$

$$\left. 2GmM^2\beta^2 \text{Ei}\left(\frac{GmM\beta}{R_1}\right) + 2GmM^2\beta^2 \log(GmM\beta) \right), \qquad (24)$$

and the specific heat reads:

$$C_v = -\frac{4GmM^2\pi^{5/2}\beta}{\mathcal{Z}T} \left(R_1 e^{\frac{GmM\beta}{R_1}} - GmM\beta + \gamma GmM\beta - \right.$$

$$\left. GmM\beta \text{Ei}\left(\frac{GmM\beta}{R_1}\right) + GmM\beta \log(GmM\beta) \right) -$$

$$\frac{\beta}{T}\left[\frac{PGmM\pi^{5/2}}{\mathcal{Z}} \left(2R_1^2 e^{\frac{GmM\beta}{R_1}} + 2R_1 GmM\beta e^{\frac{GmM\beta}{R_1}} - 3GmM^2\beta^2 + 2\gamma GmM^2\beta^2 - \right.\right.$$

$$\left. 2GmM^2\beta^2 \mathrm{Ei}\left(\frac{GmM\beta}{R_1}\right) + 2GmM^2\beta^2 \log(GmM\beta)\right)\right]^2. \tag{25}$$

We plot below these equations in Figures 7–9.

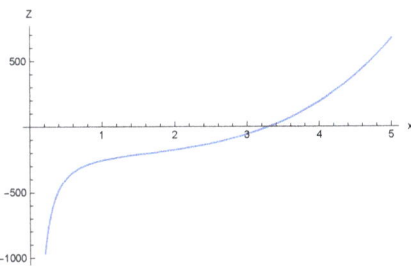

Figure 7. Plot of $\mathcal{Z}(x)$ versus $x = \beta$ for $G = m = R_1 = M = 1$. Results are unphysical for $x < 3.35$. At the ensuing very high (big-bang like) temperatures (at which matter does not exist), $\mathcal{Z}(x)$ is negative, and thus unphysical.

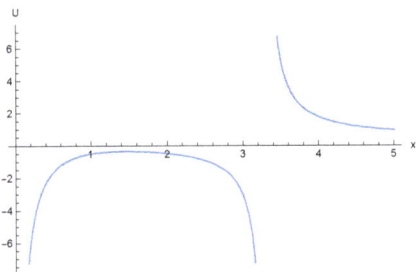

Figure 8. Plot of $<\mathcal{U}(x)>$ for $x = \beta$ $G = m = R_1 = M = 1$. Remember that the results are unphysical for $x < 3.35$. The kinetic energy diminishes as x grows and so does the energy. However, R_1 is too large to allow for statistical bounding in the region here considered.

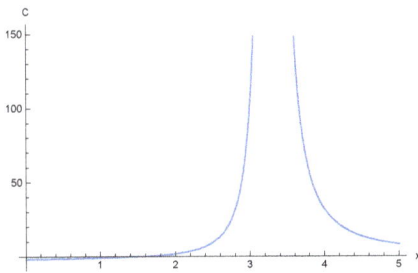

Figure 9. Plot of $C(x)$ for $x = \beta$. One has $G = m = R_1 = M = 1$. Again, the results are unphysical for $x < 3.35$. The system tends to comply with the third law. It is unbound, since $C > 0$.

5. Partition Function for $R_0 \leq r < \infty$

Finally, we confront the twin case of the precedent one (under a y to $1/y$ transform). We thus consider the case of a spherical mass M of radius R_0 interacting with a punctual mass m Accordingly, the distance between the two masses has a lower bound but no upper bound:

$$\mathcal{Z} = \left[\frac{2\pi^{\frac{3}{2}}}{\Gamma(\frac{3}{2})}\right]^2 \int_0^\infty dp \int_{R_0}^\infty dr (rp)^2 e^{-\beta\left(\frac{p^2}{2m} - \frac{GmM}{r}\right)} \tag{26}$$

Evaluating the momenta integral, we obtain:

$$\mathcal{Z} = 4\pi^{\frac{5}{2}}\left(\frac{2m}{\beta}\right)^{\frac{3}{2}}\int_{R_0}^{\infty} dr\, r^2 e^{\frac{\beta GmM}{r}}. \qquad (27)$$

This integral is divergent and can be evaluated as prescribed in reference [4], that is:

$$\mathcal{Z} = 4\pi^{\frac{5}{2}}\left(\frac{2m}{\beta}\right)^{\frac{3}{2}} \sum_{n=0}^{\infty} \frac{(\beta GmM)^n}{n!}\int_{R_0}^{\infty} r^{2-n} dr, \qquad (28)$$

leading to:

$$\int_{R_0}^{\infty} r^{2-s} dr = -\frac{R_0^{3-s}}{3-s} \quad ; \quad s \neq 3$$

$$\int_{R_0}^{\infty} r^{-1} dr = -\ln R_0 \quad ; \quad s = 3. \qquad (29)$$

Using again the result (20), we obtain:

$$\mathcal{Z} = 4\pi^{\frac{5}{2}}\left(\frac{2m}{\beta}\right)^{\frac{3}{2}}\left\{\frac{(\beta GmM)^3}{3!}\left[\psi(4) + E_i\left(\frac{\beta GmM}{R_0}\right) - \ln(\beta GmM)\right] - \right.$$

$$\left. \frac{R_0^3 e^{\frac{\beta GmM}{R_0}}}{3!}\left[\left(\frac{\beta GmM}{R_0}\right)^2 + \frac{\beta GmM}{R_0} + 2\right]\right\}. \qquad (30)$$

For the mean energy, the result is (here $a = R_0$):

$$<\mathcal{U}> = -\frac{1}{3\beta\mathcal{Z}}2\sqrt{2}\pi^{5/2}\left(\frac{m}{x}\right)^{3/2}\left(6a^3 e^{\frac{GmMx}{a}} - 3a^2 GmMx e^{\frac{GmMx}{a}} - aGm^2M^2x^2 e^{\frac{GmMx}{a}} - \right.$$

$$2aE_i((GmMx)/a)G^2m^2M^2x^2 - 2E_i((GmMx)/a)GGm^2mM^3x^3 - 2G^3m^3M^3x^3 +$$

$$2G^3m^3M^3x^3 e^{\frac{GmMx}{a}} + 3G^3m^3M^3x^3\mathrm{Ei}\left(\frac{GmMx}{a}\right) -$$

$$\left. 3G^3m^3M^3x^3\psi^{(0)}(z) + 3G^3m^3M^3x^3\log(GmMx)\right). \qquad (31)$$

For the specific heat, we have:

$$C_v = -\frac{1}{3axT\mathcal{Z}}\sqrt{2}\pi^{5/2}\left(\frac{m}{x}\right)^{3/2}\left(-30a^4 e^{\frac{GmMx}{a}} + 21a^3 GmMx e^{\frac{GmMx}{a}} + a^2 Gm^2M^2x^2 e^{\frac{GmMx}{a}} - \right.$$

$$4a^2 G^2 m^2 M^2 x^2 e^{\frac{GmMx}{a}} - 4aGGm^2 mM^3 x^3 e^{\frac{GmMx}{a}} -$$

$$8aG^3m^3M^3x^3 + 4aG^3m^3M^3x^3 e^{\frac{GmMx}{a}} - 4G^2Gm^2m^2M^4x^4 e^{\frac{GmMx}{a}} +$$

$$4G^4 m^4 M^4 x^4 e^{\frac{GmMx}{a}} + 3aG^3m^3M^3x^3\mathrm{Ei}\left(\frac{GmMx}{a}\right) -$$

$$\left. 3aG^3m^3M^3x^3\psi^{(0)}(z) + 3aG^3m^3M^3x^3\log(GmMx)\right) +$$

$$\frac{\beta}{T}\left[\frac{1}{3\beta\mathcal{Z}}2\sqrt{2}\pi^{5/2}\left(\frac{m}{x}\right)^{3/2}\left(6a^3 e^{\frac{GmMx}{a}} - 3a^2 GmMx e^{\frac{GmMx}{a}} - aGm^2M^2x^2 e^{\frac{GmMx}{a}} - \right.\right.$$

$$2aE_i((GmMx)/a)G^2m^2M^2x^2 - 2E_i((GmMx)/a)GGm^2mM^3x^3 - 2G^3m^3M^3x^3 +$$

$$\left. 2G^3m^3M^3x^3 e^{\frac{GmMx}{a}} + 3G^3m^3M^3x^3 \text{Ei}\left(\frac{GmMx}{a}\right) - \right.$$
$$\left. 3G^3m^3M^3x^3 \psi^{(0)}(z) + 3G^3m^3M^3x^3 \log(GmMx) \right)\Big]^2. \tag{32}$$

We plot the pertinent results below. Because of duality, our graphs closely resemble those of the precedent Section. See Figures 10–12.

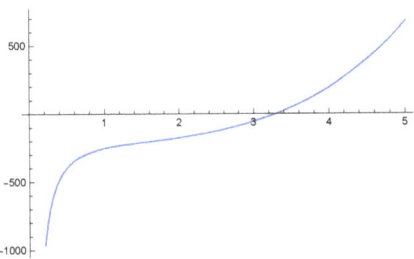

Figure 10. Plot of $\mathcal{Z}(x)$ for $x = \beta\ G = m = R_0 = M = 1$. Results turn out to be unphysical for $x < 3.35$, as in the previous case.

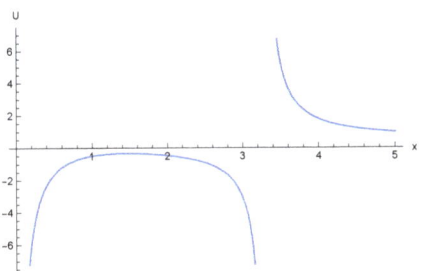

Figure 11. Plot of $<\mathcal{U}(x)>$ for $x = \beta$ and $G = m = R_1 = M = 1$. It closely resembles the companion graph of the precedent case. However, R_0 is too large to allow for statistical bounding in the region here considered.

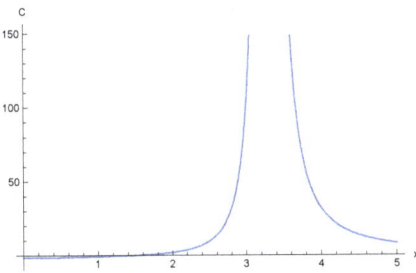

Figure 12. Plot of $C(x)$ for $x = \beta\ G = m = R_1 = M = 1$. $C > 0$ diverges in the unphysical x–region. It is positive in the physical one. The third law is complied with.

6. Conclusions

In this work, we showed how to deal with the partition function for gravitational systems in four different scenarios:

- The last two of the four scenarios envisioned here are linked by the twin transform from y to $1/y$.

- Even if our treatment is classical, the third law of thermodynamics is obeyed by the specific heat in all cases.
- It is remarkable that at a classical level, one can detect that at too high temperatures, statistical mechanics fails because the partition function becomes negative. We know now that at these Ts, matter cannot exist.
- Transformation from y to $1/y$: we might have discovered a transform that conserves the physics in the statistical mechanics of gravitation.

Author Contributions: Investigation, A.P., M.H., M.C.R. and J.Z. All authors have read and agreed to the published version of the manuscript.

Funding: This research received no external funding.

Conflicts of Interest: The authors declare no conflict of interest.

References

1. Zamora, J.; Plastino, A.; Rocca, M.C.; Ferri, G.L. Dimensionally regularized Tsallis' Statistical Mechanics and two-body Newton's gravitation. *Physica A* **2018**, *497*, 310. [CrossRef]
2. Lynden-Bell, D.; Lynden-Bell, R.M. On the negative specific heat paradox. *Mon. Not. R. Astron. Soc.* **1977**, *181*, 405. [CrossRef]
3. Gradshteyn, S.; Ryzhik, I.M. *Table of Integrals, Series and Products*; Academic Press, Inc.: Cambridge, MA, USA, 1980.
4. Gel'fand, I.M.; Shilov, G.E. *Generalized Functions*; Academic Press: Cambridge, MA, USA, 1964; Volume 1.
5. Prudnikov, A.P.; Brichkov, Y.A.; Marichev, O.I. *Integrals and Series*; Gordon and Breach Science Publishers: London, UK, 1992.

Article

Strong Interacting Internal Waves in Rotating Ocean: Novel Fractional Approach

Pundikala Veeresha [1], Haci Mehmet Baskonus [2] and Wei Gao [3,*]

[1] Department of Mathematics, CHRIST, Bengaluru 560029, India; pundikala.veeresha@christuniversity.in or viru0913@gmail.com
[2] Faculty of Education, Harran University, Sanliurfa 63050, Turkey; hmbaskonus@gmail.com
[3] School of Information Science and Technology, Yunnan Normal University, Kunming 650500, China
* Correspondence: gaowei@ynnu.edu.cn

Abstract: The main objective of the present study is to analyze the nature and capture the corresponding consequences of the solution obtained for the Gardner–Ostrovsky equation with the help of the q-homotopy analysis transform technique (q-HATT). In the rotating ocean, the considered equations exemplify strong interacting internal waves. The fractional operator employed in the present study is used in order to illustrate its importance in generalizing the models associated with kernel singular. The fixed-point theorem and the Banach space are considered to present the existence and uniqueness within the frame of the Caputo–Fabrizio (CF) fractional operator. Furthermore, for different fractional orders, the nature has been captured in plots. The realized consequences confirm that the considered procedure is reliable and highly methodical for investigating the consequences related to the nonlinear models of both integer and fractional order.

Keywords: internal waves in rotating ocean; fractional derivative; q-Homotopy analysis transform technique; fixed point theorem

MSC: 34A08; 26A33; 65L05

1. Introduction

The concept of fractional calculus (FC) has received much consideration in recent years due to its wide applicability and ability to capture more consequences of real-world problems. Even though the concept of classical calculus can exemplify most real-life problems and aid us in predicting the nature of complex phenomena, it has become a narrow subset of FC. Even though pioneers propose many new notions, many things need to be derived in order to ensure all classes of phenomena, which will be achieved by overcoming the limitations raised by mathematicians and scientists [1–7]. This is particularly true when researchers try to study, analyze and predict behaviors related to history, long-range memory, heritage, chaos, epidemiology, and other such subjects. There is always a door open for innovation, novelty, improvisation, and modifications in the research when it comes to the investigation of consequences that help us to solve real-world problems (the present pandemic, for example) and in this connection, many researchers have derived stimulating results with the aid of FC, and by using efficient techniques with the aid of fundamental results of FC [7–10].

The investigation of physical models and analysis of the associated properties are always a hot topic in applied mathematics with appropriate tools. In particular, this is the case for research on earthquakes, electrochemistry, signal processing, viscoelastic models, fluid dynamics, plasma physics, and many others. In the present investigation, we consider the nonlinear model describing the strong interacting internal waves in the ocean. In 1995, Oregon Bay experimented Coastal Ocean Probe Experiment (COPE) [11], found that the internal shallow waves are very strong, and also, it is stated that the shallow water

wave phenomena are not completely nurtured with one nonlinear term. Furthermore, there is a need for dual nonlinear terms to examine the physical behaviors of the model. These experiments on shallow-water waves aid in formulating the model with dual-power law nonlinearity.

The equation that exemplifies the above phenomena is familiarly known as the Gardner equation. Generally, the Korteweg de-Vries (KdV) equations are widely used to exemplify the behaviors of long waves and their physical significance. Here, we consider the modified KdV equation, called the Gardner–Ostrovsky (GO) equation, due to the large amplitude for long internal waves and extended rotational effects [12]. Now, we consider the following two equations respectively exemplifying Ostrovsky and Gardner-Ostrovsky (GO) equations with the isopycnic surface ($u(x, t)$):

$$\frac{d}{dx}\left(\frac{du(x,t)}{dt} + \kappa_1 u \frac{du}{dx} + \beta \frac{d^3 u}{dx^3}\right) = cu(x,t), \tag{1}$$

$$\frac{du(x,t)}{dt} + \kappa_1 u \frac{du}{dx} + \kappa_2 u^2 \frac{du}{dx} + \beta \frac{d^3 u}{dx^3} = 0, \tag{2}$$

where κ_1 and κ_2 are, respectively, quadratic and cubic nonlinearities coefficients, β is the coefficient of small-scale dispersion, and c is the velocity of dispersion-less linear waves. Here, κ_1 and κ_2 are two proportional terms involving in the above equations, and $u\frac{du}{dx}$ is due to the nonlinear hydrodynamic system, and it comes traditionally. Moreover, the term $u^2 \frac{du}{dx}$ appears when $u\frac{du}{dx}$ is a bird that is arbitrarily small. The GO equation is employed, and formally known to exemplify by large-amplitude waves with strong nonlinearity effects [13–18].

Recently, many authors illustrated the fractional operator's essence in capturing and understanding the more relative consequences of physical phenomena with higher nonlinearity and complexity with long-range memory and history-based properties [19]. In this paper, we consider the fractional-order version of Equations (1) and (2) with Caputo–Fabrizio fractional operator with order α, fractional Ostrovsky (FO) and fractional Gardner–Ostrovsky (FGO) equations given by:

$$\frac{d}{dx}\left({}_0^{CF}D_t^\alpha u(x,t) + \kappa_1 u \frac{du}{dx} + \beta \frac{d^3 u}{dx^3}\right) = cu(x,t), \tag{3}$$

$$_0^{CF}D_t^\alpha u(x,t) + \kappa_1 u \frac{du}{dx} + \kappa_2 u^2 \frac{du}{dx} + \beta \frac{d^3 u}{dx^3} = 0. \tag{4}$$

The vital and challenging job is finding the solution for nonlinear problems. Even though we have distinct schemes for finding the solution for these problems, they may possess a conversion of nonlinear to linear, partial to ordinary, perturbation, assumption of additional parameters, or discretization. Here, we consider the method proposed based on a topological concept called the homotopy analysis technique, which is nurtured to solve complicated nonlinear problems, and for 20 years, numerous scholarly works have been carried out with the aid of this [20,21]. Mankind is always searching for innovation, modernization and improvisation to increase accuracy and reduce the complexity associated with the method. In association with this, the authors in [22] came with the improvisation associated with Laplace transform (LT).

The projected scheme offers more freedom to choose the initial conditions, and the novelty is that it offers a simple solution procedure. This method is perceptible and includes all merits achieved by HAM, and also it attracted many researchers to analyze the diverse class of models and systems. The novelty of the projected scheme is the ability to solve the nonlinear problems without any assumption, perturbation, or conversion from nonlinear to linear, or partial to ordinary differential equations. Moreover, it is associated with parameters that are very helpful in converging the results attained towards a beneficial solution. Furthermore, it is related to well-posed transformation, which aims to reduce the complexity and increase the applicability and reliability of the method.

The considered technique (q-HATT) is widely used for examining diverse classes of nonlinear problems and investigating the distinct mathematical models associated with real life [23–30]. For instance, the system of nonlinear differential equations exemplifying the outbreak of COVID-19 in India is examined within the frame of a nonsingular derivative in [24] using the projected scheme; HIV infection of lymphocyte cells is numerically analyzed by researchers in [25]; some interesting results have been derived with the help of the considered algorithm by authors in [29], in which the system exemplified the wind-influenced projectile motion; and the numerical simulation is presented by authors in [28] for the coupled Korteweg-de Vries system and this derived some interesting consequences in terms of numerical simulation.

2. Preliminaries

The basic definitions are presented in this segment for the CF and Laplace transform. Specifically, we recall the notions related to Caputo–Fabrizio fractional operator [19,31].

Definition 1. *The CF fractional derivative for* $f \in H^1(a, b)$ *is presented as* [19]:

$$D_t^\alpha(f(t)) = \frac{\mathcal{M}(\alpha)}{1-\alpha} \int_a^t f'(t) \exp\left[-\alpha \frac{t-\vartheta}{1-\alpha}\right] d\vartheta, \ b > a, \tag{5}$$

where $\mathcal{M}(\alpha)$ *is a normalization function and admits* $\mathcal{M}(0) = \mathcal{M}(1) = 1$. *Further, if* $f \notin H^1(a, b)$, *then we have:*

$$D_t^\alpha(f(t)) = \frac{\alpha \mathcal{M}(\alpha)}{1-\alpha} \int_{-\infty}^t (f(t) - f(\vartheta)) \exp\left[-\alpha \frac{t-\vartheta}{1-\alpha}\right] d\vartheta. \tag{6}$$

Definition 2. *The CF fractional integral for* $f \in H^1(a, b)$ *is presented as* [31]:

$$^{CF}I^\alpha(f(t)) = \frac{2(1-\alpha)}{(2-\alpha)\mathcal{M}(\alpha)} f(t) + \frac{2\alpha}{(2-\alpha)\mathcal{M}(\alpha)} \int_0^t f(\vartheta) d\vartheta, \ 0 < \alpha < 1, \ t \geq 0. \tag{7}$$

Note: According to [31], the following must hold:

$$\frac{2(1-\alpha)}{(2-\alpha)\mathcal{M}(\alpha)} + \frac{2\alpha}{(2-\alpha)\mathcal{M}(\alpha)} = 1, \ 0 < \alpha < 1, \tag{8}$$

which gives $\mathcal{M}(\alpha) = \frac{2}{2-\alpha}$. With the help of the above equation, researchers in [31] proposed a novel Caputo derivative as follows

$$D_t^\alpha(f(t)) = \frac{1}{1-\alpha} \int_0^t f'(t) \exp\left[\alpha \frac{t-\vartheta}{1-\alpha}\right] d\vartheta, \ 0 < \alpha < 1. \tag{9}$$

Definition 3. *The LT for a CF derivative* $^{CF}_0 D_t^\alpha f(t)$ *is presented as* [19] *below:*

$$\mathcal{L}\left[^{CFC}_0 D_t^{(\alpha+n)} f(t)\right] = \frac{s^{n+1}\mathcal{L}[f(t)] - s^n f(0) - s^{n-1} f'(0) - \cdots - f^{(n)}(0)}{s + (1-s)\alpha}. \tag{10}$$

3. Fundamental Procedure of the Considered Method

Here, we hired the differential equation to present the basic algorithm of the projected method with the initial condition:

$$_0^{CF}D_t^\alpha v(x,t) + \mathcal{R}\, v(x,t) + \mathcal{N}\, v(x,t) = f(x,t), \quad 0 < \alpha \le 1, \tag{11}$$

and

$$v(x,0) = \}(x). \tag{12}$$

We obtained this by applying LT on Equation (11):

$$\mathcal{L}[v(x,t)] - \frac{\}(x)}{s} + \frac{s+(1-s)\alpha}{s}\{\mathcal{L}[\mathcal{R}v(x,t)] + \mathcal{L}[\mathcal{N}\, v(x,t)] - \mathcal{L}[f(x,t)]\} = 0. \tag{13}$$

For $\varphi(x,t;q)$, $\mathcal{N}$ is contracted as follows:

$$\begin{aligned}\mathcal{N}[\varphi(x,t;q)] &= \mathcal{L}[\varphi(x,t;q)] - \frac{\}(x)}{s} \\ &+ \frac{s+(1-s)\alpha}{s}\{\mathcal{L}[\mathcal{R}\,\varphi(x,t;q)] + L[\mathcal{N}\varphi(x,t;q)] - L[f(x,t)]\}\end{aligned} \tag{14}$$

where $q \in \left[0, \frac{1}{n}\right]$. Then, the homotopy is defined by results in:

$$(1 - nq)\mathcal{L}[\varphi(x,\, t;q) - v_0(x,t)] = \hbar q \mathcal{N}[\varphi(x,\, t;q)]. \tag{15}$$

For $q = 0$ and $q = \frac{1}{n}$, we have:

$$\varphi(x,t;0) = v_0(x,t), \quad \varphi\left(x,t;\frac{1}{n}\right) = v(x,t). \tag{16}$$

By using the Taylor theorem, we consider:

$$\varphi(x,t;q) = v_0(x,t) + \sum_{m=1}^{\infty} v_m(x,t)q^m, \tag{17}$$

where

$$v_m(x,t) = \frac{1}{m!}\frac{\partial^m \varphi(x,\, t;q)}{\partial q^m}\Big|_{q=0}. \tag{18}$$

For the appropriate chaise of $v_0(x,t)$, n and $\hbar$ the series (15) converges at $q = \frac{1}{n}$. Then:

$$v(x,t) = v_0(x,t) + \sum_{m=1}^{\infty} v_m(x,t)\left(\frac{1}{n}\right)^m. \tag{19}$$

After differentiating Equation (15) m-times with q and multiplying by $\frac{1}{m!}$ and substituting $q = 0$, one can obtain:

$$\mathcal{L}[v_m(x,t) - k_m v_{m-1}(x,t)] = \hbar \Re_m\left(\vec{v}_{m-1}\right), \tag{20}$$

where

$$\vec{v}_m = \{v_0(x,t), v_1(x,t),\ldots, v_m(x,t)\}. \tag{21}$$

Equation (20) reduces after employing inverse LT to:

$$v_m(x,t) = k_m v_{m-1}(x,t) + \hbar \mathcal{L}^{-1}\left[\Re_m\left(\vec{v}_{m-1}\right)\right], \tag{22}$$

where

$$\begin{aligned}\Re_m\left(\vec{v}_{m-1}\right) &= L[v_{m-1}(x,t)] - \left(1 - \frac{k_m}{n}\right)\left(\frac{\}(x)}{s} + \frac{s+(1-s)\alpha}{s}L[f(x,t)]\right) \\ &+ \frac{s+(1-s)\alpha}{s}L[\mathcal{R}v_{m-1} + \mathcal{H}_{m-1}],\end{aligned} \tag{23}$$

and
$$k_m = \begin{cases} 0, & m \leq 1, \\ n, & m > 1. \end{cases} \qquad (24)$$

Here, $\mathcal{H}_m$ is homotopy polynomial and presented as:

$$\mathcal{H}_m = \frac{1}{m!}\left[\frac{\partial^m \varphi(x,t;q)}{\partial q^m}\right]_{q=0} \text{ and } \varphi(x,t;q) = \varphi_0 + q\varphi_1 + q^2\varphi_2 + \ldots \qquad (25)$$

By the help of Equations (22) and (23), we found:

$$v_m(x,t) = (k_m + \hbar)v_{m-1}(x,t) - \left(1 - \frac{k_m}{n}\right)\mathcal{L}^{-1}\left(\frac{\}(x)}{s} + \frac{s+(1-s)\alpha}{s}L[f(x,t)]\right)$$
$$+ \hbar\mathcal{L}^{-1}\left\{\frac{s+(1-s)\alpha}{s}L[Rv_{m-1} + \mathcal{H}_{m-1}]\right\}. \qquad (26)$$

With the help of q-HATM, the series solution is:

$$v(x,t) = v_0(x,t) + \sum_{m=1}^{\infty} v_m(x,t)\left(\frac{1}{n}\right)^m. \qquad (27)$$

4. Solution for FO and FGO Equations

The above-illustrated scheme is employed for the two fractional-order equations (namely, FO and FGO equations).

Case I. *For FO equation*

Consider the equation defined in Equation (3):

$$^{CF}_0 D_t^\alpha u(x,t) + \kappa_1 u \frac{du}{dx} + \beta \frac{d^3 u}{dx^3} - c\int u(x,t)dx = 0, \qquad (28)$$

with
$$u(x,0) = \rho \text{sech}^2(x). \qquad (29)$$

Taking LT on Equation (28) and with the help of Equation (29), we obtain:

$$\mathcal{L}[u(x,t)] = \frac{1}{s}\left(\rho\text{sech}^2(x)\right) - \frac{s+(1-s)\alpha}{s}\mathcal{L}\left\{\kappa_1 u\frac{du}{dx} + \beta\frac{d^3u}{dx^3} - c\int u(x,t)dx\right\}. \qquad (30)$$

Now, $\mathcal{N}$ is defined as:

$$\mathcal{N}[\varphi(x,t;q)] = \mathcal{L}[\varphi(x,t;q)] - \frac{1}{s}\left(\rho\text{sech}^2(x)\right)$$
$$+ \frac{s+(1-s)\alpha}{s}\mathcal{L}\left\{\kappa_1\varphi(x,t;q)\frac{\partial\varphi(x,t;q)}{\partial x} + \beta\frac{d^3\varphi(x,t;q)}{d^3 x} - c\int\varphi(x,t;q)dx\right\}. \qquad (31)$$

At $\mathcal{H}(x,t) = 1$, the deformation equation presented as:

$$\mathcal{L}[u_m(x,t) - k_m u_{m-1}(x,t)] = \hbar\Re_m[\vec{u}_{m-1}], \qquad (32)$$

where
$$\Re_m[\vec{u}_{m-1}] = \mathcal{L}[u_{m-1}(x,t)] - \left(1 - \frac{k_m}{n}\right)\left\{\frac{1}{s}\left(\rho\text{sech}^2(x)\right)\right\}$$
$$+ \frac{s+(1-s)\alpha}{s}\mathcal{L}\left\{\kappa_1\sum_{i=0}^{m-1}u_i\frac{\partial u_{m-1-i}}{\partial x} + \beta\frac{d^3 u_{m-1}}{dx^3} - c\int u_{m-1}(x,t)dx\right\}. \qquad (33)$$

On employing inverse LT on Equation (32), it simplifies to:

$$u_m(x,t) = k_m u_{m-1}(x,t) + \hbar L^{-1}\left\{\Re_m[\vec{u}_{m-1}]\right\}. \qquad (34)$$

Now, by simplifying the above system we can evaluate the terms of the series solution:

$$u(x, t) = u_0(x, t) + \sum_{m=1}^{\infty} u_m(x, t) \left(\frac{1}{n}\right)^m. \tag{35}$$

Now, by the help of the initial condition, we can derive the terms of Equation (29) as:

$$u_1(x, t) = \hbar(1 - \alpha + \alpha t)(-\rho\tanh(x)(c - 16\beta \operatorname{sech}^4(x) + 2\rho\operatorname{sech}^4(x)\kappa_1 \\ + 8\beta\operatorname{sech}^2(x)\tanh^2(x))),$$

$$\vdots$$

4.1. Existence of Solution for Ostrovsky Equation

Here, the existence and uniqueness are illustrated with CF operator for the considered Equation (28) as:

$$^{CF}_0 D_t^\alpha [u(x, t)] = \mathcal{Q}_1(x, t, u). \tag{36}$$

Now, using Equation (36) and results derived in [31], we obtained:

$$u(x, t) - u(x, 0) = {}^{CF}_0 I_t^\alpha \left\{ -\kappa_1 u \frac{du}{dx} - \beta \frac{d^3 u}{dx^3} + c \int u(x, t)dx \right\}. \tag{37}$$

Then we have from [31] the follows:

$$u(x, t) - u(x, 0) = \frac{2(1-\alpha)}{\mathcal{M}(\alpha)} \mathcal{Q}(x, t, u) + \frac{2\alpha}{(2-\alpha)\mathcal{M}(\alpha)} \int_0^t \mathcal{Q}_1(x, \zeta, u)d\zeta. \tag{38}$$

Theorem 1. *The kernel $\mathcal{Q}$ admits the Lipschitz condition and contraction if $0 \leq (\frac{\kappa_1}{2}\Lambda(a_1 + a_2) + \beta\Lambda^3 - c\xi) < 1$ satisfies.*

Proof. Consider the two functions A and A_1 to prove the theorem, then

$$\begin{aligned}
&\|\mathcal{Q}_1(x, t, u) - \mathcal{Q}_1(x, t, u_1)\| \\
&= \left\|\kappa_1 \left[u(x,t)\frac{\partial u(x,t)}{\partial x} - u(x,t_1)\frac{\partial u(x,t_1)}{\partial x}\right] + \beta \frac{\partial^3}{\partial x^3}[u(x,t) - u(x,t_1)] - cb(x)\right\| \\
&= \left\|\frac{\kappa_1}{2}\left[\frac{\partial}{\partial x}[u^2(x,t) - u^2(x,t_1)]\right] + \beta\frac{\partial^3}{\partial x^3}[u(x,t) - u(x,t_1)] - cb(x)\right\| \\
&\leq \|\frac{\kappa_1}{2}\Lambda[u(x,t) - u(x,t_1)] + \beta\Lambda^3 - cb(x)\| \|u(x,t) - u(x,t_1)\| \\
&\leq \left(\frac{\kappa_1}{2}\Lambda(a_1 + a_2) + \beta\Lambda^3 - c\xi\right)\|u(x,t) - u(x,t_1)\|,
\end{aligned} \tag{39}$$

where $a_1 = \|u\|$ and $a_2 = \|u_1\|$ are the bounded function and $\|b(x)\| = \xi$ is also a bounded function. Set $\Psi_1 = \frac{\kappa_1}{2}\Lambda(a_1 + a_2) + \beta\Lambda^3 - c\xi$ in Equation (39), then:

$$\|\mathcal{Q}_1(x, t, u) - \mathcal{Q}_1(x, t, u_1)\| \leq \Psi_1 \|u(x, t) - u(x, t_1)\|. \tag{40}$$

Equation (40) provides the Lipschitz condition for $\mathcal{Q}_1$. Similarly, we can see that if $0 \leq \frac{\kappa_1}{2}\Lambda(a_1 + a_2) + \beta\Lambda^3 - c\xi < 1$, then it implies the contraction. With the help of the above equations, Equation (38) simplifies to:

$$u(x, t) = u(x, 0) + \frac{2(1-\alpha)}{(2-\alpha)\mathcal{M}(\alpha)} \mathcal{Q}_1(x, t, u) + \frac{2\alpha}{(2-\alpha)\mathcal{M}(\alpha)} \int_0^t \mathcal{Q}_1(x, \zeta, u)d\zeta. \tag{41}$$

$$u_n(x, t) = \frac{2(1-\alpha)}{(2-\alpha)\mathcal{M}(\alpha)} \mathcal{Q}_1(x, t, u_{n-1}) + \frac{2\alpha}{(2-\alpha)\mathcal{M}(\alpha)} \int_0^t \mathcal{Q}_1(x, \zeta, u_{n-1})d\zeta, \tag{42}$$

Now, between the terms the successive difference is defined as:

$$\phi_{1n}(x, t) = u_n(x, t) - u_{n-1}(x, t)$$
$$= \frac{2(1-\alpha)}{(2-\alpha)\mathcal{M}(\alpha)}(Q_1(x,t,u_{n-1}) - Q_1(x,t,u_{n-2})) \tag{43}$$
$$+ \frac{2\alpha}{(2-\alpha)\mathcal{M}(\alpha)} \int_0^t (Q_1(x,t,u_{n-1}) - Q_1(x,t,u_{n-2}))d\zeta.$$

Notice that:

$$u_n(x, t) = \sum_{i=1}^{n} \phi_{1i}(x, t). \tag{44}$$

Then we have:

$$\|\phi_{1n}(x, t)\| = \|u_n(x, t) - u_{n-1}(x, t)\|$$
$$= \|\frac{2(1-\alpha)}{(2-\alpha)\mathcal{M}(\alpha)}(Q_1(x, t, u_{n-1}) - Q_1(x, t, u_{n-2})) \tag{45}$$
$$+ \frac{2\alpha}{(2-\alpha)\mathcal{M}(\alpha)} \int_0^t (Q_1(x, t, u_{n-1}) - Q_1(x, t, u_{n-2}))d\zeta\|.$$

Application of the triangular inequality, Equation (45) reduces to:

$$\|\phi_{1n}(x, t)\| = \|u_n(x, t) - u_{n-1}(x, t)\|$$
$$= \frac{2(1-\alpha)}{(2-\alpha)\mathcal{M}(\alpha)} \|(Q_1(x, t, u_{n-1}) - Q_1(x, t, u_{n-2}))\| \tag{46}$$
$$+ \frac{2\alpha}{(2-\alpha)\mathcal{M}(\alpha)} \| \int_0^t (Q_1(x, t, u_{n-1}) - Q_1(x, t, u_{n-2}))d\zeta\|.$$

The Lipschitz condition satisfied by the kernel t_1, then:

$$\|\phi_{1n}(x, t)\| = \|u_n(x, t) - u_{n-1}(x, t)\| \leq \frac{2(1-\alpha)}{(2-\alpha)\mathcal{M}(\alpha)} \Psi_1 \|\phi_{(n-1)}(x, t)\|$$
$$+ \frac{2\alpha}{(2-\alpha)\mathcal{M}(\alpha)} \Psi_1 \int_0^t \|\phi_{(n-1)}(x, t)\| d\zeta. \tag{47}$$

□

By the aid of the above result, we state the following result:

Theorem 2. *If we have specific t_0, then the solution for Equation (28) will exist and be unique. Further, we have:*

$$\frac{2(1-\alpha)}{(2-\alpha)\mathcal{M}(\alpha)} \Psi_1 + \frac{2\alpha}{(2-\alpha)\mathcal{M}(\alpha)} \Psi_1 t_0 < 1.$$

Proof. Let $u(x, t)$ be the bounded functions admitting the Lipschitz condition. Then, we obtain, using Equations (46) and (47):

$$\|\phi_{1i}(x, t)\| \leq \|u_n(x, 0)\| \left[\frac{2(1-\alpha)}{(2-\alpha)\mathcal{M}(\alpha)} \Psi_1 + \frac{2\alpha}{(2-\alpha)\mathcal{M}(\alpha)} \Psi_1 t \right]^n. \tag{48}$$

Therefore, for the obtained solution, continuity and existence are verified. Now, to prove the Equation (48) is a solution for Equation (28), we consider:

$$u(x, t) - u(x, 0) = u_n(x, t) - \mathcal{K}_{1n}(x, t). \tag{49}$$

Let us consider:

$$\|\mathcal{K}_{1n}(x,t)\| = \|\frac{2(1-\alpha)}{(2-\alpha)\mathcal{M}(\alpha)}(Q_1(x, t, u) - Q_1(x, t, u_{n-1}))$$
$$+ \frac{2\alpha}{(2-\alpha)\mathcal{M}(\alpha)} \int_0^t (Q_1(x, \zeta, u) - Q_1(x, \zeta, u_{n-1}))d\zeta\|$$
$$\leq \frac{2(1-\alpha)}{(2-\alpha)\mathcal{M}(\alpha)} \|(Q_1(x, t, u) - Q_1(x, t, u_{n-1}))\| \tag{50}$$
$$+ \frac{2\alpha}{(2-\alpha)\mathcal{M}(\alpha)} \int_0^t \|(Q_1(x, \zeta, u) - Q_1(x, \zeta, u_{n-1}))\| d\zeta$$
$$\leq \frac{2(1-\alpha)}{(2-\alpha)\mathcal{M}(\alpha)} \Psi_1 \|u - u_{n-1}\| + \frac{2\alpha}{(2-\alpha)\mathcal{M}(\alpha)} \Psi_1 \|u - u_{n-1}\| t.$$

This process gives:

$$\|\mathcal{K}_{1n}(x,t)\| \leq \left(\frac{2(1-\alpha)}{(2-\alpha)\mathcal{M}(\alpha)} + \frac{2\alpha}{(2-\alpha)\mathcal{M}(\alpha)}t\right)^{n+1}\Psi_1^{n+1}M$$

Similarly, at t_0 we can obtain:

$$\|\mathcal{K}_{1n}(x,t)\| \leq \left(\frac{2(1-\alpha)}{(2-\alpha)\mathcal{M}(\alpha)} + \frac{2\alpha}{(2-\alpha)\mathcal{M}(\alpha)}t_0\right)^{n+1}\Psi_1^{n+1}M. \tag{51}$$

Next, for the solution of the projected model, we prove its uniqueness. Suppose $u^*(x,t)$ is another solution, then:

$$u(x,t) - u^*(x,t) = \frac{2(1-\alpha)}{(2-\alpha)\mathcal{M}(\alpha)}(\mathcal{Q}_1(x,t,u) - \mathcal{Q}_1(x,t,u^*)) \\ + \frac{2\alpha}{(2-\alpha)\mathcal{M}(\alpha)}\int_0^t(\mathcal{Q}_1(x,\zeta,u) - \mathcal{Q}_1(x,\zeta,u^*))d\zeta. \tag{52}$$

Now, employing the norm on the above equation, we obtain:

$$\begin{aligned}\|u(x,t) - u^*(x,t)\| &= \|\frac{2(1-\alpha)}{(2-\alpha)\mathcal{M}(\alpha)}(\mathcal{Q}_1(x,t,u) - \mathcal{Q}_1(x,t,u^*)) \\ &+ \frac{2\alpha}{(2-\alpha)\mathcal{M}(\alpha)}\int_0^t(\mathcal{Q}_1(x,\zeta,u) - \mathcal{Q}_1(x,\zeta,u^*))d\zeta\| \\ &\leq \frac{2(1-\alpha)}{(2-\alpha)\mathcal{M}(\alpha)}\Psi_1\|u(x,t)-u^*(x,t)\| + \frac{2\alpha}{(2-\alpha)\mathcal{M}(\alpha)}\Psi_1 t\|u(x,t)-u^*(x,t)\|.\end{aligned} \tag{53}$$

On simplification, this becomes:

$$\|u(x,t) - u^*(x,t)\|\left(1 - \frac{2(1-\alpha)}{(2-\alpha)\mathcal{M}(\alpha)}\Psi_1 - \frac{2\alpha}{(2-\alpha)\mathcal{M}(\alpha)}\Psi_1 t\right) \leq 0. \tag{54}$$

From the above condition, it is clear that $u(x,t) = u^*(x,t)$, if:

$$\left(1 - \frac{2(1-\alpha)}{(2-\alpha)\mathcal{M}(\alpha)}\Psi_1 - \frac{2\alpha}{(2-\alpha)\mathcal{M}(\alpha)}\Psi_1 t\right) \geq 0. \tag{55}$$

Hence, Equation (55) proves our required result. □

Case II. *For the FGO equation*

Consider the equation defined in Equation (4):

$$_0^{CF}D_t^\alpha u(x,t) + \kappa_1 u\frac{du}{dx} + \kappa_2 u^2\frac{du}{dx} + \beta\frac{d^3u}{dx^3} = 0, \tag{56}$$

with condition defined in Equation (29). Now, taking LT on Equation (56) and by the assist of Equation (29), we obtain:

$$\mathcal{L}[u(x,t)] = \frac{1}{s}\left(\rho\operatorname{sech}^2(x)\right) - \frac{s+(1-s)\alpha}{s}\mathcal{L}\left\{\kappa_1 u\frac{du}{dx} + \kappa_2 u^2\frac{du}{dx} + \beta\frac{d^3u}{dx^3}\right\}. \tag{57}$$

Now, $\mathcal{N}$ is defined as:

$$\mathcal{N}[\varphi(x,t;q)] = \mathcal{L}[\varphi(x,t;q)] - \frac{1}{s}\left(\rho\operatorname{sech}^2(x)\right) \\ + \frac{s+(1-s)\alpha}{s}\mathcal{L}\left\{\kappa_1\varphi(x,t;q)\frac{\partial\varphi(x,t;q)}{\partial x} + \kappa_2\varphi^2(x,t;q)\frac{\partial\varphi(x,t;q)}{\partial x} + \beta\frac{d^3\varphi(x,t;q)}{d^3 x}\right\}. \tag{58}$$

Here:

$$\Re_m\left[\vec{u}_{m-1}\right] = \mathcal{L}[u_{m-1}(x,t)] - \left(1 - \frac{k_m}{n}\right)\left\{\frac{1}{s}\left(\rho\operatorname{sech}^2(x)\right)\right\}$$
$$+ \frac{s+(1-s)\alpha}{s}\mathcal{L}\left\{\kappa_1 \sum_{i=0}^{m-1} u_i \frac{\partial u_{m-1-i}}{\partial x} + \kappa_1 \sum_{i=0}^{m-1} \sum_{j=0}^{i} u_{i-j} u_i \frac{\partial u_{m-1-i}}{\partial x} + \beta \frac{d^3 u_{m-1}}{dx^3}\right\}. \tag{59}$$

4.2. Existence of Solution for Fractional Gardner's Ostrovsky Equation

Here, the existence and uniqueness are illustrated with CF operator for the considered Equation (56) as:

$$^{CF}_{\ 0}D_t^\alpha[u(x,t)] = \mathcal{Q}_2(x,t,u). \tag{60}$$

Now, using Equation (56) and results derived in [31], we obtained:

$$u(x,t) - u(x,0) = {}^{CF}_{\ 0}I_t^\alpha\left\{-\kappa_1 u \frac{du}{dx} - \kappa_2 u^2 \frac{du}{dx} - \beta \frac{d^3 u}{dx^3}\right\}. \tag{61}$$

Then, we have from [31] the following:

$$u(x,t) - u(x,0) = \frac{2(1-\alpha)}{\mathcal{M}(\alpha)}\mathcal{Q}_2(x,t,u) + \frac{2\alpha}{(2-\alpha)\mathcal{M}(\alpha)}\int_0^t \mathcal{Q}_2(x,\zeta,u)d\zeta. \tag{62}$$

4.3. Existence of Solution for the Fractional Gardner–Ostrovsky Equation

Here, we will state the following results in a similar manner carried out for the above case.

Theorem 3. *The kernel $\mathcal{Q}$ admits the Lipschitz condition and contraction if $0 \leq (\frac{\kappa_1}{2}\Lambda(a_1 + a_2) + \kappa_2\Lambda(a_1 + a_2 + a_1 a_2) + \beta\Lambda^3) < 1$ satisfies for Equation (56).*

Proof. Consider the two functions A and A_1 to prove the theorem, then:

$$\begin{aligned}
&\|\mathcal{Q}_2(x,t,u) - \mathcal{Q}_2(x,t,u_1)\| \\
&= \left\|\kappa_1\left[u(x,t)\frac{\partial u(x,t)}{\partial x} - u(x,t_1)\frac{\partial u(x,t_1)}{\partial x}\right] + \kappa_2\left[u^2(x,t)\frac{\partial u(x,t)}{\partial x} - u^2(x,t_1)\frac{\partial u(x,t_1)}{\partial x}\right] + \beta\frac{\partial^3}{\partial x^3}[u(x,t) - u(x,t_1)]\right\| \\
&= \left\|\frac{\kappa_1}{2}\left[\frac{\partial}{\partial x}[u^2(x,t) - u^2(x,t_1)]\right] + \kappa_2\left[u^2(x,t)\frac{\partial u(x,t)}{\partial x} - u^2(x,t_1)\frac{\partial u(x,t_1)}{\partial x}\right] + \beta\frac{\partial^3}{\partial x^3}[u(x,t) - u(x,t_1)]\right\| \\
&\leq \left(\frac{\kappa_1}{2}\Lambda(a_1 + a_2) + \kappa_2\Lambda(a_1 + a_2 + a_1 a_2) + \beta\Lambda^3\right)\|u(x,t) - u(x,t_1)\|,
\end{aligned} \tag{63}$$

where $a_1 = \|u\|$ and $a_2 = \|u_1\|$ is the bounded function. Set $\Psi_2 = \frac{\kappa_1}{2}\Lambda(a_1 + a_2) + \beta\Lambda^3 - c\xi$ in Equation (63), then:

$$\|\mathcal{Q}_2(x,t,u) - \mathcal{Q}_2(x,t,u_1)\| \leq \Psi_2\|u(x,t) - u(x,t_1)\|. \tag{64}$$

Equation (64) provides the Lipschitz condition for $\mathcal{Q}_2$. Similarly, we can see that if $0 \leq \frac{\kappa_1}{2}\Lambda(a_1 + a_2) + \kappa_2\Lambda(a_1 + a_2 + a_1 a_2) + \beta\Lambda^3 < 1$, then it implies the contraction. By the assist of the above equation, Equation (62) simplifies to

$$u(x,t) = u(x,0) + \frac{2(1-\alpha)}{(2-\alpha)\mathcal{M}(\alpha)}\mathcal{Q}_2(x,t,u) + \frac{2\alpha}{(2-\alpha)\mathcal{M}(\alpha)}\int_0^t \mathcal{Q}_2(x,\zeta,u)d\zeta. \tag{65}$$

Then we obtain the recursive form as follows:

$$u_n(x,t) = \frac{2(1-\alpha)}{(2-\alpha)\mathcal{M}(\alpha)}\mathcal{Q}_2(x,t,u_{n-1}) + \frac{2\alpha}{(2-\alpha)\mathcal{M}(\alpha)}\int_0^t \mathcal{Q}_2(x,\zeta,u_{n-1})d\zeta, \tag{66}$$

Now, between the terms, the successive difference is defined as:

$$\phi_{2n}(x, t) = u_n(x, t) - u_{n-1}(x, t)$$
$$= \frac{2(1-\alpha)}{(2-\alpha)\mathcal{M}(\alpha)}(Q_2(x,t,u_{n-1}) - Q_2(x,t,u_{n-2}))$$
$$+ \frac{2\alpha}{(2-\alpha)\mathcal{M}(\alpha)} \int_0^t (Q_2(x,t,u_{n-1}) - Q_2(x,t,u_{n-2})) d\zeta. \quad (67)$$

Notice that:

$$u_n(x, t) = \sum_{i=1}^{n} \phi_{2i}(x, t). \quad (68)$$

Then, we have:

$$\|\phi_{2n}(x, t)\| = \|u_n(x, t) - u_{n-1}(x, t)\|$$
$$= \|\frac{2(1-\alpha)}{(2-\alpha)\mathcal{M}(\alpha)}(Q_2(x, t, u_{n-1}) - Q_2(x, t, u_{n-2}))$$
$$+ \frac{2\alpha}{(2-\alpha)\mathcal{M}(\alpha)} \int_0^t (Q_2(x, t, u_{n-1}) - Q_2(x, t, u_{n-2})) d\zeta\| \quad (69)$$

Application of the triangular inequality, Equation (69) reduces to:

$$\|\phi_{2n}(x, t)\| = \|u_n(x, t) - u_{n-1}(x, t)\|$$
$$= \frac{2(1-\alpha)}{(2-\alpha)\mathcal{M}(\alpha)} \|(Q_2(x, t, u_{n-1}) - Q_2(x, t, u_{n-2}))\|$$
$$+ \frac{2\alpha}{(2-\alpha)\mathcal{M}(\alpha)} \|\int_0^t (Q_2(x, t, u_{n-1}) - Q_2(x, t, u_{n-2})) d\zeta\|. \quad (70)$$

The Lipschitz condition satisfied by the kernel t_1, then:

$$\|\phi_{2n}(x, t)\| = \|u_n(x, t) - u_{n-1}(x, t)\| \leq \frac{2(1-\alpha)}{(2-\alpha)\mathcal{M}(\alpha)} \Psi_2 \|\phi_{(n-1)}(x, t)\|$$
$$+ \frac{2\alpha}{(2-\alpha)\mathcal{M}(\alpha)} \Psi_1 \int_0^t \|\phi_{(n-1)}(x, t)\| d\zeta. \quad (71)$$

□

By the aid of the above result, we state the following result:

Theorem 4. *If we have specific t_0, then the solution for Equation (56) will exist and be unique. Further, we have:*

Proof. Let $u(x, t)$ be the bounded functions admitting the Lipschitz condition. Then, from Equations (70) and (71), we obtain:

$$\|\phi_{2i}(x, t)\| \leq \|u_n(x, 0)\| \left[\frac{2(1-\alpha)}{(2-\alpha)\mathcal{M}(\alpha)} \Psi_2 + \frac{2\alpha}{(2-\alpha)\mathcal{M}(\alpha)} \Psi_2 t\right]^n \quad (72)$$

Therefore, for the obtained solution, continuity and existence are verified. Now, to prove that Equation (72) is a solution for Equation (56), we consider:

$$u(x, t) - u(x, 0) = u_n(x, t) - \mathcal{K}_{1n}(x, t). \quad (73)$$

Let us consider:

$$\|\mathcal{K}_{1n}(x, t)\| = \|\frac{2(1-\alpha)}{(2-\alpha)\mathcal{M}(\alpha)}(Q_2(x, t, u) - Q_2(x, t, u_{n-1}))$$
$$+ \frac{2\alpha}{(2-\alpha)\mathcal{M}(\alpha)} \int_0^t (Q_2(x, \zeta, u) - Q_2(x, \zeta, u_{n-1})) d\zeta\|$$
$$\leq \frac{2(1-\alpha)}{(2-\alpha)\mathcal{M}(\alpha)} \|(Q_2(x, t, u) - Q_2(x, t, u_{n-1}))\|$$
$$+ \frac{2\alpha}{(2-\alpha)\mathcal{M}(\alpha)} \int_0^t \|(Q_2(x, \zeta, u) - Q_2(x, \zeta, u_{n-1}))\| d\zeta$$
$$\leq \frac{2(1-\alpha)}{(2-\alpha)\mathcal{M}(\alpha)} \Psi_2 \|u - u_{n-1}\| + \frac{2\alpha}{(2-\alpha)\mathcal{M}(\alpha)} \Psi_2 \|u - u_{n-1}\| t. \quad (74)$$

This process gives:

$$\|\mathcal{K}_{2n}(x,\,t)\| \leq \left(\frac{2(1-\alpha)}{(2-\alpha)\mathcal{M}(\alpha)} + \frac{2\alpha}{(2-\alpha)\mathcal{M}(\alpha)}t\right)^{n+1} \Psi_2^{n+1} M$$

Similarly, at t_0 we can obtain:

$$\|\mathcal{K}_{2n}(x,\,t)\| \leq \left(\frac{2(1-\alpha)}{(2-\alpha)\mathcal{M}(\alpha)} + \frac{2\alpha}{(2-\alpha)\mathcal{M}(\alpha)}t_0\right)^{n+1} \Psi_2^{n+1} M. \tag{75}$$

Next, for the solution of the projected model, we prove the uniqueness. Suppose $u^*(x,\,t)$ is another solution, then:

$$u(x,t) - u^*(x,t) = \frac{2(1-\alpha)}{(2-\alpha)\mathcal{M}(\alpha)}(Q_2(x,t,u) - Q_2(x,t,u^*)) \\ + \frac{2\alpha}{(2-\alpha)\mathcal{M}(\alpha)} \int_0^t (Q_2(x,\zeta,u) - Q_2(x,\zeta,u^*))d\zeta. \tag{76}$$

Now, employing the norm on the above equation we obtain:

$$\|u(x,t) - u^*(x,t)\| = \|\frac{2(1-\alpha)}{(2-\alpha)\mathcal{M}(\alpha)}(Q_2(x,t,u) - Q_2(x,t,u^*)) \\ + \frac{2\alpha}{(2-\alpha)\mathcal{M}(\alpha)} \int_0^t (Q_2(x,\zeta,u) - Q_2(x,\zeta,u^*))d\zeta\| \\ \leq \frac{2(1-\alpha)}{(2-\alpha)\mathcal{M}(\alpha)}\Psi_2\|u(x,t)-u^*(x,t)\| + \frac{2\alpha}{(2-\alpha)\mathcal{M}(\alpha)}\Psi_2 t\|u(x,t)-u^*(x,t)\|. \tag{77}$$

On simplification:

$$\|u(x,t)-u^*(x,t)\|\left(1-\frac{2(1-\alpha)}{(2-\alpha)\mathcal{M}(\alpha)}\Psi_2 - \frac{2\alpha}{(2-\alpha)\mathcal{M}(\alpha)}\Psi_2 t\right) \leq 0. \tag{78}$$

From the above condition, it is clear that $u(x,t) = u^*(x,t)$, if

$$\left(1-\frac{2(1-\alpha)}{(2-\alpha)\mathcal{M}(\alpha)}\Psi_2 - \frac{2\alpha}{(2-\alpha)\mathcal{M}(\alpha)}\Psi_2 t\right) \geq 0. \tag{79}$$

Hence, Equation (79) proves our required result. □

5. Results and Discussion

Here, we capture the physical nature of two cases with different fractional order, small-scale dispersion, and homotopy parameters. It is essential to illustrate the effect of the fractional operator incorporated in the considered model. For Case I, we present the surfaces of fractional values of the order in Figure 1, and then it is illustrated in Figure 2 with a particular value of time ($t = 0.5$). In the Ostrovsky equation, the dispersion term plays a pivotal role and its effect is presented in Figure 3 with attained results. The main advantage of the considered scheme is the association of homotopy parameters (n and $\hbar$). They play a critical role in the archiving solution for the modulation of convergence. For Case I we presented $\hbar$-curves in Figure 4.

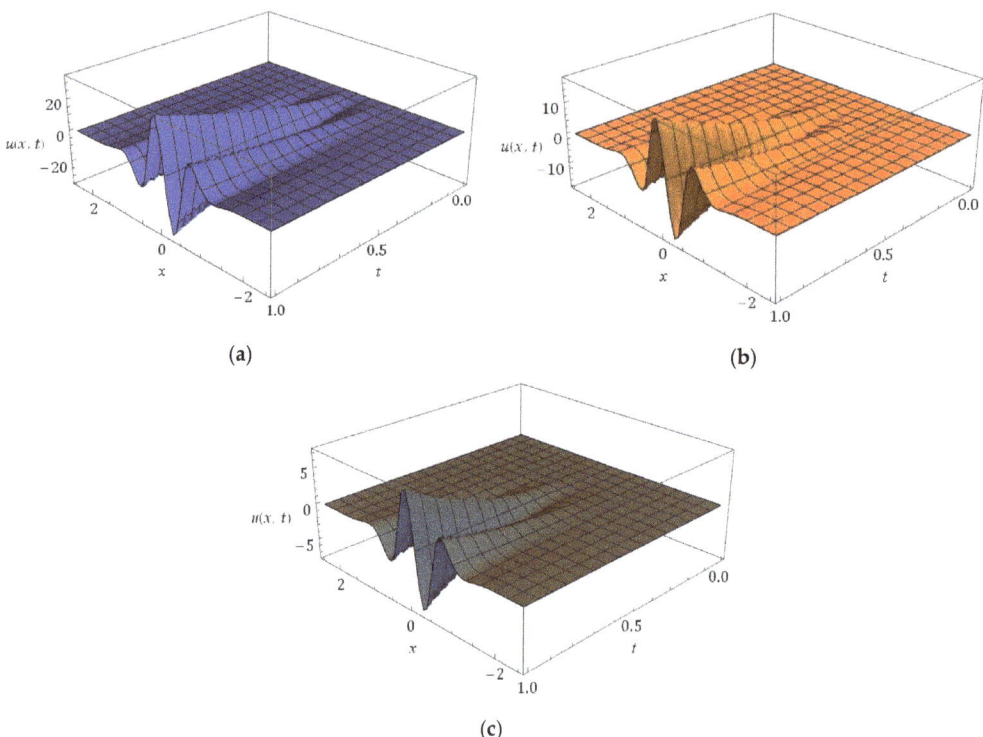

Figure 1. Surfaces of q-HATT solution for (**a**) $\alpha = 0.50$, (**b**) $\alpha = 0.75$ and (**c**) $\alpha = 1$ at $n = 1$, $\hbar = -1$, $\beta = 1$, $c = -1$, $\kappa_1 = 1$ and $\rho = 0.001$ for Case I.

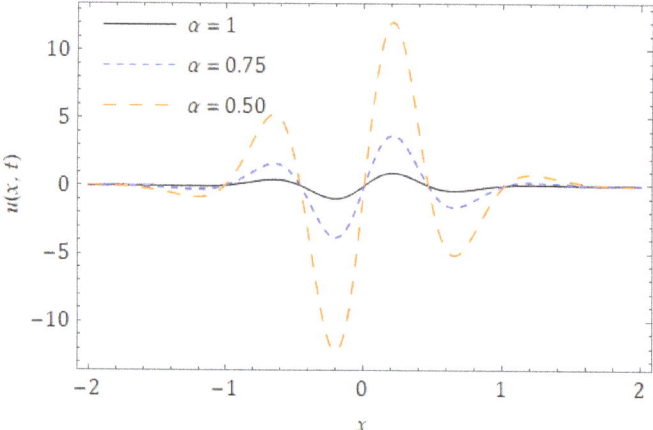

Figure 2. Response of the obtained solution with distinct α and time at $n = 1$, $t = 0.5$, $\hbar = -1$, $\beta = 1, c = -1$, $\kappa_1 = 1$ and $\rho = 0.001$ for Case I.

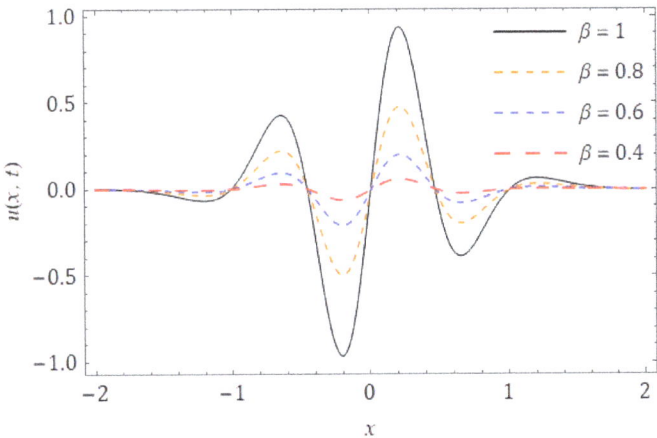

Figure 3. Response of the obtained solution with distinct β and time at $n = 1$, $t = 0.5$, $\hbar = -1$, $\alpha = 1$, $c = -1$, $\kappa_1 = 1$ and $\rho = 0.001$ for Case I.

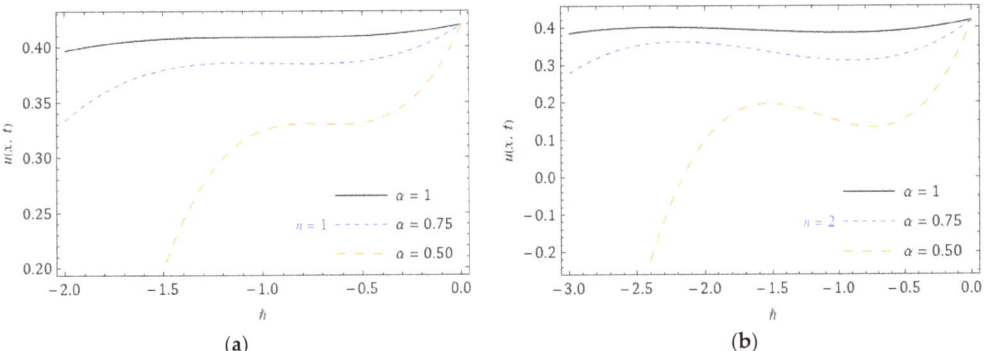

Figure 4. $\hbar$-curves for Case I at $x = 1$, $t = 0.01$, $\hbar = -1$, $\beta = 1$, $c = -1$, $\kappa_1 = 1$ and $\rho = 0.001$ for (**a**) $n = 1$ and (**b**) $n = 2$.

Similarly, we present Case II (i.e., fractional Gardner–Ostrovsky equation). For the change in fractional values of the order, we capture the response in Figures 5 and 6. The effect of the small dispersion term in the FGO equation is cited in Figure 7. In Figure 2, we have a diverse change in between the range of -1 and 1 with x. With the change in β within the range of 0.4 and 1 for this case, we can observe peak variations with the same values that we observe for x with fractional order in Figure 3, and similarly for Case II in Figures 6 and 7. For Case I and II, we respectively presented $\hbar$-curves in Figure 8. These types of studies can influence researchers to investigate physical phenomena.

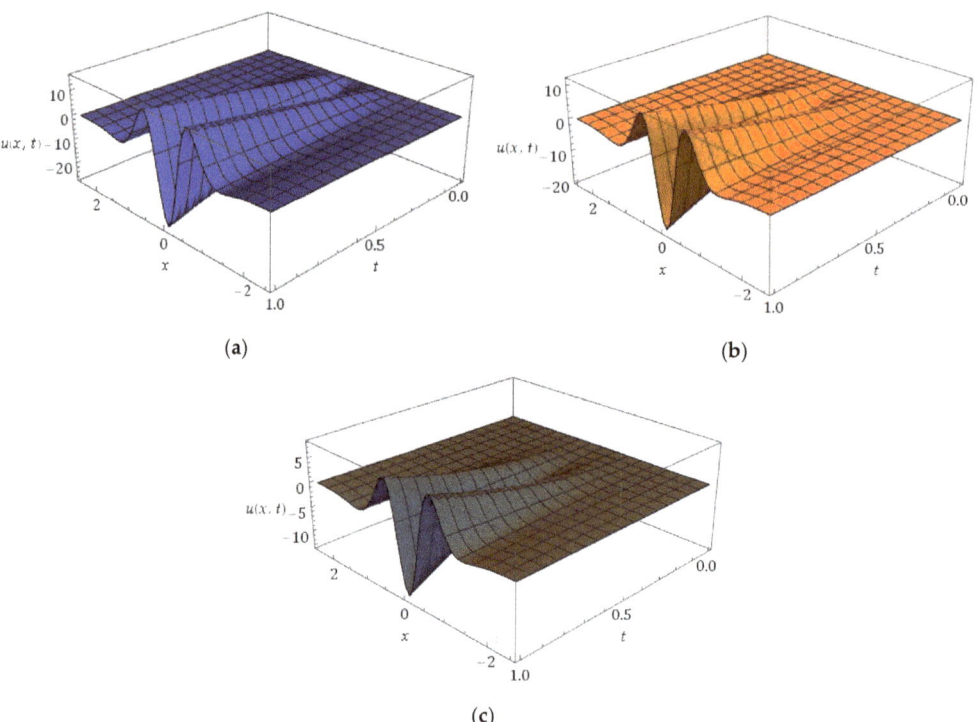

Figure 5. Surfaces of q-HATT solution for (**a**) $\alpha = 0.50$, (**b**) $\alpha = 0.75$ and (**c**) $\alpha = 1$ at $n = 1$, $\hbar = -1$, $\beta = 1$, $c = -1$, $\kappa_1 = 1$ and $\rho = 0.001$ for Case II.

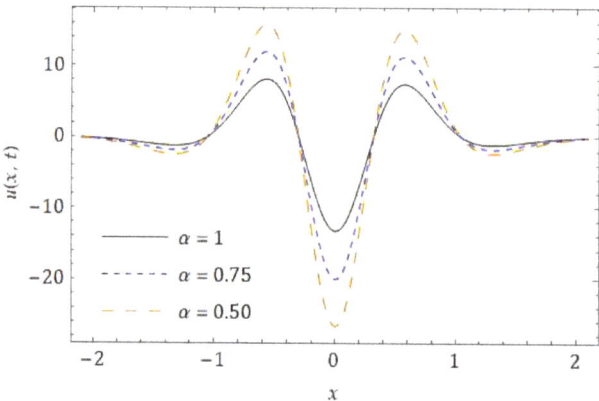

Figure 6. Response of the obtained solution with distinct α and time at $n = 1$, $t = 0.5$, $\hbar = -1$, $\beta = 1$, $\kappa_1 = \kappa_2 = 1$ and $\rho = 0.001$ for Case II.

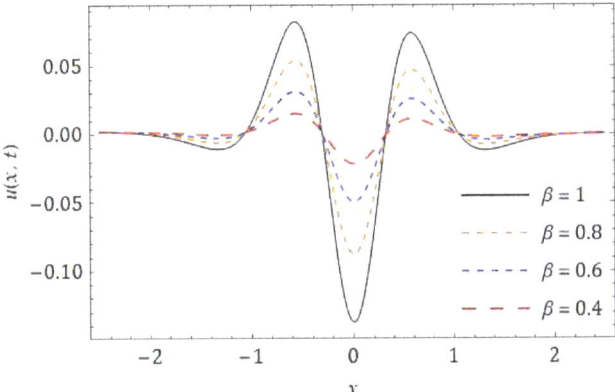

Figure 7. Response of the obtained solution with distinct β and time at $n = 1$, $t = 0.5$, $\hbar = -1$, $\alpha = 1, \kappa_1 = \kappa_2 = 1$ and $\rho = 0.001$ for Case II.

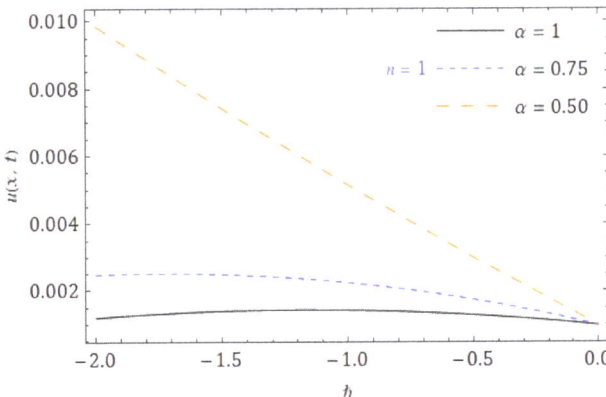

Figure 8. Curves for Case II at $x = 1$, $t = 0.01$, $\hbar = -1$, $\beta = 1$, $\kappa_1 = \kappa_2 = 1$, $n = 1$ and $\rho = 0.001$.

6. Conclusions

In this study, we investigated the equations exemplifying the rotating ocean with strong interacting internal waves with the fractional operator using q-HATT. With the help of the fixed-point hypothesis, the existence of the solution of the two cases is illustrated associated with the Caputo–Fabrizio operator. Furthermore, the effect of the coefficients of small-scale dispersion is illustrated with change in space x for both the cases, and we found noticeable variations in the archived results. The considered scheme is free from many limitations, including conversion from PDE to ODE, nonlinear to linear, or discretization. Finally, the present study demonstrates the effect of fractional order, parameters associated with models, and methods with their corresponding consequences.

Author Contributions: Conceptualization, methodology, software, writing: P.V.; validation, formal analysis: H.M.B.; investigation, resources: W.G. All authors have read and agreed to the published version of the manuscript.

Funding: This research received no external funding.

Institutional Review Board Statement: Not applicable.

Informed Consent Statement: Not applicable.

Data Availability Statement: Not applicable.

Acknowledgments: This projected work was partially (not financial) supported by the Scientific Research Project Fund of Harran University with the project number HUBAP:21132.

Conflicts of Interest: The authors declare no conflict of interest.

References

1. Liouville, J. Memoire surquelques questions de geometrieet de mecanique, et sur un nouveau genre de calcul pour resoudreces questions. *J. Ec. Polytech.* **1832**, *13*, 1–69.
2. Riemann, G.F.B. *Versuch Einer Allgemeinen Auffassung der Integration und Differentiation, Gesammelte Mathematische Werke*; Leipzig, Germany, 1896.
3. Caputo, M. *Elasticita e Dissipazione*; Zanichelli: Bologna, Italy, 1969.
4. Miller, K.S.; Ross, B. *An Introduction to Fractional Calculus and Fractional Differential Equations*; A Wiley: New York, NY, USA, 1993.
5. Podlubny, I. *Fractional Differential Equations*; Academic Press: New York, NY, USA, 1999.
6. Kilbas, A.A.; Srivastava, H.M.; Trujillo, J.J. *Theory and Applications of Fractional Differential Equations*; Elsevier: Amsterdam, The Netherlands, 2006.
7. Baleanu, D.; Guvenc, Z.B.; Tenreiro Machado, J.A. *New Trends in Nanotechnology and Fractional Calculus Applications*; Springer: Dordrecht, The Netherlands; Heidelberg, Germany; London, UK; New York, NY, USA, 2010.
8. Baishya, C.; Achar, S.J.; Veeresha, P.; Prakasha, D.G. Dynamics of a fractional epidemiological model with disease infection in both the populations. *Chaos* **2021**, *31*. [CrossRef]
9. Akinyemi, L.; Veeresha, P.; Senol, M. Numerical solutions for coupled nonlinear schrodinger-Korteweg-de Vries and Maccari's systems of equations. *Mod. Phys. Lett. B* **2021**. [CrossRef]
10. Baishya, C.; Jaipala. Numerical solution of fractional predator-prey model by trapezoidal based homotopy perturbation method. *Int. J. Math. Arch.* **2018**, *9*, 252–259.
11. Antonova, M.; Biswas, A. Adiabatic parameter dynamics of perturbed solitary waves. *Commun. Nonlinear Sci. Numer. Simul.* **2009**, *14*, 734–748. [CrossRef]
12. Holloway, P.; Pelinovsky, E.; Talipova, T. A generalised Korteweg-de Vries model of internal tide transformation in the coastal zone. *J. Geophys.* **1999**, *104*, 333–350. [CrossRef]
13. Wazwaz, A.M. The variational iteration method for solving linear and nonlinear ODEs and scientific models with variable coefficients. *Cent. Eur. J. Eng.* **2014**, *4*, 64–71. [CrossRef]
14. Apel, J.; Ostrovsky, L.A.; Stepanyants, Y.A.; Lynch, J.F. Internal solitons in the ocean and their effect on underwater sound. *J. Acoust. Soc. Am.* **2007**, *121*, 695–722. [CrossRef]
15. Ostrovsky, L.A.; Pelinovsky, E.N.; Shrira, V.I.; Stepanyants, Y.A. Beyond the KDV: Post-explosion development. *Chaos* **2015**, *25*, 097620. [CrossRef]
16. Biswas, A.; Krishnan, E.V. Exact solutions for Ostrovsky equation. *Indian J. Phys.* **2011**, *85*, 1513–1521. [CrossRef]
17. Grimshaw, R.; Stepanyants, Y.; Alias, A. Formation of wave packets in the Ostrovsky equation for both normal and anomalous dispersion. *Proc. R. Soc. A* **2015**, *472*. [CrossRef]
18. Stepanyants, Y.A. Nonlinear Waves in a rotating ocean the Ostrovsky equation and its generalizations and Applications. *Atmos. Ocean. Phys.* **2020**, *56*, 16–32. [CrossRef]
19. Caputo, M.; Fabrizio, M. A new definition of fractional derivative without singular kernel. *Prog. Fract. Differ. Appl.* **2015**, *1*, 73–85.
20. Liao, S.J. Homotopy analysis method and its applications in mathematics. *J. Basic Sci. Eng.* **1997**, *5*, 111–125.
21. Liao, S.J. Homotopy analysis method: A new analytic method for nonlinear problems. *Appl. Math. Mech.* **1998**, *19*, 957–962.
22. Singh, J.; Kumar, D.; Swroop, R. Numerical solution of time- and space-fractional coupled Burgers' equations via homotopy algorithm. *Alex. Eng. J.* **2016**, *55*, 1753–1763. [CrossRef]
23. Srivastava, H.M.; Kumar, D.; Singh, J. An efficient analytical technique for fractional model of vibration equation. *Appl. Math. Model.* **2017**, *45*, 192–204. [CrossRef]
24. Safare, K.M.; Betageri, V.S.; Prakasha, D.G.; Veeresha, P.; Kumar, S. A mathematical analysis of ongoing outbreak COVID-19 in India through nonsingular derivative. *Numer. Methods Partial Differ. Equ.* **2021**, *37*, 1282–1298. [CrossRef]
25. Bulut, H.; Kumar, D.; Singh, J.; Swroop, R.; Baskonus, H.M. Analytic study for a fractional model of HIV infection of CD4+T lymphocyte cells. *Math. Nat. Sci.* **2018**, *2*, 33–43. [CrossRef]
26. Prakasha, D.G.; Malagi, N.S.; Veeresha, P.; Prasannakumara, B.C. An efficient computational technique for time-fractional Kaup-Kupershmidt equation. *Numer. Methods Partial Differ. Equ.* **2021**, *37*, 1299–1316. [CrossRef]
27. Kumar, D.; Agarwal, R.P.; Singh, J. A modified numerical scheme and convergence analysis for fractional model of Lienard's equation. *J. Comput. Appl. Math.* **2018**, *399*, 405–413. [CrossRef]
28. Akinyemi, L.; Huseen, S.N. A powerful approach to study the new modified coupled Korteweg-de Vries system. *Math. Comput. Simul.* **2021**, *177*, 556–567. [CrossRef]
29. Veeresha, P.; Ilhan, E.; Baskonus, H.M. Fractional approach for analysis of the model describing wind-influenced projectile motion. *Phys. Scr.* **2021**, *96*, 075209. [CrossRef]
30. Akinyemi, L.; Şenol, M.; Huseen, S.N. Modified homotopy methods for generalized fractional perturbed Zakharov–Kuznetsov equation in dusty plasma. *Adv. Differ. Equ.* **2021**, *45*. [CrossRef]
31. Losada, J.; Nieto, J.J. Properties of the new fractional derivative without singular Kernel. *Prog. Fract. Differ. Appl.* **2015**, *1*, 87–92.

Article

On a Class of Isoperimetric Constrained Controlled Optimization Problems

Savin Treanţă

Department of Applied Mathematics, University Politehnica of Bucharest, 060042 Bucharest, Romania; savin.treanta@upb.ro

Abstract: In this paper, we investigate the Lagrange dynamics generated by a class of isoperimetric constrained controlled optimization problems involving second-order partial derivatives and boundary conditions. More precisely, we derive necessary optimality conditions for the considered class of variational control problems governed by path-independent curvilinear integral functionals. Moreover, the theoretical results presented in the paper are accompanied by an illustrative example. Furthermore, an algorithm is proposed to emphasize the steps to be followed to solve a control problem such as the one studied in this paper.

Keywords: controlled second-order Lagrangian; Euler–Lagrange equations; isoperimetric constraints; curvilinear integral; differential 1-form

MSC: 49K15; 49K20; 49K21; 65K10

Citation: Treanţă, S. On a Class of Isoperimetric Constrained Controlled Optimization Problems. *Axioms* **2021**, *10*, 112. https://doi.org/10.3390/axioms10020112

Academic Editors: Davron Aslonqulovich Juraev, Samad Noeiaghdam and Hsien-Chung Wu

Received: 26 April 2021
Accepted: 1 June 2021
Published: 3 June 2021

Publisher's Note: MDPI stays neutral with regard to jurisdictional claims in published maps and institutional affiliations.

Copyright: © 2021 by the authors. Licensee MDPI, Basel, Switzerland. This article is an open access article distributed under the terms and conditions of the Creative Commons Attribution (CC BY) license (https://creativecommons.org/licenses/by/4.0/).

1. Introduction

In the last decade, several researchers (see, for instance, Treanţă [1–8], Jayswal et al. [9] and Mititelu and Treanţă [10]) have studied several controlled processes by considering some integral functionals with PDE, PDI, or mixed constraints. More specifically, these researchers have introduced and investigated new classes of optimization problems governed by multiple and path-independent curvilinear integral functionals with mixed constraints involving first-order PDEs of m-flow type, partial differential inequations and boundary conditions. In this regard, quite recently, Treanţă [11] established the optimality conditions for a class of constrained interval-valued optimization problems governed by path-independent curvilinear integral (mechanical work) cost functionals. More exactly, he formulated and proved a minimal criterion of optimality such that a local LU-optimal solution of the considered constrained optimization problem to be its global LU-optimal solution. On the other hand, due to their importance in the applied sciences and engineering, the isoperimetric constrained optimization problems have been introduced, studied and analyzed by many researchers. In this respect, by using the Pontryagin's principle, Schmitendorf [12] established necessary optimality conditions for a class of isoperimetric constrained control problems with inequality constraints at the terminal time. Further, Forster and Long [13] have studied the same isoperimetric constrained optimization problem formulated in Schmitendorf [12] (see, also, Schmitendorf [14]). They have established the associated necessary conditions of optimality by considering an alternative transformation technique. Recently, Benner et al. [15] investigated bang-bang control strategies corresponding to periodic trajectories with isoperimetric constraints for a control problem, with application to nonlinear chemical reactions. For other different but connected ideas on this subject, the reader is directed to the following reasearch works [16–20].

In this paper, motivated and inspired by the research works conducted by Hestenes [21], Lee [22], Schmitendorf [12] and Treanţă [4], we introduce a new class of isoperimetric constrained controlled optimization problems governed by path-independent curvilinear integral functionals which involves second-order partial derivatives and boundary conditions.

Concretely, in comparison with other related research papers, without restrict our analysis to linear systems having convex cost (see Lee [22]), we build a mathematical framework that is more general than in Hestenes [21] and Schmitendorf [12], both by the presence of path-independent curvilinear integrals as isoperimetric constraints but also by the inclusion of second-order partial derivatives and the new proof associated with the main result. Furthermore, besides totally new elements mentioned above, due to the physical meaning of the integral functionals used (as is well-known the path-independent curvilinear integrals represent the mechanical work performed by a variable force in order to move its point of application along a given piecewise smooth curve), this paper becomes a fundamental work for researchers in the field of applied mathematics and ingineering.

The paper is divided as follows. Section 2 introduces the controlled optimization problem under study, and includes the main result of the current paper, namely, Theorem 1. This result establishes the necessary conditions of optimality for the considered isoperimetric constrained variational control problem. Furthermore, an illustrative example is presented in the second part of Section 2. Moreover, to emphasize the steps to be followed to solve a control problem such as the one studied in this paper, an algorithm is presented. Section 3 contains the conclusions of the paper.

2. Isoperimetric Constrained Controlled Optimization Problem

In the following, let $\mathcal{L}_\zeta\big(s(t), s_\gamma(t), s_{\alpha\beta}(t), u(t), t\big)$, $\zeta = \overline{1,m}$, be C^3-class functions, called *multi-time controlled second-order Lagrangians*, where $t = (t^\alpha) = (t^1, \cdots, t^m) \in \Lambda_{t_0,t_1} \subset \mathbb{R}_+^m$, $s = (s^i) = \big(s^1, \cdots, s^n\big) : \Lambda_{t_0,t_1} \to \mathbb{R}^n$ is a C^4-class function (called the *state variable*) and $u = (u^\vartheta) = \big(u^1, \cdots, u^k\big) : \Lambda_{t_0,t_1} \to \mathbb{R}^k$ is a piecewise continuous function (called the *control variable*). Furthermore, denote $s_\alpha(t) := \dfrac{\partial s}{\partial t^\alpha}(t)$, $s_{\alpha\beta}(t) := \dfrac{\partial^2 s}{\partial t^\alpha \partial t^\beta}(t)$, $\alpha, \beta \in \{1, ..., m\}$, and consider $\Lambda_{t_0,t_1} = [t_0, t_1]$ (*multi-time interval* in $\mathbb{R}_+^m$) is a hyper-parallelepiped determined by the diagonally opposite points $t_0, t_1 \in \mathbb{R}_+^m$. Moreover, we assume that the previous multi-time controlled second-order Lagrangians determine a controlled closed (complete integrable) Lagrange 1-form

$$\mathcal{L}_\zeta\big(s(t), s_\gamma(t), s_{\alpha\beta}(t), u(t), t\big) dt^\zeta$$

(see summation over the repeated indices, Einstein summation), which generates the following controlled path-independent curvilinear integral functional

$$J(s(\cdot), u(\cdot)) = \int_{\Upsilon_{t_0,t_1}} \mathcal{L}_\zeta\big(s(t), s_\gamma(t), s_{\alpha\beta}(t), u(t), t\big) dt^\zeta, \tag{1}$$

where Υ_{t_0,t_1} is a smooth curve, included in Λ_{t_0,t_1}, joining the points $t_0, t_1 \in \mathbb{R}_+^m$.

Isoperimetric constrained controlled optimization problem. *Find the pair (s^*, u^*) that minimizes the above controlled path-independent curvilinear integral functional (1), among all the pair functions (s, u) satisfying*

$$s(t_0) = s_0, \quad s(t_1) = s_1, \quad s_\gamma(t_0) = \tilde{s}_{\gamma 0}, \quad s_\gamma(t_1) = \tilde{s}_{\gamma 1}$$

and the isoperimetric constraints (constant level sets of some controlled curvilinear integral functionals) defined as follows:

$$\int_{\Upsilon_{t_0,t_1}} g_\zeta^a\big(s(t), s_\gamma(t), s_{\alpha\beta}(t), u(t), t\big) dt^\zeta = l^a, \quad a = 1, 2, \cdots, r \leq n,$$

where

$$g_\zeta^a\big(s(t), s_\gamma(t), s_{\alpha\beta}(t), u(t), t\big) dt^\zeta, \quad a = 1, 2, \cdots, r$$

are (C^1-class functions) complete integrable differential 1-forms, that is, $D_\gamma g_\zeta = D_\zeta g_\gamma$, $\gamma, \zeta \in \{1, \cdots, m\}$, $\gamma \neq \zeta$, where $D_\gamma := \frac{\partial}{\partial t^\gamma}$, $\gamma \in \{1, \cdots, m\}$.

In order to formulate the necessary optimality conditions of the above controlled optimization problem (1), associated with the aforementioned isoperimetric constraints, we introduce the curve $Y_{t_0,t} \subset Y_{t_0,t_1}$ and the auxiliary variables

$$y^a(t) = \int_{Y_{t_0,t}} g^a_\zeta(s(\tau), s_\gamma(\tau), s_{\alpha\beta}(\tau), u(\tau), \tau) d\tau^\zeta, \quad a = 1, 2, \cdots, r,$$

which satisfy $y^a(t_0) = 0$, $y^a(t_1) = l^a$. It results that the functions y^a fulfil the following controlled complete integrable first-order PDEs

$$\frac{\partial y^a}{\partial t^\zeta}(t) = g^a_\zeta(s(t), s_\gamma(t), s_{\alpha\beta}(t), u(t), t), \quad y^a(t_1) = l^a.$$

Now, under the Abadie constraint qualifications, considering the Lagrange multiplier $p = (p_a(t))$ and by denoting $y = (y^a(t))$, we build new multi-time controlled second-order Lagrangians

$$\mathcal{L}_{1\zeta}(s(t), s_\gamma(t), s_{\alpha\beta}(t), u(t), y(t), y_\zeta(t), p(t), t) = \mathcal{L}_\zeta(s(t), s_\gamma(t), s_{\alpha\beta}(t), u(t), t)$$

$$+ p_a(t) \left(g^a_\zeta(s(t), s_\gamma(t), s_{\alpha\beta}(t), u(t), t) - \frac{\partial y^a}{\partial t^\zeta}(t) \right), \quad \zeta = \overline{1, m},$$

which change the initial controlled optimization problem (with isoperimetric constraints defined by controlled path-independent curvilinear integral functionals) into an unconstrained controlled optimization problem

$$\min_{(s(\cdot), u(\cdot), y(\cdot), p(\cdot))} \int_{Y_{t_0,t_1}} \mathcal{L}_{1\zeta}(s(t), s_\gamma(t), s_{\alpha\beta}(t), u(t), y(t), y_\zeta(t), p(t), t) dt^\zeta \quad (2)$$

$$s(t_q) = s_q, \quad s_\gamma(t_q) = \tilde{s}_{\gamma q}, \quad q = 0, 1$$

$$y(t_0) = 0, \quad y(t_1) = l.$$

According to Lagrange theory (Treanţă [4]), a minimum point of (1) is found among the minimum points of (2).

A *multi-index* (see Saunders [23]) is an m-tuple U of natural numbers. The components of U are denoted $U(\alpha)$, where α is an ordinary index, $1 \leq \alpha \leq m$. The multi-index 1_α is defined by $1_\alpha(\alpha) = 1$, $1_\alpha(\beta) = 0$ for $\alpha \neq \beta$. The addition and the substraction of the multi-indexes are defined componentwise (although the result of a substraction might not be a multi-index): $(U \pm V)(\alpha) = U(\alpha) \pm V(\alpha)$. The length of a multi-index is $|U| = \sum_{\alpha=1}^{m} U(\alpha)$, and its factorial is $U! = \prod_{\alpha=1}^{m}(U(\alpha))!$. The number of distinct indices represented by $\{\alpha_1, \alpha_2, ..., \alpha_k\}$, $\alpha_j \in \{1, 2, ..., m\}$, $j = \overline{1, k}$, is

$$\mu(\alpha_1, \alpha_2, ..., \alpha_k) = \frac{|1_{\alpha_1} + 1_{\alpha_2} + ... + 1_{\alpha_k}|!}{(1_{\alpha_1} + 1_{\alpha_2} + ... + 1_{\alpha_k})!}.$$

The following theorem represents the main result of this paper. It establishes the necessary conditions of optimality associated with the considered isoperimetric constrained controlled optimization problem.

Theorem 1. *If $(s^*(\cdot), u^*(\cdot), y^*(\cdot), p^*(\cdot))$ is solution for (2), then*

$$(s^*(\cdot), u^*(\cdot), y^*(\cdot), p^*(\cdot))$$

is solution of the following Euler–Lagrange system of PDEs

$$\frac{\partial \mathcal{L}_{1\zeta}}{\partial s^i} - D_\gamma \frac{\partial \mathcal{L}_{1\zeta}}{\partial s^i_\gamma} + \frac{1}{\mu(\alpha,\beta)} D^2_{\alpha\beta} \frac{\partial \mathcal{L}_{1\zeta}}{\partial s^i_{\alpha\beta}} = 0, \quad i = \overline{1,n}, \; \zeta = \overline{1,m}$$

$$\frac{\partial \mathcal{L}_{1\zeta}}{\partial u^\vartheta} - D_\gamma \frac{\partial \mathcal{L}_{1\zeta}}{\partial u^\vartheta_\gamma} + \frac{1}{\mu(\alpha,\beta)} D^2_{\alpha\beta} \frac{\partial \mathcal{L}_{1\zeta}}{\partial u^\vartheta_{\alpha\beta}} = 0, \quad \vartheta = \overline{1,k}, \; \zeta = \overline{1,m}$$

$$\frac{\partial \mathcal{L}_{1\zeta}}{\partial y^a} - D_\zeta \frac{\partial \mathcal{L}_{1\zeta}}{\partial y^a_\zeta} + \frac{1}{\mu(\alpha,\beta)} D^2_{\alpha\beta} \frac{\partial \mathcal{L}_{1\zeta}}{\partial y^a_{\alpha\beta}} = 0, \quad a = \overline{1,r}, \; \zeta = \overline{1,m}$$

$$\frac{\partial \mathcal{L}_{1\zeta}}{\partial p_a} - D_\gamma \frac{\partial \mathcal{L}_{1\zeta}}{\partial p_{a,\gamma}} + \frac{1}{\mu(\alpha,\beta)} D^2_{\alpha\beta} \frac{\partial \mathcal{L}_{1\zeta}}{\partial p_{a,\alpha\beta}} = 0, \quad a = \overline{1,r}, \; \zeta = \overline{1,m}$$

where $p_{a,\gamma} := \frac{\partial p_a}{\partial t^\gamma}$, $p_{a,\alpha\beta} := \frac{\partial^2 p_a}{\partial t^\alpha \partial t^\beta}$, $u^\vartheta_{\alpha\beta} := \frac{\partial^2 u^\vartheta}{\partial t^\alpha \partial t^\beta}$, $y^a_{\alpha\beta} := \frac{\partial^2 y^a}{\partial t^\alpha \partial t^\beta}$, $\alpha, \beta, \gamma, \zeta \in \{1, 2, ..., m\}$.

Proof. Let $(s(t), u(t), y(t), p(t))$ be a solution for (2) and $s(t) + \varepsilon h(t)$ is a variation of $s(t)$, with $h(t_0) = h(t_1) = 0$, $h_\eta(t_0) = h_\eta(t_1) = 0$, $\eta \in \{1, 2, ..., m\}$ (see $h_\eta := \frac{\partial h}{\partial t^\eta}$). Furthermore, let $p(t) + \varepsilon f(t)$ be a variation of $p(t)$, with $f(t_0) = f(t_1) = 0$. In the same manner, consider $u(t) + \varepsilon m(t)$, $y(t) + \varepsilon n(t)$ be a variation of $u(t)$ and $y(t)$, respectively, with $m(t_0) = m(t_1) = n(t_0) = n(t_1) = 0$. The functions h, f, m, n represent some "small" variations and ε is a "small" parameter used in our variational arguments. By considering the aforementioned variations, the controlled curvilinear integral functional becomes a function depending by ε, that is, a controlled curvilinear integral with parameter

$$I(\varepsilon) = \int_{\Upsilon_{t_0,t_1}} \mathcal{L}_{1\zeta}(s(t) + \varepsilon h(t), s_\gamma(t) + \varepsilon h_\gamma(t), s_{\alpha\beta}(t) + \varepsilon h_{\alpha\beta}(t), u(t) + \varepsilon m(t),$$

$$y(t) + \varepsilon n(t), y_\zeta(t) + \varepsilon n_\zeta(t), p(t) + \varepsilon f(t), t) dt^\zeta.$$

By hypothesis, we must have the following relation

$$0 = \frac{d}{d\varepsilon} I(\varepsilon)|_{\varepsilon=0} = \int_{\Upsilon_{t_0,t_1}} \left(\frac{\partial \mathcal{L}_{1\zeta}}{\partial s^j} h^j + \frac{\partial \mathcal{L}_{1\zeta}}{\partial s^j_\gamma} h^j_\gamma + \frac{1}{\mu(\alpha,\beta)} \frac{\partial \mathcal{L}_{1\zeta}}{\partial s^j_{\alpha\beta}} h^j_{\alpha\beta} + \frac{\partial \mathcal{L}_{1\zeta}}{\partial u^\vartheta} m^\vartheta \right.$$

$$\left. + \frac{\partial \mathcal{L}_{1\zeta}}{\partial y^a} n^a + \frac{\partial \mathcal{L}_{1\zeta}}{\partial y^a_\zeta} n^a_\zeta + \frac{\partial \mathcal{L}_{1\zeta}}{\partial p_a} f^a \right) dt^\zeta$$

$$= BT + \int_{\Upsilon_{t_0,t_1}} \left(\frac{\partial \mathcal{L}_{1\zeta}}{\partial s^j} - D_\gamma \frac{\partial \mathcal{L}_{1\zeta}}{\partial s^j_\gamma} + \frac{1}{\mu(\alpha,\beta)} D^2_{\alpha\beta} \frac{\partial \mathcal{L}_{1\zeta}}{\partial s^j_{\alpha\beta}} \right) h^j dt^\zeta$$

$$+ \int_{\Upsilon_{t_0,t_1}} \left(\frac{\partial \mathcal{L}_{1\zeta}}{\partial y^a} - D_\zeta \frac{\partial \mathcal{L}_{1\zeta}}{\partial y^a_\zeta} + \frac{1}{\mu(\alpha,\beta)} D^2_{\alpha\beta} \frac{\partial \mathcal{L}_{1\zeta}}{\partial y^a_{\alpha\beta}} \right) n^a dt^\zeta$$

$$+ \int_{\Upsilon_{t_0,t_1}} \left(\frac{\partial \mathcal{L}_{1\zeta}}{\partial u^\vartheta} - D_\gamma \frac{\partial \mathcal{L}_{1\zeta}}{\partial u^\vartheta_\gamma} + \frac{1}{\mu(\alpha,\beta)} D^2_{\alpha\beta} \frac{\partial \mathcal{L}_{1\zeta}}{\partial u^\vartheta_{\alpha\beta}} \right) m^\vartheta dt^\zeta$$

$$+ \int_{Y_{t_0,t_1}} \left(\frac{\partial \mathcal{L}_{1\zeta}}{\partial p_a} - D_\gamma \frac{\partial \mathcal{L}_{1\zeta}}{\partial p_{a,\gamma}} + \frac{1}{\mu(\alpha,\beta)} D^2_{\alpha\beta} \frac{\partial \mathcal{L}_{1\zeta}}{\partial p_{a,\alpha\beta}} \right) f^a dt^\zeta.$$

Taking into account the formula of integration by parts, we find the following equalities

$$\frac{\partial \mathcal{L}_{1\zeta}}{\partial s^j_\gamma} h^j_\gamma = -h^j D_\gamma \frac{\partial \mathcal{L}_{1\zeta}}{\partial s^j_\gamma} + D_\gamma \left(\frac{\partial \mathcal{L}_{1\zeta}}{\partial s^j_\gamma} h^j \right),$$

$$\frac{\partial \mathcal{L}_{1\zeta}}{\partial y^a_\zeta} n^a_\zeta = -n^a D_\zeta \frac{\partial \mathcal{L}_{1\zeta}}{\partial y^a_\zeta} + D_\zeta \left(\frac{\partial \mathcal{L}_{1\zeta}}{\partial y^a_\zeta} n^a \right),$$

$$\frac{1}{\mu(\alpha,\beta)} \frac{\partial \mathcal{L}_{1\zeta}}{\partial s^j_{\alpha\beta}} h^j_{\alpha\beta} = \frac{1}{\mu(\alpha,\beta)} \left[h^j D^2_{\alpha\beta} \frac{\partial \mathcal{L}_{1\zeta}}{\partial s^j_{\alpha\beta}} - D_\alpha \left(h^j D_\beta \frac{\partial \mathcal{L}_{1\zeta}}{\partial s^j_{\alpha\beta}} \right) + D_\beta \left(\frac{\partial \mathcal{L}_{1\zeta}}{\partial s^j_{\alpha\beta}} h^j_\alpha \right) \right].$$

The boundary terms BT vanish (see, also, $h(t_q) = m(t_q) = n(t_q) = f(t_q) = 0$, $h_\eta(t_q) = 0$, $q = 0, 1$), by considering the following equalities

$$D_\gamma \left(\frac{\partial \mathcal{L}_{1\zeta}}{\partial s^j_\gamma} h^j \right) = D_\zeta \left(\frac{\partial \mathcal{L}_{1\gamma}}{\partial s^j_\gamma} h^j \right),$$

$$D_\alpha \left(h^j D_\beta \frac{\partial \mathcal{L}_{1\zeta}}{\partial s^j_{\alpha\beta}} \right) = D_\zeta \left(h^j D_\beta \frac{\partial \mathcal{L}_{1\alpha}}{\partial s^j_{\alpha\beta}} \right),$$

$$D_\beta \left(\frac{\partial \mathcal{L}_{1\zeta}}{\partial s^j_{\alpha\beta}} h^j_\alpha \right) = D_\zeta \left(\frac{\partial \mathcal{L}_{1\beta}}{\partial s^j_{\alpha\beta}} h^j_\alpha \right).$$

In addition, we assume that the solution $(s(t), u(t), y(t), p(t))$ in (2) fulfils the following complete integrability conditions (closeness conditions) of Lagrange 1-form $L_{1\zeta}$, that is,

$$\frac{\partial \mathcal{L}_{1\zeta}}{\partial s^i} \frac{\partial s^i}{\partial t^\alpha} + \frac{\partial \mathcal{L}_{1\zeta}}{\partial s^i_\gamma} \frac{\partial s^i_\gamma}{\partial t^\alpha} + \frac{1}{\mu(\alpha,\beta)} \frac{\partial \mathcal{L}_{1\zeta}}{\partial s^i_{\alpha\beta}} \frac{\partial s^i_{\alpha\beta}}{\partial t^\alpha} + \frac{\partial \mathcal{L}_{1\zeta}}{\partial p_a} \frac{\partial p_a}{\partial t^\alpha} + \frac{\partial \mathcal{L}_{1\zeta}}{\partial t^\alpha}$$

$$+ \frac{\partial \mathcal{L}_{1\zeta}}{\partial u^\theta} \frac{\partial u^\theta}{\partial t^\alpha} + \frac{\partial \mathcal{L}_{1\zeta}}{\partial y^a} \frac{\partial y^a}{\partial t^\alpha} + \frac{\partial \mathcal{L}_{1\zeta}}{\partial y^a_\zeta} \frac{\partial y^a_\zeta}{\partial t^\alpha}$$

$$= \frac{\partial \mathcal{L}_{1\alpha}}{\partial s^i} \frac{\partial s^i}{\partial t^\zeta} + \frac{\partial \mathcal{L}_{1\alpha}}{\partial s^i_\gamma} \frac{\partial s^i_\gamma}{\partial t^\zeta} + \frac{1}{\mu(\alpha,\beta)} \frac{\partial \mathcal{L}_{1\alpha}}{\partial s^i_{\alpha\beta}} \frac{\partial s^i_{\alpha\beta}}{\partial t^\zeta} + \frac{\partial \mathcal{L}_{1\alpha}}{\partial p_a} \frac{\partial p_a}{\partial t^\zeta} + \frac{\partial \mathcal{L}_{1\alpha}}{\partial t^\zeta}$$

$$+ \frac{\partial \mathcal{L}_{1\alpha}}{\partial u^\theta} \frac{\partial u^\theta}{\partial t^\zeta} + \frac{\partial \mathcal{L}_{1\alpha}}{\partial y^a} \frac{\partial y^a}{\partial t^\zeta} + \frac{\partial \mathcal{L}_{1\alpha}}{\partial y^a_\zeta} \frac{\partial y^a_\zeta}{\partial t^\zeta}.$$

Furthermore, we assume that the variation functions h, f, m, n satisfy the closeness conditions of the 1-form

$$\mathcal{L}_{1\zeta}(s(t) + \varepsilon h(t), s_\gamma(t) + \varepsilon h_\gamma(t), s_{\alpha\beta}(t) + \varepsilon h_{\alpha\beta}(t), u(t) + \varepsilon m(t),$$

$$y(t) + \varepsilon n(t), y_\zeta(t) + \varepsilon n_\zeta(t), p(t) + \varepsilon f(t), t)dt^\zeta.$$

This condition adds the following PDEs

$$\frac{\partial \mathcal{L}_{1\zeta}}{\partial s^i} \frac{\partial h^i}{\partial t^\alpha} + \frac{\partial \mathcal{L}_{1\zeta}}{\partial s^i_\gamma} \frac{\partial h^i_\gamma}{\partial t^\alpha} + \frac{1}{\mu(\alpha,\beta)} \frac{\partial \mathcal{L}_{1\zeta}}{\partial s^i_{\alpha\beta}} \frac{\partial h^i_{\alpha\beta}}{\partial t^\alpha} + \frac{\partial \mathcal{L}_{1\zeta}}{\partial p_a} \frac{\partial f^a}{\partial t^\alpha}$$

$$+ \frac{\partial \mathcal{L}_{1\zeta}}{\partial u^\vartheta} \frac{\partial m^\vartheta}{\partial t^\alpha} + \frac{\partial \mathcal{L}_{1\zeta}}{\partial y^a} \frac{\partial n^a}{\partial t^\alpha} + \frac{\partial \mathcal{L}_{1\zeta}}{\partial y^a_\zeta} \frac{\partial n^a_\zeta}{\partial t^\alpha}$$

$$= \frac{\partial \mathcal{L}_{1\alpha}}{\partial s^i} \frac{\partial h^i}{\partial t^\zeta} + \frac{\partial \mathcal{L}_{1\alpha}}{\partial s^i_\gamma} \frac{\partial h^i_\gamma}{\partial t^\zeta} + \frac{1}{\mu(\alpha,\beta)} \frac{\partial \mathcal{L}_{1\alpha}}{\partial s^i_{\alpha\beta}} \frac{\partial h^i_{\alpha\beta}}{\partial t^\zeta} + \frac{\partial \mathcal{L}_{1\alpha}}{\partial p_a} \frac{\partial f^a}{\partial t^\zeta}$$

$$+ \frac{\partial \mathcal{L}_{1\alpha}}{\partial u^\vartheta} \frac{\partial m^\vartheta}{\partial t^\zeta} + \frac{\partial \mathcal{L}_{1\alpha}}{\partial y^a} \frac{\partial n^a}{\partial t^\zeta} + \frac{\partial \mathcal{L}_{1\alpha}}{\partial y^a_\zeta} \frac{\partial n^a_\zeta}{\partial t^\zeta}.$$

Finally, we get

$$0 = \int_{\Upsilon_{t_0,t_1}} \left(\frac{\partial \mathcal{L}_{1\zeta}}{\partial s^j} - D_\gamma \frac{\partial \mathcal{L}_{1\zeta}}{\partial s^j_\gamma} + \frac{1}{\mu(\alpha,\beta)} D^2_{\alpha\beta} \frac{\partial \mathcal{L}_{1\zeta}}{\partial s^j_{\alpha\beta}} \right) h^j dt^\zeta$$

$$+ \int_{\Upsilon_{t_0,t_1}} \left(\frac{\partial \mathcal{L}_{1\zeta}}{\partial y^a} - D_\zeta \frac{\partial \mathcal{L}_{1\zeta}}{\partial y^a_\zeta} + \frac{1}{\mu(\alpha,\beta)} D^2_{\alpha\beta} \frac{\partial \mathcal{L}_{1\zeta}}{\partial y^a_{\alpha\beta}} \right) n^a dt^\zeta$$

$$+ \int_{\Upsilon_{t_0,t_1}} \left(\frac{\partial \mathcal{L}_{1\zeta}}{\partial u^\vartheta} - D_\gamma \frac{\partial \mathcal{L}_{1\zeta}}{\partial u^\vartheta_\gamma} + \frac{1}{\mu(\alpha,\beta)} D^2_{\alpha\beta} \frac{\partial \mathcal{L}_{1\zeta}}{\partial u^\vartheta_{\alpha\beta}} \right) m^\vartheta dt^\zeta$$

$$+ \int_{\Upsilon_{t_0,t_1}} \left(\frac{\partial \mathcal{L}_{1\zeta}}{\partial p_a} - D_\gamma \frac{\partial \mathcal{L}_{1\zeta}}{\partial p_{a,\gamma}} + \frac{1}{\mu(\alpha,\beta)} D^2_{\alpha\beta} \frac{\partial \mathcal{L}_{1\zeta}}{\partial p_{a,\alpha\beta}} \right) f^a dt^\zeta$$

and, since the smooth curve Υ_{t_0,t_1} is arbitrary, we obtain the Euler–Lagrange system of PDEs formulated in theorem. □

Remark 1. *The Euler–Lagrange system of PDEs in Theorem 1 can be rewritten as follows*

$$\frac{\partial \mathcal{L}_{1\zeta}}{\partial s^i} - D_\gamma \frac{\partial \mathcal{L}_{1\zeta}}{\partial s^i_\gamma} + \frac{1}{\mu(\alpha,\beta)} D^2_{\alpha\beta} \frac{\partial \mathcal{L}_{1\zeta}}{\partial s^i_{\alpha\beta}} = 0, \quad i = \overline{1,n}, \ \zeta = \overline{1,m}$$

$$\frac{\partial \mathcal{L}_{1\zeta}}{\partial u^\vartheta} - D_\gamma \frac{\partial \mathcal{L}_{1\zeta}}{\partial u^\vartheta_\gamma} + \frac{1}{\mu(\alpha,\beta)} D^2_{\alpha\beta} \frac{\partial \mathcal{L}_{1\zeta}}{\partial u^\vartheta_{\alpha\beta}} = 0, \quad \vartheta = \overline{1,k}, \ \zeta = \overline{1,m}$$

$$\frac{\partial p_a}{\partial t^\zeta} = 0, \quad a = \overline{1,r}, \ \zeta = \overline{1,m}$$

$$\frac{\partial y^a}{\partial t^\zeta}(t) = g^a_\zeta(s(t), s_\gamma(t), s_{\alpha\beta}(t), u(t), t), \quad a = \overline{1,r}, \ \zeta = \overline{1,m}.$$

In consequence, the Lagrange multiplier p is constant. Moreover, it is well determined only if the optimal solution is not an extrem for at least one of the following controlled path-independent curvilinear integral functionals

$$\int_{\Upsilon_{t_0,t_1}} g^a_\zeta(s(t), s_\gamma(t), s_{\alpha\beta}(t), u(t), t) dt^\zeta, \quad a = \overline{1,r}.$$

Illustrative example. Let us find the minimum for the following controlled curvilinear integral functional

$$J(s(\cdot), u(\cdot)) = \int_{Y_{0,1}} \left(s^2(t) + u^2(t)\right) dt^1 + \left(s^2(t) + u^2(t)\right) dt^2$$

subject to: $\int_{Y_{0,1}} s_{t^1}(t) dt^1 + s_{t^2}(t) dt^2 = 0$ (path-independent curvilinear integral) and the boundary conditions $s(0,0) = 0$, $s(1,1) = 0$, where $Y_{0,1}$ is a C^1-class curve, included in $[0,1]^2$, joining the points $(0,0)$, $(1,1)$.

Solution. The path-independence associated with the cost functional $J(s(\cdot), u(\cdot))$ gives the relation

$$s\left(\frac{\partial s}{\partial t^2} - \frac{\partial s}{\partial t^1}\right) = u\left(\frac{\partial u}{\partial t^1} - \frac{\partial u}{\partial t^2}\right).$$

Furthermore, the associated Lagrange 1-form has the following components

$$\mathcal{L}_{11} = s^2(t) + u^2(t) + p(y_{t^1}(t) - s_{t^1}(t)),$$

$$\mathcal{L}_{12} = s^2(t) + u^2(t) + p(y_{t^2}(t) - s_{t^2}(t))$$

and the extremals are described by the following system of Euler–Lagrange PDEs

$$2s + \frac{\partial p}{\partial t^1} = 0, \quad 2s + \frac{\partial p}{\partial t^2} = 0,$$

$$2u = 0,$$

$$y_{t^1}(t) - s_{t^1}(t) = 0, \quad y_{t^2}(t) - s_{t^2}(t) = 0,$$

implying that $(s^\star, u^\star) = (0,0)$ is the optimal solution of the considered isoperimetric constrained controlled optimization problem.

Further, taking into account the above illustrative example and the theory developed in the paper, we formulate an algorithm. The main intention of the next algorithm is to synthesize the concrete steps to be followed to solve a control problem such as those studied in the paper. In particular, for a controlled path-independent curvilinear integral cost functional and a set of mixed (isoperimetric and boundary conditions) restrictions and self or normal data, the main goal is to find $(s^\star, u^\star)$ (satisfying the set of mixed constraints and normal data) such that $J(s^\star, u^\star) \leq J(s, u)$, for all feasible points (s, u). For this purpose, we start with a feasible point (s, u). If the pair (s, u) fulfils the necessary optimality conditions formulated in Theorem 1, then the "Generating Stage" (see below) is satisfied and we go to the next step, namely "Detecting Stage"; else, the algorithm stops. If the set of self or normal data is fulfilled, then the "Detecting Stage" is satisfied and we go to the next step, namely "Deciding Stage" (see below); else, the algorithm stops. For $(s^\star, u^\star)$ derived in "Detecting Stage", if $J(s^\star, u^\star) \leq J(s, u)$ holds for all feasible points (s, u), then $(s^\star, u^\star)$ is an optimal solution; else, the Algorithm 1 stops.

Algorithm 1:
DATA:
• controlled path-independent curvilinear integral cost functional $$\min_{(s,u)} J(s,u) = \int_{\Upsilon_{t_0,t_1}} \mathcal{L}_\zeta\big(s(t),s_\gamma(t),s_{\alpha\beta}(t),u(t),t\big)dt^\zeta;$$ • set of mixed constraints $$\int_{\Upsilon_{t_0,t_1}} g^a_\zeta\big(s(t),s_\gamma(t),s_{\alpha\beta}(t),u(t),t\big)dt^\zeta = l^a, \quad a=1,2,\cdots,r \leq n$$ and $$s(t_q) = s_q, \quad s_\gamma(t_q) = \tilde{s}_{\gamma q}, \quad q=0,1;$$ • set of self or normal data - the differential 1-form $g = \left(g^a_\zeta\right)$ satisfies the closeness conditions; RESULT: $$S = \{(s^\star,u^\star) \mid J(s^\star,u^\star) \leq J(s,u),$$ with $(s^\star,u^\star)$ satisfying the set of ; mixed ; constraints ; and ; normal ; data$\}$; BEGIN • Generating Stage: consider (s,u) a feasible point if the necessary optimality conditions (see Theorem 1) are not compatible with respect to (s,u) then STOP else GO to the next step   • Detecting Stage: monitoring of Lagrange multipliers if the set of self or normal data is not fulfilled then STOP else GO to the next step   • Deciding Stage: let $(s^\star,u^\star)$ be derived in Detecting Stage if $J(s,u) \geq J(s^\star,u^\star)$ holds for all feasible points (s,u) then $(s^\star,u^\star)$ is an optimal solution else STOP
END

3. Conclusions

In this paper, we have studied a new class of isoperimetric constrained controlled optimization problems. In accordance with Lagrange Theory, necessary optimality conditions have been formulated and proved for the considered class of variational control problems governed by path-independent curvilinear integrals and second-order partial derivatives. The theoretical mathematical results developed in the paper have been highlighted by an illustrative example and an algorithm.

As a new research direction on the class of problems introduced in this paper, we mention, for example, the study of well-posedness.

Funding: This research received no external funding.

Institutional Review Board Statement: Not applicable.

Informed Consent Statement: Not applicable.

Data Availability Statement: Not applicable.

Conflicts of Interest: The author declares no conflict of interest.

References

1. Treanţă, S. A necessary and sufficient condition of optimality for a class of multidimensional control problems. *Optim. Control Appl. Methods* **2020**, *41*, 2137–2148. [CrossRef]
2. Treanţă, S. On a global efficiency criterion in multiobjective variational control problems with path-independent curvilinear integral cost functionals. *Ann. Oper. Res.* **2020**, 1–9. [CrossRef]
3. Treanţă, S. Saddle-point optimality criteria in modified variational control problems with PDE constraints. *Optim. Control Appl. Methods* **2020**, *41*, 1160–1175. [CrossRef]
4. Treanţă, S. Constrained variational problems governed by second-order Lagrangians. *Appl. Anal.* **2020**, *99*, 1467–1484. [CrossRef]
5. Treanţă, S.; Mititelu, Ş. Efficiency for variational control problems on Riemann manifolds with geodesic quasiinvex curvilinear integral functionals. *Rev. Real Acad. Cienc. Exactas Físicas Nat. Ser. Matemáticas* **2020**, *114*, 113. [CrossRef]
6. Treanţă, S.; Arana-Jiménez, M.; Antczak, T. A necessary and sufficient condition on the equivalence between local and global optimal solutions in variational control problems. *Nonlinear Anal.* **2020**, *191*, 111640. [CrossRef]
7. Treanţă, S. On a modified optimal control problem with first-order PDE constraints and the associated saddle-point optimality criterion. *Eur. J. Control* **2020**, *51*, 1–9. [CrossRef]
8. Treanţă, S. Efficiency in generalized V-KT-pseudoinvex control problems. *Int. J. Control* **2020**, *93*, 611–618. [CrossRef]
9. Jayswal, A.; Antczak, T.; Jha, S. Modified objective function approach for multitime variational problems. *Turk. J. Math.* **2018**, *42*, 1111–1129.
10. Mititelu, Ş.; Treanţă, S. Efficiency conditions in vector control problems governed by multiple integrals. *J. Appl. Math. Comput.* **2018**, *57*, 647–665. [CrossRef]
11. Treanţă, S. On a class of constrained interval-valued optimization problems governed by mechanical work cost functionals. *J. Optim. Theory Appl.* **2021**, *188*, 913–924. [CrossRef]
12. Schmitendorf, W.E. Pontryagin's principle for problems with isoperimetric constraints and for problems with inequality terminal constraints. *J. Optim. Theory Appl.* **1976**, *18*, 561–567. [CrossRef]
13. Forster, B.A.; Long, N.V. Pontryagin's principle for problems with isoperimetric constraints and for problems with inequality terminal constraints: Comment. *J. Optim. Theory Appl.* **1978**, *25*, 317–322. [CrossRef]
14. Schmitendorf, W.E. Pontryagin's principle for problems with isoperimetric constraints and for problems with inequality terminal constraints: Reply. *J. Optim. Theory Appl.* **1978**, *25*, 323. [CrossRef]
15. Benner, P.; Seidel-Morgenstern, A.; Zuyev, A. Periodic switching strategies for an isoperimetric control problem with application to nonlinear chemical reactions. *Appl. Math. Model.* **2019**, *69*, 287–300. [CrossRef]
16. Bildhauer, M.; Fuchs, M.; Muller, J. A reciprocity principle for constrained isoperimetric problems and existence of isoperimetric subregions in convex sets. *Calc. Var. Partial. Differ. Equ.* **2018**, *57*, 60. [CrossRef]
17. Curtis, J.P. Complementary extremum principles for isoperimetric optimization problems. *Optim. Eng.* **2004**, *5*, 417–430. [CrossRef]
18. Demyanov, V.F.; Tamasyan, G.S. Exact penalty functions in isoperimetric problems. *Optimization* **2011**, *60*, 153–177. [CrossRef]
19. Harper, L.H. *Global Methods for Combinatorial Isoperimetric Problems*; Cambridge University Press: Cambridge, UK, 2010.
20. Urziceanu, S.A. Necessary optimality conditions in isoperimetric constrained optimal control problems. *Symmetry* **2019**, *11*, 1380. [CrossRef]
21. Hestenes, M. *Calculus of Variations and Optimal Control Theory*; John Wiley and Sons: New York, NY, USA, 1966.
22. Lee, E.B. Linear Optimal Control Problems with Isoperimetric Constraints. *IEEE Trans. Autom. Control* **1967**, *12*, 87–90. [CrossRef]
23. Saunders, D.J. *The Geometry of Jet Bundles*; London Mathematical Society Lecture Notes Series 142; Cambridge University Press: Cambridge, UK, 1989.

Article

Monitoring and Recognizing Enterprise Public Opinion from High-Risk Users Based on User Portrait and Random Forest Algorithm

Tinggui Chen [1,*], Xiaohua Yin [1], Lijuan Peng [1], Jingtao Rong [1], Jianjun Yang [2] and Guodong Cong [3]

1. School of Statistics and Mathematics, Zhejiang Gongshang University, Hangzhou 310018, China; yinxh0213@163.com (X.Y.); Cherrylijuanpeng@163.com (L.P.); rjt323@126.com (J.R.)
2. Department of Computer Science and Information Systems, University of North Georgia, Oakwood, GA 30566, USA; Jianjun.Yang@ung.edu
3. School of Tourism and Urban-Rural Planning, Zhejiang Gongshang University, Hangzhou 310018, China; cgd@mail.zjsu.edu.cn
* Correspondence: ctgsimon@mail.zjgsu.edu.cn

Abstract: With the rapid development of "We media" technology, netizens can freely express their opinions regarding enterprise products on a network platform. Consequently, online public opinion about enterprises has become a prominent issue. Negative comments posted by some netizens may trigger negative public opinion, which can have a significant impact on an enterprise's image. From the perspective of helping enterprises deal with negative public opinion, this paper combines user portrait technology and a random forest algorithm to help enterprises identify high-risk users who have posted negative comments and thus may trigger negative public opinion. In this way, enterprises can monitor the public opinion of high-risk users to prevent negative public opinion events. Firstly, we crawled the information of users participating in discussions of product experience, and we constructed a portrait of enterprise public opinion users. Then, the characteristics of the portraits were quantified into indicators such as the user's activity, the user's influence, and the user's emotional tendency, and the indicators were sorted. According to the order of the indicators, the users were divided into high-risk, moderate-risk, and low-risk categories. Next, a supervised high-risk user identification model for this classification was established, based on a random forest algorithm. In turn, the trained random forest identifier can be used to predict whether the authors of newly published public opinion information are high-risk users. Finally, a back propagation neural network algorithm was used to identify users and compared with the results of model recognition in this paper. The results showed that the average recognition accuracy of the back propagation neural network is only 72.33%, while the average recognition accuracy of the model constructed in this paper is as high as 98.49%, which verifies the feasibility and accuracy of the proposed random forest recognition method.

Keywords: product user experience; enterprise network public opinion; identification of high-risk users; random forest algorithm; user portrait

1. Introduction

At present, with the popularization and development of media technology, a growing number of netizens are frequently posting opinions about enterprise products on social platforms, including some comments about poor experiences. However, this open transmission of information sometimes leads to comments about negative experiences, which are likely to cause negative public opinion. Some users may post negative comments via Weibo, for example, to compromise an enterprise, bringing huge damage to the business and resulting in a negative impact on business operations. However, due to the large number of users who post comments online, enterprises cannot monitor the public opinion of all users. Therefore, personalized classification and identification of users can help enterprises develop more targeted solutions for public opinion and can also greatly reduce the cost of controlling public opinion for enterprises. Based on this, the monitoring and identification of users' online comments are particularly critical to improving the efficiency of managing public opinion and maintaining the corporate image.

Generally speaking, monitoring online public opinion for enterprises mainly refers to capturing the relevant public opinion information through a series of technical means to realize the monitoring and tracking of public opinion [1]. At present, the academic research on enterprises' online public opinion focuses on both macro and micro levels. With regard to research on the macro level, most scholars carry out a theoretical analysis on the communication characteristics of enterprise public opinion and propose suggestions to deal with negative public opinion. However, much of the macro research ignores the characteristics of the event content itself and is short of user tracking of negative public opinion events because such research is focused on the overall perspective. On the micro level, many scholars have introduced natural language processing tools to identify the emotional polarity of user comments and described the emotional distribution of product users. Although such studies are helpful for enterprises to understand the emotional tendencies of users, they lack personalized identification and monitoring of users in enterprises' negative public opinion events. In addition, in the study of recognition and detection algorithms, many scholars use neural network algorithms for classification and prediction, but this often requires manual classification of training samples in advance, and the recognition accuracy needs to be improved.

In this context, this paper combines user portrait technology and random forest algorithm to create a supervised model to recognize and monitor high-risk users expressing a public opinion on enterprises. The traditional technique of manually classifying training samples is abandoned in this model, while user portrait technology is adopted to classify the public opinion data of users. Further, such data are used as the training samples of the random forest algorithm, which lowers the workload of manual labeling and eliminates subjectivity, making it more scientific and objective. Specifically, Python (3.8 32-bit) is firstly used to crawl the information of users participating in the discussion of product experience. In turn, indicators are quantified according to the user's portrait technology. Consequently, the users are divided into high-risk, moderate-risk, and low-risk types in terms of the size of quantized indicators. Moreover, a high-risk user identification model based on a supervised random forest algorithm is established, which can learn and train the random forest classifier and predict whether newly released public opinion information users are high-risk users. Finally, by analyzing the public opinion of Hive Box and optimizing the parameters of the random forest algorithm, the parameters applicable to "Hive Box Charges" are selected to demonstrate the feasibility of the model. Compared with the BP (Back Propagation) neural network, the proposed model is verified by recognition results with high accuracy.

2. Literature Review

Negative public opinion caused by negative product experience has a significant impact on corporate image, and this has attracted the attention of many scholars. In this paper, user portrait technology and a random forest algorithm are used to monitor enterprise public opinion. Therefore, the applications of these two methods in enterprise public opinion are summarized.

Many scholars have studied the public opinion of enterprises from the perspective of user portraits. For example, Wang et al. [2] presented a novel learning model called the personal service ecosystem (PSE) to delineate user preferences and interests naturally. Virginia et al. [3] explored the relationship between the user's perceived quality, the club service dimensions, and the golf club performance. The results showed that the strategy to increase user satisfaction should be quite different depending on whether users were beginners or advanced golf players. Martínez-Cevallos et al. [4] analyzed the segmentation of participants in a sports event according to their perceived quality, perceived value, satisfaction, and future intentions. Tiwari et al. [5] leveraged the power of the categorical information stored in the Wikipedia database to assign relative weights to entities that a user followed on Twitter. Zhang et al. [6] explored the application of user portrait technology in agriculture, and they constructed a situational recommendation system of agricultural science and technology resources based on user portraits. Sun and Chai [7] classified the portrait of online learners into three dimensions, and they constructed a tag system of learner portraits based on the data fields of an online learning platform. You et al. [8] excavated user characteristics through identified user behaviors and constructed user portraits based on behavior perception according to actual cases. Widiyaningtyas et al. [9] proposed a new similarity algorithm—so-called "user profile correlation-based similarity", which examined genre data and user profile data, namely, age, gender, occupation, and location. Ni et al. [10] designed a new user portrait construction architecture based on knowledge database construction and fingerprint matching technology. Chen et al. [11] modeled the public opinion reversal process, taking into consideration external intervention information and individual internal characteristics. Their simulation results showed that the intensity of external intervention information affected the direction and degree of public opinion reversal. Although the above literature works have promoted the development of user portrait theory, few scholars have applied this method to the research of enterprise network public opinion. Most scholars have only established an indicator system of user characteristics, while few scholars have quantified the indicators within it.

On the other hand, many scholars use random forests to study public opinion. For example, Yelkanat [12] investigated the performance of the random forest machine learning algorithm in estimating near-future case numbers for 190 countries in the world, and it was mapped in comparison with actual confirmed cases' results. Chen et al. [13] focused on studying the polarization phenomenon and established a model of public opinion dissemination with the polarization process considering individual heterogeneity. Simsekler et al. [14] used a tree-based machine learning algorithm as well as random forests to estimate accurate and stable associations. The results of our analysis showed that safety perception, management support, and supervisor/manager expectations were the leading drivers of patient safety grade. Chen et al. [15] introduced the social preference theory and revealed the micro-interaction mechanism of public opinion polarization. Different social preferences held by individuals had different influences on the public opinion polarization effects. Xiao et al. [16] held that the evolution of individual opinions was not only influenced by the interactions between neighboring individuals but was also updated naturally due to individual factors themselves in the absence of interactions. Li and Xiao [17] combined social judgment theory with the multi-agent model and proposed a multidimensional opinion evolution model for studying the dynamics of opinion polarization. The results demonstrated that polarization was influenced by the average degree of the network, and the polarization process was affected by the parameters of the assimilation effect and contrast effect. Although the above studies have promoted the development of the random forest algorithm, few scholars have applied it to the detection and identification of enterprise network public opinion users, and few scholars have combined it with user portrait technology.

This paper takes enterprise network public opinion as the research background. The current academic research in this field mainly focuses on the macro and micro perspectives. For research on the macro level, Regester and Larkin [18] hold that enterprise reputation is an important component of intangible assets, and they also put forward the 3T principle for enterprises to deal with a reputation crisis. Taylor and Perry [19] proposed that enterprises should establish a crisis response mechanism combining traditional media and network media when dealing with crises. By analyzing specific cases, Li and Dong [20] proposed the communication process model of enterprises' online public opinion and believe that the attention of netizens plays a leading role in the development direction of events. At the micro level, Yu [21] proposed the Leader Rank algorithm, which recognizes the emotional characteristics of users by commenting on them and infers their positive, negative, or neutral attitude towards the author of an article. Yin et al. [22] examined how a perceived locus of the crisis was caused and used a real case study to perform a quantitative content analysis with a sample of 503 comments under online articles. Zhang and Zhang [23] used a sentiment analysis to calculate the emotional strength of defect events and explored the effectiveness of the trust repair strategy adopted by enterprises at different evolution stages. Bella et al. [24] selected the comment text of the official Twitter of an enterprise and used SVM (Support Vector Machines) to classify the text to construct an opinion leader style model. Liu et al. [25] used natural language processing technology to study the word segmentation of text information on a network and applied it for risk detection of corporate public opinion. Aggarwal and Singh [26] monitored user review data from different social network regions and optimized corporate advertising strategies. However, the described objects of the aforementioned research are products or services provided by enterprises, which lack the monitoring of users during negative public opinion events of enterprises.

3. Model Construction

After the outbreak of negative public opinions of an enterprise due to poor user experiences, monitoring high-risk users involves monitoring and recognizing those users participating in negative public opinion discussions. Firstly, user data in the network are crawled and classified by user portrait technology. Then, supervised random forest algorithms are adopted to identify and monitor high-risk groups. The overall framework of the paper is shown in Figure 1.

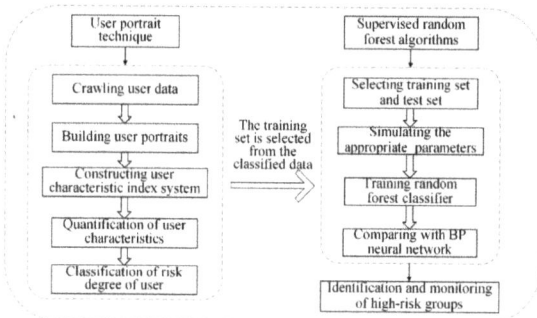

Figure 1. Research framework.

The variables and parameters involved in this paper are shown in Table 1.

Table 1. Relevant parameters.

Parameter	Description
Y	User's influence
Q	User's emotional tendency
S	User's base properties
H	User's activation
A	The number of thumbs up
B	The number of secondary comments
C	The number of fans
D	The amount of attention
F	Total number of Weibo posts shared
$Gini$	$Gini$ coefficient
c	Sample category
$nTree$	The number of decision trees
$maxf$	The maximum number of features
$leaf$	The number of leaf nodes

3.1. Crawling User Data

The product name is used as a keyword to search on various social network platforms, and public opinion topics related to the product experience are selected. Subsequently, the Python (3.8 32-bit) crawler is used to crawl user information under the relevant topic, which is mainly for three kinds of information: (1) user information, such as the number of followers, address, and other account information; (2) user experience, comments, and other text information; and (3) interactive information, such as the number of reposts/comments, etc. In order to facilitate the subsequent emotional segmentation, pre-processing of the comment text data is required. Firstly, the symbols, expressions, and punctuations in the comments are removed. Secondly, comments that are not relevant to the company, such as advertisements and comments with very few words, are also deleted.

3.2. Building the User Portraits

A user portrait refers to the characterization of users through direct or indirect data, which is widely used in enterprise precision marketing, crime prevention, financial risk prediction, and other fields. In addition, by studying users' behaviors on the internet, user portraits can also be used for online public opinion governance [27]. At present, the user portrait system construction methods are varied, and user portrait construction methods with statistical analysis or rules are based on the practice of business knowledge combined with the understanding of the business, scenes, and problems. A characteristic index based on scenes is put forward to construct a user portrait framework and deal with real problems. In this paper, this method is adopted to construct an enterprise network public opinion user portrait. Based on the theoretical knowledge of enterprise public opinion, the user characteristic labels are crawled, and the data are quantified as indicators affecting the risk degree to build a user characteristic index system of enterprise public opinion and take it as the training data of the random forest.

3.2.1. Constructing the User Characteristic Index System

After the outbreak of negative public opinions on an enterprise, the user can participate in discussions and make comments on Weibo. While constructing the portrait of users participating in the discussions of negative public opinion, the characteristics can be divided into dynamic and static categories. The former is the attributes of users when they participate in social activities, including the time they participate in discussions and the opinions they hold. The latter describes basic properties, including an individual's ID, gender, site, number of followers, number of fans, number of Weibo posts shared, and other characteristics. Based on this, the specific user characteristic index system is shown in Table 2.

Table 2. User characteristic index system.

The First Index	The Second Index	The Third Index	The Fourth Index
User portrait	User's dynamic characteristics	User's emotional tendency Q	Comment text content
		User's influence Y	The number of thumbs up A The number of secondary comments B The number of fans C
	User's static characteristics	User's base properties S	Weibo ID Gender Site
		User's activation H	The amount of attention D Total number of Weibo posts shared F

3.2.2. Quantification of User Characteristics

In the index system constructed above, the static characteristics of users can be directly obtained according to the crawled Weibo user data. However, the dynamic characteristics of users change with the content posted by users, so it is necessary to quantify these characteristics.

For the quantification of emotional tendency, the Jieba segmentation tool [28] is adopted, and word frequency is counted. SnowNLP is used to score the emotion so as to obtain the emotional value of each posted text comment as a user's emotional tendency.

For the quantification of user influence, the influence of users in public opinion events is comprehensively judged according to the number of users' comments by likes, the number of secondary comments, and the number of fans. The calculation is shown in the following Formula (1):

$$Y_i = \frac{\frac{A_i}{A_{max}} + \frac{B_i}{B_{max}} + \frac{C_i}{C_{max}}}{3} \quad (1)$$

where Y_i represents the relative influence of the ith Weibo user; A_i represents the likes on a comment by user i, and A_{max} represents the maximum number of likes by all the users participating in the public opinion event; B_i represents the number of secondary comments of i, and B_{max} represents the highest number of secondary comments of all the users participating in the public opinion event; C_i represents the number of fans of i, and C_{max} represents the highest number of fans of all the users participating in the public opinion event.

With regard to the quantification of user activation, the participation in public opinion events can be comprehensively judged according to the amount of users' attention and the total number of posts on Weibo. The calculation is shown in the following Formula (2):

$$H_i = \frac{\frac{D_i}{D_{max}} + \frac{F_i}{F_{max}}}{2} \quad (2)$$

where H_i represents the relative activation of user i; D_i represents the number of followers of i, and D_{max} represents the highest number of followers of all users participating in the public opinion event; F_i represents the total number of Weibo posts shared by i, while F_{max} represents the maximum total number of Weibo posts shared by all users participating in the public opinion event.

3.2.3. Classification of Risk Degree of User

In our classification, those who post negative comments on Weibo and can easily bring great losses to a corporate image are defined as high-risk groups. On this basis, the risk levels of users who post comments on the internet are divided into three categories: high-risk users, medium-risk users, and low-risk users. According to the above-quantified user portrait index system, the risk degree of each user is positively correlated with the influence and activity of the user and negatively correlated with the affective tendency value of the user. Therefore, the user influence and user activity are sorted in descending order, while the emotional tendency is sorted in ascending order. High-risk users are defined as those who rank in the top 10% for influence, activity, and emotional propensity, and their label is set as 2. After the high-risk users are excluded from the top 20% of the users in the three indicators, the remaining users are defined as moderate-risk users, and the label is set as 1. After excluding high-risk and moderate-risk users, the remaining users shall be low-risk users, and the label shall be set as 0. The specific judgment criteria are shown in Table 3.

Table 3. Criteria for classification of user portraits.

Classification	Classification Label	User's Emotional Tendency Q	User's Influence Y	User's Activation H
High-risk type	2	$Q = 0$	$Y \in (0.1, Y_{max})$	$H \in (0.1, Y_{max})$
Moderate-risk type	1	$Q \in (0, 0.05)$	$Y \in (0.05, 0.1)$	$H \in (0.05, 0.1)$
Low-risk type	0	$Q \in (0.05, 1)$	$Y \in (0, 0.05)$	$H \in (0, 0.05)$

3.3. Supervised Random Forest Algorithms

After the outbreak of negative public opinion, monitoring and identifying high-risk users involves classifying those high-risk users participating in discussions of the product/experience, so as to identify the high-risk users with label 2. In recent years, the use of machine learning algorithms to classify all kinds of data has extended to a very wide range of applications, and there are various machine learning algorithms. Since the involved data belong to the supervised classification type, the random forest algorithm with strong generalization ability and fast training speed is adopted. The random forest algorithm was first proposed by Leo [29] in 2001, which was a classifier with a decision tree as the basic unit. It is equivalent to consisting of multiple decision tree classifiers. Each decision tree produces a result, and the final classification result is determined by voting. As such, the principle of random forest is introduced based on decision trees, and the specific steps of the algorithm are described as well.

3.3.1. Decision Tree

A decision tree algorithm is a supervised learning algorithm, which is mainly used for regression and classification. There are three optimal feature selection methods in the decision tree—namely, information gain, information gain rate, and Gini coefficient. By calculating the Gini coefficient selection characteristics, the user set $S = \{s_1, s_2, s_3, \ldots s_n\}$ is obtained, meaning that there are n different users participating in a public opinion discussion, and each user can be represented by $V = \{v_1, v_2, v_3, \ldots v_M\}$. It is assumed that each user in set S can be divided into k subsets, and each subset represents a class. The Gini coefficient [30] of the attribute v in dataset S is as follows:

$$Gini(S, v) \sum_{j=1}^{v} p(v_j) \times gini(S_j | V = v_j) \quad (3)$$

$$Gini(S_j | V = v_j) = \sum \sum_{i=1}^{n} p_i (1 - p_i) \quad (4)$$

In Formula (4), p_i is the probability that any user belongs to a specific category. By calculating the information gain of each feature, the feature with the lowest Gini coefficient in V is obtained, which is taken as the best feature.

3.3.2. Steps of Random Forest Algorithm

The random forest algorithm involves the integration of multiple decision tree classifiers. The tree classifier is set as an independent identically distributed random vector $\{h(x, \theta_k), k = 1, \ldots\}$. Given the dataset $T = \{(x_1, y_1), \ldots, (x_l, y_l)\}, l = 1, 2, \ldots, n\}$, the algorithm flow is as follows:

(1) Firstly, sampling with a replacement is adopted, and each sub-training set is extracted from n samples of the original training set by the replacement, namely $\{\theta_k\}$;
(2) Based on the decision tree, a binary tree corresponding to each sub-training set is formed. The process is as follows:
   (a) In the decision tree for constructing each node, select $m(M \geq m)$ features out of M as the candidate attribute characteristics for prediction or classification;

(b) Calculate the Gini coefficients of the selected m characteristics, and select the smallest attribute characteristic of the Gini coefficient as the optimal classification characteristic attribute;

(c) Classify the nodes according to the selected optimal attribute characteristics, and select the sub-characteristic attributes next to the optimal attribute characteristics from the remaining ones to ensure that each binary tree is a full binary tree.

(3) Repeat steps (1) and (2) until the generated tree can accurately classify the samples in the training set and combine all tree models in the operation process to form a random forest model;

(4) For any test sample, the final classification result of the sample is usually decided by simple voting. The specific formula is as follows:

$$c = argmax_c \left(\frac{1}{ntree} \sum_{k=1}^{ntree} I(h(x, \{\theta_k\}) = c) \right) \quad (5)$$

where $I(\cdot)$ is the indicator function, and c is the sample category that receives the most votes.

The specific algorithm diagram is shown in Figure 2.

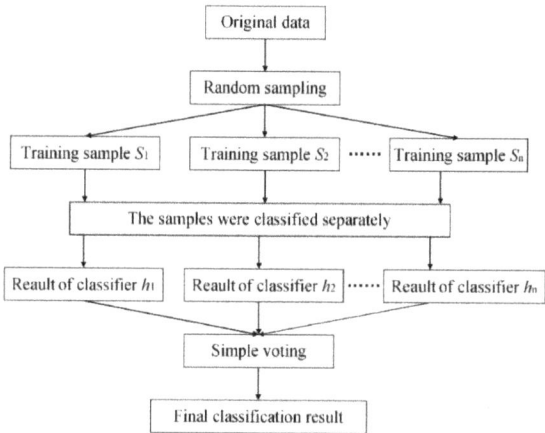

Figure 2. Random forest algorithm diagram.

3.4. Combination of the User Portrait and Random Forest Algorithm

Based on the theory of user portraits, first, the data of user characteristics are crawled, and the user characteristic index system is built up. Then, the quantitative characteristics after the indicators are considered as independent variables, and the user's degree of risk is considered as the dependent variable. According to the characteristics of the target, the degrees of risk to the user classification are sorted out. Furthermore, data from the classification of the dataset are selected as a training set for the random forest algorithm through continuously generating a decision tree.

4. The Empirical Analysis

In this section, Hive Box (one of the largest express enterprises in China), a typical representative of the logistics industry, is selected as the research object to conduct an empirical analysis of the recent outbreak of a "Hive Box charges" public opinion event.

On 30 April 2020, Hive Box launched a membership system in which ordinary users could keep a package for free within 12 h, for CNY 0.5 for 12 h after overtime, and for CNY 3 for capping. After the implementation of this system, most users' experiences of Hive

Box were very poor, and they made negative comments on various social platforms. On 8 May, some communities in Hangzhou and Shanghai posted notices to stop using Hive Box, and the negative public opinion continued to spread online. As of 11 May, the number of discussions about Hive Box on Weibo had reached 3.5 million. In view of the negative public opinion of Hive Box caused by poor user experiences, identification and monitoring of the online public opinion of the aforementioned enterprise were applied to depict the user portrait characteristics of Hive Box enterprise and to identify high-risk groups, so as to take targeted measures to improve user experience.

4.1. Crawling User Data

"Hive Box" and "Hive Box charges" were chosen as the main key words, the related Weibo comments starting from "30 April 2020" were selected, and the Python (3.8 32-bit) software crawl was adopted to collect user data—mainly user information, comment information for experience, and interactive information. In total, 48,267 data pieces were processed by Python (3.8 32-bit); a piece of data refers to a single comment posted by a user on the network, and it can include many bytes of varying lengths. As such, about 44,775 valid pieces of data were obtained after removing useless punctuations and emojis.

4.2. Building the User Portrait

Firstly, the characteristics of the users were quantified according to the above-built model. User comment text was segmented using the Jieba tool and was scored through the natural language processing tool SnowNLP. The text emotional value Q was quantified and mapped to [0,1] as the user's emotional tendency. The specific distribution is shown in Figure 3.

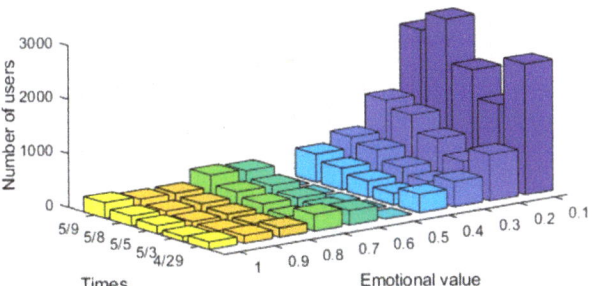

Figure 3. Distribution of emotional value for "Hive Box charges".

Figure 3 shows that in the initial stage of the "Hive Box charges", most users' emotional tendency was less than 0.1, a few users' emotional value was 0.5, and a small number of users' emotional value was more than 0.5. According to the calculation, 86.19% of users' emotional value was less than 0.5, which shows that most users held a negative emotional tendency towards this event, while only a small number of users held a positive attitude, and almost no users held a neutral attitude. Subsequently, on 8 May, a number of communities in Hangzhou and Shanghai jointly boycotted Hive Box, and most of the negative emotions of users also reached a peak. It can be seen that this event had a great impact on the corporate image, and the user experience was very poor. In such a case, Hive Box enterprise would need to take immediate strategies to improve the user experience.

4.3. Classification of Risk Degree of User

With regard to the quantification of influence and activation, the influence $Y \in (0, 0.666709)$ and the activation $H \in (0, 0.83015)$ were calculated using the above Formulas (1) and (2), and then, the users participating in the public opinion discussion were classified according to the above classification criteria. The final results are shown in Table 4.

Table 4. Classification of high-risk groups.

User Type	High-Risk	Moderate-Risk	Low-Risk
User number	26	151	44,598
Proportion	0.058%	0.34%	99.6%

Table 4 shows that most users were low-risk, and only a small number were high-risk, accounting for only 0.058% ratio, which indicates that although the majority of users held a negative opinion, most of them were ordinary users in the network with limited influence and were not significant enough to become high-risk users. Based on this, in order to prevent the continuous spread of negative public opinions, Hive Box enterprise could focus on monitoring the high-risk users and improving user experience.

4.4. Selecting Training Set and Test Data for Random Forest Algorithm

According to the classification of data, influence Y, activation H, and emotional value Q were deemed as attributes, and a training set was randomly selected from the data classified using the user portrait technique. A random forest classifier was trained using the random forest algorithm. From the data excluding the training samples, 1000 pieces of data were randomly selected as the test sample to identify the corresponding category of each sample, and the recognition accuracy of the algorithm was calculated. The training set and test set were disjoint.

4.5. Simulating the Appropriate Parameters

As different parameters have a great influence on the random forest algorithm's identification precision, different datasets generally use different values of parameters. Selecting appropriate parameters not only improves the identification precision of the algorithm but also speeds up the training. Therefore, four parameters in the random forest algorithm were simulated and analyzed: the number of decision trees (*nTree*), the size of the training sample, the maximum number of features (*maxf*), and the number of leaf nodes (*leaf*). The suitable parameters for this case were determined by simulation analysis. The specific results are shown in Figures 4 and 5.

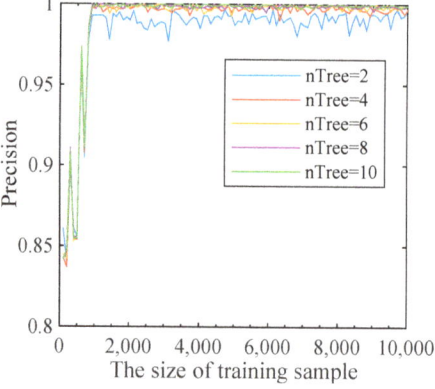

Figure 4. Simulation of *nTree* and the size of training sample.

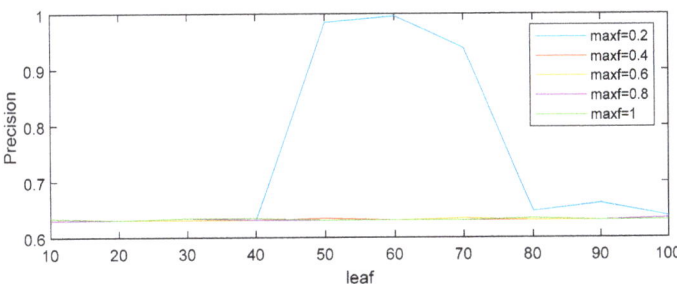

Figure 5. Simulation of *leaf* and *maxf*.

As can be seen from Figure 4, when the training sample is small, the accuracy of recognition of the high-risk population fluctuates greatly. However, when the training sample is larger than 1000, the recognition accuracy fluctuates only slightly. Therefore, in reality, the high-risk population monitoring and identification model requires at least 1000 pieces of data. Since there are enough data crawled in this paper, 10,000 pieces of data were selected as the training sample. In addition, with the increase in the number of decision trees in the random forest, the recognition accuracy is further improved, but at the same time, the running speed needs to be taken into account. Therefore, *nTree* = 10 was selected for training. As shown in Figure 5, only when *maxf* = 0.2 will the prediction accuracy change, and when *leaf* =60, the identification accuracy will be the highest. Therefore, in this case, our simulation set the parameters *maxf* = 0.2 and *leaf* = 60. Based on the above simulation results, the parameter settings are shown in Table 5 below.

Table 5. Parameter settings.

Parameter	The Size of Training Sample	nTree	Maxf	Leaf
setvalue	>1000	10	0.2	60

4.6. Training Random Forest Classifier and Comparing the Results with BP Neural Network

The trained classifier was used to identify the test samples. In order to improve the reliability of the results, 2000 pieces of data were randomly selected as the test set and divided into two groups. BP neural network and random forest were used to identify the samples, respectively, and the identification accuracy is shown in Table 6.

Table 6. Comparison of identification accuracy.

Test Dataset	Recognition Accuracy of BP Neural Network	Recognition Accuracy of Random Forest
Dataset 1	62%	98.98%
Dataset 2	84%	97.8%
Dataset 3	71%	98.7%
Average identification accuracy	72.33%	98.49%

In Table 6, the identification accuracy of random forest in the three datasets is much higher than that of BP neural network. The average identification accuracy of random forest is 98.49%, while the average accuracy of BP neural network is only 72.33%. Since 1000 pieces of data are too large an amount, in order to better visualize the identification results, 100 pieces of data were selected from each dataset for display. The results are shown in Figures 6–11.

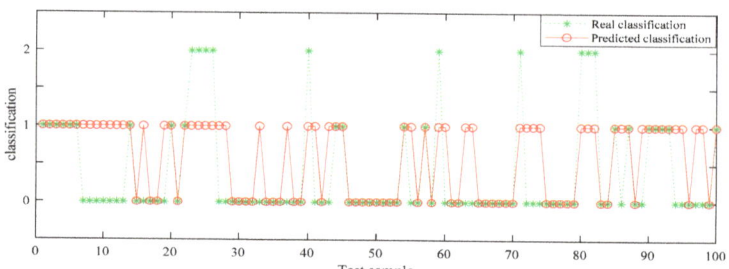

Figure 6. BP (back propagation) neural network recognition results from dataset 1.

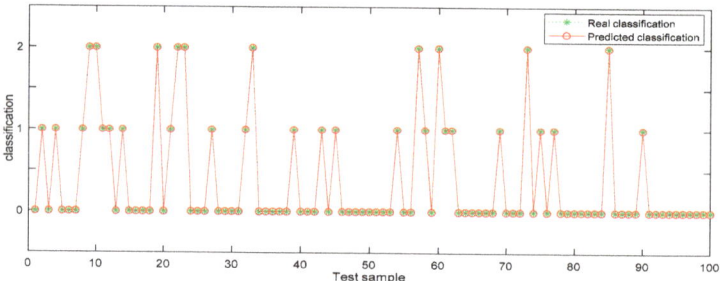

Figure 7. Random forest identification results from dataset 1.

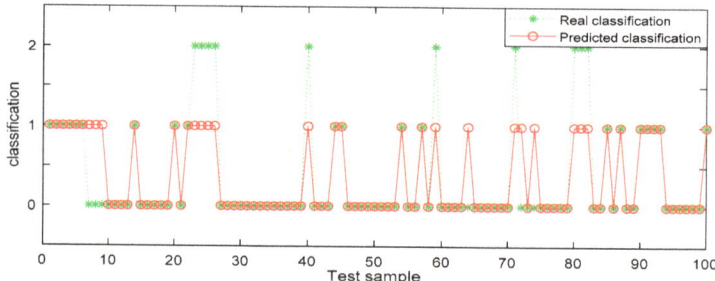

Figure 8. BP (back propagation) neural network recognition results from dataset 2.

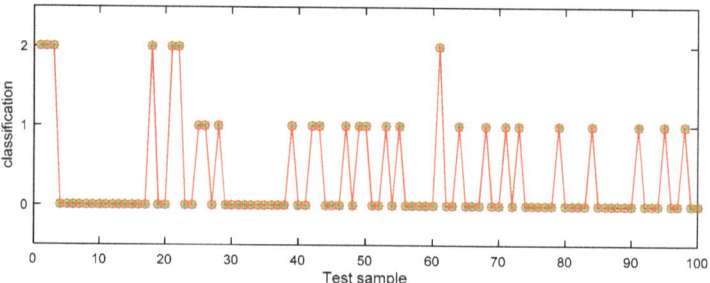

Figure 9. Random forest identification results from dataset 2.

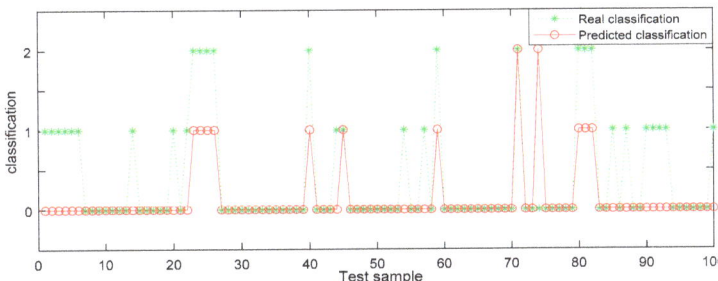

Figure 10. BP (back propagation) neural network recognition results from dataset 3.

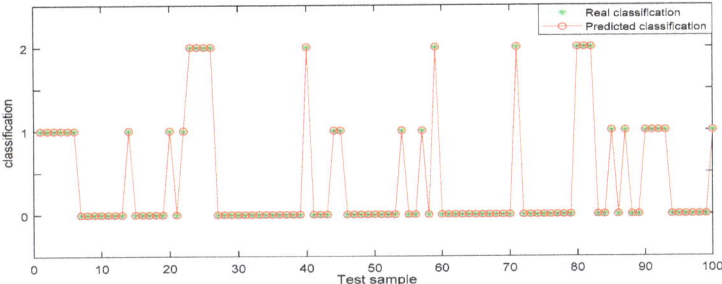

Figure 11. Random forest identification results from dataset 3.

In the three datasets, most of the users who made comments are low-risk users, while only a small number of users are high-risk users, which is also consistent with the data characteristics in Section 4.3 above. In the 100 test samples of Figures 6–11, the category of users identified by random forest is consistent with the real category, but the recognition results from BP (back propagation) neural network are only partially accurate. This suggests that the random forest identification method has higher accuracy and feasibility, which can be used for identifying high-risk users in public opinion discussions of an enterprise network.

4.7. Discussion

User portrait technology and the random forest algorithm were used to monitor and identify the public opinion events related to "Hive Box charges". The related discussions are as follows:

(1) When the sample size is big enough, the results of training indicate a dependence of precision on the training sample size. However, if the sample size is small, the result of training will be influenced by the size of the sample. This may be due to contingency. Therefore, we tested different sample sizes and demonstrated that the required sample size is at least above 1000 in the "Hive Box charges" public opinion event, and the identification accuracy will not be affected by the sample size;

(2) With regard to the "Hive Box charges" public opinion event, although 86.19% of the users participating in the discussion held negative emotions, most of them were low-risk users and only 0.058% were high-risk users. Therefore, monitoring the public opinion of high-risk groups could greatly save the costs of public opinion control of the enterprise;

(3) In the random forest algorithm, the final result is the average output result of each decision tree. Generally speaking, the larger the number of decision trees is, the stronger the robustness and the higher the accuracy of the random forest algorithm are. However, if the number of decision trees is too large, it is easy to increase

the correlation between the trees and increase the running time of the algorithm. Therefore, determining the appropriate number of decision trees is helpful to improve accuracy and efficiency. Our simulation showed that in the "Hive Box charges" public opinion event, the appropriate number of decision trees is 10;

(4) Compared with the BP neural network algorithm, the random forest algorithm used in this paper has a higher identification accuracy and is better for identifying high-risk groups of enterprise public opinion.

5. Conclusions

From the perspective of improving a user's experience with an enterprise, this paper proposes a monitoring and identification method combining user portrait technology and random forest algorithm to find high-risk users holding a negative enterprise public opinion. Accordingly, the traditional way of manual classification of training samples is abandoned. The proposed scheme can help enterprises to identify high-risk users that have had a poor experience and who may trigger negative public opinion. The main contributions of this paper are as follows:

(1) The traditional supervised machine learning algorithm is abandoned, while user portrait technology is adopted to classify the public opinion data of enterprise users. Furthermore, such data are used as the input data for the random forest algorithm, which lowers the workload of manual labeling and is not subjective, making it more scientific and objective;

(2) Combining the user portrait technique and the random forest algorithm, a model to identify public opinions on enterprises from high-risk users in terms of user experience of a product was established, which could allow enterprises to identify and monitor high-risk groups that may threaten their network image. The user portrait technique can identify the characteristics of users, providing a new perspective for the management of enterprise public opinion;

(3) In analyzing the public opinion of Hive Box, after optimizing the parameters of the random forest algorithm, the parameters applicable to "Hive Box charges" were selected to demonstrate the feasibility of the model. Compared with the BP neural network recognition results, the proposed model was verified by the recognition results with high accuracy.

However, there are still some limitations in this paper that need to be further explored. In this paper, only users who participated in the public discussion of opinions and experiences were identified, and no specific countermeasures were proposed. Therefore, further research could carry out a specific analysis based on the comments of high-risk users' experience perception to help enterprises come up with targeted measures.

Author Contributions: T.C. described the proposed framework and wrote the whole manuscript; X.Y. implemented the simulation experiments; L.P. and J.R. collected data; J.Y. and G.C. revised the manuscript. All authors have read and agreed to the published version of the manuscript.

Funding: This research is supported by the National Social Science Foundation of China (Grant No. 20BTQ059), the China (Hangzhou) cross-border electricity business school, Contemporary Business and Trade Research Center and Center for Collaborative Innovation Studies of Modern Business of Zhejiang Gongshang University of China (Grant No. 14SMXY05YB), Zhejiang Federation of Humanities and Social Sciences funded project, China (Grant No. 2019N21), Research Topics Project in higher education of Zhejiang Gongshang University (Grant No. Xgy20034), Discipline Construction and Management Project of Zhejiang Gongshang University (Grant No. XXK2019007), as well as First Class Discipline of Zhejiang-A (Zhejiang Gongshang University-Statistics).

Institutional Review Board Statement: Not Applicable.

Informed Consent Statement: Informed consent was obtained from all subjects involved in the study.

Data Availability Statement: The data used to support the findings of this study are available from the corresponding author upon request.

Conflicts of Interest: The authors declare that they have no competing interests.

References

1. Zhang, J. Research summary of Network public opinion information mining in China. *Intell. Sci.* **2016**, *34*, 167–172.
2. Wang, H.; Tu, Z.; Fu, Y.; Wang, Z.; Xu, X. Time-aware user profiling from personal service ecosystem. *Neural Comput. Appl.* **2020**, *33*, 3597–3619. [CrossRef]
3. Serrano-Gómez, V.; García-García, Ó.; Pinasa, I.V.G.; Fernández-Liporace, M.; Hernández-Mendo, A.; Rial-Boubeta, A. Measuring perceived service quality and its impact on golf courses performance according to types of facilities and user profile. *Sustainability* **2020**, *12*, 5746. [CrossRef]
4. Martínez-Cevallos, D.; Proao-Grijalva, A.; Alguacil, M.; Duclos-Bastias, D.; Parra-Camacho, D. Segmentation of participants in a sports event using cluster analysis. *Sustainability* **2020**, *12*, 5641. [CrossRef]
5. Tiwari, S.; Saini, A.; Paliwal, V.; Singh, A.; Mattoo, R. Implicit preferences discovery for biography recommender system using twitter. *Procedia Comput. Sci.* **2020**, *167*, 1411–1420. [CrossRef]
6. Zhang, H.; Qin, X.; Zheng, H. Research on contextual recommendation system of agricultural science and technology resource based on user portrait. *J. Phys. Conf. Ser.* **2020**, *1693*, 012186. [CrossRef]
7. Sun, Y.; Chai, R. An early-warning model for online learners based on user portrait. *Ingénierie Systèmesd' Inf.* **2020**, *25*, 535–541. [CrossRef]
8. You, M.; Yin, Y.; Lu, S. User portrait based on behavior perception. *J. Zhejiang Univ.* **2021**, *4*, 1–8.
9. Widiyaningtyas, T.; Hidayah, I.; Adji, T.B. User profile correlation-based similarity algorithm in movie recommendation system. *J. Big Data* **2021**, *8*, 52. [CrossRef]
10. Ni, J.; Li, W.; Dong, W.; Zhuang, Y. User behavior detection and portrait construction system based on web log. *Comput. Era* **2021**, *11*, 42–46.
11. Chen, T.; Wang, Y.; Yang, J.; Cong, G. Modeling multidimensional public opinion polarization process under the context of derived topics. *Int. J. Environ. Res. Public Health* **2021**, *18*, 472. [CrossRef]
12. Yelkanat, C.M. Spatio-temporal estimation of the daily cases of COVID-19 in worldwide using random forest machine learning algorithm. *Chaos Solitons Fractals* **2020**, *140*, 110210. [CrossRef]
13. Chen, T.; Rong, J.; Yang, J.; Cong, G.; Li, G. Combining public opinion dissemination with polarization process considering individual heterogeneity. *Healthcare* **2021**, *9*, 176. [CrossRef] [PubMed]
14. Simsekler, M.C.E.; Qazi, A.; Alalami, M.; Ellahham, S.; Ozonoff, A. Evaluation of patient safety culture using a random forest algorithm. *Reliab. Eng. Syst. Saf.* **2020**, *204*, 107186. [CrossRef]
15. Chen, T.; Li, Q.; Fu, P.; Yang, J.; Xu, C.; Cong, G.; Li, G. Public opinion polarization by individual revenue from the social preference theory. *Int. J. Environ. Res. Public Health* **2020**, *17*, 946. [CrossRef]
16. Xiao, R.; Yu, T.; Hou, J. Modeling and simulation of opinion natural reversal dynamics with opinion leader based on HK bounded confidence model. *Complexity* **2020**, *2020*, 7360302. [CrossRef]
17. Li, J.; Xiao, R. Agent-based modelling approach for multidimensional opinion polarization in collective behaviour. *J. Artif. Soc. Soc. Simul.* **2017**, *20*, 14. [CrossRef]
18. Regester, M.; Larkin, J. Risk issues and crisis management in public relations: A casebook of best practice. *Res. Nurs. Health* **2008**, *5*, 93–101.
19. Taylor, M.; Perry, D.C. Diffusion of traditional and new media tactics in crisis communication. *Public Relat. Rev.* **2005**, *31*, 209–217. [CrossRef]
20. Li, G.; Dong, Q. Research and empirical analysis on the communication process of enterprises' online public opinion under Web2.0 environment. *Intell. Sci.* **2011**, *29*, 1810–1814.
21. Yu, X.; Wei, X.U.; Lin, X. Networking groups opinion leader identification algorithms based on sentiment analysis. *Comput. Sci.* **2012**, *39*, 34–37.
22. Yin, C.; Liang, C.; Yen, F. Determinants of public attitude towards a social enterprise crisis in the digital era: Lessons learnt from THINX. *Public Relat. Rev.* **2018**, *44*, 784–793.
23. Zhang, L.; Zhang, N. Effectiveness of trust repair strategies in the crisis of corporate internet public opinion. *Am. J. Manag. Sci. Eng.* **2020**, *5*, 10. [CrossRef]
24. Bella, A.L.; Colladon, A.F.; Battistoni, E.; Castellan, S.; Francucci, M. Assessing perceived organizational leadership styles through twitter text mining. *J. Assoc. Inf. Sci. Technol.* **2018**, *69*, 21–31. [CrossRef]
25. Liu, D.; Su, J.; Song, L.; Qiu, Z. Application of internet segmentation research based on natural language processing technology in enterprise public opinion risk monitoring. *J. Phys. Conf. Ser.* **2019**, *1187*, 42007. [CrossRef]
26. Aggarwal, A.; Singh, A. Geo-localized public perception visualization using GLOPP for social media. In Proceedings of the Annual Information Technology, Electronics and Mobile Communication Conference (IEMCON), Vancouver, BC, Canada, 3–5 October 2017; pp. 439–445.

27. Chen, Z.; Hu, Z. UGC user portrait research. *Comput. Syst. Appl.* **2017**, *26*, 24–30.
28. Chen, T.; Peng, L.; Yin, X.; Jing, B.; Yang, J.; Cong, G.; Li, G. A policy category analysis model for tourism promotion in china during the COVID-19 pandemic based on data mining and binary regression. *Risk Manag. Healthc. Policy* **2020**, *13*, 3211–3233. [CrossRef] [PubMed]
29. Breiman, L. *Classification and Regression Trees*; CRC Press: Boca Raton, FL, USA, 2017.
30. Everitt, B.S. Classification and regression trees. In *Encyclopedia of Statistics in Behavioral Science*; John Wiley & Sons, Ltd.: Hoboken, NJ, USA, 2005.

Article

Comments on the Navier–Stokes Problem

Alexander G. Ramm

Department of Mathematics, Kansas State University, Manhattan, KS 66506, USA; ramm@ksu.edu

Abstract: The aim of this paper is to explain for broad audience the author's result concerning the Navier–Stokes problem (NSP) in $\mathbb{R}^3$ without boundaries. It is proved that the NSP is contradictory in the following sense: if one assumes that the initial data $v(x,0) \not\equiv 0$, $\nabla \cdot v(x,0) = 0$ and the solution to the NSP exists for all $t \geq 0$, then one proves that the solution $v(x,t)$ to the NSP has the property $v(x,0) = 0$. This paradox shows that the NSP is not a correct description of the fluid mechanics problem and the NSP does not have a solution. In the exceptional case, when the data are equal to zero, the solution $v(x,t)$ to the NSP exists for all $t \geq 0$ and is equal to zero, $v(x,t) \equiv 0$. Thus, one of the millennium problems is solved.

Keywords: hyper-singular integrals; Navier–Stokes problem

MSC: 44A10; 45A05; 45H05; 35Q30; 76D05

1. Introduction

The results of this paper are proved in detail in the monograph [1]. In the author's papers, listed in the References (see [2–5]), some preliminary results are obtained. In paper [6] some of the results are summarized. These results are stated in the abstract. The aim of this paper is to explain for broad audience the author's result concerning the Navier–Stokes problem (NSP) in $\mathbb{R}^3$ without boundaries. The result, proven in detail in the book [1], can now be stated:

If the exterior force $f(x,t) = 0$, the initial velocity $v_0(x) := v(x,0) \not\equiv 0$, $\nabla \cdot v_0(x) = 0$ and the solution $v(x,t)$ of the NSP exists for all $t \geq 0$, then $v_0(x) = 0$.

This result, that we call **the NSP paradox** (or just paradox), shows that

The NSP is not a correct statement of the problem of motion of viscous incompressible fluid. The NSP is neither physically nor mathematically correct statement of the dynamics of incompressible viscous fluid.

Let us explain the steps of our proof. The NSP consists of solving the equations:

$$v' + (v, \nabla)v = -\nabla p + \nu \Delta v\text{s.} + f, \quad x \in \mathbb{R}^3, \; t \geq 0, \quad \nabla \cdot v = 0, \quad v(x,0) = v_0(x), \quad (1)$$

see, for example, books [1,7]. Here $v = v(x,t)$ is the velocity of incompressible viscous fluid, $p = p(x,t)$ is the pressure, $f = f(x,t)$ is the exterior force, $\nu = const > 0$ is the viscocuity coefficient, $v_0 = v_0(x)$ is the initial velocity, $\nabla \cdot v_0 = 0$. The data v_0 and f are given, the v and p are to be found. The fluid's density $\rho = 1$.

(a) First we reduce the NSP to an equivalent integral equation.

$$v(x,t) = F - \int_0^t ds \int_{\mathbb{R}^3} G(x-y, t-s)(v, \nabla)v dy, \quad (2)$$

where $F = F(x,t)$ depends only on the data f and v_0, see [1]. We assume that $f = 0$. This is done for simplicity only. Under this assumption one has (see [1]):

$$F(x,t) := \int_{\mathbb{R}^3} g(x-y,t)v_0(y)dy, \quad (3)$$

where
$$g(x,t) = \frac{e^{-\frac{|x|^2}{4\nu t}}}{4\nu\pi t}, \quad t > 0; \quad g(x,t) = 0, \quad t \leq 0. \tag{4}$$

The tensor $G = G(x,t) = G_{jm}(x,t)$ is calculated explicitly in [1]:

$$G(x,t) = (2\pi)^{-3} \int_{\mathbb{R}^3} e^{i\xi \cdot x}\left(\delta_{pm} - \frac{\xi_p \xi_m}{\xi^2}\right) e^{-\nu\xi^2 t} d\xi. \tag{5}$$

Let us define the Fourier transform:

$$\tilde{v} := \tilde{v}(\xi,t) := (2\pi)^{-3} \int_{\mathbb{R}^3} v(x,t) e^{-i\xi \cdot x} dx. \tag{6}$$

Take the Fourier transform of Equation (2) and get the integral equation:

$$\tilde{v}(\xi,t) = \tilde{F}(\xi,t) - \int_0^t ds \tilde{G}(\xi, t-s)\tilde{v} \star (i\xi\tilde{v}), \tag{7}$$

where $\star$ denotes the convolution in $\mathbb{R}^3$. The following inequality, that follows from the Cauchy inequality, is useful:

$$|\tilde{v} \star (i\xi\tilde{v})| \leq \|\tilde{v}\| \|\xi|\tilde{v}\|. \tag{8}$$

One proves a priori estimate (see [1]):

$$\sup_{t \geq 0} \|\tilde{v}\| < c. \tag{9}$$

By c here and throughout the paper various positive constants, independent of t, are denoted. We denote by $c_1 := |\Gamma(-\frac{1}{4})| > 0$ the special constant from Equation (27), see below.

Let us prove inequality (9).

We denote $v_{j,m} := \frac{\partial v_j}{\partial x_m}$, $\int := \int_{\mathbb{R}^3}$, write Equation (1) as

$$v'_j + v_m v_{j,m} = f_j - p_{,j} + \nu v_{,jj}, \quad v_{j,j} = 0, \tag{10}$$

where over the repeated indices summation is understood, $1 \leq j \leq 3$. We assume that $v = v(x,t)$ is real-valued and

$$\|v_0\| + \int_0^\infty \|f(x,t)\| dt < c. \tag{11}$$

Here $\|\cdot\|$ is $L^2(\mathbb{R}^3)$ norm.

Multiply Equation (10) by v_j, integrate over $\mathbb{R}^3$ and sum up over j to get

$$\frac{1}{2}(\|v\|^2)_{,t} \leq |(f,v)| \leq \|f\|\|v\|, \tag{12}$$

where $z_{,t} := \frac{\partial z}{\partial t}$. In deriving inequality (12) we have used integration by parts: $-\int p_{,j} v_j dx = \int p v_{j,j} dx = 0$, $\int \nu v_{,jj} v_j dx = -\nu \int v_{,j} v_{,j} dx \leq 0$ and $\int v_m v_{j,m} v_j dx = -\frac{1}{2} \int v_{m,m} v_j v_j dx = 0$. From inequality (12) it follows that $\|v\|_{,t} \leq \|f\|$. Consequently,

$$\|v\| \leq \|v_0\| + \int_0^\infty \|f\| dt.$$

This and our assumption (11) imply estimate $\sup_{t \geq 0} \|v\|$. By Parseval equality the desired estimate (9) follows. Estimate (9) is proved.

Inequalities (8) and (9) imply

$$|\tilde{v} \star (i\xi\tilde{v})| \leq c\|\xi|\tilde{v}\|. \tag{13}$$

Therefore, Equation (7) implies inequality (22), see below.
From (5) it follows that
$$|\tilde{G}(\xi, t-s)| \leq c e^{-\nu(t-s)\xi^2}, \tag{14}$$
because $\left|\left(\delta_{pm} - \frac{\xi_p \xi_m}{\xi^2}\right)\right| < c$.

(b) Secondly, we prove that any solution to Equation (7) satisfies integral inequality (15), see below. The integral in this inequality is a convolution with the kernel that is hyper-singular; this integral diverges classically, that is, from the classical point of view. This inequality is derived in Section 2 (see also [1]):
$$b(t) \leq b_0(t) + c \int_0^t (t-s)^{-\frac{5}{4}} b(s) ds, \tag{15}$$
where
$$b_0(t) := \|\tilde{F}\|, \quad b(t) := \||\xi|\tilde{v}\| \geq 0. \tag{16}$$

The norm here and below is $L^2(\mathbb{R}^3)$ norm. Since the convolution integral in (15) diverges classically, we give a new definition to this integral in Section 3 and estimate the solution $b(t)$ to integral inequality (15) by the solution $q(t)$ to an integral equation with the same hyper-singular kernel:
$$q(t) = b_0(t) + c \int_0^t (t-s)^{-\frac{5}{4}} q(s) ds. \tag{17}$$

Namely, we prove the following inequality
$$b(t) \leq q(t). \tag{18}$$

The term $b_0(t)$ depends on the data only (on v_0 since we assume $f = 0$) and we may assume that $b_0(t)$ is smooth and rapidly decaying as $t \to \infty$.

(c) We prove a priori estimate
$$\sup_{t \geq 0}(\|\nabla v\| + \|v\|) < c, \tag{19}$$
part of which is inequality (9). One has
$$(2\pi)^{3/2}\|\tilde{v}\| = \|v\|, \quad (2\pi)^3 \||\xi|\tilde{v}\|^2 = \|\nabla v\|^2, \tag{20}$$
by the Parseval equality. We prove that Equation (17) has a solution in the space $C(\mathbb{R}_+)$, $\sup_{t \geq 0} q(t) < c$, provided that the data $v_0(x)$ is smooth and rapidly decaying at infinity. Moreover, this solution is unique and
$$q(0) = 0. \tag{21}$$

(d) We prove that any solution $b(t) \geq 0$ of inequality (15) with $b_0(t)$ a smooth rapidly decaying function satisfies inequality (18). Since $q(0) = 0$ and $0 \leq b(t) \leq q(t)$ it follows that $b(0) = 0$. This yields the NSP paradox mentioned at the beginning of this section. Indeed, the initial data $v_0(x) \not\equiv 0$, so $b(0) > 0$ and we prove that $b(0) = 0$.

The NSP paradox impies the conclusions we have made:

The NSP is physically not a correct description of motion of incompressible viscous fluid in $\mathbb{R}^3$ and the NSP does not have a solution on the interval $[0, \infty)$ unless the data are equal to zero. In this case the solution to the NSP does exist on the whole inteval $[0, \infty)$ and is identically equal to zero.

The uniqueness of the solution to NSP is proved in Section 4, see Theorem 3.

2. Derivation of the Integral Inequality

Take the absolute value of both sides of Equation (7), then use inequalities (9) and (14) to get

$$u \leq \mu + c \int_0^t e^{-\nu(t-s)\xi^2} \|u\| \||\xi|u\| ds \leq \mu + c \int_0^t e^{-\nu(t-s)\xi^2} b(s) ds, \quad b(s) := \|\,|\xi|u\,\|, \quad (22)$$

where the Parseval formula $(2\pi)^{3/2}\|\tilde{v}\| = \|v\| < c$ was used, and we denoted

$$|\tilde{v}(\xi,t)| := u, \quad |\tilde{F}| := \mu(\xi,t) := \mu. \quad (23)$$

In this paper, by c various constants, independent of t, are denoted. Multiply inequality (22) by $|\xi|$, take the norm $\|\cdot\|$ of both sides of the resulting inequality and get inequality (15). In this calculation one uses the formulas

$$\|e^{-\nu(t-s)\xi^2}\| = \frac{c}{(t-s)^{3/4}}, \quad \|\,|\xi|e^{-\nu(t-s)\xi^2}\,\| = \frac{c}{(t-s)^{5/4}}, \quad 0 \leq s < t, \quad (24)$$

which are easy to derive. The c are different in these formulas.

To study integral Equation (17) and integral inequality (15) we need to define the hyper-singular integral in this equation.

To do this, one needs some auxiliary material.

Let us define the function

$$\Phi_\lambda := \frac{t^{\lambda-1}}{\Gamma(\lambda)}, \quad (25)$$

where $\Gamma(\lambda)$ is the gamma function. Here and throughout $t = t_+$, that is, $t = 0$ for $t < 0$, $t := t$ for $t \geq 0$. It is known (see [8]) that $\Gamma(\lambda)$ is an analytic function of $\lambda \in \mathbb{C}$ except for the points $\lambda = 0, -1, -2, \ldots$, at which it has simple poles. The function $\frac{1}{\Gamma(\lambda)}$ is entire function of λ.

Consider the convolution operator

$$\Phi_\lambda \star b := \int_0^t \Phi_\lambda(t-s)b(s)ds. \quad (26)$$

One has

$$\int_0^t (t-s)^{-\frac{5}{4}} b(s) ds = \Gamma(-\frac{1}{4}) \Phi_{-\frac{1}{4}} \star b = -c_1 \Phi_{-\frac{1}{4}} \star b,$$

where $\star$ denotes the convolution on $\mathbb{R}_+$ and $c_1 := |\Gamma(-\frac{1}{4})|$. Inequality (15) can be written as

$$b(t) \leq b_0(t) - cc_1 \Phi_{-\frac{1}{4}} \star b. \quad (27)$$

Consider also the corresponding integral equation:

$$q(t) = b_0(t) - cc_1 \Phi_{-\frac{1}{4}} \star q. \quad (28)$$

3. Investigation of Integral Equations and Inequalities with Hyper-Singular Kernel

In this section, we solve Equation (28) analytically and prove estimate (18). First, let us define the hyper-singular integral $\psi := \Phi_\lambda \star q$. We are especially interested in the value $\lambda = -\frac{1}{4}$ because it appears in Equation (28). For $\lambda > 0$ the convolution $\Phi_\lambda \star q$ is defined classically for $q \in L^2(\mathbb{R}_+)$ and one has $L(\psi) = L(q)p^{-\lambda}$, where L is the Laplace transform operator defined as

$$Q(p) := L(q) := \int_0^\infty e^{-pt} q(t) dt, \quad (29)$$

which is analytic in the region $\operatorname{Re} p > 0$ if $q \in L^2(\mathbb{R}_+)$. If $q(t)e^{-at} \in L^2(\mathbb{R}_+)$ and $a = const > 0$, then $L(q)$ is an analytic function of p in the region $\operatorname{Re} p > a$. The Laplace transform is injective on any domain of its definition. Therefore the inverse operator L^{-1} is well defined on the range of L. The inversion formula is known:

$$q(t) := L^{-1}L(q) := \frac{1}{2\pi i}\int_{C_\sigma} e^{pt}L(q)dp, \qquad (30)$$

where C_σ is the straight line $\sigma = const > a$, $p = \sigma + i\omega$, ω changes from $-\infty$ to ∞ and $L(q)$ is a function of p. In Appendix 3 of [1] one finds information on the Laplace transform used in this paper. In particular, the following Lemmas 1–3 will be used. Their proofs can be found in Appendix 3 of [1].

Lemma 1. *Example text of a Lemma.*
  If $Q(p)$ is analytic in the region $\mathrm{Re}\, p > 0$ and

$$|Q(p)| < c(1+|p|)^{-b}, \quad b > 1/2, \quad |p| \gg 1, \quad \mathrm{Re}\, p > 0, \qquad (31)$$

then $q(t) = \frac{1}{2\pi}\int_{-\infty}^{\infty} e^{i\omega t}Q(i\omega)d\omega$, $q(t) \in L^2(\mathbb{R}_+)$ and $L(q) = Q(p)$.

In Lemma 1 sufficient conditions are given for a function, analytic in the region $\mathrm{Re}\, p > 0$ to be the Laplace transform of an $L^2(\mathbb{R}_+)$ function.

Lemma 2. *Let the assumptions of Lemma 1 hold with $b > 1$. Then $q(0) = 0$.*

Lemma 3. *One has*

$$L(\Phi_\lambda \star q) = L(q)p^{-\lambda}. \qquad (32)$$

Here we have used the known result (see [1] or [9]):

$$L(\Phi_\lambda) = p^{-\lambda}, \qquad (33)$$

and the known formula

$$L(\Phi_\lambda \star q) = L(\Phi_\lambda)L(q). \qquad (34)$$

For $\lambda > 0$ and q smooth and decaying at infinity this formula can be understood classically. For $\lambda < 0$ it is defined by the analytic continuation with respect to $\lambda \in \mathbb{C}$ where $L(\Phi_\lambda)$ is given in formula (33). Formula (33) is valid for all $\lambda \in \mathbb{C}$ by the analytic continuation from the region $\mathrm{Re}\,\lambda > 0$, where it is valid classically.

Let us define convolution $\psi := \Phi_\lambda \star q$ by the formula:

$$\psi(t) := L^{-1}(L(q)p^{-\lambda}). \qquad (35)$$

The expression under the sign L^{-1} is an entire function of λ. For $\lambda > 0$ the $\psi(t)$ is well defined classically if $q \in C(\mathbb{R}_+) \cap L^2(\mathbb{R}_+)$. The function $L(\psi)$ admits analytic continuation with respect to λ to the whole complex plane $\mathbb{C}$. Therefore, the convolution ψ is defined for all $\lambda \in \mathbb{C}$. We are especially interested in the value $\lambda = -\frac{1}{4}$ because it appears in Equation (27).

To illustrate the argument with analytic continuation, consider a simple example:

$$\int_0^\infty t^{z-1}e^{-pt}dt = \int_0^\infty s^{z-1}e^{-s}ds\, p^{-z} = \Gamma(z)p^{-z}, \qquad (36)$$

where $s = pt$. Formula (36) is valid classically for $\mathrm{Re}\, z > 0$, but remains valid for all $z \in \mathbb{C}$, $z \neq 0, -1, -2, \ldots\ldots$, by the analytic continuation with respect to z because $\Gamma(z)$ is analytic for $z \in \mathbb{C}$, $z \neq 0, -1, -2, \ldots\ldots$, and p^{-z} is an entire function of z. Formula (33) follows from (36) immediately: just divide both sides of (36) by $\Gamma(z)$. The integral (36) diverges classically for $\mathrm{Re}\, z \leq 0$, but formula (36) is valid by analytic continuation for all $z \in \mathbb{C}$

except for $z \neq 0, -1, -2, \ldots\ldots$. In [10], a regularization method is described for defining divergent integrals. By this method one writes

$$\int_0^\infty s^{z-1} e^{-s} ds = \int_0^1 s^{z-1}(e^{-s} - 1 - s) ds + \int_1^\infty s^{z-1} e^{-s} ds + (\frac{s^z}{z} + 0.5 \frac{s^{z+1}}{z+1})|_0^1, \quad (37)$$

and uses analytic continuation with respect to z. The third term of the right side of Equation (37) for $\text{Re} z > 0$ can be written as $\frac{1}{z} + 0.5\frac{1}{z+1}$. The first integral on the right side of (37) is analytic with respect to z in the region $\text{Re} z > -2$, the second integral is also analytic with respect to z in this region and the third term, $\frac{1}{z} + 0.5\frac{1}{z+1}$, admits analytic continuation with respect to z from the region $\text{Re} z > 0$ to the complex plane $\mathbb{C}$ except for the points $z = 0$ and $z = -1$. Thus, the right side of Equation (37) admits analytic continuation with respect to z to the region $\text{Re} z > -2$, except for the points $z = 0$ and $z = -1$ at which it has simple poles. So, this right side is well defined at $z = -\frac{1}{4}$.

However, the right side of (37) is much less convenient than $\Gamma(z)$, the expression we use. If one deals with the integral $\Phi_\lambda \star q$, then the advantage of our definition, based on the Laplace transform, is even greater because the three terms, analogous to the terms on the right side of Equation (37), will depend on p and on z and there is no separation of z-dependence similar to the one we have in Equation (36). Furthermore, these three terms are not all the Laplace transforms. Consequently, it is wrong to use the regularization procedure from [10] in our problem.

Lemma 4. *One has*

$$\Phi_\lambda \star \Phi_\mu = \Phi_{\lambda+\mu}. \quad (38)$$

for any $\lambda, \mu \in \mathbb{C}$. If $\lambda + \mu = 0$ then

$$\Phi_0(t) = \delta(t), \quad (39)$$

where $\delta(t)$ is the Dirac distribution.

Proof. By formulas (32) and (33) one gets

$$L(\Phi_\lambda \star \Phi_\mu) = \frac{1}{p^{\lambda+\mu}}. \quad (40)$$

By formula (33) one has

$$L^{-1}\left(\frac{1}{p^{\lambda+\mu}}\right) = \Phi_{\lambda+\mu}. \quad (41)$$

This proves formula (38).
If $\lambda + \mu = 0$ then

$$p^{-(\lambda+\mu)} = 1, \quad L^{-1}1 = \delta(t). \quad (42)$$

This proves formula (39).
Lemma 4 is proved. □

This proof is taken from [1].

Our plan is to prove that Equation (28) has a solution $q(t) \in C(\mathbb{R}_+)$ provided that $b_0(t)$ is smooth and rapidly decaying as $t \to \infty$. Moreover, this solution is unique in $C(\mathbb{R}_+)$ and $q(0) = 0$. Any solution $b(t) \geq 0$ to inequality (27) satisfies the relation $b(t) \leq q(t)$.

In particular, $b(0) = 0$. This is the NSP paradox because a priori $b(0) \neq 0$.

To realize this plan, let us investigate Equation (28). First, let us apply to (28) the operator $\Phi_{1/4}\star$ and use Lemma 4 to get

$$\Phi_{1/4} \star q = \Phi_{1/4} \star b_0 - cc_1 q. \quad (43)$$

This implies
$$q = c_3(\Phi_{1/4} \star b_0 - \Phi_{1/4} \star q), \quad c_3 := (cc_1)^{-1}, \quad c_3 > 0. \tag{44}$$

Take the Laplace transform of (44) to get
$$L(q) = c_3 L(b_0) p^{-\frac{1}{4}} - c_3 L(q) p^{-\frac{1}{4}}. \tag{45}$$

Therefore
$$L(q) = \frac{c_3 L(b_0)}{p^{1/4} + c_3}. \tag{46}$$

The function $p^{1/4} := |p|e^{i\phi}$ is analytic function of p in the region $-\pi/2 \leq \phi \leq \pi/2$, where ϕ is the argument of p. One can check that the function $\frac{1}{p^{1/4}+c_3}$, $c_3 > 0$, is an analytic function of p in the region $\mathrm{Re}\, p > 0$ and is bounded in this region. To check this, denote $r := |p|$ and write
$$|re^{i\phi/4} + c_3|^2 = r^2 + 2rc_3 \cos(\phi/4) + c_3^2 \geq c_3^2(1 - \cos^2(\phi/4)) + (r\cos(\phi/4) + c_3)^2 > c > 0.$$

This inequality is valid for all $-\pi/2 \leq \phi \leq \pi/2$. The function $L(b_0)$ is also analytic in this region. Therefore, the function $L(q)$ in formula (46) is analytic in this region. We assumed that $b_0(t)$ is smooth and rapidly decaying as $t \to \infty$. Thus, $L(b_0)$ is analytic in the region $\mathrm{Re}\, p > 0$ and
$$|L(b_0)| < c(1 + |p|)^{-1}, \quad \mathrm{Re}\, p > 0, \quad |p| \gg 1. \tag{47}$$

Therefore, $L(q)$ is analytic in the region $\mathrm{Re}\, p > 0$ and
$$|L(q)| < c(1 + |p|)^{-\frac{5}{4}}, \quad \mathrm{Re}\, p > 0, \quad |p| \gg 1. \tag{48}$$

By Lemma 1, the function $L(q)$ is the Laplace transform of the function $q(t) \in C(\mathbb{R}_+)$ and $q(0) = 0$. We have proved the following result.

Theorem 1. *Assume that $v_0(x)$ is smooth and rapidly decaying as $|x| \to \infty$, $f(x,t) = 0$ and $x \in \mathbb{R}^3$. Then estimate (48) holds, Equation (28) is solvable in $C(\mathbb{R}_+)$, its solution $q(t)$ is unique in this space and $q(0) = 0$.*

Let us now prove that $b(t) \leq q(t)$, where $b(t) \geq 0$ solves inequality (27).

Theorem 2. *Any solution $b(t) \geq 0$ of inequality (27) satisfies the inequality $b(t) \leq q(t)$.*

Proof of Theorem 2 requires the following lemma.

Lemma 5. *The operator $Af := \int_0^t (t-s)^a f(s) ds$ in the space $X := C(0,T)$ for any fixed $T \in [0, \infty)$ and $a > -1$ has spectral radius $r(A)$ equal to zero. The equation $f = Af + g$ is uniquely solvable in X. Its solution can be obtained by iterations*
$$f_{n+1} = Af_n + g, \quad f_0 = g; \quad \lim_{n \to \infty} f_n = f, \quad f = \sum_{j=0}^{\infty} A^j g, \tag{49}$$

for any $g \in X$ and the convergence holds in X.

Proof. The spectral radius of a linear operator A is defined by the formula
$$r(A) = \lim_{n \to \infty} \|A^n\|^{1/n}.$$

By induction one proves that

$$|A^n f| \leq t^{n(p+1)} \frac{\Gamma^n(p+1)}{\Gamma(n(p+1)+1)} \|f\|_X, \quad n \geq 1. \tag{50}$$

From this formula and the known asymptotic of the gamma function $\Gamma(z)$ for $z \to \infty$ (see [8]) the conclusion $r(A) = 0$ follows. If $r(A) = 0$ then the solution to equation $f = Af + g$ is unique and can be calculated by the iterative process (49).

This proof is taken from [1] where more details are provided.

Lemma 5 is proved. □

By Lemma 5 the solution to Equation (44) can be obtained as

$$q = \sum_{j=0}^{\infty} (-c_3 \Phi_{1/4} \star)^j c_3 \Phi_{1/4} \star b_0, \tag{51}$$

and any solution to inequality (27) satisfies the inequality

$$b \leq \sum_{j=0}^{\infty} (-c_3 \Phi_{1/4} \star)^j c_3 \Phi_{1/4} \star b_0, \tag{52}$$

which is checked by iterations.

Proof of Theorem 2. From (51) and (52) the inequality $b(t) \leq q(t)$ follows.

Theorem 2 is proved. □

It follows from Theorems 1 and 2 that $\sup_{t \geq 0} q(t) < c$, $b(t) \leq q(t)$. This and the Parseval equality implies $\sup_{t \geq 0} \|\nabla \cdot v\| < c$. Together with the estimate (9) this proves the a priori estimate (19). So, solutions to Equation (7) belong to $W_2^1(\mathbb{R}^3) \times C(\mathbb{R}_+)$, where $W_2^1(\mathbb{R}^3)$ is the Sobolev space.

4. Uniqueness of the Solution to the NSP

Theorem 3. *There is no more than one solution to the NSP in the space $W_2^1(\mathbb{R}^3) \times C(\mathbb{R}_+)$.*

Proof. Let there be two solutions $\tilde{v}$ and $\tilde{w}$ to (7) and $z := \tilde{v} - \tilde{w}$. Then, subtracting from the first equation the second, one gets:

$$z = -\int_0^t ds \tilde{G}(\xi, t-s) \left(z \star (i\xi \tilde{v}) + \tilde{w} \star (i\xi z) \right). \tag{53}$$

Using estimate (13) and (19), one obtains from (53) the following inequality:

$$|z| \leq c \int_0^t e^{-\nu(t-s)\xi^2} \eta(s) ds, \quad \eta := \|z\| + \||\xi|z\|. \tag{54}$$

From (54), taking the norm $\|\cdot\|$ and using (24), one obtains:

$$\|z\| \leq c \int_0^t (t-s)^{-\frac{3}{4}} \eta(s) ds. \tag{55}$$

Multiply (54) by $|\xi|$ and take the norm $\|\cdot\|$. One gets:

$$\||\xi|z\| \leq c \int_0^t (t-s)^{-\frac{5}{4}} \eta(s) ds. \tag{56}$$

Taking the Laplace transform of (55) and of (56) and summing the results yields:

$$L(\eta) \leq c \left(\Gamma(-\tfrac{1}{4}) p^{\frac{1}{4}} + p^{-\frac{1}{4}} \Gamma(1/4) \right) L(\eta) = c \left(-c_1 p^{\frac{1}{4}} + p^{-\frac{1}{4}} \Gamma(1/4) \right) L(\eta), \quad c_1 > 0. \tag{57}$$

Since $L(\eta) \geq 0$, one concludes that $1 \leq c(-c_1 p^{\frac{1}{4}} + p^{-\frac{1}{4}}\Gamma(1/4))$. If $L(\eta) \not\equiv 0$, then one has a contradiction: take $p \to +\infty$, then the above inequality yields $1 \leq -\infty$. This contradiction proves that $L(\eta) = 0$, so $z = 0$. Theorem 3 is proved. □

Theorem 3 is not used in the derivation of our basic conclusions. This theorem is new. Earlier uniqueness theorems were proved under different assumptions on the spaces to which the solution to the NSP belongs, see [2,11].

5. Conclusions

From Theorems 1 and 2 the NSP paradox follows. From the NSP paradox we conclude that the NSP is physically and mathematically contradictive and is not a correct description of the dynamics of incompressible viscous fluid.

Thus, one of the millennium problems is solved.

Funding: This research received no external funding.

Conflicts of Interest: The author declares no conflict of interest.

References

1. Ramm, A.G. *The Navier–Stokes Problem*; Morgan & Claypool Publishers: San Rafael, CA, USA, 2021.
2. Ramm, A.G. Concerning the Navier–Stokes problem. *Open J. Math. Anal. (OMA)* **2020**, *4*, 89–92. [CrossRef]
3. Ramm, A.G. Existence of the solutions to convolution equations with distributional kernels. *Glob. J. Math. Anal.* **2018**, *6*, 1–2
4. Ramm, A.G. On a hyper-singular equation. *Open J. Math. Anal.* **2020**, *4*, 8–10. [CrossRef]
5. Ramm, A.G. On hyper-singular integrals. *Open J. Math. Anal.* **2020**, *4*, 101–103. [CrossRef]
6. Ramm, A.G. Theory of Hyper-Singular Integrals and Its Application to the Navier–Stokes Problem. *Contrib. Math.* **2020**, *2*, 47–54.
7. Landau, L.; Lifshitz, E. *Fluid Mechanics*; Pergamon Press: New York, NY, USA, 1964.
8. Lebedev, N. *Special Functions and Their Applications*; Dover: New York, NY, USA, 1972.
9. Brychkov, Y.; Prudnikov, A. *Integral Tranforms of Generalized Functions*; Nauka: Moskow, Russia, 1977. (In Russian)
10. Gel'fand, I.; Shilov, G. *Generalized Functions*; GIFML: Moscow, Russia, 1959; Volume 1. (In Russian)
11. Ladyzhenskaya, O. *The Mathematical Theory of Viscous Incompressible Fluid*; Gordon and Breach: New York, NY, USA, 1969.

Article

Degenerate Canonical Forms of Ordinary Second-Order Linear Homogeneous Differential Equations

Dimitris M. Christodoulou [1,*,†], Eric Kehoe [2,†] and Qutaibeh D. Katatbeh [3,†]

[1] Department of Mathematical Sciences, University of Massachusetts Lowell, Lowell, MA 01854, USA
[2] Department of Mathematics, Colorado State University, Fort Collins, CO 80523, USA; ekehoe@colostate.edu
[3] Department of Mathematics & Statistics, Jordan University of Science & Technology, Irbid 22110, Jordan; qutaibeh@yahoo.com
* Correspondence: dimitris_christodoulou@uml.edu
† These authors contributed equally to this work.

Abstract: For each fundamental and widely used ordinary second-order linear homogeneous differential equation of mathematical physics, we derive a family of associated differential equations that share the same "degenerate" canonical form. These equations can be solved easily if the original equation is known to possess analytic solutions, otherwise their properties and the properties of their solutions are de facto known as they are comparable to those already deduced for the fundamental equation. We analyze several particular cases of new families related to some of the famous differential equations applied to physical problems, and the degenerate eigenstates of the radial Schrödinger equation for the hydrogen atom in N dimensions.

Keywords: ordinary differential equations; analytical methods; mathematical models; Riccati equation; radial Schrödinger equation; transformations

MSC: 34A25; 34A30

Citation: Christodoulou, D.M.; Kehoe, E.; Katatbeh, Q.D. Degenerate Canonical Forms of Ordinary Second-Order Linear Homogeneous Differential Equations. *Axioms* **2021**, *10*, 94. https://doi.org/10.3390/axioms10020094

Academic Editors: Davron Aslonqulovich Juraev and Samad Noeiaghdam

Received: 23 April 2021
Accepted: 18 May 2021
Published: 19 May 2021

Publisher's Note: MDPI stays neutral with regard to jurisdictional claims in published maps and institutional affiliations.

Copyright: © 2021 by the authors. Licensee MDPI, Basel, Switzerland. This article is an open access article distributed under the terms and conditions of the Creative Commons Attribution (CC BY) license (https://creativecommons.org/licenses/by/4.0/).

1. Introduction

The ordinary second-order linear homogeneous (OSLH) differential equations of mathematical physics have the general form [1–4]

$$y_0'' + b_0(x)y_0' + c_0(x)y_0 = 0,\qquad(1)$$

where primes denote derivatives with respect to the independent variable x and $b_0(x)$ and $c_0(x)$ are functions of x. Equation (1) can be transformed to the canonical form [5–8]

$$u_0'' + q_0(x)u_0 = 0,\qquad(2)$$

where

$$q_0 \equiv c_0 - \frac{1}{4}\left(2b_0' + b_0^2\right),\qquad(3)$$

and then the solutions $y_0(x)$ are given by

$$y_0(x) = u_0(x)\exp\left(-\frac{1}{2}\int b_0(x)dx\right).\qquad(4)$$

Equation (2) is degenerate in the sense that it can also be obtained from another equation of the form

$$y'' + b(x)y' + c(x)y = 0,\qquad(5)$$

in which the functions $b \neq b_0$ and c obey the condition that

$$q \equiv c - \frac{1}{4}\left(2b' + b^2\right) = q_0, \tag{6}$$

and then the solutions $y(x)$ of Equation (5) are given by

$$y(x) = u_0(x) \exp\left(-\frac{1}{2}\int b(x)dx\right). \tag{7}$$

Therefore, the original transformation $(b_0, c_0) \to q_0$ is not uniquely invertible as there exist an infinite number of function pairs (b, c) that result in the same q_0 coefficient in Equation (2). The solutions $y_0(x)$ and $y(x)$ of the two differential equations still differ in their exponential factors, but the $u_0(x)$ function is the same in Equations (4) and (7) and generally ascribes similar qualitative properties to the solutions.

The degeneracy of the canonical form (2) effectively provides a new method of solution or at least of investigation of an enormous number of potentially useful OSLH differential equations. In what follows, we determine some of these families of associated equations that may prove to be of current or future interest in applied mathematics and in physics applications. In Section 2, we describe the general theory and some notable special cases derived from degenerate canonical forms. In Sections 3 and 4, we analyze specific examples of such families with closely related properties and solutions. In particular, we revisit 15 fundamental OSLH equations of mathematical physics listed in [3] and the degeneracies of the radial Schrödinger equation across $N \geq 1$ spatial dimensions. In Section 5, we summarize and discuss our results.

2. Exploiting the Degeneracy of the Canonical Form

We consider Equations (1) and (5) with $b \neq b_0$ and/or $c \neq c_0$ leading to the same canonical form (2) with coefficient $q_0(x)$. Ibragimov [5] calls $q_0(x)$ the invariant function and the associated equations equivalent by function (his Theorem 3.3.2, page 112) in the Lie symmetry group of second-order linear equations [6], but he does not pursue the classification further, as we do. We assume that the solutions (or at least their properties) are known for Equation (1) and we determine all other OSLH equations of the form (5) that are closely related due to the appearance of the same $u_0(x)$ function in their solutions (7). Combining Equations (3) and (6), we find that

$$\left(2b' + b^2\right) - \left(2b_0' + b_0^2\right) = 4(c - c_0). \tag{8}$$

The coefficients $b_0(x)$ and $c_0(x)$ are known functions of x, whereas $b(x)$ and $c(x)$ are generally unknown functions to be determined. If $b = b_0$, then $c = c_0$ also, in which case there is no family of associated equations. If $c = c_0$, then $b = b_0$ is only a particular solution of Equation (8). We examine this case in Section 2.1, two special cases with $c \neq c_0$ in Section 2.2, and the general case for arbitrary $b(x)$ and $c(x)$ in Section 2.3 below.

Written as a Riccati equation for $b(x)$, Equation (8) takes the form

$$b' = 2(c - q_0) - \frac{1}{2}b^2, \tag{9}$$

where q_0 is known by virtue of Equation (3). A given $c(x)$ and the general solution $b(x)$ of the Riccati equation determine together a family of coefficients for the associated Equation (5); some examples of important differential equations from mathematical physics with $c = c_0$ are analyzed in Section 3 below. Furthermore, two chosen functions $b(x)$ and $c(x)$ such that they satisfy Equation (9) identically (i.e., $q \equiv q_0$) produce additional (and generally more complicated) members of the same family; a physically interesting problem from multidimensional quantum mechanics is analyzed in Section 4 below.

2.1. The Case for $b(x)$ When $c = c_0$

When the Riccati Equation (9) is solved to obtain $b(x)$, a particular solution $b_P(x)$ is needed [7,8]. In the case with $c = c_0$, we already know that $b_P = b_0$. In this case:

Theorem 1. *The general solution of Equation (9) is given by*

$$b = b_0 + \frac{1}{z}, \tag{10}$$

where $z(x)$ is the general solution of the linear differential equation

$$z' - b_0(x)z = \frac{1}{2}. \tag{11}$$

Proof. See Procedure 2 in page 392 of [8]. □

This result appears to be important for physics applications using equations of the form (1) with predetermined coefficients $b_0(x)$ and $c_0(x)$. It shows that when the new term $1/z(x)$ is added to the coefficient $b_0(x)$ of the first derivative (Equation (10)), the complexity of the mathematical problem does not increase at all; and the new problem remains just as mathematically tractable as the original problem since the two equations share the exact same canonical form (Equation (2)).

2.2. Additional Riccati Cases with Particular Solutions $b_P = b_0$

(a) For $c = Kb$ and $c_0 = Kb_0$, where K is a constant, the Riccati Equation (9) takes the form

$$b' = -2q_0(x) + 2Kb - \frac{1}{2}b^2, \tag{12}$$

for which $b_P = b_0$ is a particular solution. Then Equation (10) is the general solution, where $z(x)$ is the general solution of the linear equation

$$z' + [2K - b_0(x)]z = \frac{1}{2}. \tag{13}$$

Example 1. *In the special case with $b_0 = c_0 = 0$, the method generates a family of damped harmonic oscillators (associated with the basic equation $y_0'' = 0$ [5]) whose simplest member has constant coefficients $b = 4K$ and $c = 4K^2$ in Equation (5).*

(b) For $c = Kb^2$ and $c_0 = Kb_0^2$, where K is a constant, the Riccati Equation (9) takes the form

$$b' = -2q_0(x) + \frac{1}{2}(4K - 1)b^2, \tag{14}$$

for which $b_P = b_0$ is a particular solution. Then Equation (10) is again the general solution, where $z(x)$ is the general solution of the linear equation

$$z' + (4K - 1)b_0(x)z = \frac{1}{2}(1 - 4K). \tag{15}$$

Example 2. *In the special case with $K = 1/4$, then $z = 1/C = $ constant, and the known function $b_0(x)$ is shifted vertically in order to produce the family of associated coefficients, i.e., $b = b_0 + C$ and $c = \frac{1}{4}(b_0 + C)^2$, in Equation (5).*

Example 3. *On the other hand, for $K \neq 1/4$ and for $b_0 = c_0 = 0$, the method generates a family of Cauchy–Euler equations (associated with $y_0'' = 0$ [5]) whose simplest member has coefficients $b = B_0/x$ and $c = K(B_0/x)^2$ in Equation (5), where $B_0 = 2/(1 - 4K) = $ constant.*

By comparing the associated families in Examples 1 and 3 above, we see how complexity is being built up into the coefficients of the general OSLH form (5), starting merely from the simplest possible OSLH equation $y_0'' = 0$; but without causing any serious difficulties to the investigations of properties or solutions of the associated equations (see also related examples in [5], pages 112 and 114).

2.3. The General Case for $b(x)$ and $c(x)$

2.3.1. Solving a Riccati Equation

For arbitrary coefficients $c(x)$ and $c_0(x)$ (not related to b and b_0, respectively), Equation (8) or (9) can be written as a Riccati equation without a linear b-term, viz.

$$b' = p(x) - \frac{1}{2}b^2, \tag{16}$$

where

$$p \equiv 2(c - q_0) = 2(c - c_0) + b_0' + \frac{1}{2}b_0^2, \tag{17}$$

is a function of x with no particular dependencies among the functions involved or any special symmetries. This function does not appear explicitly in the calculations that follow, but it does affect the determination of the sought-after particular solution. The general solution of Equation (16) from Theorem 1 is

$$b = b_P + \frac{1}{z}, \tag{18}$$

where $b_P(x)$ is a particular solution and $z(x)$ is the general solution of the linear equation

$$z' - b_P(x)z = \frac{1}{2}. \tag{19}$$

The particular solution $b_P(x)$ cannot be specified in general terms. Its form will depend on the details of the given fundamental differential Equation (1) and on the coefficient $c(x)$ that will be chosen for the family of the associated Equation (5).

2.3.2. Solving a Canonical Equation

If a particular solution $b_P(x)$ cannot be found, then there is one more transformation that one can try ([8], Section 86, page 392):

Theorem 2. *Equation (16) can be recast as an OSLH equation in canonical form (since there is no linear b-term, and the coefficient of b^2 is a constant), viz.*

$$v'' - \frac{1}{2}p(x)v = 0, \tag{20}$$

where p is given by Equation (17), and $b(x)$ will then be determined from the general solution $v(x)$, viz.

$$b = \frac{2v'}{v}. \tag{21}$$

Proof. See Procedure 1 in page 392 of [8]. □

It is important that this $b(x)$ coefficient will finally contain only one arbitrary constant, just as the solution (18). The two integration constants in the solution of Equation (20) will always combine into one constant in Equation (21), thus the solutions (18) and (21) are equivalent, as shown following Example 4.

Example 4. *An example of such a reduction to one arbitrary constant is provided by the simplest case with $p = 0$. In this case, $v(x)$ is a linear function of x, i.e., $v = C_1 x + C_2$, where C_1 and C_2 are the integration constants, and then Equation (21) gives*

$$b = \frac{2C_1}{C_1 x + C_2} = \frac{2}{x + C}, \quad (22)$$

where $C \equiv C_2/C_1$. Thus, Equation (21) produces a function $b(x)$ that depends on only one arbitrary constant C.

In the general case, $v = C_1 v_1 + C_2 v_2$, where $v_1(x)$ and $v_2(x)$ are two nontrivial linearly-indepenent particular solutions of Equation (20). Then Equation (21) gives

$$b = \frac{2(C_1 v_1' + C_2 v_2')}{C_1 v_1 + C_2 v_2} = \frac{2(v_1' + C v_2')}{v_1 + C v_2}, \quad (23)$$

where, again, $C \equiv C_2/C_1$. In this case as well, the determined $b(x)$ coefficient depends on only one arbitrary constant C.

Example 5. *A simple choice that results in complicated associated equations is $b_0 = 0$ and $c - c_0 = x \geq 0$. Then, $p = 2x$ from Equation (17), and Equation (16) gives $b' = 2x - b^2/2$, a Riccati equation for which a particular solution b_P cannot be readily found. Thus, we turn to Equation (20) which takes the form of Airy's differential equation $v'' - xv = 0$ with particular solutions $v_1 = \mathcal{A}i(x)$ and $v_2 = \mathcal{B}i(x)$, where $\mathcal{A}i$ and $\mathcal{B}i$ are the Airy functions [3]; and the general solution of $b(x)$ is then given by Equation (23), where C is an arbitrary constant. For $C = 0$, the principal solution is $b = 2(\ln \mathcal{A}i)'$, which is much more involved as compared to the initial choice of $b_0 = 0$.*

3. Families of Associated Differential Equations with $c = c_0$

We analyze several examples of families of associated OSLH differential equations of the form (5) that are closely related to well-known and widely used equations of mathematical physics that take the form of Equation (1). In this section, we limit ourselves to families with

$$c = c_0, \quad (24)$$

hence the methodology of Section 2.1 is applicable. The new differential equations have significantly more complicated coefficients $b(x)$ due to the addition of nontrivial terms $1/z$ (see Equation (10)) for which $z(x)$ is determined by solving the first-order linear differential Equation (11).

In physics applications of the standard form (1), the term $b_0 y_0'$ usually represents damping due to friction or other resisting forces [7,9], unless it was created by the specific choice of a curvilinear coordinate system [4], as for example the inertial term y_0'/x in the cylindrical Bessel differential equation [4,10]. The new coefficient $b = b_0 + 1/z$ then generally represents a significantly more sophisticated model of resistance to motion that surprisingly has a similar effect on the dynamics of the physical system as the original simpler damping coefficient b_0 (see Table 1 for a summary). The similarity is not precise however because the solutions (4) and (7) also contain differing exponential factors. The differences in the exponential factors, $\exp(-\int [b(x) - b_0(x)]dx/2)$, are also summarized in Table 1.

Table 1. Exponential factors and $1/z$ terms that appear in the solutions (4) and (7) of the OSLH differential Equations (1) and (5) with $c = c_0$, due to transformations to the canonical form (2).

Differential Equation(s) (1)	Section (2)	$b_0(x)$ $1/z(x) = b(x) - b_0(x)$ (3)	$\exp(-\int b_0(x)dx/2)$ $\exp(-\int [b(x) - b_0(x)]dx/2)$ (4)
Canonical Equations ($b_0 = 0$)	3.1	0 $2/(x+C)$	1 $1/\|x+C\|$
Damped Harmonic Oscillator	3.2	$2k$ $4k/[C\exp(2kx) - 1]$	$\exp(-kx)$ $1/\|\exp(-2kx) - C\|$
Cauchy–Euler ($B_0 \neq 1$)	3.3	B_0/x $2(1 - B_0)/(x + Cx^{B_0})$	$\|x\|^{-B_0/2}$ $1/\|C + x^{1-B_0}\|$
Cauchy–Euler ($B_0 = 1$) and (Modified) Bessel	3.3 3.4	$1/x$ $2/(x \ln\|Cx\|)$	$1/\sqrt{\|x\|}$ $1/\|\ln\|Cx\|\|$
Legendre and Associated Legendre	3.5	$-2x/(1-x^2)$ $4/[(1-x^2)(C + \ln[(1+x)/(1-x)])]$	$(1-x^2)^{-1/2}$ $1/\|C + \ln[(1+x)/(1-x)]\|$
Chebyshev	3.6	$-x/(1-x^2)$ $2/[\sqrt{1-x^2}(C + \sin^{-1}x)]$	$(1-x^2)^{-1/4}$ $1/\|C + \sin^{-1}x\|$
Hermite (Physics)	3.7	$-2x$ $2/[C\exp(-x^2) + \mathcal{D}i(x)]$	$\exp(x^2/2)$ $1/\|C + \exp(x^2)\mathcal{D}i(x)\|$
Hermite (Probability)	3.7	$-x$ $\sqrt{2}/[C\exp(-x^2/2) + \mathcal{D}i(x/\sqrt{2})]$	$\exp(x^2/4)$ $1/\|C + \exp(x^2/2)\mathcal{D}i(x/\sqrt{2})\|$
Laguerre	3.8	$(1-x)/x$ $2\exp(x)/[x(C + \mathcal{E}i(x))]$	$x^{-1/2}\exp(x/2)$ $1/\|C + \mathcal{E}i(x)\|$
Associated Laguerre	3.8	$(\nu + 1 - x)/x$ $2\exp(x)/[x^{\nu+1}(C + (-1)^{\nu+1}\Gamma(-\nu, -x))]$	$x^{-(\nu+1)/2}\exp(x/2)$ $1/\|C + (-1)^{\nu+1}\Gamma(-\nu, -x)\|$
3-D Radial Schrödinger (Hydrogen Atom)	3.9	$2/x$ $2/[x(Cx - 1)]$	$1/x$ $x/\|Cx - 1\|$
3-D Radial Schrödinger (Kummer's Form)	3.9	$2(\ell+1)/x - 1$ $2\exp(x)/[x^{2(\ell+1)}(C + \Gamma(-2\ell - 1, -x))]$	$x^{-(\ell+1)}\exp(x/2)$ $1/\|C + \Gamma(-2\ell - 1, -x)\|$
3-D Radial Schrödinger (Whittaker's Form)	3.9	1 $2/[C\exp(x) - 1]$	$\exp(-x/2)$ $1/\|\exp(-x) - C\|$

Notes: (a) To obtain the coefficient $b(x)$, add the two functions in column (3) in each case. (b) To obtain the factor $\exp(-\int b(x)dx/2)$, multiply the two functions in column (4) in each case. (c) In Sections 3.5 and 3.6, $|x| < 1$. In Sections 3.8 and 3.9, $x > 0$ and $-\nu, \ell \geq 0$ are integers. (d) Whittaker's form with $b_0 = 1 =$ constant is a form of damped harmonic oscillator with $k = \frac{1}{2}$. Definitions (Ref. [3]): (1) Dawson's Integral: $\mathcal{D}i(x) = \int_0^x dt \exp(t^2 - x^2)$ (2) Exponential Integral: $\mathcal{E}i(x) = -\int_{-x}^\infty dt \exp(-t)/t$, $x > 0$ (3) Upper Incomplete Gamma Function: $\Gamma(a, x) = \int_x^\infty dt\, t^{a-1}\exp(-t)$.

3.1. Canonical Equations of Physics with $b_0 = 0$

There are quite a few OSLH equations of mathematical physics that lack a first derivative term ($b_0 = 0$ in Equation (1)) [3,4] and their properties and solutions depend only on the single remaining coefficient $c_0(x)$. For such equations, we find that the associated Equation (5) admit nonzero terms of the form $b(x)y'$ that complicate their appearances but not their studies. For $b_0 = 0$ and $c = c_0$, Equation (11) reduces to $z' = 1/2$ and Equation (10) provides a nonzero coefficient $b(x)$ of the form (22) (since $p = 0$ from Equation (17)), viz.

$$b = 2/(x + C), \qquad (25)$$

where C is an arbitrary constant. This is not a trivial result. The principal ($C = 0$) particular solution $b = 2/x$ is ubiquitous in physical models [1,4,9] and the degeneracy of the canonical form was first discovered in this case: transformations of equations with $b_0 = 2/x$ to their canonical forms would eliminate the b_0-terms from q_0, thus leading to $q_0 \equiv c_0$ in such models (Equation (3) with $b_0 = 2/x$; see also Section 6.2 in [4]).

3.2. Damped Harmonic Oscillator

The damped harmonic oscillator [7] is described by Equation (1) with $b_0 = 2k =$ constant and $c_0 = \omega_0^2 =$ constant. A family of associated differential equations is obtained from Equations (10) and (11). We find that the family members with $c = c_0$ have coefficients $b(x)$ of the form

$$b = 2k \left[\frac{C \exp(2kx) + 1}{C \exp(2kx) - 1} \right], \qquad (26)$$

where C is an arbitrary constant. The result can also be written in terms of hyperbolic functions (Appendix 2 in [11]). It may be surprising that such a complicated damping coefficient can be introduced to the harmonic oscillator, yet the problem remains analytically solvable. We have seen analogous "harmless" complications in the past (hyperbolic tangents in $b(x)$; Equations (56) and (59) in [4]) when we solved analytically the CDOS differential equation [11,12].

3.3. Cauchy–Euler Equation

The Cauchy–Euler equation [2,7] is described by Equation (1) with $b_0 = B_0/x$ and $c_0 = C_0/x^2$, where B_0 and C_0 are constants. We find that its family members with $c = c_0$ have

$$b = \begin{cases} 1/x + 2/(x \ln |Cx|), & \text{for } B_0 = 1 \\ B_0/x + 2(1 - B_0)/(x + Cx^{B_0}), & \text{for } B_0 \neq 1 \end{cases}, \qquad (27)$$

where C is an arbitrary constant. As with the Bessel differential equation [10], the $1/x$ term in the $B_0 = 1$ case does not represent damping if x is a cylindrical radial coordinate [4]. This must be the case for the new term as well, because b and b_0 lead to the same canonical form with $q_0 = (C_0 + 1/4)/x^2$, which implies that $q_0 > 1/(4x^2)$ for $C_0 > 0$; thus, the solutions are oscillatory in $x > 0$ for any positive value of the constant C_0 (see [4] for details).

Example 6. *The cases with $B_0 = 0$ and $B_0 = 2$ are also notable and consistent with the results obtained in Section 3.1 above and in Section 3.9 below, respectively:*

(a) For $B_0 = 0$ (i.e., $b_0 = 0$), then $b = 2/(x + C)$, a renowned coefficient [1,4,9].
(b) For $B_0 = 2$ (i.e., $b_0 = 2/x$), then $b = 2C/(Cx + 1)$, a coefficient that includes the special forms $b = 0$ (for $C = 0$) and $b = 2/(x + C)$ (for $C \to 1/C$).

It is important to note here that both Cauchy–Euler special cases with $B_0 = 0$ and $B_0 = 2$ include the ubiquitous result that $b = 2/(x + C)$.

3.4. Bessel Equations

The Bessel equation of order n [10,13] is described by Equation (1) with $b_0 = 1/x$ and $c_0 = 1 - n^2/x^2$, where n is a constant. We find that its family members with $c = c_0$ have

$$b = 1/x + 2/(x \ln |Cx|), \qquad (28)$$

where C is an arbitrary constant. In this case too, the new coefficient $b(x)$ does not represent damping in a cylindrical coordinate frame (see also equation (77) in [4]).

The modified Bessel equation of order n [10,13] also has $b_0 = 1/x$, but it differs in the form of $c_0 = -(1 + n^2/x^2)$. Members of this family are described by the same coefficient $b(x)$ as that in Equation (28) and they are distinguished from the corresponding Bessel family members only because of their "modified" coefficient $c(x) = -(1 + n^2/x^2)$.

3.5. Legendre Equations

The Legendre ($m = 0$) and associated Legendre ($m \neq 0$) equations [13] are described by Equation (1) with $b_0 = -2x/(1 - x^2)$ and $c_0 = \ell(\ell + 1)/(1 - x^2) - m^2/(1 - x^2)^2$, where $|x| < 1$ and ℓ, m are constants. We find that their family members with $c = c_0$ have

$$b = -2x/(1 - x^2) + 4/\left[(1 - x^2)(C + \ln[(1 + x)/(1 - x)])\right], \tag{29}$$

where C is an arbitrary constant other than zero. The condition $C \neq 0$ eliminates a singularity at $x = 0$ where $b(0) = 4/C$.

3.6. Chebyshev Equation

The Chebyshev equation [13] is described by Equation (1) with $b_0 = -x/(1 - x^2)$ and $c_0 = n^2/(1 - x^2)$, where $|x| < 1$ and n is a constant. We find that its family members with $c = c_0$ have

$$b = -x/(1 - x^2) + 2/[\sqrt{1 - x^2}\left(C + \sin^{-1} x\right)], \tag{30}$$

where C is an arbitrary constant other than zero. The condition $C \neq 0$ eliminates a singularity at $x = 0$ where $b(0) = 2/C$. The Chebyshev equation and the associated differential equations can all be solved analytically by a transformation to their degenerate canonical form [4,11].

3.7. Hermite Equations

The Hermite differential equation [13] for the so-called $H_\lambda(x)$ polynomials in physics applications is described by Equation (1) with $b_0 = -2x$ and $c_0 = 2\lambda$, where $\lambda \geq 0$ is an integer. We find that its family members with $c = c_0$ have

$$b = -2x + 2/[C \exp(-x^2) + \mathcal{D}i(x)], \tag{31}$$

where $C \neq 0$ is an arbitrary constant and $\mathcal{D}i(x)$ is Dawson's integral [3,14].

In probability applications, the Hermite differential equation for the so-called $He_\lambda(x)$ polynomials is written with $b_0 = -x$ and an integer $c_0 = \lambda \geq 0$ [3]. In this case, we find that family members with $c = c_0$ have

$$b = -x + \sqrt{2}/[C \exp(-x^2/2) + \mathcal{D}i(x/\sqrt{2})], \tag{32}$$

where, again, $C \neq 0$ is an arbitrary constant and $\mathcal{D}i(x/\sqrt{2})$ is Dawson's integral [3,14]. In both of the above $b(x)$ coefficients, the condition that $C \neq 0$ eliminates the singularity at $x = 0$ introduced by $\mathcal{D}i(0) = 0$.

3.8. Laguerre Equations

The Laguerre equation [13] is described by Equation (1) with $b_0 = (1 - x)/x$ and $c_0 = \lambda/x$, where $x > 0$ and $\lambda \geq 0$ is a constant. We find that its family members with $c = c_0$ have

$$b = (1 - x)/x + 2 \exp(x)/[x(C + \mathcal{E}i(x))], \tag{33}$$

where C is an arbitrary constant and $\mathcal{E}i(x)$ is the exponential integral [3,11].

The associated Laguerre equation [13] is described by Equation (1) with $b_0 = (\nu + 1 - x)/x$ and $c_0 = \lambda/x$, where $x > 0$ and $\lambda \geq 0$, ν are real constants. Here we take ν to be a negative integer so that the coefficients $b(x)$ will be real (on the other hand, $\nu = 0$ leads back to Equation (33)). We find that family members with $c = c_0$ have

$$b = (\nu + 1 - x)/x + 2\exp(x)/[x^{\nu+1}(C + (-1)^{\nu+1}\Gamma(-\nu, -x))], \quad (34)$$

where C is an arbitrary constant and $\Gamma(-\nu, -x)$ is the upper incomplete Gamma function [3]. We note that the coefficient $b(x)$ in Equation (34) is not a real function of $x > 0$ if ν is taken to be a real number other than a negative integer or zero.

3.9. Radial Schrödinger Equation in Three Dimensions

The radial Schrödinger equation for the hydrogen atom [1–3,15–17] is described by Equation (1) with $b_0 = 2/x$ and $c_0 = n/x - \ell(\ell+1)/x^2 - 1/4$, where $x > 0$ is a spherical radial coordinate and the integers $n \geq 1$ and $0 \leq \ell \leq n - 1$ are the principal and secondary quantum numbers, respectively. It is often written in alternative forms such as in Kummer's form of the confluent hypergeometric equation (Section 67 in [15]) with $b_0 = 2(\ell+1)/x - 1$ and $c_0 = (n - \ell - 1)/x$; and as Whittaker's differential equation (Section 16.1 in [1]) with $b_0 = 1$, $c_0 = (-m^2 + 1/4)/x^2$, and $m = \ell + 1/2$. All three equations share the same canonical form (2) with $q_0 = n/x - \ell(\ell+1)/x^2 - 1/4$ [16].

For $c = c_0$, the above forms produce three distinct families of associated differential equations having $b(x)$ coefficients (Equation (10))

$$b = \frac{2}{x+C}, \quad (35)$$

$$b = 2(\ell+1)/x - 1 + 2\exp(x)/[x^{2(\ell+1)}(C + \Gamma(-2\ell - 1, -x))], \quad (36)$$

and

$$b = \frac{C\exp(x) + 1}{C\exp(x) - 1}, \quad (37)$$

respectively, where C is an arbitrary constant and $\Gamma(-2\ell - 1, -x)$ is the upper incomplete Gamma function [3]. The coefficient (35) with $C = 0$ is ubiquitous in mathematical physics [1,4,9]. On the other hand, we find that, as in Equation (34) above with integer $-\nu < 0$, the coefficient (36) here is not a real function of $x > 0$ since $-2\ell - 1 < 0$ in the Gamma function for all quantum numbers $\ell \geq 0$. Finally, $b(x)$ in Equation (37) (derived from the original $b_0 = 1$) corresponds to the associated coefficient (26) of a damped harmonic oscillator derived from an original constant damping of $b_0 = 2k = 1$.

4. Radial Schrödinger Equations in N Dimensions

Here we consider the eigenvalue problem posed by the radial Schrödinger equation in N dimensions with quantum numbers $n \geq 1$ and $0 \leq \ell \leq n - 1$ and radial scale $x > 0$. The fundamental N-dimensional equation [17] takes the form (1) with $b_0 = (N-1)/x$ and $c_0 = E_{n_r\ell}^N - V(x) - \ell(\ell + N - 2)/x^2$, where V is the potential and $E_{n_r\ell}^N$ is the discrete spectrum of the eigenvalues with radial quantum numbers $n_r = n - \ell - 1$ such that $0 \leq n_r \leq n - 1$.

The corresponding eigenfunctions $\psi(x) \in L^2(\mathbb{R}^+)$, $\psi(0) = 0$, $\psi(x) \to 0$ as $x \to \infty$, and they have n_r radial nodes, not counting the boundary node at $x = 0$. For $N = 3$ and $V = -k/x$, where $k > 0$ is a constant, and with the proper normalization of variables, the main differential equation in [17] reduces to the spherical form discussed at the top of Section 3.9 above and in [16] for the hydrogen atom. In this transformation, the eigenvalues (usually denoted by E_n) are absorbed by the scaling (Section 67 in [15]) and they can be obtained from $E_n = -1/(2n^2)$ in atomic units (Section 3.9.1 in [18]) or, more commonly, from $E_n \simeq -13.6/n^2$ in electron-volts, where $n \geq 1$. (We note that, in the metric system of units [19], 13.6 eV = 2.18×10^{-18} J.)

Our interest in this differential equation stems from the comparison theorems of Hall and Katatbeh [17] who showed that the eigenvalues and the corresponding eigenstates with the same number of radial nodes n_r are related across different dimensions because the associated differential equations share effectively the same canonical form. Using our formulation, we recover and extend their Theorem 2 that quantifies the degeneracies between eigenvalues across dimensions N and $M \neq N$ (and within the 1-dimensional case itself) for the same potential function $V(x)$ and with quantum numbers n_r, ℓ and n_r, ℓ', respectively. We note that, although n_r is taken to be the same in degenerate eigenstates, their principal quantum numbers may still differ since n depends also on ℓ [15], viz.

$$n \equiv n_r + \ell + 1 \quad (0 \leq n_r, \ell \leq n-1). \tag{38}$$

On the other hand, it is the number of radial nodes that determines the number of oscillations in the corresponding eigenfunctions, causing thus the appearance of similar qualitative characteristics in the degenerate eigenstates [16].

4.1. The Case with $\ell' = 0$

For the given b_0 and c_0 functions, the coefficient of the N-dimensional canonical form (2) is

$$q_0 = E^N_{n_r \ell} - V(x) - \frac{1}{4x^2}\left[(N-1)(N-3) + 4\ell(\ell + N - 2)\right]. \tag{39}$$

Degeneracy occurs between these eigenstates with discrete eigenvalues $E^N_{n_r \ell}$ and the families of the corresponding eigenstates in M dimensions with eigenvalues $E^M_{n_r 0}$ and canonical coefficients

$$q = E^M_{n_r 0} - V(x) - \frac{1}{4x^2}(M-1)(M-3), \tag{40}$$

in which the secondary quantum number is $\ell' = 0$ [17]. The condition $q = q_0$ then results in two intersecting sets of degenerate solutions with eigenvalues $E^M_{n_r 0} = E^N_{n_r \ell}$: (i) $M = N + 2\ell$ and (ii) $M = 4 - (N + 2\ell)$. Set (ii) is finite (since $M \geq 1$ requires that $N + 2\ell \leq 3$ and $M + N \geq 2$ requires that $\ell \leq 1$) and its elements are also contained in set (i), except for one particular solution: $M = 1$ for $N = 3$ and $\ell = 0$. This solution indicates that in three-dimensional and in one-dimensional spaces, the corresponding coefficients $b_0 = 2/x$ and $b_0 = 0$ result in the same canonical form for $\ell = \ell' = 0$. This occurs because the s-orbitals effectively respond to the same radial potential $V(x)$ in one and three dimensions. The same property does not extend to the s-orbitals in two dimensions because the electron sees a different effective potential, $V(x) - 1/(4x^2)$, when we restrict its motion to be on a plane (Equation (39) with $N = 2$ and $\ell = 0$).

We note that Equations (39) and (40) allow for more sets of solutions with $\ell = 2 - N$ and/or $\ell' = 2 - M$ for $1 \leq M, N \leq 2$. These sets are finite and their solutions are included in the fundamental set (i). We conclude that in the N-dimensional radial Schrödinger equation, given an eigenstate with eigenvalue $E^N_{n_r \ell}$ for potential $V(x)$, the degenerate eigenstates with $\ell' = 0$ are described by the conditions

$$E^{N+2\ell}_{n_r 0} = E^N_{n_r \ell} \quad \text{and} \quad E^1_{n_r 0} = E^3_{n_r 0}, \tag{41}$$

where the integers $N \geq 1$, $0 \leq n_r \leq n-1$, and $1 \leq \ell \leq n-1$ (because using $\ell = 0$ for s-orbitals in the first condition leads to a tautology).

4.2. The Case with $c = c_0$

For $c = E^M_{n_r \ell'} - V(x) - \ell'(\ell' + M - 2)/x^2 = c_0$, the inferred equation

$$\ell'(\ell' + M - 2) = \ell(\ell + N - 2), \tag{42}$$

can be rewritten in the convenient form

$$\left[(M + N - 4) + 2(\ell' + \ell)\right]\left[(M - N) + 2(\ell' - \ell)\right] = (M + N - 4)(M - N), \tag{43}$$

which has three nontrivial solution sets with degenerate eigenvalues $E_{n_r\ell'}^M = E_{n_r\ell}^N$ in the case $M + N = 4$: (iii) $M = 1$ for $N = 3$, $\ell' = \ell + 1$ (where $\ell' \geq 1$); (iv) $M = 3$ for $N = 1$, $\ell' = \ell - 1$ (where $\ell \geq 1$); and (v) $\ell' + \ell = 0$ and $M \neq N$ (which is identical to the second condition in Equation (41) obtained in Section 4.1). Sets (iii) and (iv) are equivalent, thus the degenerate eigenstates with $c = c_0$, $M + N = 4$, and $\ell' + \ell > 0$ are described by the condition

$$E_{n_r\ell+1}^1 = E_{n_r\ell}^3 \quad (0 \leq n_r, \ell \leq n-1), \tag{44}$$

that associates the $(\ell + 1)$-orbitals in 1 dimension with the corresponding ℓ-orbitals in three dimensions and the same number of radial nodes.

Equation (42) also has a solution set (vi) for $M = N = 1$: In one dimension, we find that $\ell' + \ell = 1$, which gives the degeneracy condition

$$E_{n_r1}^1 = E_{n_r0}^1 \quad (0 \leq n_r \leq n-1). \tag{45}$$

This condition shows that the s- and p-orbitals are degenerate in one dimension, if they have the same number of radial nodes n_r (i.e., if their principal quantum numbers differ by 1). Finally, combining Equations (44) and (45) and with $\ell = 0$, we infer the second condition in Equation (41) which also results from set (v) above.

4.3. The General Case for Any $c(x)$ Function

For the same potential function $V(x)$, the degeneracy condition $q = q_0$ in the general case takes the form

$$(M-1)(M-3) + 4\ell'(\ell' + M - 2) = (N-1)(N-3) + 4\ell(\ell + N - 2), \tag{46}$$

which can be recast as a quadratic equation for $M + 2\ell'$ in terms of $N + 2\ell$, viz.

$$(M + 2\ell' - 2)^2 = (N + 2\ell - 2)^2, \tag{47}$$

that has two sets of solutions: (I) $M + 2\ell' = N + 2\ell$ and (II) $M + 2\ell' = 4 - (N + 2\ell)$.

The two sets are intesecting and the combinations $(M \pm N)$ are even integers in all solutions, just as in the subsets of solutions with $\ell' = 0$ studied in Section 4.1. Similarly here, set (II) is finite and small in size since its solutions are valid only for $2 \leq M + N \leq 4$ and $0 \leq \ell + \ell' \leq 1$. Sets (I) and (II) include all special cases found in Sections 4.1 and 4.2 above:

(a) From set (II) and for $M + N = 2$, we recover the solution set (45);
(b) whereas for $M + N = 4$ in set (II), we recover the second condition (41).
(c) Finally, condition (44) is recovered here from set (I) for $N - M = 2$;
(d) and the first condition (41) is recovered also from set (I) for $\ell' = 0$.

5. Summary and Discussion

For OSLH differential equations of the form (1), we have determined entire families of associated differential Equation (5) of the same form, but with generally different coefficients $b(x)$ and/or $c(x)$, that exhibit comparable qualitative properties in their solutions. All such equations belonging to the same family share the same canonical form (see Equations (2) and (6)) and their general solutions $y(x)$ differ only by the introduction of exponential factors in Equations (4) and (7), such as those listed in the $\exp(-\int [b(x) - b_0(x)] dx/2)$ entries of the summarizing Table 1. Given an original well-studied and widely used differential equation, the methods for determining associated equations with comparable qualitative properties were described in Section 2, and several examples known from physics applications were analyzed in Section 3 ($c = c_0$; see also Table 1) and Section 4 (generally $b \neq b_0$ and $c \neq c_0$ in $L^2(\mathbb{R}^+)$ Hilbert spaces with different spatial dimensions).

Although one may generally create arbitrarily complicated differential equations (as in Sections 2.2 and 2.3), we focused here on the "tip of the iceberg," that is, on the multidimensional radial Schrödinger equations of quantum mechanics (Section 4), as well as on other physically-important OSLH differential equations (Section 3) in which the y-coefficients of Equations (1) and (5) remain the same (Equation (24)) within each family of associated equations. In the latter case, the transformations of coefficients $b_0(x) \to b(x)$ that we carried out are not iterative: If the derived function $b(x)$ is used in place of the original $b_0(x)$, then the new derived function is equivalent to the input $b(x)$, that is, repeated transformations produce the sequence $b_0 \to b \to b$ and only one general solution $b(x)$.

Example 7. *For instance, in the canonical case with $b_0 = 0$ (Section 3.1):*

$$b_0 = 0 \;\to\; b = 2/(x+C) \;\to\; b = 2/(x+\overline{C});$$

and similarly in the Bessel case with $b_0 = 1/x$ (Section 3.4):

$$b_0 = 1/x \;\to\; b = 1/x + 2/(x \ln |Cx|) \;\to\; b = 1/x + 2/(x \ln |\overline{C}x|),$$

where C and $\overline{C}$ are arbitrary constants.

This property arises from the method of solution of the Riccati Equation (9). For any choice of the arbitrary constant $C = C_1$, $b(x)$ becomes a particular solution and if it is used in place of $b_0(x)$ in Equation (11), then this equation will produce the same general solution (10) for $b(x)$ that will contain yet another arbitrary constant $\overline{C}$ which absorbs both C_1 and the new integration constant C_2. In particular, in the two cases of Example 7, we have $\overline{C} = C_1 + C_2$ and $\overline{C} = C_1 C_2$, respectively.

The results listed in Table 1 indicate that $b \to b_0$ as $C \to \pm\infty$, and then the listed $\exp(-\int [b(x) - b_0(x)]dx/2)$ entries are not applicable; as $b \to b_0$, these exponential factors tend to 1. On the other hand, for $C = 0$, the principal solutions $b(x)$ are described mostly by elementary functions and by three notable special functions (Dawson's integral $\mathcal{D}i(x)$, the exponential integral $\mathcal{E}i(x)$, and the upper incomplete Gamma function $\Gamma(a,x)$; their standard definitions are given in [3] and in the notes to Table 1). Because of their appearance in the corresponding families of associated differential equations, these special functions have just grown somewhat in importance to mathematical physics. Of the three special functions appearing in Table 1, $\mathcal{E}i(x)$ and $\Gamma(-\nu, -x)$ (for $x > 0$ and $-\nu > 0$ an even integer) contain singular points other than the familiar $x = 0$ in the coefficients $b(x)$ of the (associated) Laguerre equation for $C = 0$ (Equations (33) and (34), respectively, in Section 3.8). In particular, the only root of $\mathcal{E}i(x) = 0$ is $x \approx 0.372507$ and it lies in the domain $x > 0$ of the Laguerre equation; and the root of $\Gamma(2, -x) = 0$ is $x = 1$ and it lies in the domain $x > 0$ of the associated Laguerre equation; similarly, the real roots of $\Gamma(-\nu, -x) = 0$ for $-\nu = 4, 6, 8, 10$ are $x \approx 1.596072, 2.180607, 2.759003, 3.333551$, respectively.

The coefficients $b(x)$ derived from Equation (10) for $c = c_0$ and listed in Table 1 (one has to add up the two b-entries in each case) generally describe damping of motion due to friction ([20], Section 3.4, page 172), or air resistance ([20], Section 2.3, page 93), or other dissipative processes (e.g., [21], Section 17.9, page 603) in physics applications (unless the $b(x)y'$ term is inertial created by the curvature of the coordinate system; see [4]). At present, there is no general theory of friction or such resisting forces [22]. Then, these new functions $b(x)$ would potentially represent more complicated and more sophisticated models of resisting forces acting on the corresponding dynamical systems. Despite their intimidating look at first sight (owing to the overly complicated $b(x)y'$ terms), the associated differential equations of the various families are quite easily mathematically tractable, provided that the original models involving simpler damping terms of the form $b_0(x)y_0'$ in Equation (1) are already well-studied and their qualitative properties are fully understood.

Author Contributions: Conceptualization, D.M.C.; methodology, D.M.C. and Q.D.K.; formal analysis and investigation, D.M.C., E.K., and Q.D.K.; writing—original draft preparation, D.M.C.;

writing—review and editing, E.K. and Q.D.K. All authors have read and agreed to the published version of the manuscript.

Funding: This research received no external funding.

Data Availability Statement: All data generated in this study are included in this published article.

Acknowledgments: During this research project, D.M.C. was supported by the University of Massachusetts Lowell, E.K. was supported by Colorado State University, and Q.D.K. was supported by the Jordan University of Science and Technology.

Conflicts of Interest: The authors declare no conflict of interest.

Abbreviations

The following abbreviations are used in this manuscript:

OSLH	Ordinary Second-order Linear Homogeneous
CDOS	Chuaqui, Duren, Osgood, Stowe (Ref. [12])
$\mathcal{A}i, \mathcal{B}i$	Airy Functions (Ref. [3])
$\mathcal{D}i$	Dawson's Integral (Refs. [3,14])
$\mathcal{E}i$	Exponential Integral (Refs. [3,11])
Γ	Upper Incomplete Gamma Function (Ref. [3])

References

1. Whittaker, E.T.; Watson, G.N. *A Course of Modern Analysis*, 3rd ed.; Cambridge University Press: Cambridge, UK, 1920.
2. Hartman, P. *Ordinary Differential Equations*; Wiley: New York, NY, USA, 1964.
3. Abramowitz, M.; Stegun, I.A. *Handbook of Mathematical Functions with Formulas, Graphs, and Mathematical Tables*; Dover: New York, NY, USA, 1972.
4. Christodoulou, D.M.; Graham-Eagle, J.; Katatbeh, Q.D. A program for predicting the intervals of oscillations in the solutions of ordinary second-order linear homogeneous differential equations. *Adv. Differ. Equ.* **2016**, *48*. [CrossRef]
5. Ibragimov, N.H. *A Practical Course in Differential Equations and Mathematical Modeling*; World Scientific: Singapore, 2010; pp. 112–114.
6. Mahomed, F.M. Symmetry group classification of ordinary differential equations: Survey of some results. *Math. Meth. Appl. Sci.* **2007**, *30*, 1995–2012. [CrossRef]
7. Boyce, W.E.; DiPrima, R.C. *Elementary Differential Equations*, 10th ed.; Wiley: New York, NY, USA, 2012.
8. Zwillinger, D. *Handbook of Differential Equations*, 3rd ed.; Academic Press: Boston, MA, USA, 1997.
9. Agarwal, R.P.; Grace, S.R.; O'Regan, D. *Oscillation Theory for Second Order Linear, Half-Linear, Superlinear and Sublinear Dynamic Equations*; Kluwer Academic: Dordrecht, The Netherlands, 2002.
10. Watson, G.N. *A Treatise on the Theory of Bessel Functions*; Cambridge University Press: Cambridge, UK, 1922.
11. Christodoulou, D.M.; Katatbeh, Q.D. A powerful diagnostic tool of analytic solutions of ordinary second-order linear homogeneous differential equations. *Adv. Differ. Equ.* **2017**, *208*. [CrossRef]
12. Chuaqui, M.; Duren, P.; Osgood, B.; Stowe, D. Oscillation of solutions of linear differential equations. *Bull. Aust. Math. Soc.* **2009**, *79*, 161–169. [CrossRef]
13. Bell, W.W. *Special Functions for Scientists and Engineers*; Dover: New York, NY, USA, 2004.
14. McCabe, J.H. A continued fraction expansion with a truncation error estimate for Dawson's integral. *Math. Comput.* **1974**, *28*, 811–816. [CrossRef]
15. Flügge, S. *Practical Quantum Mechanics*; Springer: Berlin, Germany, 1994.
16. Katatbeh, Q.D.; Christodoulou, D.M.; Graham-Eagle, J. The intervals of oscillations in the solutions of the radial Schrödinger differential equation. *Adv. Differ. Equ.* **2016**, *47*. [CrossRef]
17. Hall, R.L.; Katatbeh, Q.D. Generalized comparison theorems in quantum mechanics. *J. Phys. A* **2002**, *35*, 8727–8742. [CrossRef]
18. Cohen, E.R.; Cvitas, T.; Frey, J.G.; Holmström, B.; Kuchitsu, K.; Marquardt, R.; Mills, I.; Pavese, F.; Quack, M.; Stohner, J.; et al. *Quantities, Units and Symbols in Physical Chemistry, IUPAC Green Book*, 3rd ed.; IUPAC & RSC Publishing: Cambridge, UK, 2008.
19. Newell, D.B.; Tiesinga, E. (Eds.) *The International System of Units (SI)*, 9th ed.; Special Publication 330; National Institute of Standards and Technology: Gaithersburg, MD, USA, 2019.
20. Edwards, C.H.; Penney, D.E.; Calvis, D. *Differential Equations and Boundary Value Problems*, 5th ed.; Pearson: Boston, MA, USA, 2015.
21. Jackson, J.D. *Classical Electrodynamics*; Wiley: New York, NY, USA, 1962.
22. Tabor, D. Friction as a Dissipative Process. In *Fundamentals of Friction: Macroscopic and Microscopic Processes*; Singer, I.L., Pollock, H.M., Eds.; NATO ASI Series; Springer: Dordrecht, The Netherlands, 1992; Volume 220, pp. 3–24.

Article

Ranking Road Sections Based on MCDM Model: New Improved Fuzzy SWARA (IMF SWARA)

Sabahudin Vrtagić [1], Edis Softić [2], Marko Subotić [3], Željko Stević [3,*], Milan Dordevic [1] and Mirza Ponjavic [4]

1. College of Engineering and Technology, American University of the Middle East, Street 250, Block 6, Egaila 54200, Kuwait; sabahudin.vrtagic@aum.edu.kw (S.V.); milan.dordevic@aum.edu.kw (M.D.)
2. Technical Faculty, University of Bihac, Irfana Ljubijankića bb, 77000 Bihać, Bosnia and Herzegovina; edis.softic@unbi.ba
3. Faculty of Transport and Traffic Engineering, University of East Sarajevo, Vojvode Mišića 52, 74000 Doboj, Bosnia and Herzegovina; marko.subotic@sf.ues.rs.ba
4. Faculty of Engineering and Natural Sciences, International Burch University, Francuske revolucije bb, Ilidža, 71210 Sarajevo, Bosnia and Herzegovina; mirza.ponjavic@ibu.edu.ba
* Correspondence: zeljkostevic88@yahoo.com or zeljko.stevic@sf.ues.rs.ba

Citation: Vrtagić, S.; Softić, E.; Subotić, M.; Stević, Ž.; Dordevic, M.; Ponjavic, M. Ranking Road Sections Based on MCDM Model: New Improved Fuzzy SWARA (IMF SWARA). *Axioms* **2021**, *10*, 92. https://doi.org/10.3390/axioms 10020092

Academic Editors: Davron Aslonqulovich Juraev and Samad Noeiaghdam

Received: 21 April 2021
Accepted: 13 May 2021
Published: 15 May 2021

Publisher's Note: MDPI stays neutral with regard to jurisdictional claims in published maps and institutional affiliations.

Copyright: © 2021 by the authors. Licensee MDPI, Basel, Switzerland. This article is an open access article distributed under the terms and conditions of the Creative Commons Attribution (CC BY) license (https:// creativecommons.org/licenses/by/ 4.0/).

Abstract: Traffic management is a significantly difficult and demanding task. It is necessary to know the main parameters of road networks in order to adequately meet traffic management requirements. Through this paper, an integrated fuzzy model for ranking road sections based on four inputs and four outputs was developed. The goal was to determine the safety degree of the observed road sections by the methodology developed. The greatest contribution of the paper is reflected in the development of the improved fuzzy step-wise weight assessment ratio analysis (IMF SWARA) method and integration with the fuzzy measurement alternatives and ranking according to the compromise solution (fuzzy MARCOS) method. First, the data envelopment analysis (DEA) model was applied, showing that three road sections have a high traffic risk. After that, IMF SWARA was applied to determine the values of the weight coefficients of the criteria, and the fuzzy MARCOS method was used for the final ranking of the sections. The obtained results were verified through a three-phase sensitivity analysis with an emphasis on forming 40 new scenarios in which input values were simulated. The stability of the model was proven in all phases of sensitivity analysis.

Keywords: road section; IMF SWARA; traffic safety; fuzzy MARCOS; DEA

MSC: 90B20; 90B50; 90C08

1. Introduction

Roads represent considerable resources and are some of the most important public investments of a country, with significant funds allocated for their construction and maintenance with two main tasks: that they are efficient and safe. In a real traffic flow, almost all functional dependencies given by international road standards are based on the correlation between flow, speed, density, number of traffic accidents, etc., as traffic parameters and longitudinal gradient, minimum radius of horizontal curve, etc., as road parameters. The analysis of functional dependence requires special attention and is particularly evident for two-lane roads, since these roads make up the largest percentage of a road network of a country. Based on the research [1] conducted on 3450 km stretch of two-lane roads in Valencia, it was shown that traffic risk is significantly affected by the density of access points, average sight distance, average speed limit, and the proportion of no-passing zones. The impact of a larger number of access points also results in a large number of conflict points on the road, which significantly affects both speed and flow, as well as a negative trend in safety for traffic participants.

Deviation from speed limits is most often associated with an increased probability of traffic accidents. In addition, any credible deviation from the speed limit of real traffic flow is an imperative in the analysis of speeding, and thus the potential occurrence of incidents. Speed credibility analysis is a method of traffic engineering management which is often neglected and can lead to worsening traffic risk. Any deviation from speed limits is often related to both technical and operational road characteristics, as well as to the psychophysical abilities of drivers. For example, across slope ranges of −5.5% to 4.50%, it was found that speed increases with where the slope increases, and with a decrease, stress while driving increases too [2]. Within this research, measured sections on the slope were identified as potential places with a high percentage of traffic accidents. According to the research [3], about 40% of drivers drove above the permitted speed limit, and this percentage of speeding varies from location to location. It was concluded that the speed limit of 60 km/h was not appropriate for most of the investigated locations.

In traffic and operational analyzes conducted on two-lane roads, unadjusted speed is one of key indicators for the occurrence of traffic accidents. The importance of the 85th percentile speed is especially emphasized in the scientific literature, since it is a representative speed in a traffic flow of road network users [3,4]. Commonly, the number of traffic accidents and an increase in risk are related to the exploitation speed. Exploitation speeds have been shown to be higher than design speeds for a speed limit of about 55 mph or less. Therefore, it is important to present speed potentially through five specific indicators of speed dependence on geometric road characteristics [5].

In addition to the importance of the research field and motivation for conducting this study, it is very important to emphasize that in addition to the professional contribution reflected in assessing the safety level of considered sections, a significant scientific contribution was made too. It is reflected through forming an integrated fuzzy multi-criteria decision-making (MCDM) model with an emphasis on defining the IMF SWARA method that eliminates the shortcomings of the previously developed fuzzy SWARA method, which is explained in detail in the Materials and Methods section. It is also important to emphasize that Dombi and Bonferroni aggregators were used to average specific values of inputs and outputs.

After the introduction in Section 1, the rest of the paper is structured as follows. Section 2 provides an overview of the literature related to the field of application. Section 3 presents the preliminaries regarding the development of the methodology. Section 4 presents the creation of the methodology that is described and explained in detail. Section 5 provides a case study with a detailed calculation given for each approach of the methodology proposed. Section 6 demonstrates the stability of the model through a three-phase sensitivity analysis, while Section 7 presents a conclusion with guidelines for further research.

2. Literature Review

Contemporary HCM methodology [6], depending on the speed limit, can classify all two-lane roads in Bosnia and Herzegovina in class II of two-lane roads (where limit speed does not exceed 80 km/h), and the assessment of the qualitative measure level of service (LOS) is based on determining the percentage of time losses, but not on the mean value of speed or deviation from the speed limit. The German methodology HBS 2001 [7] mainly expresses the problem of the functional dependence of travel speed on the width of traffic lanes. This method defines a minimum lane width of 2.75 m, while the lane width can be 2.50 m on some sections of road in Bosnia and Herzegovina. Chapter 10 of the Highway Safety Manual [8] (HSM, 2010) defines a method for predicting traffic accidents for suburban two-lane, two-way roads. This prediction method provides a procedure consisting of 18 steps for estimating the "expected average number of traffic accidents" (according to the total number of accidents, the severity of accidents, or the type of collision) on road networks, facilities, or locations.

However, some authors have investigated the functional dependence of speed, length of individual geometric road elements, radius of curvature of transverse/longitudinal slopes, and traffic accidents [9–12]. Additionally, in some research the problem of reducing speed due to geometric road characteristics arises. The reduction in speed of heavy vehicles was specifically analyzed in a study where a longitudinal slope of 9.0% at 1.20 km of road was included, showing a significant reduction in the speed of heavy vehicles. According to this study, in order to increase speed due to influential road factors, power-to-mass ratio must be improved [13]. Based on a report conducted in Texas [14], regression analysis determined that the following variables affect the prediction of traffic accidents: AADT, lane width, shoulder width, and section length. The use of wider longitudinal road markings leads to a reduction in the speed of vehicles, and thus a reduction in the number of traffic accidents. The analysis of speed reduction in day and night driving conditions at average traffic volume showed the following reduction in vehicle speed: 2.24% during the day and 1.96% at night for light vehicles, and 2.46% during the day and 2.15% at night for heavy vehicles [15]. Using the Bayesian network analysis for predicting the probability of the influence of traffic and road factors on the occurrence of traffic accidents, it is especially emphasized that vehicle speed, horizontal curve radius, vehicle type, adhesion coefficient, and longitudinal slope are important factors [11]. Additionally, the results of the study [12] show that the continuous use of several boundary indices and the excessive average gradient of long and steep road sections are some of the main causes of frequent accidents on the Hexi section.

In a study conducted according to data on 465 traffic accidents in Chongqing, vehicle stability is analyzed as a function of a combination of steep slopes and the sharp radii of curves, where a relevant safety model of a combined section of a steep slope and sharp curve was established. The simulation iteration frequency at 100 Hz (using the target speed control mode) was performed on steep downhills where safety factors were analyzed downhill at speeds of 40 km/h, 60 km/h, and 80 km/h [16]. Additionally, in a study researching the validation of mean speeds collected in a driving simulator, speed was monitored on curves corresponding to real-world road sections. A road in Iran was simulated in two experiments. In the first experiment, the speeds of 30 participants were collected at the beginning, middle, and end of five curves, and in the second experiment, the speeds of 40 different participants were collected at specified points of five curves on the real road using radar. In most curves, the mean speeds in the simulator were higher than on the real road, and the trend in changing speed from the beginning to the end of the curve was similar in both experiments [17]. Additionally, the dynamic simulation procedure analyzed several factors (longitudinal grades, minimum radius of horizontal curve, vehicle speed, changes in vehicle length and weight, and differences in changes in driver behavior) where the level of traffic safety was determined across large radius curves. This research demonstrates the need for the introduction of clothoid transition curves, and also proposed a methodology for the analysis of lateral friction for speeds of 50, 80, 110, and 130 km/h [18]. At the same time, the analysis of lateral friction, through the reduction of the coefficient of friction by 5% in vertical curves, increased the probability of a traffic accident by 20% for rescue vehicles. By increasing the speed of these vehicles in curves by 10%, the probability of a crash increases by 25% [19]. Some researchers have analyzed roads in rural areas in order to determine the functional correlation between vehicle speed at different horizontal radii of curves in order to link this to traffic safety. Within this research, a linear model for predicting operating speeds depending on the radius and preceding tangent length of the curve was developed [20].

The implementation of different approaches in one integrated model is a practice that very often ensures more accurate results. Such models are preferable, thus many researchers have brought new ideas and have implemented them in the field of transport and traffic. For example, an approach based on an individual DEA for determining the efficiency of 197 municipalities containing two inputs and 14 outputs was applied in [21]. The obtained results showed that due to the weights of the input, it was possible for a more efficient municipality to be ranked lower. In the study [22] the analytic hierarchy process (AHP) method for determining the influence of traffic factor interaction on the rate of traffic accidents was applied. The MCDM model was also applied in [23] for the identification of priority black spots in order to decrease traffic risk, while in [24] authors have implemented the AHP method in order to evaluate and rank road section designs. In the research [25], a new multi-criteria and simultaneous multi-objective optimization (MOO) model using the AHP method for evaluating and ranking traffic and geometric elements was created.

For transport projects, policies, or policy measures, various multi-criteria methods have been used. The study [26] defines why MCDM models play such an important role in dealing with various categories of decision problems that arise in mass transit systems. Additionally, the importance of MCDM models in this field has been proven with two case studies. In the study [27], a novel hybrid model which combines the fuzzy step-wise weight assessment ratio analysis (FSWARA) and the fuzzy best-worst method (FBWM) was developed for selection of equipment in a container terminal. By applying MCDM, the quality of life in urban environments was assessed at three spatial levels (socioeconomic, environmental, and accessibility) [28]. For transport projects, policies, or the evaluation of policy measures, from 1982 to 2019 we have seen the development and effective application of various multi-criteria methods to complement conventional cost effectiveness and cost benefit analyses [29]. The determination of transportation and traffic risk is a very popular and interesting activity that can help participants avoid risk and conflict situations. The authors in [30] applied a fuzzy pivot pairwise relative criteria importance assessment (Fuzzy PIPRECIA) method to determine and rank the road transportation risk factors in the Giresun province.

3. Preliminaries

3.1. Preliminaries—Operations with Fuzzy Numbers

A fuzzy number $\overline{A}$ on R is to be a TFN if its membership function $\mu_{\overline{A}}(x)$: R→[0,1] is equal to following Equation (1) [31,32]:

$$\mu_{\overline{A}}(x) = \begin{cases} \frac{x-l}{m-l} & l \leq x \leq m \\ \frac{u-x}{u-m} & m \leq x \leq u \\ 0 & \text{otherwise} \end{cases} \quad (1)$$

In Equation (1), l and u indicated the lower and upper bounds of the fuzzy number $\overline{A}$, and m is the modal value for $\overline{A}$. The TFN can be denoted by $\overline{A} = (l, m, u)$.

The operational laws of TFN $\overline{A} = (l_1, m_1, u_1)$ and $\overline{A} = (l_2, m_2, u_2)$ are displayed as the following equations.

Addition:

$$\overline{A_1} + \overline{A_2} = (l_1, m_1, u_1) + (l_2, m_2, u_2) = (l_1 + l_2, m_1 + m_2, u_1 + u_2) \quad (2)$$

Multiplication:

$$\overline{A_1} \times \overline{A_2} = (l_1, m_1, u_1) \times (l_2, m_2, u_2) = (l_1 \times l_2, m_1 \times m_2, u_1 \times u_2) \quad (3)$$

Subtraction:

$$\overline{A_1} - \overline{A_2} = (l_1, m_1, u_1) - (l_2, m_2, u_2) = (l_1 - u_2, m_1 - m_2, u_1 - l_2) \quad (4)$$

Division:

$$\frac{\overline{A_1}}{\overline{A_2}} = \frac{(l_1, m_1, u_1)}{(l_2, m_2, u_2)} = \left(\frac{l_1}{u_2}, \frac{m_1}{m_2}, \frac{u_1}{l_2}\right) \quad (5)$$

Reciprocal:

$$\overline{A_1}^{-1} = (l_1, m_1, u_1)^{-1} = \left(\frac{1}{u_1}, \frac{1}{m_1}, \frac{1}{l_1}\right) \quad (6)$$

3.2. Preliminaries—Dombi Aggregator

The Dombi aggregator [33] was used to determine the value of AADT, since data for a total of 12 years were considered.

$$(b_j) = \frac{\sum_{j=1}^{n}(b_j)}{1+\left(\sum_{j=1}^{n} w_j \left(\frac{1-f(b_j)}{f(b_j)}\right)\right)}$$

$$f(b_j) = \frac{b_j}{\sum_{j=1}^{n} b_j} \quad (7)$$

where w_j represents the weight of each considered year of data separately.

3.3. Preliminaries—Bonferroni Aggregator

Since the previous Dombi aggregator is not suitable for averaging the value of traffic accidents due to the occurrence of the value of zero, the Bonferroni aggregator was used [34].

$$a_j = \left(\frac{1}{e(e-1)} \sum_{\substack{i,j=1 \\ i \neq j}}^{e} a_i^p \otimes a_j^q\right)^{\frac{1}{p+q}} \quad (8)$$

In this research, e represents the number of years for traffic accidents, while $p, q \geq 0$ are a set of non-negative numbers.

4. Materials and Methods

This section presents the main phases of the research with the methodology created for determining the safety degree of road sections. All phases and the overall methodology are described below and shown in Figure 1, consisting of a total of 10 steps arranged in three phases which are causally connected.

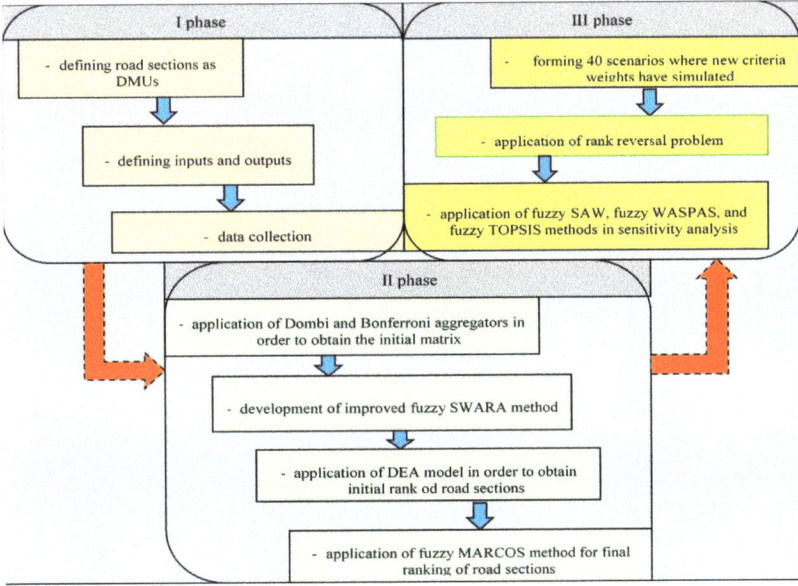

Figure 1. Research flow diagram with the applied methodology.

4.1. The First Phase

The first phase is to define sections of road infrastructure, determine entrances and exits, and collect data. A total of six sections of the road network on the territory of the Republic of Srpska were considered: Vrhovi-Šešlije I, Vrhovi-Šešlije II, Rudanka-Doboj, Šepak-Karakaj 3, Donje Caparde-Karakaj 1, and Border (RS/FBIH)-Donje Caparde. The input parameters for the given sections were as follows: section length—I1, road slope—I2, deviation from the speed limit—I3, and average annual daily traffic (AADT) —I4. The classifications of traffic accidents with fatalities, severe injuries, minor injuries, and material damage are defined as output parameters, O_1, O_2, O_3, and O_4, respectively.

4.2. The Second Phase

The second phase is the most important part of this research as it develops a methodology for determining the safety degree of certain road sections. After collecting all data from the first phase, it is necessary to determine an initial matrix that synthesizes the input and output parameters of the model. In terms of input data, AADT in the last 12 years has been considered as a very important factor in both this analysis and others. Since the data refer to a large number of years in order to obtain unique values, the Dombi aggregator described in the previous section was applied. The traffic parameters that occurred in the last five years on the presented road sections were considered as output parameters. Since no traffic accidents with fatalities occurred on certain sections during the year, i.e., their value was zero, the Bonferroni aggregator, also described in the previous section, was used.

4.2.1. Improved Fuzzy SWARA Method (IMF SWARA)

Step 1: After defining all the criteria on the basis of which the decision was made, it is necessary to arrange them in descending order based on their expected significance. For example, the most significant criterion is placed in first position and the least significant criterion is in the last position.

Step 2: Starting from the previously determined rank, the relatively smaller significance of the criterion (criterion C_j) was determined in relation to the previous one (C_j-1), and this was repeated for each subsequent criterion. This relation, i.e., comparative significance of the average value, is denoted with $\overline{s_j}$. A key problem in the original fuzzy SWARA [35] method is the application of an inadequate scale to determine the comparative significance of criteria, as demonstrated in detail below. Therefore, in this paper, the key point that improves the fuzzy SWARA method is the development of an adequate TFN scale (Table 1) that enables the precise and good quality determination of the significance of criteria using improved fuzzy SWARA (IMF SWARA).

Table 1. New linguistics and the TFN scale for the evaluation of the criteria in the improved fuzzy SWARA (IMF SWARA) method.

Linguistic Variable	Abbreviation	TFN Scale		
Absolutely less significant	ALS	1.000	1.000	1.000
Dominantly less significant	DLS	$\frac{1}{2}$	2/3	1.000
Much less significant	MLS	2/5	1/2	2/3
Really less significant	RLS	1/3	2/5	1/2
Less significant	LS	2/7	1/3	2/5
Moderately less significant	MDLS	$\frac{1}{4}$	2/7	1/3
Weakly less significant	WLS	2/9	1/4	2/7
Equally significant	ES	0.000	0.000	0.000

Step 3: Determining the fuzzy coefficient $\overline{k_j}$ (9):

$$\overline{k_j} = \begin{cases} \overline{1} & j=1 \\ \overline{s_j} & j>1 \end{cases} \tag{9}$$

Step 4: Determining the calculated weights $\overline{q}_j$ (10):

$$\overline{q}_j = \begin{cases} \overline{1} & j = 1 \\ \frac{\overline{q}_{j-1}}{\overline{k}_j} & j > 1 \end{cases} \tag{10}$$

Step 5: Calculation of the fuzzy weight coefficients using the following Equation (11):

$$\overline{w}_j = \frac{\overline{q}_j}{\sum_{j=1}^{m} \overline{q}_j} \tag{11}$$

where w_j represents the fuzzy relative weight of the criteria j, and m represents the total number of criteria.

We will now present the reasoning behind the need to develop the improved fuzzy SWARA (IMF SWARA) method, i.e., why the original fuzzy SWARA method [35] was not well conceived. To do this we have taken two examples, both from the paper in which the original extension of the fuzzy SWARA method was performed. The authors considered four main criteria: economic, environmental, social, and risk, applying the scale developed by Chang in 1996 [36] for the purposes of calculating the weights of criteria using the fuzzy AHP method, as shown in Table 2.

Table 2. Inadequate scale used by the authors in [35] in fuzzy SWARA.

Linguistic Scale	Response Scale
Equally important	(1, 1, 1)
Moderately less important	(2/3, 1, 3/2)
Less important	(2/5, 1/2, 2/3)
Very less important	(2/7, 1/3, 2/5)
Much less important	(2/9, 1/4, 2/7)

Therefore, the authors in developing the fuzzy SWARA method in [35] should not have used the scale created for the fuzzy AHP method, since the procedure for solving and thus evaluating the criteria is different for these two methods. The authors obtained the values of the criteria as shown in Table 3. As can be seen, the evaluation was performed as follows: less important, much less important, and moderately less important, respectively.

Table 3. Calculation process and weights of the main criteria applying fuzzy SWARA in [35].

	$\overline{s}_j$			$\overline{k}_j$			$\overline{q}_j$			$\overline{w}_j$			Crisp Value
C1				1.000	1.000	1.000	1.000	1.000	1.000	0.377	0.405	0.444	0.407
C2	0.400	0.500	0.667	1.400	1.500	1.667	0.600	0.667	0.714	0.226	0.270	0.317	0.271
C3	0.222	0.250	0.286	1.222	1.250	1.286	0.467	0.533	0.584	0.176	0.216	0.259	0.217
C4	0.667	1.000	1.500	1.667	2.000	2.500	0.187	0.267	0.351	0.070	0.108	0.156	0.110
						SUM	2.253	2.467	2.649				

The calculation below instead applies the improved fuzzy SWARA (IMF SWARA) (Table 4) method developed in this paper using the same linguistic variables: LS, MLS, and MDLS in order to show differences in the obtained fuzzy weights.

Table 4. Calculation process and weights of the main criteria applying improved fuzzy SWARA (IMF SWARA) in the same example.

	$\overline{s_j}$			$\overline{k_j}$			$\overline{q_j}$			$\overline{w_j}$			Crisp Value
C1				1.000	1.000	1.000	1.000	1.000	1.000	0.360	0.379	0.406	0.380
C2	2/7	1/3	2/5	1.286	1.333	1.400	0.714	0.750	0.778	0.257	0.284	0.316	0.285
C3	2/5	1/2	2/3	1.400	1.500	1.667	0.429	0.500	0.556	0.154	0.189	0.225	0.190
C4	1/4	2/7	1/3	1.250	1.286	1.333	0.321	0.389	0.444	0.116	0.147	0.180	0.148
						SUM	2.464	2.639	2.778				

Considering the comparative analysis on the basis of the same example by applying both the fuzzy SWARA and improved fuzzy SWARA (IMF SWARA) method, three key points can be identified:

(1) Using the fuzzy SWARA method, it is impossible to obtain results in which two criteria have equal fuzzy weights. By applying the improved fuzzy SWARA method, two or more criteria can have equal values.

(2) On the contrary, applying the inadequate TFN scale shown in Table 2, where decision-makers indicate that two criteria have the same value by assigning TFN (1,1,1), the criterion Cj in relation to $Cj-1$ received a value that is twice less than Cj. By applying the improved fuzzy SWARA method, assigning the value (0,0,0), equal values are obtained and not values twice as large.

(3) By increasing the number of criteria in the model, the least significant criteria receive values that can be negligible, i.e., with a tendency to zero. By applying the improved fuzzy SWARA method, less significant criteria have higher values and can play a greater role in the decision-making process.

We prove the above through a comparative analysis of the second example, which is a continuation of the first one. Namely, it is further considered the sub-criteria of the main economic criterion, which contains a total of eight criteria. A comparative analysis from which the inadequacy of the results of the fuzzy SWARA method can be concluded and the advantages of the improved fuzzy SWARA (IMF SWARA) method are shown in Table 5.

Table 5. Comparative analysis of fuzzy SWARA and improved fuzzy SWARA (IMF SWARA) in another example.

Fuzzy SWARA													
	$\overline{s_j}$			$\overline{k_j}$			$\overline{q_j}$			$\overline{w_j}$			Crisp Value
C1				1.000	1.000	1.000	1.000	1.000	1.000	0.292	0.319	0.351	0.320
C2	0.286	0.333	0.400	1.286	1.333	1.400	0.714	0.750	0.778	0.209	0.239	0.273	0.240
C3	0.222	0.250	0.286	1.222	1.250	1.286	0.556	0.600	0.636	0.162	0.191	0.223	0.192
C4	0.400	0.500	0.667	1.400	1.500	1.667	0.333	0.400	0.455	0.097	0.127	0.160	0.128
C5	0.667	1.000	1.500	1.667	2.000	2.500	0.133	0.200	0.273	0.039	0.064	0.096	0.065
C6	1.000	1.000	1.000	2.000	2.000	2.000	0.067	0.100	0.136	0.019	0.032	0.048	0.032
C7	0.667	1.000	1.500	1.667	2.000	2.500	0.027	0.050	0.082	0.008	0.016	0.029	0.017
C8	0.286	0.333	0.400	1.286	1.333	1.400	0.019	0.038	0.064	0.006	0.012	0.022	0.013
						SUM	2.849	3.138	3.423				
IMF SWARA													
	$\overline{s_j}$			$\overline{k_j}$			$\overline{q_j}$			$\overline{w_j}$			Crisp Value
C1				1.000	1.000	1.000	1.000	1.000	1.000	0.243	0.263	0.292	0.265
C2	2/7	1/3	2/5	1.286	1.333	1.400	0.714	0.750	0.778	0.174	0.198	0.227	0.199
C3	2/9	1/4	2/7	1.222	1.250	1.286	0.556	0.600	0.636	0.135	0.158	0.186	0.159
C4	2/5	1/2	2/3	1.400	1.500	1.667	0.333	0.400	0.455	0.081	0.105	0.133	0.106
C5	1/4	2/7	1/3	1.250	1.286	1.333	0.250	0.311	0.364	0.061	0.082	0.106	0.082
C6	0	0	0	1.000	1.000	1.000	0.250	0.311	0.364	0.061	0.082	0.106	0.082
C7	1/4	2/7	1/3	1.250	1.286	1.333	0.188	0.242	0.291	0.046	0.064	0.085	0.064
C8	2/7	1/3	2/5	1.286	1.333	1.400	0.134	0.181	0.226	0.033	0.048	0.066	0.048
						SUM	3.425	3.796	4.113				

In Table 5, we prove that all three key points were enhanced through the development of the improved fuzzy SWARA (IMF SWARA) method. As can be observed using the fuzzy SWARA method, there is no case where two criteria have the same weight, although in the process of the criterion evaluation it is indicated that criteria C5 and C6 should have equal values. If, using the fuzzy SWARA method, we see that criterion C6 is assigned TFN (1,1,1), this should mean that it has the same value as the preceding C5 criterion. However, the results show that C5 is twice as important as C6, which is an inconsistency in the process of evaluating the criteria and its weights. This shortcoming is eliminated by applying the improved fuzzy SWARA (IMF SWARA) method, which can be seen in Table 5. Additionally, it is evident that less significant criteria with an increase in the total number of criteria do not have a maximum tendency to zero.

Some examples of the application of the fuzzy SWARA method with an inadequate scale were given in studies [37–40]. It should be noted that the future application of the fuzzy SWARA method should be replaced by the improved fuzzy SWARA (IMF SWARA) method or the application of another scale in fuzzy SWARA that eliminates these previously observed shortcomings.

4.2.2. DEA Model

The CCR model is the most basic model of DEA [41,42]. Here, two DEA CCR models have been formed according to an input-oriented model (max) and according to an output-oriented model (min). The DEA CCR input-oriented model (max) is:

$$DEA_{input} = \max \sum_{i=1}^{m} w_i x_{i-input}$$

$st:$

$$\sum_{i=1}^{m} w_i x_{ij} - \sum_{i=m+1}^{m+s} w_i y_{ij} \leq 0, \ j = 1, \ldots, n \qquad (12)$$

$$\sum_{i=m+1}^{m+s} w_i y_{i-output} = 1$$

$$w_i \geq 0, \ i = 1, \ldots, m+s$$

The decision-making units (DMUs) are presented as m inputs for each alternative x_{ij}, through s which represents outputs for each alternative y_{ij}. Additionally, the weights of the parameters w_i are taken into account [43]. The total number of DMUs is denoted by n.

The DEA CCR output-oriented model (min) is:

$$DEA_{output} = \min \sum_{i=m+1}^{m+s} w_i y_{i-output}$$

$st:$

$$-\left(\sum_{i=1}^{m} w_i x_{ij}\right) + \sum_{i=m+1}^{m+s} w_i y_{ij} \geq 0, \ j = 1, \ldots, n \qquad (13)$$

$$\sum_{i=1}^{m} w_i x_{i-input} = 1$$

$$w_i \geq 0, \ i = 1, \ldots, m+s$$

4.2.3. Fuzzy MARCOS Method

The fuzzy MARCOS method was developed by Stanković et al. [44] and consists of the following steps [45]:

Step 1. Forming an initial fuzzy decision matrix.

Step 2. Expanding the initial fuzzy decision matrix with an anti-ideal solution (AAI):

$$\widetilde{A}(AI) = \min_i \widetilde{x}_{ij} \ if \ j \in B \ and \ \max_i \widetilde{x}_{ij} \ if \ j \in C \qquad (14)$$

and the ideal solution (AI):

$$\tilde{A}(ID) = \max_i \tilde{x}_{ij} \text{ if } j \in B \text{ and } \min_i \tilde{x}_{ij} \text{ if } j \in C \qquad (15)$$

Step 3. Normalizing the initial fuzzy decision matrix.

$$\tilde{n}_{ij} = \left(n^l_{ij}, n^m_{ij}, n^u_{ij}\right) = \left(\frac{x^l_{id}}{x^u_{ij}}, \frac{x^l_{id}}{x^m_{ij}}, \frac{x^l_{id}}{x^l_{ij}}\right) \text{ if } j \in C \qquad (16)$$

$$\tilde{n}_{ij} = \left(n^l_{ij}, n^m_{ij}, n^u_{ij}\right) = \left(\frac{x^l_{ij}}{x^u_{id}}, \frac{x^m_{ij}}{x^u_{id}}, \frac{x^u_{ij}}{x^u_{id}}\right) \text{ if } j \in B \qquad (17)$$

Step 4. Weighting the normalized decision matrix.

$$\tilde{v}_{ij} = \left(v^l_{ij}, v^m_{ij}, v^u_{ij}\right) = \tilde{n}_{ij} \otimes \tilde{w}_j = \left(n^l_{ij} \times w^l_j, n^m_{ij} \times w^m_j, n^u_{ij} \times w^u_j\right) \qquad (18)$$

Step 5. Calculating the S_i matrix:

$$\tilde{S}_i = \sum_{i=1}^{n} \tilde{v}_{ij} \qquad (19)$$

Step 6. Calculating the degree of usefulness K_i.

$$\tilde{K}_i^- = \frac{\tilde{S}_i}{\tilde{S}_{ai}} = \left(\frac{s^l_i}{s^u_{ai}}, \frac{s^m_i}{s^m_{ai}}, \frac{s^u_i}{s^l_{ai}}\right) \qquad (20)$$

$$\tilde{K}_i^+ = \frac{\tilde{S}_i}{\tilde{S}_{id}} = \left(\frac{s^l_i}{s^u_{id}}, \frac{s^m_i}{s^m_{id}}, \frac{s^u_i}{s^l_{id}}\right) \qquad (21)$$

Step 7. Calculating the fuzzy matrix $\tilde{T}_i$.

$$\tilde{T}_i = \tilde{t}_i = \left(t^l_i, t^m_i, t^u_i\right) = \tilde{K}_i^- \oplus \tilde{K}_i^+ = \left(k_i^{-l} + k_i^{+l}, k_i^{-m} + k_i^{+m}, k_i^{-u} + k_i^{+u}\right) \qquad (22)$$

Determining the fuzzy number $\tilde{D}$:

$$\tilde{D} = \left(d^l, d^m, d^u\right) = \max_i \tilde{t}_{ij} \qquad (23)$$

Step 8. The de-fuzzification of fuzzy numbers:

$$df_{crisp} = \frac{l + 4m + u}{6} \qquad (24)$$

Step 9. Determining the utility function $f\left(\tilde{K}_i\right)$:
Utility function according to the anti-ideal solution.

$$f\left(\tilde{K}_i^+\right) = \frac{\tilde{K}_i^-}{df_{crisp}} = \left(\frac{k_i^{-l}}{df_{crisp}}, \frac{k_i^{-m}}{df_{crisp}}, \frac{k_i^{-u}}{df_{crisp}}\right) \qquad (25)$$

Utility function according to the ideal solution.

$$f\left(\tilde{K}_i^-\right) = \frac{\tilde{K}_i^+}{df_{crisp}} = \left(\frac{k_i^{+l}}{df_{crisp}}, \frac{k_i^{+m}}{df_{crisp}}, \frac{k_i^{+u}}{df_{crisp}}\right) \qquad (26)$$

Step 10. Calculating the final utility function:

$$f(K_i) = \frac{K_i^+ + K_i^-}{1 + \frac{1-f(K_i^+)}{f(K_i^+)} + \frac{1-f(K_i^-)}{f(K_i^-)}}; \quad (27)$$

Step 11. Ranking alternatives.

4.3. The Third Phase

In the third phase of this research, a sensitivity analysis consisting of three parts was conducted. The first part presents the change of input values through 40 newly formed scenarios in which their weights ware simulated using Equation (28).

$$W_{n\beta} = (1 - W_{n\alpha}) \frac{W_\beta}{(1 - W_n)} \quad (28)$$

$\widetilde{W}_{n\alpha}$ indicates the reduced value of the input the weight of which changes, $\widetilde{W}_\beta$ indicates the real value of the input considered, while $\widetilde{W}_n$ indicates the original value of the input the value of which increases. In the second part of the verification of the results, the ranking reversal problem is applied [46], in which the size of the initial fuzzy matrix is changed. This is followed by a comparative analysis with three other fuzzy MCDM methods: fuzzy WASPAS [47], fuzzy SAW [48], and fuzzy TOPSIS [49].

5. Case Study

5.1. Formation of Input-Output Parameters and Averaging Using Dombi and Bonferroni Aggregators

The input parameters for the given sections are: section length—I1, road slope—I2, deviation from the speed limit—I3, and average annual daily traffic (AADT)—I4. The classifications of traffic accidents with fatalities, severe injuries, minor injuries, and material damage are defined as output parameters, O_1, O_2, O_3, and O_4, respectively. These parameters are defined on the basis of the authors' practical experiences, similar studies, and dialogue with other experts. Table 6 shows the values of all DMUs according to the input-output parameters.

Table 6. Road section parameters in relation to input-output parameters.

	I_1	I_2	I_3	I_4	O_1	O_2	O_3	O_4
DMU$_1$	14.07	5.00	10.16	4578.95	0.63	2.49	3.70	7.26
DMU$_2$	14.07	1.92	11.67	4578.95	0.63	2.49	3.70	7.26
DMU$_3$	7.41	0.02	4.88	13,179.39	1.38	4.81	14.42	49.72
DMU$_4$	20.95	1.00	9.61	5988.48	1.38	5.67	19.38	53.18
DMU$_5$	15.35	3.00	6.33	3367.41	0.55	3.16	6.26	21.09
DMU$_6$	3.14	7.00	6.29	3871.79	0.00	0.32	0.84	1.95

A total of six sections of the road network on the territory of the Republic of Srpska: Vrhovi-Šešlije I, Vrhovi-Šešlije II, Rudanka-Doboj, Šepak-Karakaj 3, Donje Caparde-Karakaj 1, and Border (RS/FBIH)-Donje Caparde, were considered.

Figure 2 shows the lengths of six analyzed sections, the slopes on the measured sections (upgrade/downgrade), and the arithmetic mean of the speed deviation of passenger cars from the speed limit on the measured sections. The measured sections were determined for different values of longitudinal gradients. At 1000 m in front of the measured cross section, the value of the average downgrade/upgrade was determined (−5.00%, −1.92%, −0.017%, + 1.00%, 3.00%, and 7.00%). The largest section length is Šepak-Karakaj 3 (20.95 km), and the smallest is Border (RS-FBIH)-Donje Caparde (3.14 km). The sections Vrhovi-Šešlije and Border (RS-FBIH)-Donje Caparde can be classified as hilly sections according to their terrain configuration. The arithmetic means of deviations from the speed limit were determined on the measured sections. The credibility of the speed limit on the section Vrhovi-Šešlije for both values of the measured longitudinal downgrades is especially endangered, with exceedances of over 10 km/h. In this case, the speeding of passenger cars as representative vehicles was analyzed, and it was determined that the credibility of the speed limit was significantly endangered for the downgrade. Additionally, with the increase of the ascent on the measured sections, a deviation from the speed limit which did not endanger the credibility of the speed limit was determined.

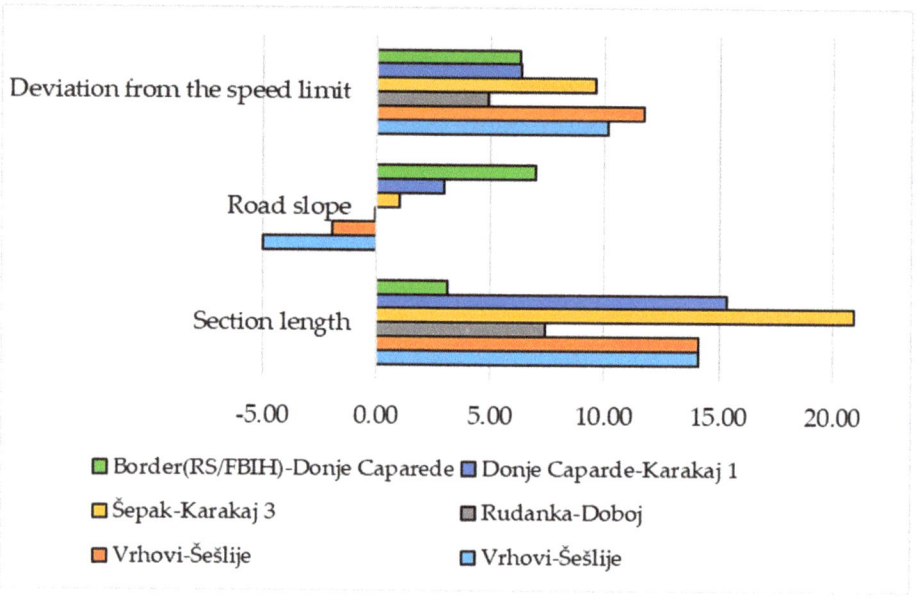

Figure 2. Display of the section length, road slope, and deviation from the speed limit.

Figure 3 shows the AADT values for the available period of 12 years, from 2005 to 2016, where the value of AADT on the plain section Rudanka-Doboj deviates significantly (0.017% ≈ 0.00%), while, on the other five sections, the arithmetic mean of AADT does not exceed the value of 6000 vehicles/day, except at the section Rudanka-Doboj, where the AADT is 13,179.39 vehicles/day. On all sections, the slight increase/decrease of AADT by years is not balanced.

Figure 4 shows the average number of traffic accidents in the period from 2015 to 2019 by the type of accident and the specified sections. Figure 4 shows the total number of traffic accidents, with the section DMU3, Rudanka-Doboj, standing out negatively, considering that there are also accidents with fatalities in the period from 2015 to 2019. On average, 1.4 persons were killed annually on this section. When observing the stated time period, the number of traffic accidents with fatalities is 2, 2, 1, 1, and 1, respectively. Additionally, on the section DMU4, Šepak 3-Karakaj, 1.4 persons were killed annually, with the number of fatalities by years 2, 1, 1, 2, and 1, respectively. On average, 53.40 traffic accidents with material damage occurred on the same section in five years, making this section especially risky. The lowest number of accidents was recorded on the shortest section DMU6, Border (RS-BIH)-Donje Caparde, where in the period of five analyzed years, the number of total accidents per year does not exceed five, and there are no fatalities on this section. These indicators of the number of accidents per year were analyzed and used in further data synthesis. It is clearly noticeable that DMU4 and DMU3 stand out in terms of the number of traffic accidents by all classes.

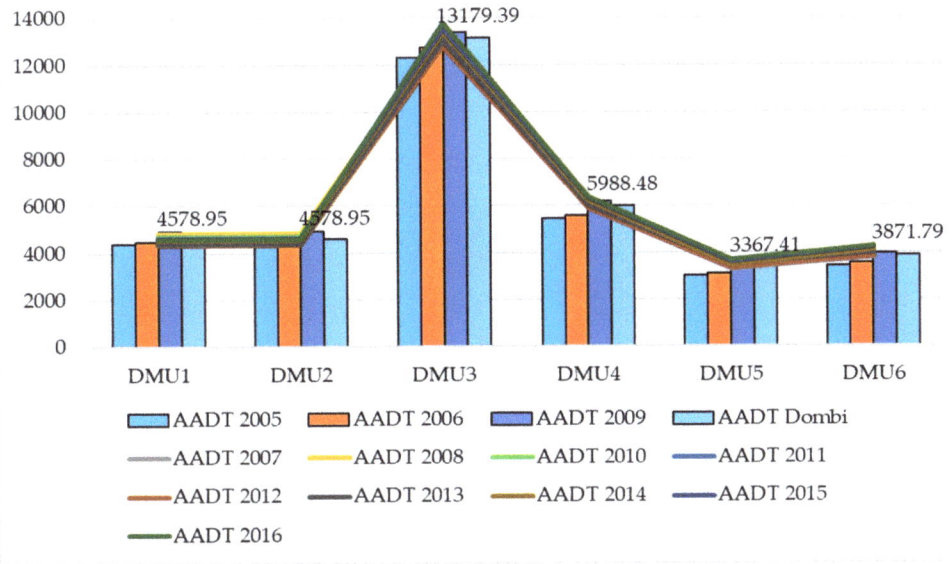

Figure 3. AADT for the period 2005–2016.

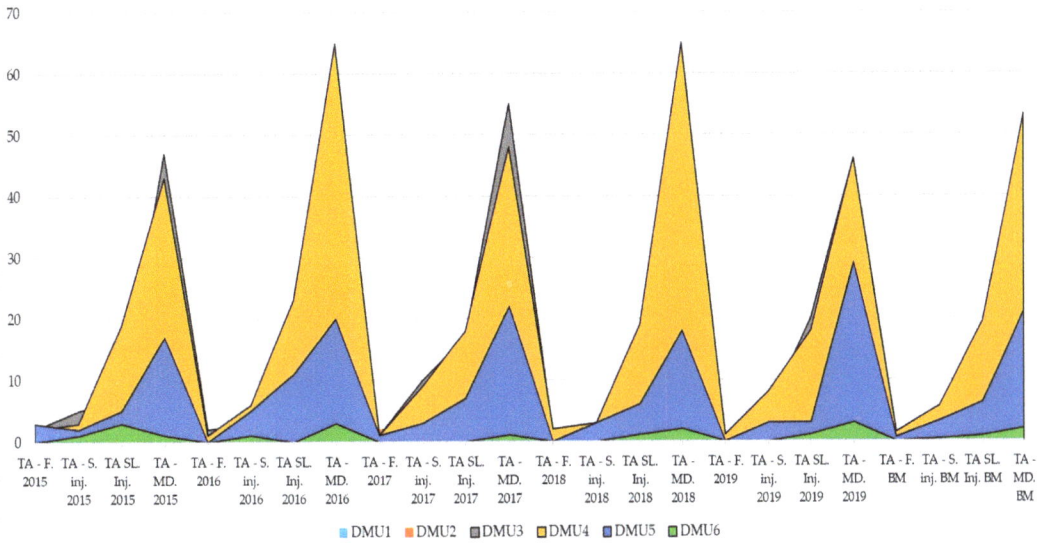

Figure 4. The total number of traffic accidents on all road sections in the period 2015–2019.

In order to obtain the final data that will represent the input, i.e., output of the DEA model, it is necessary to use the Dombi aggregator for averaging the historical data for the AADT as inputs, and the Bonferroni aggregator for averaging all outputs of the DEA model. An example of the application of both aggregators and the averaging of the stated values is shown below. The AADT values for the first and second DMU are obtained using a Dombi aggregator as follows:

$$(b_1) = \frac{\sum_{j=1}^{n}(b_1)}{1+\left(\sum_{j=1}^{n} w_j \left(\frac{1-f(b_j)}{f(b_j)}\right)\right)} = \frac{55033}{1+\left(\frac{1}{12}\left(\frac{1-0.0795}{0.0795}\right)+\frac{1}{12}\left(\frac{1-0.081}{0.081}\right)+\ldots+\frac{1}{12}\left(\frac{1-0.083}{0.083}\right)\right)} = \frac{55033}{12.019} = 4578.95$$

$$f(b_1) = \frac{4375}{4375+4470+4782+4829+4931+4676+4547+4334+4366+4548+4600+4575} = \frac{4375}{55033} = 0.0795$$

All AADT values shown in Figure 3 were obtained in the same way.

The values of all outputs are obtained using the Bonferroni aggregator as follows. For the first DMU1 for the first output O1—traffic accidents with fatalities:

$$BM^{p=1,q=1} = (2,0,2,0,0) = \omega_{DMU_{1(1)}} = \left(\frac{1}{5(5-1)} \sum_{\substack{i,j=1 \\ i \neq j}}^{5} \omega_{DMU_{1(1)}i} \omega_{DMU_{1(1)}j}\right)^{\frac{1}{1+1}} =$$

$$(0.050(2^1 \cdot 0^1 + 2^1 \cdot 2^1 + 2^1 \cdot 0^1 + 2^1 \cdot 0^1 + 0^1 \cdot 2^1 + 0^1 \cdot 2^1 + 0^1 \cdot 0^1 + 0^1 \cdot 0^1 + \ldots + 0^1 \cdot 2^1 + 0^1 \cdot 0^1 + 0^1 \cdot 2^1 + 0^1 \cdot 0^1))^{\frac{1}{1+1}} = 0.632$$

In this research, e represents the year number for traffic accidents, which means individually each year in the interval 2015–2019, while $p, q \geq 0$ are a set of non-negative numbers. In the same way, averaged values are obtained for all outputs shown in Figure 4.

5.2. Determination of Weight Values Using the Improved Fuzzy SWARA (IMF SWARA) Method

In the previous section of the paper, the validity of the developed IMF SWARA method was proved, so the results of weight coefficients for the criteria, i.e., the inputs and outputs of the considered study, are presented below. The weight coefficients of the criteria obtained using the IMF SWARA method are shown in Table 7.

Table 7. Values of input and output significance in the considered model obtained by applying the IMF SWARA method.

							IMF SWARA						
		$\bar{s}_j$			$\bar{k}_j$			$\bar{q}_j$			$\bar{w}_j$		Crisp Value
C5				1.000	1.000	1.000	1.000	1.000	1.000	0.203	0.213	0.225	0.213
C6	2/9	1/4	2/7	1.222	1.250	1.286	0.778	0.800	0.818	0.158	0.170	0.184	0.170
C3	0.000	0.000	0.000	1.000	1.000	1.000	0.778	0.800	0.818	0.158	0.170	0.184	0.170
C4	2/9	1/4	2/7	1.222	1.250	1.286	0.605	0.640	0.669	0.123	0.136	0.151	0.136
C1	1/4	2/7	1/3	1.250	1.286	1.333	0.454	0.498	0.536	0.092	0.106	0.121	0.106
C2	1/4	2/7	1/3	1.250	1.286	1.333	0.340	0.387	0.428	0.069	0.082	0.096	0.082
C7	2/7	1/3	2/5	1.286	1.333	1.400	0.243	0.290	0.333	0.049	0.062	0.075	0.062
C8	0.000	0.000	0.000	1.000	1.000	1.000	0.243	0.290	0.333	0.049	0.062	0.075	0.062
						SUM	4.441	4.706	4.936				

The most significant criterion from the set of observed criteria is O1 with a value of 0.213, which represents the number of traffic accidents with fatalities. According to this criterion, the prominent sections, DMU3 and DMU4, are evaluated especially negatively. Additionally, a significant influential criterion is the number of accidents with severe injuries, O2, and the deviation from the speed limit, I3, with a value of 0.170. The values of the input parameters in a small range are slightly higher compared to the less significant ones. Compared to other inputs, there is no input with a significantly higher advantage.

5.3. Application of DEA Model

Further in the paper, the DEA output-oriented model is presented in order to determine the sections of two-lane roads, i.e., DMUs, which show a rather satisfactory safety degree in a certain manner. It is these sections that are implemented further into the model and ranked using the fuzzy MARCOS method. The DEA algorithms are defined and solved using the Lingo 17 software, an example of which is given below (DMU1).

$DEA_{ouput-DMU1} = \text{MIN} = 0.63 \cdot w5 + 2.49 \cdot w6 + 3.7 \cdot w7 + 7.26 \cdot w8;$
$-14.07 \cdot w1 - 5.00 \cdot w2 - 10.16 \cdot w3 - 4578.95 \cdot w4 + (0.63 \cdot w5 + 2.49 \cdot w6 + 3.7 \cdot w7 + 7.26 \cdot w8) >= 0;$
$-14.07 \cdot w1 - 1.92 \cdot w2 - 11.67 \cdot w3 - 4578.95 \cdot w4 + (0.63 \cdot w5 + 2.49 \cdot w6 + 3.7 \cdot w7 + 7.26 \cdot w8) >= 0;$
$-7.41 \cdot w1 - 0.02 \cdot w2 - 4.88 \cdot w3 - 13179.39 \cdot w4 + (1.38 \cdot w5 + 4.81 \cdot w6 + 14.42 \cdot w7 + 49.72 \cdot w8) >= 0;$
$-20.95 \cdot w1 - 1.00 \cdot w2 - 9.61 \cdot w3 - 5988.48 \cdot w4 + (1.38 \cdot w5 + 5.67 \cdot w6 + 19.38 \cdot w7 + 53.18 \cdot w8) >= 0;$
$-15.35 \cdot w1 - 3.00 \cdot w2 - 6.33 \cdot w3 - 3367.41 \cdot w4 + (0.55 \cdot w5 + 3.16 \cdot w6 + 6.26 \cdot w7 + 21.09 \cdot w8) >= 0;$
$-3.14 \cdot w1 - 7.00 \cdot w2 - 6.29 \cdot w3 - 3871.79 \cdot w4 + (0.00 \cdot w5 + 0.32 \cdot w6 + 0.84 \cdot w7 + 1.95 \cdot w8) >= 0;$
$14.07 \cdot w1 + 5.00 \cdot w2 + 10.16 \cdot w3 + 4578.95 \cdot w4 = 1;$
$w1 > 0; w2 > 0; w3 > 0; w4 > 0; w5 > 0; w6 > 0; w7 > 0; w8 > 0;$

After solving the set algorithm, the objective function is 1.000. After that, the algorithms for the other DMUs according to inputs are formed. Regardless of the fact that it is not necessary to perform the calculation using both models (input and output), the output-oriented model is performed for internal calculation control in this study. The final results using the DEA model are shown in Table 8.

Table 8. Results of the safety situation on road sections after the application of the DEA model.

		DEA-Input	DEA-Output	DEA-Final
DMU1	Vrhovi-Šešlije I	1.00	1.00	1.00
DMU2	Vrhovi-Šešlije II	1.00	1.00	1.00
DMU3	Rudanka-Doboj	0.23	4.42	19.54
DMU4	Šepak-Karakaj 3	0.38	2.66	7.064
DMU5	Donje Caparde-Karakaj 1	0.65	1.54	2.362
DMU6	Border (RS/FBIH)-Donje Caparde	1.00	1.00	1.00

Road sections marked DMU1, DMU2, and DMU6 with a displayed value of 1.000 are further implemented in the model and ranked using the fuzzy MARCOS method. Other sections, DMU3, DMU4, and DMU5, with the values of 19.540, 7.064, and 2.362, respectively, are eliminated further from the model because they show an inadequate safety degree and a corrective action in the implementation of certain measures is necessary.

5.4. Application of the Fuzzy MARCOS Method in Order to Make Final Ranking of Road Sections

In the previous section of the paper, the DEA model was applied, and it was determined which road sections should be included further in the model. Additionally, the initial fuzzy decision matrix was defined on the basis of the scale shown in Table 9.

Table 9. Fuzzy linguistic scale for evaluating alternatives [41].

		Benefit	Cost
Linguistic term	Mark	TFN	TFN
Extremely poor	EP	(1,1,1)	(7,9,9)
Very poor	VP	(1,1,3)	(7,7,9)
Poor	P	(1,3,3)	(5,7,7)
Medium poor	MP	(3,3,5)	(5,5,7)
Medium	M	(3,5,5)	(3,5,5)
Medium good	MG	(5,5,7)	(3,3,5)
Good	G	(5,7,7)	(1,3,3)
Very good	VG	(7,7,9)	(1,1,3)
Extremely good	EG	(7,9,9)	(1,1,1)

Taking into account the given linguistic scale, the initial fuzzy decision matrix was formed, after which Equations (14) and (15) were applied in order to determine ideal and anti-ideal solutions. The extended fuzzy initial matrix is shown in Table 10, and it is important to note that criteria C1, C2, and C4 are beneficial and thus for whom the maximum value is desired, while the others belong to the group of cost criteria with the preference of the minimum value.

Table 10. Extended fuzzy initial decision matrix.

	C1	C2	C3	C4	C5	C6	C7	C8
AAI	(1,3,3)	(3,3,5)	(5,5,7)	(3,5,5)	(3,3,5)	(5,5,7)	(5,5,7)	(7,7,9)
DMU1	(7,9,9)	(5,7,7)	(3,5,5)	(5,7,7)	(3,3,5)	(5,5,7)	(5,5,7)	(7,7,9)
DMU2	(7,9,9)	(3,3,5)	(5,5,7)	(5,7,7)	(3,3,5)	(5,5,7)	(5,5,7)	(7,7,9)
DMU6	(1,3,3)	(7,7,9)	(1,1,3)	(3,5,5)	(1,1,3)	(1,1,3)	(1,3,3)	(3,3,5)
ID	(7,9,9)	(7,7,9)	(1,1,3)	(5,7,7)	(1,1,3)	(1,1,3)	(1,3,3)	(3,3,5)

The next step involves performing the normalization of the matrix shown in Table 10, taking into account the type of criteria. Equation (16) is applied to the cost group of criteria as follows: $\tilde{n}_{13} = \left(\frac{1.000}{5.000}, \frac{1.000}{5.000}, \frac{1.000}{3.000}\right) = (0.200, 0.200, 0.333)$, while for the group of criteria that prefer the maximum value, Equation (17) is applied: $\tilde{n}_{11} = \left(\frac{7.000}{9.000}, \frac{9.000}{9.000}, \frac{9.000}{9.000}\right) = (0.778, 1.000, 1.000)$. The complete fuzzy normalized matrix is shown in Table 11.

Table 11. Fuzzy normalized decision matrix.

	C1	C2	C3	C4
AAI	(0.111,0.333,0.333)	(0.333,0.333,0.556)	(0.143,0.2,0.2)	(0.429,0.714,0.714)
DMU1	(0.778,1,1)	(0.556,0.778,0.778)	(0.2,0.2,0.333)	(0.714,1,1)
DMU2	(0.778,1,1)	(0.333,0.333,0.556)	(0.143,0.2,0.2)	(0.714,1,1)
DMU6	(0.111,0.333,0.333)	(0.778,0.778,1)	(0.333,1,1)	(0.429,0.714,0.714)
ID	(0.778,1,1)	(0.778,0.778,1)	(0.333,1,1)	(0.714,1,1)
	C5	C6	C7	C8
AAI	(0.2,0.333,0.333)	(0.143,0.2,0.2)	(0.143,0.2,0.2)	(0.333,0.429,0.429)
DMU1	(0.2,0.333,0.333)	(0.143,0.2,0.2)	(0.143,0.2,0.2)	(0.333,0.429,0.429)
DMU2	(0.2,0.333,0.333)	(0.143,0.2,0.2)	(0.143,0.2,0.2)	(0.333,0.429,0.429)
DMU6	(0.333,1,1)	(0.333,1,1)	(0.333,0.333,1)	(0.6,1,1)
ID	(0.333,1,1)	(0.333,1,1)	(0.333,0.333,1)	(0.6,1,1)

In the next step, the previous matrix is weighted by multiplying the values from Table 11 by the weight coefficients obtained using the IMF SWARA method by Equation (18) as follows:

$$\tilde{v}_{11} = \left(n_{11}^l \times w_1^l, n_{11}^m \times w_1^m, n_{11}^u \times w_1^u\right) = (0.778 \times 0.213, 1.000 \times 0.213, 1.000 \times 0.213) = (0.166, 0.213, 0.213)$$

The complete weighted fuzzy matrix is given in Table 12.

Table 12. Fuzzy weighted normalized decision matrix.

	C1	C2	C3	C4
AAI	(0.02,0.07,0.07)	(0.06,0.06,0.09)	(0.02,0.03,0.03)	(0.06,0.1,0.1)
DMU1	(0.17,0.21,0.21)	(0.09,0.13,0.13)	(0.03,0.03,0.06)	(0.1,0.14,0.14)
DMU2	(0.17,0.21,0.21)	(0.06,0.06,0.09)	(0.02,0.03,0.03)	(0.1,0.14,0.14)
DMU6	(0.02,0.07,0.07)	(0.13,0.13,0.17)	(0.06,0.17,0.17)	(0.06,0.1,0.1)
ID	(0.17,0.21,0.21)	(0.13,0.13,0.17)	(0.06,0.17,0.17)	(0.1,0.14,0.14)
	C5	C6	C7	C8
AAI	(0.02,0.04,0.04)	(0.01,0.02,0.02)	(0.01,0.01,0.01)	(0.02,0.03,0.03)
DMU1	(0.02,0.04,0.04)	(0.01,0.02,0.02)	(0.01,0.01,0.01)	(0.02,0.03,0.03)
DMU2	(0.02,0.04,0.04)	(0.01,0.02,0.02)	(0.01,0.01,0.01)	(0.02,0.03,0.03)
DMU6	(0.04,0.11,0.11)	(0.03,0.08,0.08)	(0.02,0.02,0.06)	(0.04,0.06,0.06)
ID	(0.04,0.11,0.11)	(0.03,0.08,0.08)	(0.02,0.02,0.06)	(0.04,0.06,0.06)

The fuzzy matrix $\tilde{S}_i$ is obtained by applying Equation (19),

$$\tilde{S}_{ai} = (0.226, 0.350, 0.388), \quad \tilde{S}_1 = (0.454, 0.606, 0.629)$$
$$\tilde{S}_2 = (0.406, 0.531, 0.569), \quad \tilde{S}_6 = (0.392, 0.742, 0.821),$$
$$\tilde{S}_{id} = (0.573, 0.923, 1.002)$$

as follows:

$$\tilde{S}_{ai} = \begin{pmatrix} 0.024 + 0.057 + 0.024 + 0.058 + 0.021 + 0.012 + 0.009 + 0.021 \\ 0.071 + 0.057 + 0.034 + 0.097 + 0.035 + 0.016 + 0.012 + 0.027 \\ 0.071 + 0.095 + 0.034 + 0.097 + 0.035 + 0.016 + 0.012 + 0.027 \end{pmatrix} = (0.226, 0.350, 0.388)$$

Using Equation (20), the matrix $\tilde{K}_i^-$ is obtained,

$$\tilde{k}_1^- = (1.171, 1.734, 2.789)$$
$$\tilde{k}_2^- = (1.049, 1.517, 2.521)$$
$$\tilde{k}_6^- = (1.011, 2.121, 3.640)$$

as follows:

$$\tilde{k}_1^- = \frac{\tilde{S}_1}{\tilde{S}_{ai}} = \left(\frac{s_1^l}{s_{ai}^u}, \frac{s_1^m}{s_{ai}^m}, \frac{s_1^u}{s_{ai}^l} \right) = \left(\frac{0.454}{0.388}, \frac{0.606}{0.350}, \frac{0.629}{0.226} \right) = (1.171, 1.734, 2.789)$$

Using Equation (21), the matrix $\tilde{K}_i^+$ is obtained,

$$\tilde{k}_1^+ = (0.453, 0.657, 1.098)$$
$$\tilde{k}_2^+ = (0.406, 0.575, 0.993)$$
$$\tilde{k}_6^+ = (0.391, 0.804, 1.434)$$

as follows:

$$\tilde{k}_1^+ = \frac{\tilde{S}_1}{\tilde{S}_{id}} = \left(\frac{s_1^l}{s_{id}^u}, \frac{s_1^m}{s_{id}^m}, \frac{s_1^u}{s_{id}^l} \right) = \left(\frac{0.454}{1.002}, \frac{0.606}{0.923}, \frac{0.629}{0.573} \right) = (0.453, 0.657, 1.098)$$

The matrix $\tilde{T}_i$ is calculated using Equation (22):

$$\tilde{t}_1 = (1.624, 2.391, 3.888), \quad \tilde{t}_2 = (1.454, 2.092, 3.513),$$
$$\tilde{t}_6 = (1.402, 2.925, 5.073)$$

in the following way:

$$\tilde{t}_1 = (1.171 + 0.453, 1.734 + 0.657, 2.789 + 1.098) = (1.624, 2.391, 3.888)$$

After that, a fuzzy number $\tilde{D} = (1.624, 2.925, 5.073)$ using Equation (23) is calculated. Defuzzification is done by Equation (24) obtaining the number $df_{crisp} = 3.066$. The final results calculated using the fuzzy MARCOS method are shown in Table 13.

Table 13. Results of the integrated IMF SWARA–fuzzy MARCOS model.

	$f(\tilde{K}_i^-)$	$f(\tilde{K}_i^+)$	K-	K+	fK-	fK+	Ki	Rank
DMU1	(0.148,0.214,0.358)	(0.382,0.565,0.91)	1.816	0.697	0.227	0.592	0.494	2
DMU2	(0.132,0.188,0.324)	(0.342,0.495,0.822)	1.606	0.616	0.201	0.524	0.378	3
DMU6	(0.128,0.262,0.468)	(0.33,0.692,1.187)	2.189	0.840	0.274	0.714	0.748	1

Utility functions $f\left(\tilde{K}_i^+\right)$ and $f\left(\tilde{K}_i^-\right)$ are calculated applying Equations (25) and (26).

$$f\left(\tilde{K}_1^+\right) = \frac{\tilde{K}_1^-}{df_{crisp}} = \left(\frac{1.171}{3.066}, \frac{1.734}{3.066}, \frac{2.789}{3.066}\right), f\left(\tilde{K}_1^-\right) = \frac{\tilde{K}_1^+}{df_{crisp}} = \left(\frac{0.453}{3.066}, \frac{0.657}{3.066}, \frac{1.098}{3.066}\right)$$

Then defuzzification if performed for $\tilde{K}_i^-, \tilde{K}_i^+, f\left(\tilde{K}_i^+\right), f\left(\tilde{K}_i^-\right)$. The calculation of the utility function of alternatives fK_i is obtained using Equation (27).

$$f(K_1) = \frac{K_1^+ + K_1^-}{1 + \frac{1-f(K_1^+)}{f(K_1^+)} + \frac{1-f(K_1^-)}{f(K_1^-)}} = \frac{0.453 + 1.171}{1 + \frac{1-0.382}{0.382} + \frac{1-0.148}{0.148}} = 0.494$$

After applying the integrated DEA—IMF SWARA—fuzzy MARCOS methodology based on the application of Dombi and Bonferroni aggregators, the final results are obtained. The section of the two-lane road DMU6 Border (RS-FBIH)-Donje Caparde is ranked with the highest level of traffic safety functionally dependent on the technical and exploitation indicators of the road network. The value of the rank size is 0.748, while the other two sections, DMU1 and DMU2, have incomparably lower values of 0.494 and 0.378, which shows a higher traffic risk on the measured parts of these sections.

6. Sensitivity Analysis

6.1. Testing the Change in Weights of Inputs

One way to test the sensitivity of the model is to simulate the weight coefficients of the criteria by applying the aforementioned Equation (28). It is important to note that a total of 40 new scenarios were formed in which the value of the input was simulated through 10 scenarios individually. In the first 10 scenarios, the value of the first input C1 was reduced by 10% for each scenario, meaning that in scenario S10, the first input had no significance, i.e., the value was zero. The other 30 scenarios for criteria C2, C3, and C4 were formed in the same way. Thus, the input values were corrected in the range of 10–100%. The values of input weight coefficients through the newly formed 40 scenarios are given in Figure 5.

Each scenario was observed separately, so recalculation was performed with the fuzzy MARCOS method and new results were obtained that confirmed the original rank. Thus, there is no change in any rank, while the final values of the DMUs certainly change. In the first 10 scenarios where the significance of the first input is reduced, the value of the best DMU6 increases, while the other two decrease. The same is the case with the change in the influence of the fourth input, i.e., in scenarios S31–S40. When the value of the second input changes, the value of the best-ranked DMU6 decreases, while the value of DMU2 increases and approaches DMU1. As the value of the third input changes, the value of DMU6 decreases, and the other two increase, with DMU1 approaching DMU6. This means that the model is completely insensitive to changes in input significance.

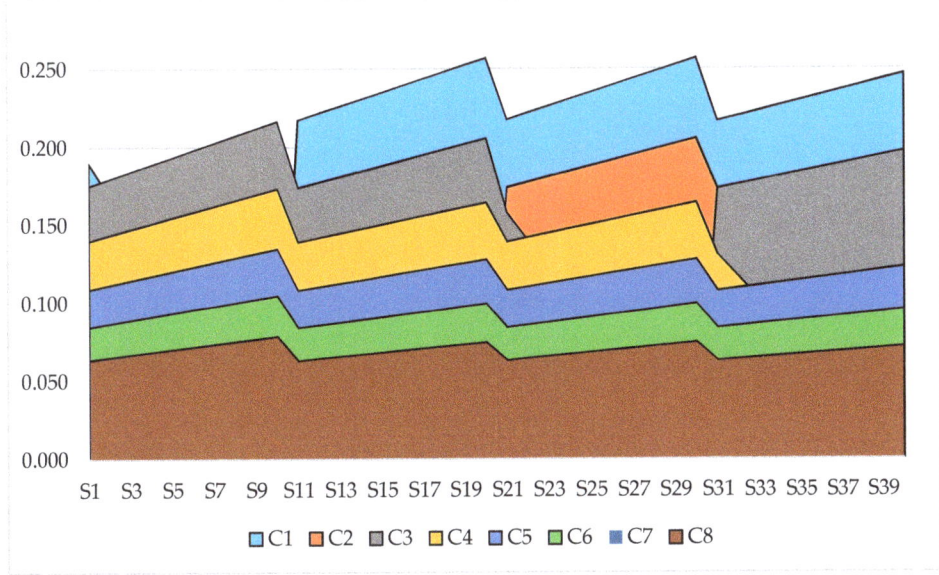

Figure 5. Input weight coefficient values through 40 scenarios.

6.2. Comparison with Other MCDM Methods in a Fuzzy Form

As a second part of this analysis, a comparative analysis with three other methods in a fuzzy form was performed, as explained in detail in the third phase of the methodology section. Figure 6 shows the values and ranks of the DMUs using the fuzzy SAW, fuzzy WASPAS, and fuzzy TOPSIS methods.

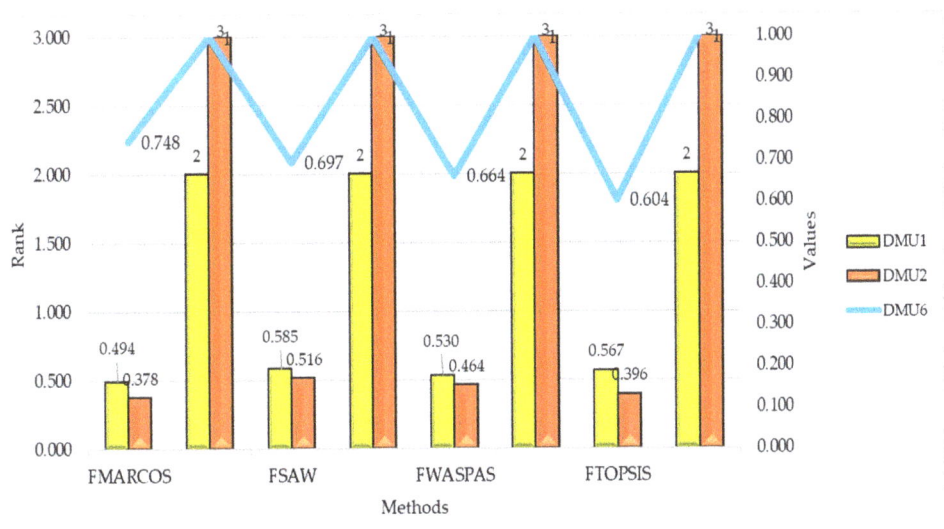

Figure 6. Values and ranks of DMUs using the fuzzy SAW, fuzzy WASPAS, and fuzzy TOPSIS methods.

Figure 6 shows that the ranks are fully correlated with the initial rank obtained by applying the integrated IMF SWARA—fuzzy MARCOS model. In addition to the ranks, the values of all DMUs are shown in order to provide a better insight into the comparative analysis. Once again, the stability of the model and the more precise data obtained with the fuzzy MARCOS method can be proven through the observation of the intervals of all values using all methods.

6.3. Influence of Dynamic Initial Matrix Formation

Figure 7 shows the results after applying the sensitivity analysis related to the change in the size of the initial fuzzy matrix. Two new scenarios have been defined in which the sizes of the initial matrix are changed in such a way that the worst-ranked alternative is eliminated from the initial matrix.

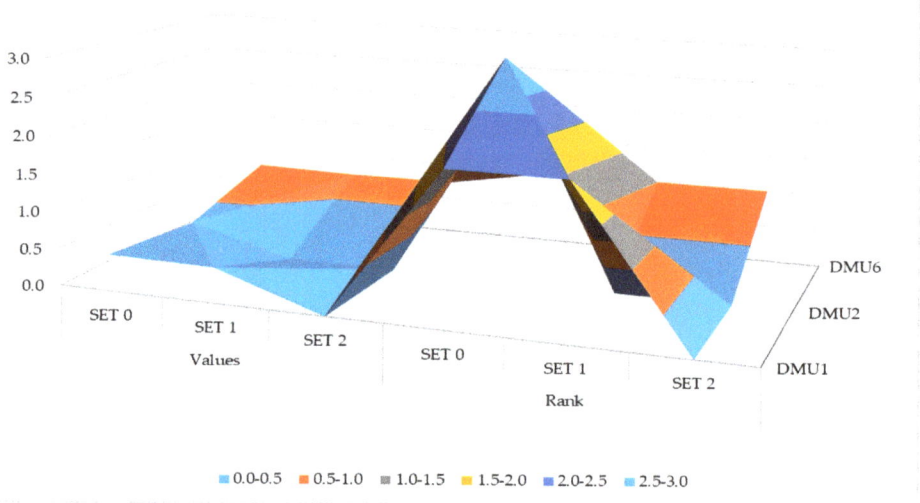

Figure 7. Results after resizing the fuzzy initial matrix.

Observing the results in Figure 7, was can see that the change in the size of the fuzzy initial matrix has absolutely no effect on the rankings, as the rankings do not change in either of the two scenarios formed.

7. Conclusions

Through this study we defined a new integrated fuzzy model that combines several different approaches in the field of decision making, and a combination of this with the DEA model for initial determination of road section safety based on the geometric exploitation of road parameters: section length, road slope, speed deviation from the speed limit, and AADT as input parameters. The output parameters in the DEA model represent different classifications of traffic accidents. The overall research was previously presented in detail through three phases. The greatest contributions of this research can be observed from two aspects: scientific and professional. From a scientific aspect, the main, and probably the most important contribution, is the development of the improved fuzzy SWARA method (IMF SWARA), overcomes the disadvantages of the fuzzy SWARA method in the following ways:

(1) Using the fuzzy SWARA method, it is impossible to obtain results in which two criteria have equal fuzzy weights. By applying the improved fuzzy SWARA method, two or more criteria can have equal values.

(2) On the contrary, applying the inadequate TFN scale shown in Table 2, where decision-makers indicate that two criteria have the same value by assigning TFN (1,1,1), the criterion C_j in relation to C_{j-1} received a value that is twice less than C_j. By applying the improved fuzzy SWARA method, assigning the value (0,0,0), equal values are obtained and not values twice as large.

(3) By increasing the number of criteria in the model, the least significant criteria receive values that can be negligible, i.e., with a tendency to zero. By applying the improved fuzzy SWARA method, less significant criteria have higher values and can play a greater role in the decision-making process.

Then, the integration of the developed IMF SWARA method with the fuzzy MARCOS method was performed, which is also a contribution of this paper. In order to achieve greater precision in the creation of input data, the originality of the given model was manifested by the application of two aggregators: Dombi and Bonferroni, which were used to average the input-output parameters. The creation and verification of such a model certainly represents a contribution to the overall literature that considers multi-criteria problems. If we take into account that according to Salabun et. al. [50] in many studies, the problem of selecting the proper method and parameters for decision-making is raised, we can conclude that our integrated fuzzy model represents a good solution. This has been proved through the aforementioned scientific contributions of the developed model.

From a professional perspective, the contribution of this research is reflected through the quantification of the safety degree of certain sections of road infrastructure, which is an important parameter in practice. Taking into account the previously stated parameters with a focus on passenger cars exceeding the speed limit, the road sections from the considered set that represent risk have been defined, indicating that a corrective measure in traffic management is necessary.

The limitations of this study can be manifested through the small number of considered road sections and the inclusion of input/output parameters without the performance principal component analysis. The results calculated using the original IMF SWARA–Fuzzy MARCOS model indicate that the sixth section of the observed two-lane roads shows the highest level of traffic safety with a value of 0.748, taking into account the complexity of the set model. The next section is the DMU2 with a value of 0.494, and in the third position is section DMU2, while the remaining three sections, DMU3, DMU4, and DMU5, have a higher risk of side effects, i.e., traffic accidents. After applying the original IMF SWARA—Fuzzy MARCOS model, a sensitivity analysis was performed, through which the overall stability of the model was proven.

Future research may need to be conducted on a number of road sections with the possibility of implementing the proposed model. The existing model, with an increase of input parameters, could contribute to a much more selective level of choosing the rank of road sections. It is also possible for the number of access points on each section, the radius of the road, etc., to be added as input parameters. Additionally, by applying the existing model through observing other input parameters, an extensive analysis in ranking the safety level of the two-lane roads in terms of multi-criteria traffic analysis can be obtained. Besides, a PCA–DEA analysis [51] can be applied to determine the efficiency of the considered road sections or to implement the group MCDM model [52] along with some different uncertainty theories such multi-granular unbalanced linguistic information [53], intuitionistic 2-tuple linguistic sets [54], or grey theory [55]. Considering the approach presented in [56] can be useful in developing future integrated decision-making models.

Author Contributions: Conceptualization, S.V. and E.S.; methodology, Ž.S. and M.S.; formal analysis, M.S. and M.P.; data curation, M.S. and E.S.; writing—original draft preparation, Ž.S. and S.V.; writing—review and editing, M.D. and M.P.; supervision, M.D. and M.P.; All authors have read and agreed to the published version of the manuscript.

Funding: This research received no external funding.

Informed Consent Statement: Not applicable.

Data Availability Statement: Not applicable.

Conflicts of Interest: The authors declare no conflict of interest.

References

1. Mayora, J.P.; Rubio, R.L. Relevant Variables for Crash Rate Prediction in Spain´s Two Lane Rural Roads. In Proceedings of the 82nd Transportation Research Board Annual Meeting, Washington, DC, USA, 12–16 January 2003.
2. Qiao, J.; Wen, Y.; Yang, N.; Song, J. The research of two-lane highway longitudinal slope based on the running speed in the plateau areas. In Proceedings of the 2011 International Conference on Consumer Electronics, Communications and Networks (CECNet), Xianning, China, 16–18 April 2011; pp. 1827–1830. [CrossRef]
3. Hashim, I.H. Analysis of speed characteristics for rural two-lane roads: A field study from Minoufiya Governorate, Egypt. *Ain Shams Eng. J.* **2011**, *2*, 43–52. [CrossRef]
4. D'Andrea, A.; Carbone, F.; Salviera, S.; Pellegrino, O. The Most Influential Variables in the Determination of V85 Speed. *Procedia Soc. Behav. Sci.* **2012**, *53*, 633–644. [CrossRef]
5. Porter, R.J.; Donnell, E.T.; Mason, J.M. Geometric Design, Speed, and Safety. *Transp. Res. Rec.* **2012**, *2309*, 39–47. [CrossRef]
6. *Highway Capacity Manual 2016 (HCM 2016)*; TRB. National Research Council: Washington, DC, USA, 2016.
7. *Handbuch für die Bemessung von Straßenverkehrsanlagen (HBS) (German Highway Capacity Manual)*; Forschungsgesellschaft für Straßen-und Verkehrswesen: Köln, Germany, 2005.
8. AASHTO. *The Highway Safety Manual*; American Association of State Highway Transportation Professionals: Washington, DC, USA, 2010.
9. Aarts, L.; van Schagen, I. Driving speed and the risk of road crashes: A review. *Accid. Anal. Prev.* **2006**, *38*, 215–224. [CrossRef] [PubMed]
10. Cruzado, I.; Donnell, E.T. Factors Affecting Driver Speed Choice along Two-Lane Rural Highway Transition Zones. *J. Transp. Eng.* **2010**, *136*, 755–764. [CrossRef]
11. Cheng, G.; Cheng, R.; Pei, Y.; Xu, L. Probability of Roadside Accidents for Curved Sections on Highways. *Math. Probl. Eng.* **2020**, *2020*, 1–18. [CrossRef]
12. Zheng, Y.; Guo, H.; Wei, X. The Evaluation Analysis of Design Code About the Road Design of longitudinal gradient in the Mountain Road. In Proceedings of the 7th International Conference on Education, Management, Computer and Society (EMCS 2017), Shenyang, China, 17–19 March 2017; pp. 693–699. [CrossRef]
13. Srnová, B. A Case of Road Design in Mountainous Terrain with an Evaluation of Heavy Vehicles Performance Stockholm. Master's Thesis, KTH Royal Institute of Technology, Stockholm, Sweden, 2017.
14. Fitzpatrick, K.; Schneider, W.H.; Park, E.S. *Comparisons of Crashes on Rural Two-Lane and Four-Lane Highways in Texas*; FHWA/TX-06/0-4618-1; Report 0-4618-1; Texas Transportation Institute: Bryan, TX, USA, 2005.
15. Calvo-Poyo, F.; de Oña, J.; Garach Morcillo, L.; Navarro-Moreno, J. Influence of Wider Longitudinal Road Markings on Vehicle Speeds in Two-Lane Rural Highways. *Sustainability* **2020**, *12*, 8305. [CrossRef]
16. Yue, L.; Wang, H. An Optimization Design Method of Combination of Steep Slope and Sharp Curve Sections for Mountain Highways. *Math. Probl. Eng.* **2019**, *2019*, 1–13. [CrossRef]
17. Kazemzadehazad, S.; Monajjem, S.; Larue, G.; King, M.J. Driving simulator validation for speed research on curves of two-lane rural roads. In *Proceedings of the Institution of Civil Engineers—Transport*; Thomas Telford Ltd.: London, UK, 2018; pp. 1–18. [CrossRef]
18. Abdollahzadeh Nasiri, A.S.; Rahmani, O.; Abdi Kordani, A.; Karballaeezadeh, N.; Mosavi, A. Evaluation of Safety in Horizontal Curves of Roads Using a Multi-Body Dynamic Simulation Process. *Int. J. Environ. Res. Public Health* **2020**, *17*, 5975. [CrossRef]
19. Khabiri, M.M.; Ghaforifard, Z. The Effect of Low Friction in Pavement Due to Floods and High-speed Vehicles in Increasing the Number of Rescue Vehicles' Driving Accidents. *Health Emerg. Disasters* **2020**, *6*, 29–38. [CrossRef]
20. Sil, G.; Maji, A.; Nama, S.; Maurya, A.K. Operating speed prediction model as a tool for consistency based geometric design of four-lane divided highways. *Transport* **2019**, *34*, 425–436. [CrossRef]
21. Alper, D.; Sinuany-Stern, Z.; Shinar, D. Evaluating the efficiency of local municipalities in providing traffic safety using the Data Envelopment Analysis. *Accid. Anal. Prev.* **2015**, *78*, 39–50. [CrossRef]
22. Podvezko, V.; Sivilevičius, H. The use of AHP and rank correlation methods for determining the significance of the interaction between the elements of a transport system having a strong influence on traffic safety. *Transport* **2013**, *28*, 389–403. [CrossRef]
23. Pirdavani, A.; Brijs, T.; Wets, G. A Multiple Criteria Decision-Making Approach for Prioritizing Accident Hotspots in the Absence of Crash Data. *Transp. Rev.* **2010**, *30*, 97–113. [CrossRef]
24. Barić, D.; Pilko, H.; Strujić, J. An analytic hierarchy process model to evaluate road section design. *Transport* **2016**, *31*, 312–321. [CrossRef]
25. Pilko, H.; Mandžuka, S.; Barić, D. Urban single-lane roundabouts: A new analytical approach using multi-criteria and simultaneous multi-objective optimization of geometry design, efficiency and safety. *Transp. Res. Part C Emerg. Technol.* **2017**, *80*, 257–271. [CrossRef]
26. Zak, J. The methodology of multiple criteria decision making/aiding in public transportation. *J. Adv. Transp.* **2010**, *45*, 1–20. [CrossRef]

27. Krstić, M.D.; Tadić, S.R.; Brnjac, N.; Zečević, S. Intermodal Terminal Handling Equipment Selection Using a Fuzzy Multi-criteria Decision-making Model. *Promet Traffic Transp.* **2019**, *31*, 89–100. [CrossRef]
28. Vakilipour, S.; Sadeghi-Niaraki, A.; Ghodousi, M.; Choi, S.-M. Comparison between Multi-Criteria Decision-Making Methods and Evaluating the Quality of Life at Different Spatial Levels. *Sustainability* **2021**, *13*, 4067. [CrossRef]
29. Yannis, G.; Kopsacheili, A.; Dragomanovits, A.; Petraki, V. State-of-the-art review on multi-criteria decision-making in the transport sector. *J. Traffic Transp. Eng. (Engl. Ed.)* **2020**, *7*, 413–431. [CrossRef]
30. Memiş, S.; Demir, E.; Karamaşa, Ç.; Korucuk, S. Prioritization of road transportation risks: An application in Giresun province. *Oper. Res. Eng. Sci. Theory Appl.* **2020**, *3*, 111–126. [CrossRef]
31. Đalić, I.; Ateljević, J.; Stević, Ž.; Terzić, S. An integrated swot–fuzzy piprecia model for analysis of competitiveness in order to improve logistics performances. *Facta Univ. Ser. Mech. Eng.* **2020**, *18*, 439–451. [CrossRef]
32. Vesković, S.; Milinković, S.; Abramović, B.; Ljubaj, I. Determining criteria significance in selecting reach stackers by applying the fuzzy PIPRECIA method. *Oper. Res. Eng. Sci. Theory Appl.* **2020**, *3*, 72–88. [CrossRef]
33. Pamucar, D. Normalized weighted Geometric Dombi Bonferoni Mean Operator with interval grey numbers: Application in multicriteria decision making. *Rep. Mech. Eng.* **2020**, *1*, 44–52. [CrossRef]
34. Yager, R.R. On generalized Bonferroni mean operators for multi-criteria aggregation. *Int. J. Approx. Reason.* **2009**, *50*, 1279–1286. [CrossRef]
35. Mavi, R.K.; Goh, M.; Zarbakhshnia, N. Sustainable third-party reverse logistic provider selection with fuzzy SWARA and fuzzy MOORA in plastic industry. *Int. J. Adv. Manuf. Technol.* **2017**, *91*, 2401–2418. [CrossRef]
36. Chang, D.Y. Applications of the extent analysis method on fuzzy AHP. *Eur. J. Oper. Res.* **1996**, *95*, 649–655. [CrossRef]
37. Zarbakhshnia, N.; Soleimani, H.; Ghaderi, H. Sustainable third-party reverse logistics provider evaluation and selection using fuzzy SWARA and developed fuzzy COPRAS in the presence of risk criteria. *Appl. Soft Comput.* **2018**, *65*, 307–319. [CrossRef]
38. Ulutaş, A.; Karakuş, C.B.; Topal, A. Location selection for logistics center with fuzzy SWARA and CoCoSo methods. *J. Intell. Fuzzy Syst.* **2020**, *38*, 4693–4709. [CrossRef]
39. Ansari, Z.N.; Kant, R.; Shankar, R. Evaluation and ranking of solutions to mitigate sustainable remanufacturing supply chain risks: A hybrid fuzzy SWARA-fuzzy COPRAS framework approach. *Int. J. Sustain. Eng.* **2020**, *13*, 473–494. [CrossRef]
40. Ghasemi, P.; Mehdiabadi, A.; Spulbar, C.; Birau, R. Ranking of Sustainable Medical Tourism Destinations in Iran: An Integrated Approach Using Fuzzy SWARA-PROMETHEE. *Sustainability* **2021**, *13*, 683. [CrossRef]
41. Mitrović Simić, J.; Stević, Ž.; Zavadskas, E.K.; Bogdanović, V.; Subotić, M.; Mardani, A. A Novel CRITIC-Fuzzy FUCOM-DEA-Fuzzy MARCOS Model for Safety Evaluation of Road Sections Based on Geometric Parameters of Road. *Symmetry* **2020**, *12*, 2006. [CrossRef]
42. Blagojević, A.; Vesković, S.; Kasalica, S.; Gojić, A.; Allamani, A. The application of the fuzzy AHP and DEA for measuring the efficiency of freight transport railway undertakings. *Oper. Res. Eng. Sci. Theory Appl.* **2020**, *3*, 1–23. [CrossRef]
43. Andrejić, M.; Kilibarda, M.; Pajić, V. Measuring efficiency change in time applying malmquist productivity index: A case of distribution centres in Sserbia. *Facta Univ. Ser. Mech. Eng.* **2021**.
44. Stanković, M.; Stević, Ž.; Das, D.K.; Subotić, M.; Pamučar, D. A new fuzzy MARCOS method for road traffic risk analysis. *Mathematics* **2020**, *8*, 457. [CrossRef]
45. Bakır, M.; Atalık, Ö. Application of Fuzzy AHP and Fuzzy MARCOS Approach for the Evaluation of E-Service Quality in the Airline Industry. *Decis. Mak. Appl. Manag. Eng.* **2021**, *4*, 127–152. [CrossRef]
46. Zahir, S. Normalisation and rank reversals in the additive analytic hierarchy process: A new analysis. *Int. J. Oper. Res.* **2009**, *4*, 446–467. [CrossRef]
47. Petrović, G.; Mihajlović, J.; Ćojbašić, Ž.; Madić, M.; Marinković, D. Comparison of three fuzzy MCDM methods for solving the supplier selection problem. *Facta Univ. Ser. Mech. Eng.* **2019**, *17*, 455–469. [CrossRef]
48. Roszkowska, E.; Kacprzak, D. The fuzzy saw and fuzzy TOPSIS procedures based on ordered fuzzy numbers. *Inf. Sci.* **2016**, *369*, 564–584. [CrossRef]
49. Hassanpour, M. Evaluation of Iranian wood and cellulose industries. *Decis. Mak. Appl. Manag. Eng.* **2019**, *2*, 13–34. [CrossRef]
50. Sałabun, W.; Wątróbski, J.; Shekhovtsov, A. Are MCDA Methods Benchmarkable? A Comparative Study of TOPSIS, VIKOR, COPRAS, and PROMETHEE II Methods. *Symmetry* **2020**, *12*, 1549. [CrossRef]
51. Andrejić, M.M.; Kilibarda, M.J. Measuring global logistics efficiency using PCA-DEA approach. *Tehnika* **2016**, *71*, 733–740. [CrossRef]
52. Zhang, Z.; Gao, Y.; Li, Z. Consensus reaching for social network group decision making by considering leadership and bounded confidence. *Knowl. Based Syst.* **2020**, *204*, 106240. [CrossRef]
53. Zhang, Z.; Li, Z.; Gao, Y. Consensus reaching for group decision making with multi-granular unbalanced linguistic information: A bounded confidence and minimum adjustment-based approach. *Inf. Fusion* **2021**, *74*, 96–110. [CrossRef]
54. Faizi, S.; Sałabun, W.; Nawaz, S. Best-Worst method and Hamacher aggregation operations for intuitionistic 2-tuple linguistic sets. *Expert Syst. Appl.* **2021**, 115088. [CrossRef]
55. Badi, I.; Pamucar, D. Supplier selection for steelmaking company by using combined Grey-MARCOS methods. *Decis. Mak. Appl. Manag. Eng.* **2020**, *3*, 37–48. [CrossRef]
56. Zhang, Z.; Kou, X.; Yu, W.; Gao, Y. Consistency improvement for fuzzy preference relations with self-confidence: An application in two-sided matching decision making. *J. Oper. Res. Soc.* **2020**, 1–14. [CrossRef]

Article

Regularization of the Ill-Posed Cauchy Problem for Matrix Factorizations of the Helmholtz Equation on the Plane

Davron Aslonqulovich Juraev [1,*] and Samad Noeiaghdam [2,3]

1. Higher Military Aviation School of the Republic of Uzbekistan, Karshi City 180100, Uzbekistan
2. Industrial Mathematics Laboratory, Baikal School of BRICS, Irkutsk National Research Technical University, 664074 Irkutsk, Russia; snoei@istu.edu or noiagdams@susu.ru
3. Department of Applied Mathematics and Programming, South Ural State University, Lenin Prospect 76, 454080 Chelyabinsk, Russia
* Correspondence: juraev_davron@list.ru

Abstract: In this paper, we present an explicit formula for the approximate solution of the Cauchy problem for the matrix factorizations of the Helmholtz equation in a bounded domain on the plane. Our formula for an approximate solution also includes the construction of a family of fundamental solutions for the Helmholtz operator on the plane. This family is parameterized by function $K(w)$ which depends on the space dimension. In this paper, based on the results of previous works, the better results can be obtained by choosing the function $K(w)$.

Keywords: cauchy problem; regularization; factorization; regular solution; fundamental solution

MSC: 35J46; 65J20

Citation: Juraev, D.A.; Noeiaghdam, S. Regularization of the Ill-Posed Cauchy Problem for Matrix Factorizations of the Helmholtz Equation on the Plane. *Axioms* **2021**, *10*, 82. https://doi.org/10.3390/axioms10020082

Academic Editors: Chris Goodrich and Giampiero Palatucci

Received: 20 March 2021
Accepted: 28 April 2021
Published: 2 May 2021

Publisher's Note: MDPI stays neutral with regard to jurisdictional claims in published maps and institutional affiliations.

Copyright: © 2021 by the authors. Licensee MDPI, Basel, Switzerland. This article is an open access article distributed under the terms and conditions of the Creative Commons Attribution (CC BY) license (https://creativecommons.org/licenses/by/4.0/).

1. Introduction

The paper studies the construction of the exact and approximate solutions of the ill-posed Cauchy problem for matrix factorizations of the Helmholtz equation. Such problems naturally arise in mathematical physics and in various fields of natural science such as electro-geological exploration, cardiology, electrodynamics and so on. In general, the theory of ill-posed problems for elliptic system of equations has been sufficiently formed by Tikhonov, Ivanov, Lavrent'ev and Tarkhanov in [1–5]. Among them, the most important and applicable topic is related to the conditionally well-posed problems, characterized by stability in the presence of additional information about the nature of the problem data. One of the most effective ways to study such problems is to construct the regularizing operators. For example, it can be done by the Carleman-type formulas (as in complex analysis) or iterative processes (the Kozlov–Maz'ya–Fomin algorithm, etc.).

The work is devoted to the main problem for partial differential equations, which is the Cauchy problem. The main aim of this study is to find the regularization formulas to find the solutions of the Cauchy problem for matrix factorizations of the Helmholtz equation. The question of the existence of a solution of the problem is not considered—it is assumed a priori. At the same time, it should be noted that any regularization formula leads to an approximate solution of the Cauchy problem for all data, even if there is no solution in the usual classic sense. Moreover, for explicit regularization formulas, the optimal solution can be obtained. In this sense, exact regularization formulas are very useful for real numerical calculations. There is good reason to hope that numerous practical applications of regularization formulas are still ahead. In [6–8] some applications of the Cauchy problem and the regularization technique for solving different kinds of integral equations have been presented.

The Cauchy problem for matrix factorizations of the Helmholtz equation is among ill-posed and unstable problems. It is known that the Cauchy problem for elliptic equations is among unstable problems which by a small change in the data the problem will be incorrect [1,4,9–13]. Tarkhanov [14] has published a criterion for the solvability of a large class of boundary value problems for elliptic systems. In some cases of unstable problems, we should apply some operators for solving the problem. But the image of these operators are not closed, therefore, the solvability condition can not be written in terms of continuous linear functions. So, in the Cauchy problem for elliptic equations with data on part of the boundary of the domain the solution is usually unique and the problem is solvable for everywhere dense a set of data, but this set is not closed. Consequently, the theory of solvability of such problems is much more difficult and deeper than theory of solvability of Fredholm equations. The first results in this direction appeared only in the mid-1980s in the works of Aizenberg, Kytmanov and Tarkhanov [5].

The uniqueness of the solution follows from Holmgren's general theorem [10]. The conditional stability of the problem follows from the work of Tikhonov [1], if we restrict the class of possible solutions to a compactum.

Formulas that allow finding a solution to an elliptic equation in the case when the Cauchy data are known only on a part of the boundary of the domain are called Carleman type formulas. In [13], Carleman established a formula giving a solution to the Cauchy–Riemann equations in a a special form of a domain. Developing his idea, Goluzin and Krylov [15] derived a formula to determine the values of analytic functions from known data. A multidimensional analogue of Carleman's formula for analytic functions of several variables was constructed in [11]. The Carleman formula to find the solution of the differential operator with special properties can be found in [3,4]. Yarmukhamedov [16–19] applied this method to construct the Carleman functions for the Laplace and Helmholtz operators for special form and domain. In [5] an integral formula was proved for the first order elliptic type system of equations with constant coefficients in a bounded domain. In [20], Ikehata applied the presented methodologies in [16–19] to consider the probe method and Carleman functions for the Laplace and Helmholtz equations in the three-dimensional domain. In [21], a formula for solving the Helmholtz equation with a variable coefficient for regions in space where the unknown data are located on a section of the hypersurface $\{x \cdot s = t\}$ has been presented by Ikehata.

Carleman type formulas for various elliptic equations and systems were also obtained in [5,14–28] and [29–41]. In [22] the Cauchy problem for the Helmholtz equation in an arbitrary bounded plane domain with Cauchy data which is known only on the boundary region was discussed. The solvability criterion of the Cauchy problem for the Laplace equation in $\mathbb{R}^m$ was considered by Shlapunov [25]. In [42], the continuation of the Helmholtz equation was investigated and the results of the numerical experiments were presented.

The construction of the Carleman matrix for elliptic systems was carried out by Yarmukhamedov, Tarkhanov, Shlapunov, Niyozov, Juraev and others [5,14,16–19,23–41]. The system considered in this paper was introduced by Tarkhanov. For this system, he studied correct boundary value problems and found an analogue of the Cauchy integral formula in a bounded domain (see, for instance [5]).

In many well-posed problems of the system of equations of the first order elliptic type with constant coefficients that factorize the Helmholtz operator, calculating the values of the vector function on the entire boundary is not possible. Therefore, the problem of reconstructing the solution of system of equations of the first order elliptic type with constant coefficients and factorizing the Helmholtz operator [29–41] are among the more challenging problems in the theory of differential equations.

Additionally, the ill-posed problems of mathematical physics have been investigated by many researchers. The properties of solutions of the Cauchy problem for the Laplace equation were studied in [3,4,16–19] and subsequently developed in [5,14,15,20–41].

Let $\mathbb{R}^2$ be the two-dimensional real Euclidean space,

$$x = (x_1, x_2) \in \mathbb{R}^2, \ y = (y_1, y_2) \in \mathbb{R}^2.$$

$G \subset \mathbb{R}^2$ is a bounded simply-connected domain with piecewise smooth boundary consisting of the plane $T: y_2 = 0$ and some smooth curve S lying in the half-space $y_2 > 0$, i.e., $\partial G = S \bigcup T$.

We introduce the following notation:

$$r = |y - x|, \alpha = |y_1 - x_1|, w = i\sqrt{u^2 + \alpha^2} + y_2, u \geq 0,$$

$$\frac{\partial}{\partial x} = \left(\frac{\partial}{\partial x_1}, \frac{\partial}{\partial x_2}\right)^T, \frac{\partial}{\partial x} \to \xi^T, \xi^T = \begin{pmatrix} \xi_1 \\ \xi_2 \end{pmatrix} \text{ which is a transposed vector of } \xi,$$

$$U(x) = (U_1(x), ..., U_n(x))^T, u^0 = (1, ..., 1) \in \mathbb{R}^n, n = 2^m, m = 2,$$

$$E(z) = \begin{Vmatrix} z_1 ... 0 \\ ...... \\ 0 ... z_n \end{Vmatrix} - \text{diagonal matrix}, z = (z_1, ..., z_n) \in \mathbb{R}^n.$$

Let $D(\xi^T)$ be a $(n \times n)$-dimensional matrix with elements consisting of a set of linear functions with constant coefficients of the complex plane which is satisfied in the following condition

$$D^*(\xi^T)D(\xi^T) = E((|\xi|^2 + \lambda^2)u^0),$$

where $D^*(\xi^T)$ is the Hermitian conjugate matrix $D(\xi^T)$ and λ is a real number.

We consider the following system of differential equations

$$D\left(\frac{\partial}{\partial x}\right)U(x) = 0, \tag{1}$$

in the region G where $D\left(\frac{\partial}{\partial x}\right)$ is the matrix of first-order differential operators.

We denote the class of vector functions in the domain G by $A(G)$ which is continuous on $\overline{G} = G \bigcup \partial G$ and satisfy in the system (1).

2. Construction of the Carleman Matrix and the Cauchy Problem

Formulation of the problem: suppose $U(y) \in A(G)$ and

$$U(y)|_S = f(y), y \in S, \tag{2}$$

where $f(y)$ is a given continuous vector-function on S. It is required to note that the vector function $U(y)$ is in the domain G, based on $f(y)$ on S.

If $U(y) \in A(G)$, then the following Cauchy type integral formula

$$U(x) = \int_{\partial G} N(y, x; \lambda)U(y)ds_y, \quad x \in G, \tag{3}$$

is valid and

$$N(y, x; \lambda) = \left(E\left(\varphi_2(\lambda r)u^0\right)D^*\left(\frac{\partial}{\partial x}\right)\right)D(t^T),$$

where $t = (t_1, t_2)$ shows the unit exterior normal which is drawn at point y on curve ∂G and $\varphi_2(\lambda r)$ denotes the fundamental solution of the Helmholtz equation in $\mathbb{R}^2$, which is defined in the following form

$$\varphi_2(\lambda r) = -\frac{i}{4}H_0^{(1)}(\lambda r). \tag{4}$$

Here $H_0^{(1)}(\lambda r)$ is the the Hankel function of the first kind [12].

We introduce $K(w)$ as an entire function which takes real values as real part of w, ($w = u + iv$, u, v—real numbers) and satisfies in the following conditions:

$$K(u) \neq 0, \ \sup_{v \geq 1} \left| v^p K^{(p)}(w) \right| = B(u, p) < \infty, \tag{5}$$

$$-\infty < u < \infty, \ p = 0, 1, 2.$$

We define the function $\Phi(y, x; \lambda)$ at $y \neq x$ by the following equality

$$\Phi(y, x; \lambda) = -\frac{1}{2\pi K(x_2)} \int_0^\infty \operatorname{Im} \left[\frac{K(w)}{w - x_2} \right] \frac{u I_0(\lambda u)}{\sqrt{u^2 + \alpha^2}} du, \tag{6}$$

where $I_0(\lambda u) = J_0(i\lambda u)$—is the zero order Bessel function of the first kind [10].
In the Formula (6), choosing

$$K(w) = \exp(\sigma w^2), \ K(x_2) = \exp(\sigma x_2^2), \ \sigma > 0, \tag{7}$$

we get

$$\Phi_\sigma(y, x; \lambda) = -\frac{e^{-\sigma x_2^2}}{2\pi} \int_0^\infty \operatorname{Im} \left[\frac{\exp(\sigma w^2)}{w - x_2} \right] \frac{u I_0(\lambda u)}{\sqrt{u^2 + \alpha^2}} du. \tag{8}$$

Substituting

$$\Phi_\sigma(y, x; \lambda) = \varphi_2(\lambda r) + g_\sigma(y, x; \lambda), \tag{9}$$

in Equation (3) instead of $\varphi_2(\lambda r)$, the formula will be correct where $g_\sigma(y, x)$ is the regular solution of the Helmholtz equation with respect to the variable y, including the point $y = x$.
Then the integral formula can written in the follwoing form:

$$U(x) = \int_{\partial G} N_\sigma(y, x; \lambda) U(y) ds_y, \ x \in G, \tag{10}$$

where

$$N_\sigma(y, x; \lambda) = \left(E \left(\Phi_\sigma(y, x; \lambda) u^0 \right) D^* \left(\frac{\partial}{\partial x} \right) \right) D(t^T).$$

3. The Continuation Formula and Regularization According to M.M. Lavrent'ev's

Theorem 1. *Let $U(y) \in A(G)$ satisfy in the following inequality*

$$|U(y)| \leq M, \ y \in T. \tag{11}$$

If

$$U_\sigma(x) = \int_S N_\sigma(y, x; \lambda) U(y) ds_y, \ x \in G, \tag{12}$$

then the following estimations are correct:

$$|U(x) - U_\sigma(x)| \leq C(\lambda, x) \sigma M e^{-\sigma x_2^2}, \ \sigma > 1, \ x \in G, \tag{13}$$

$$\left| \frac{\partial U(x)}{\partial x_j} - \frac{\partial U_\sigma(x)}{\partial x_j} \right| \leq C(\lambda, x) \sigma M e^{-\sigma x_2^2}, \ \sigma > 1, \ x \in G, \ j = 1, 2, \tag{14}$$

where $C(\lambda, x)$ shows the bounded functions on compact subsets of the domain G.

Proof. Let us first estimate inequality (13). Using the integral Formula (10) and the equality (12), we obtain

$$U(x) = \int_S N_\sigma(y, x; \lambda) U(y) ds_y + \int_T N_\sigma(y, x; \lambda) U(y) ds_y$$

$$= U_\sigma(x) + \int_T N_\sigma(y, x; \lambda) U(y) ds_y, \quad x \in G.$$

Taking into account the inequality (11), we estimate the following

$$|U(x) - U_\sigma(x)| \le \left| \int_T N_\sigma(y, x; \lambda) U(y) ds_y \right| \tag{15}$$

$$\le \int_T |N_\sigma(y, x; \lambda)||U(y)| ds_y \le M \int_T |N_\sigma(y, x; \lambda)| ds_y, \quad x \in G.$$

For this aim, we estimate the integrals $\int_T |\Phi_\sigma(y, x; \lambda)| ds_y$, $\int_T \left| \dfrac{\partial \Phi_\sigma(y, x; \lambda)}{\partial y_1} \right| ds_y$, and

$\int_T \left| \dfrac{\partial \Phi_\sigma(y, x; \lambda)}{\partial y_2} \right| ds_y$ on the part T of the plane $y_2 = 0$.

Separating the imaginary part of (8), we obtain

$$\Phi_\sigma(y, x; \lambda) = \frac{e^{\sigma(y_2^2 - x_2^2)}}{2\pi} \left[\int_0^\infty \frac{e^{-\sigma(u^2 + \alpha^2)} \cos \sigma \sqrt{u^2 + \alpha^2}}{u^2 + r^2} u I_0(\lambda u) du \right. \tag{17}$$

$$\left. - \int_0^\infty \frac{e^{-\sigma(u^2 + \alpha^2)} (y_2 - x_2) \sin \sigma \sqrt{u^2 + \alpha^2}}{u^2 + r^2} \frac{u I_0(\lambda u)}{\sqrt{u^2 + \alpha^2}} du \right], \quad x_2 > 0.$$

Given (16) and the inequality

$$I_0(\lambda u) \le \sqrt{\frac{2}{\lambda \pi u}}, \tag{17}$$

we have

$$\int_T |\Phi_\sigma(y, x; \lambda)| ds_y \le C(\lambda, x) \sigma e^{-\sigma x_2^2}, \quad \sigma > 1, \ x \in G. \tag{18}$$

To estimate the second integral, we use the equality

$$\frac{\partial \Phi_\sigma(y, x; \lambda)}{\partial y_1} = \frac{\partial \Phi_\sigma(y, x; \lambda)}{\partial s} \frac{\partial s}{\partial y_1} = 2(y_1 - x_1) \frac{\partial \Phi_\sigma(y, x; \lambda)}{\partial s}, \tag{19}$$

$$s = \alpha^2.$$

Given equality (16), inequality (17) and equality (19), we obtain

$$\int_T \left| \frac{\partial \Phi_\sigma(y, x; \lambda)}{\partial y_1} \right| ds_y \le C(\lambda, x) \sigma e^{-\sigma x_2^2}, \quad \sigma > 1, \ x \in G, \tag{20}$$

Now, we estimate the integral $\int_T \left| \dfrac{\partial \Phi_\sigma(y, x; \lambda)}{\partial y_2} \right| ds_y$.

Taking into account equality (16) and inequality (17), we obtain

$$\int_T \left|\frac{\partial \Phi_\sigma(y,x;\lambda)}{\partial y_2}\right| ds_y \leq C(\lambda,x)\sigma e^{-\sigma x_2^2}, \ \sigma > 1, \ x \in G, \qquad (21)$$

From inequalities (17), (20) and (21), bearing in mind (15), we obtain an estimate (13).
Now the inequality (14) can be proved. To do this, we take the derivatives from equalities (10) and (12) with respect to x_j, $(j=1,2)$ then we get:

$$\frac{\partial U(x)}{\partial x_j} = \int_S \frac{\partial N_\sigma(y,x;\lambda)}{\partial x_j} U(y) ds_y + \int_T \frac{\partial N_\sigma(y,x;\lambda)}{\partial x_j} U(y) ds_y,$$

$$\frac{\partial U_\sigma(x)}{\partial x_j} = \int_S \frac{\partial N_\sigma(y,x;\lambda)}{\partial x_j} U(y) ds_y, \ x \in G, \ j = 1,2. \qquad (22)$$

Taking into account the (22) and inequality (11), we estimate the following

$$\left|\frac{\partial U(x)}{\partial x_j} - \frac{\partial_\sigma U(x)}{\partial x_j}\right| \leq \left|\int_T \frac{\partial N_\sigma(y,x;\lambda)}{\partial x_j} U(y) ds_y\right|$$

$$\leq \int_T \left|\frac{\partial N_\sigma(y,x;\lambda)}{\partial x_j}\right| |U(y)| ds_y \leq M \int_T \left|\frac{\partial N_\sigma(y,x;\lambda)}{\partial x_j}\right| ds_y, \qquad (23)$$

$$x \in G, \ j=1,2.$$

To do this, we estimate the integrals $\int_T \left|\frac{\partial \Phi_\sigma(y,x;\lambda)}{\partial x_1}\right| ds_y$ and $\int_T \left|\frac{\partial \Phi_\sigma(y,x;\lambda)}{\partial x_2}\right| ds_y$ on the part T of the plane $y_2 = 0$.

To estimate the first integrals, we use the equality

$$\frac{\partial \Phi_\sigma(y,x;\lambda)}{\partial x_1} = \frac{\partial \Phi_\sigma(y,x;\lambda)}{\partial s} \frac{\partial s}{\partial x_1} = -2(y_1-x_1)\frac{\partial \Phi_\sigma(y,x;\lambda)}{\partial s}, \qquad (24)$$

$$s = \alpha^2.$$

Applying equality (16), inequality (17) and equality (24), we obtain

$$\int_T \left|\frac{\partial \Phi_\sigma(y,x;\lambda)}{\partial x_1}\right| ds_y \leq C(\lambda,x)\sigma e^{-\sigma x_2^2}, \ \sigma > 1, \ x \in G. \qquad (25)$$

Now, we estimate the integral $\int_T \left|\frac{\partial \Phi_\sigma(y,x;\lambda)}{\partial x_2}\right| ds_y$.

Taking into account equality (16) and inequality (17), we obtain

$$\int_T \left|\frac{\partial \Phi_\sigma(y,x;\lambda)}{\partial x_2}\right| ds_y \leq C(\lambda,x)\sigma e^{-\sigma x_2^2}, \ \sigma > 1, \ x \in G. \qquad (26)$$

From inequalities (25) and (26), bearing in mind (23), we obtain an estimate of (14). Theorem 1 is proved. □

Corollary 1. *For each $x \in G$, the equalities are true*

$$\lim_{\sigma \to \infty} U_\sigma(x) = U(x), \ \lim_{\sigma \to \infty} \frac{\partial U_\sigma(x)}{\partial x_j} = \frac{\partial U(x)}{\partial x_j}, \ j = 1,2.$$

We define $\overline{G}_\varepsilon$ as

$$\overline{G}_\varepsilon = \left\{ (x_1, x_2) \in G,\ a > x_2 \geq \varepsilon,\ a = \max_T \psi(x_1),\ 0 < \varepsilon < a \right\}.$$

Here $\psi(x_1)$-is a curve. It is easy to see that the set $\overline{G}_\varepsilon \subset G$ is compact.

Corollary 2. *If $x \in \overline{G}_\varepsilon$, then the families of functions $\{U_\sigma(x)\}$ and $\left\{\dfrac{\partial U_\sigma(x)}{\partial x_j}\right\}$ converge uniformly for $\sigma \to \infty$, i.e.:*

$$U_\sigma(x) \rightrightarrows U(x),\quad \frac{\partial U_\sigma(x)}{\partial x_j} \rightrightarrows \frac{\partial U(x)}{\partial x_j},\ j = 1, 2.$$

We should note that the set $E_\varepsilon = G \backslash \overline{G}_\varepsilon$ serves as a boundary layer for this problem, as in the theory of singular perturbations, where there is no uniform convergence.

4. Estimation of the Stability of the Solution to the Cauchy Problem

Suppose that the curve S is given by the equation

$$y_2 = \psi(y_1),\ y_1 \in \mathbb{R},$$

where $\psi(y_1)$ is a single-valued function satisfying the Lyapunov conditions.

We put

$$a = \max_T \psi(y_1),\quad b = \max_T \sqrt{1 + \psi'^2(y_1)}.$$

Theorem 2. *Let $U(y) \in A(G)$ satisfies in the condition (20), and on a smooth curve S the inequality*

$$|U(y)| \leq \delta,\ 0 < \delta \leq M e^{-\sigma a^2}. \tag{27}$$

Then the following relations are true

$$|U(x)| \leq C(\lambda, x) \sigma M^{1 - \frac{x_2^2}{a^2}} \delta^{\frac{x_2^2}{a^2}},\ \sigma > 1,\ x \in G. \tag{28}$$

$$\left|\frac{\partial U(x)}{\partial x_j}\right| \leq C(\lambda, x) \sigma M^{1 - \frac{x_2^2}{a^2}} \delta^{\frac{x_2^2}{a^2}},\ \sigma > 1,\ x \in G, \tag{29}$$

$$j = 1, 2.$$

Proof. Let us first estimate inequality (28). Using the integral formula (10), we have

$$U(x) = \int_S N_\sigma(y, x; \lambda) U(y) ds_y + \int_T N_\sigma(y, x; \lambda)) U(y) ds_y,\ x \in G. \tag{30}$$

We estimate the following

$$|U(x)| \leq \left|\int_S N_\sigma(y, x; \lambda) U(y) ds_y\right| + \left|\int_T N_\sigma(y, x; \lambda) U(y) ds_y\right|,\ x \in G. \tag{31}$$

Given inequality (27), we estimate the first integral of inequality (31).

$$\left| \int_S N_\sigma(y, x; \lambda) U(y) ds_y \right| \leq \int_S |N_\sigma(y, x; \lambda)| |U(y)| ds_y$$
$$\leq \delta \int_S |N_\sigma(y, x; \lambda)| ds_y, \ x \in G. \tag{32}$$

To do this, we estimate the integrals $\int_S |\Phi_\sigma(y, x; \lambda)| ds_y$, $\int_S \left| \frac{\partial \Phi_\sigma(y, x; \lambda)}{\partial y_1} \right| ds_y$ and $\int_S \left| \frac{\partial \Phi_\sigma(y, x; \lambda)}{\partial y_2} \right| ds_y$ on a smooth curve S.

Given equality (16) and the inequality (17), we have

$$\int_S |\Phi_\sigma(y, x; \lambda)| ds_y \leq C(\lambda, x) \sigma e^{\sigma(a^2 - x_2^2)}, \ \sigma > 1, \ x \in G. \tag{33}$$

To estimate the second integral, using equalities (16) and (19) as well as inequality (17), we obtain

$$\int_S \left| \frac{\partial \Phi_\sigma(y, x; \lambda)}{\partial y_1} \right| ds_y \leq C(\lambda, x) \sigma e^{\sigma(a^2 - x_2^2)}, \ \sigma > 1, \ x \in G. \tag{34}$$

To find the integral $\int_S \left| \frac{\partial \Phi_\sigma(y, x; \lambda)}{\partial y_2} \right| ds_y$, using equality (16) and inequality (17), we obtain

$$\int_S \left| \frac{\partial \Phi_\sigma(y, x; \lambda)}{\partial y_2} \right| ds_y \leq C(\lambda, x) \sigma e^{\sigma(a^2 - x_2^2)}, \ \sigma > 1, \ x \in G. \tag{35}$$

From (33)–(35) and applying (32), we obtain

$$\left| \int_S N_\sigma(y, x; \lambda) U(y) ds_y \right| \leq C(\lambda, x) \sigma \delta \, e^{\sigma(a^2 - x_2^2)}, \ \sigma > 1, \ x \in G. \tag{36}$$

The following is known

$$\left| \int_T N_\sigma(y, x; \lambda) U(y) ds_y \right| \leq C(\lambda, x) \sigma M e^{-\sigma x_2^2}, \ \sigma > 1, \ x \in G. \tag{37}$$

Now taking into account (36)–(37) and using (31), we have

$$|U(x)| \leq \frac{C(\lambda, x) \sigma}{2} (\delta e^{\sigma a^2} + M) e^{-\sigma x_2^2}, \ \sigma > 1, \ x \in G. \tag{38}$$

Choosing σ from the equality

$$\sigma = \frac{1}{a^2} \ln \frac{M}{\delta}, \tag{39}$$

we obtain an estimate (28).

Now let us prove inequality (29). To do this, we find the partial derivative from the integral formula (10) with respect to the variable x_j, $j = 1, 2$:

$$\frac{\partial U(x)}{\partial x_j} = \int_S \frac{\partial N_\sigma(y, x; \lambda)}{\partial x_j} U(y) ds_y + \int_T \frac{\partial N_\sigma(y, x; \lambda)}{\partial x_j} U(y) ds_y$$

(40)

$$= \frac{\partial U_\sigma(x)}{\partial x_j} + \int_T \frac{\partial N_\sigma(y, x; \lambda)}{\partial x_j} U(y) ds_y, \quad x \in G, \ j = 1, 2.$$

Here

$$\frac{\partial U_\sigma(x)}{\partial x_j} = \int_S \frac{\partial N_\sigma(y, x; \lambda)}{\partial x_j} U(y) ds_y.$$

(41)

We estimate the following

$$\left| \frac{\partial U(x)}{\partial x_j} \right| \leq \left| \int_S \frac{\partial N_\sigma(y, x; \lambda)}{\partial x_j} U(y) ds_y \right|$$

$$+ \left| \int_T \frac{\partial N_\sigma(y, x; \lambda)}{\partial x_j} U(y) ds_y \right| \leq \left| \frac{\partial U_\sigma(x)}{\partial x_j} \right|$$

(42)

$$+ \left| \int_T \frac{\partial N_\sigma(y, x; \lambda)}{\partial x_j} U(y) ds_y \right|, \quad x \in G, \ j = 1, 2.$$

Given inequality (27), we estimate the first integral of inequality (42).

$$\left| \int_S \frac{\partial N_\sigma(y, x; \lambda)}{\partial x_j} U(y) ds_y \right| \leq \int_S \left| \frac{\partial N_\sigma(y, x; \lambda)}{\partial x_j} \right| |U(y)| ds_y$$

$$\leq \delta \int_S \left| \frac{\partial N_\sigma(y, x; \lambda)}{\partial x_j} \right| ds_y, \quad x \in G, \ j = 1, 2.$$

(43)

To do this, we estimate the integrals $\int_S \left| \frac{\partial \Phi_\sigma(y, x; \lambda)}{\partial x_1} \right| ds_y$, and $\int_S \left| \frac{\partial \Phi_\sigma(y, x; \lambda)}{\partial x_2} \right| ds_y$ on a smooth curve S.

Given equality (16), inequality (17) and equality (24), we obtain

$$\int_S \left| \frac{\partial \Phi_\sigma(y, x; \lambda)}{\partial x_1} \right| ds_y \leq C(\lambda, x) \sigma e^{\sigma(a^2 - x_2^2)}, \quad \sigma > 1, \ x \in G,$$

(44)

Now, we estimate the integral $\int_S \left| \frac{\partial \Phi_\sigma(y, x; \lambda)}{\partial x_2} \right| ds_y$.

Taking into account equality (16) and inequality (17), we obtain

$$\int_S \left| \frac{\partial \Phi_\sigma(y, x; \lambda)}{\partial x_2} \right| ds_y \leq C(\lambda, x) \sigma e^{\sigma(a^2 - x_2^2)}, \quad \sigma > 1, \ x \in G,$$

(45)

From (44) and (45), bearing in mind (43), we obtain

$$\left| \int_S \frac{\partial N_\sigma(y, x; \lambda)}{\partial x_j} U(y) ds_y \right| \leq C(\lambda, x) \sigma \delta e^{-\sigma x_2^2}, \ \sigma > 1, \ x \in G, \quad (46)$$
$$j = 1, 2.$$

The following is known

$$\left| \int_T \frac{\partial N_\sigma(y, x; \lambda)}{\partial x_j} U(y) ds_y \right| \leq C(\lambda, x) \sigma M e^{-\sigma x_2^2}, \ \sigma > 1, \ x \in G, \quad (47)$$
$$j = 1, 2.$$

Now taking into account (46)–(47), bearing in mind (42), we have

$$\left| \frac{\partial U(x)}{\partial x_j} \right| \leq \frac{C(\lambda, x) \sigma}{2} (\delta e^{\sigma a^2} + M) e^{-\sigma x_2^2}, \ \sigma > 1, \ x \in G, \quad (48)$$
$$j = 1, 2.$$

Choosing σ from the equality (39) we obtain an estimate (29). Theorem 2 is proved. □

Assume that $U(y) \in A(G)$ is defined on S and $f_\delta(y)$ is its approximation with an error $0 < \delta \leq M e^{-\sigma a^2}$ then

$$\max_S |U(y) - f_\delta(y)| \leq \delta. \quad (49)$$

We put

$$U_{\sigma(\delta)}(x) = \int_S N_\sigma(y, x; \lambda) f_\delta(y) ds_y, \ x \in G. \quad (50)$$

Theorem 3. *Let $U(y) \in A(G)$ on the part of the plane $y_2 = 0$ satisfies in the condition (11). Then the following estimates is true*

$$\left| U(x) - U_{\sigma(\delta)}(x) \right| \leq C(\lambda, x) \sigma M^{1 - \frac{x_2^2}{a^2}} \delta^{\frac{x_2^2}{a^2}}, \ \sigma > 1, \ x \in G. \quad (51)$$

$$\left| \frac{\partial U(x)}{\partial x_j} - \frac{\partial U_{\sigma(\delta)}(x)}{\partial x_j} \right| \leq C(\lambda, x) \sigma M^{1 - \frac{x_2^2}{a^2}} \delta^{\frac{x_2^2}{a^2}}, \ \sigma > 1, \ x \in G, \quad (52)$$
$$j = 1, 2.$$

Proof. From the integral formulas (10) and (50), we have

$$U(x) - U_{\sigma(\delta)}(x) = \int_{\partial G} N_\sigma(y, x; \lambda) U(y) ds_y$$

$$- \int_S N_\sigma(y, x; \lambda) f_\delta(y) ds_y = \int_S N_\sigma(y, x; \lambda) U(y) ds_y$$

$$+ \int_T N_\sigma(y, x; \lambda) U(y) ds_y - \int_S N_\sigma(y, x; \lambda) f_\delta(y) ds_y$$

$$= \int_S N_\sigma(y, x; \lambda) \{U(y) - f_\delta(y)\} ds_y + \int_T N_\sigma(y, x; \lambda) U(y) ds_y.$$

and
$$\frac{\partial U(x)}{\partial x_j} - \frac{\partial U_{\sigma(\delta)}(x)}{\partial x_j} = \int_{\partial G} \frac{\partial N_\sigma(y,x;\lambda)}{\partial x_j} U(y)ds_y$$

$$-\int_S \frac{\partial N_\sigma(y,x;\lambda)}{\partial x_j} f_\delta(y)ds_y = \int_S \frac{\partial N_\sigma(y,x;\lambda)}{\partial x_j} U(y)ds_y$$

$$+\int_T \frac{\partial N_\sigma(y,x;\lambda)}{\partial x_j} U(y)ds_y - \int_S \frac{\partial N_\sigma(y,x;\lambda)}{\partial x_j} f_\delta(y)ds_y$$

$$= \int_S \frac{\partial N_\sigma(y,x;\lambda)}{\partial x_j} \{U(y) - f_\delta(y)\}ds_y + \int_T \frac{\partial N_\sigma(y,x;\lambda)}{\partial x_j} U(y)ds_y,$$

$$j = 1,2.$$

Using conditions (11) and (49), we estimate the following:

$$\left|U(x) - U_{\sigma(\delta)}(x)\right| = \left|\int_S N_\sigma(y,x;\lambda)\{U(y) - f_\delta(y)\}ds_y\right|$$

$$+\left|\int_T N_\sigma(y,x;\lambda)U(y)ds_y\right| \le \int_S |N_\sigma(y,x;\lambda)||\{U(y) - f_\delta(y)\}|ds_y$$

$$+\int_T |N_\sigma(y,x;\lambda)||U(y)|ds_y \le \delta \int_S |N_\sigma(y,x;\lambda)|ds_y$$

$$+M \int_T |N_\sigma(y,x;\lambda)|ds_y.$$

and
$$\left|\frac{\partial U(x)}{\partial x_j} - \frac{\partial U_{\sigma(\delta)}(x)}{\partial x_j}\right| = \left|\int_S \frac{\partial N_\sigma(y,x;\lambda)}{\partial x_j}\{U(y) - f_\delta(y)\}ds_y\right|$$

$$+\left|\int_T \frac{\partial N_\sigma(y,x;\lambda)}{\partial x_j} U(y)ds_y\right| \le \int_S \left|\frac{\partial N_\sigma(y,x;\lambda)}{\partial x_j}\right||\{U(y) - f_\delta(y)\}|ds_y$$

$$+\int_T \left|\frac{\partial N_\sigma(y,x;\lambda)}{\partial x_j}\right||U(y)|ds_y \le \delta \int_S \left|\frac{\partial N_\sigma(y,x;\lambda)}{\partial x_j}\right|ds_y$$

$$+M \int_T \left|\frac{\partial N_\sigma(y,x;\lambda)}{\partial x_j}\right|ds_y, \; j = 1,2.$$

Now, repeating the proof of Theorems 1 and 2, we obtain

$$\left|U(x) - U_{\sigma(\delta)}(x)\right| \le \frac{C(\lambda,x)\sigma}{2}(\delta e^{\sigma a^2} + M)e^{-\sigma x_2^2}.$$

$$\left|\frac{\partial U(x)}{\partial x_j} - \frac{U_{\sigma(\delta)}(x)}{\partial x_j}\right| \le \frac{C(\lambda,x)\sigma}{2}(\delta e^{\sigma a^2} + M)e^{-\sigma x_2^2}, \; j = 1,2.$$

From here, choosing σ from equality (39), we obtain an estimates (51) and (52). Thus Theorem 3 is proved. □

Corollary 3. *For each $x \in G$, the following equalities are true*

$$\lim_{\delta \to 0} U_{\sigma(\delta)}(x) = U(x), \quad \lim_{\delta \to 0} \frac{\partial U_{\sigma(\delta)}(x)}{\partial x_j} = \frac{\partial U(x)}{\partial x_j}, \quad j = 1, 2.$$

Corollary 4. *If $x \in \overline{G}_\varepsilon$, then the families of functions $\left\{ U_{\sigma(\delta)}(x) \right\}$ and $\left\{ \frac{\partial U_{\sigma(\delta)}(x)}{\partial x_j} \right\}$ are convergent uniformly for $\delta \to 0$, i.e.:*

$$U_{\sigma(\delta)}(x) \rightrightarrows U(x), \quad \frac{\partial U_{\sigma(\delta)}(x)}{\partial x_j} \rightrightarrows \frac{\partial U(x)}{\partial x_j}, \quad j = 1, 2.$$

5. Conclusions

The article obtained the following results:

Using the Carleman function, a formula can be obtained for the continuation of the solution of linear elliptic systems of the first order with constant coefficients in a spatial bounded domain $\mathbb{R}^2$. The resulting formula is an analogue of the classical formula of Riemann, Voltaire and Hadamard, which they constructed to solve the Cauchy problem in the theory of hyperbolic equations. An estimate of the stability of the solution of the Cauchy problem in the classical sense for matrix factorizations of the Helmholtz equation was presented. This problem can be considered when, instead of the exact data of the Cauchy problem we have their approximations with a given deviation in the uniform metric and under the assumption that the solution of the Cauchy problem is bounded on part T, of the boundary of the domain G.

We note that for solving applicable problems, the approximate values of $U(x)$ and $\frac{\partial U(x)}{\partial x_j}$, $x \in G$, $j = 1, 2$ should be found.

In this paper, we have built a family of vector-functions $U(x, f_\delta) = U_{\sigma(\delta)}(x)$ and $\frac{\partial U(x, f_\delta)}{\partial x_j} = \frac{\partial U_{\sigma(\delta)}(x)}{\partial x_j}$, $(j = 1, 2)$ depend on a parameter σ. Also, we prove that under certain conditions and a special choice of the parameter $\sigma = \sigma(\delta)$, at $\delta \to 0$, the family $U_{\sigma(\delta)}(x)$ and $\frac{\partial U_{\sigma(\delta)}(x)}{\partial x_j}$ are convergent to a solution $U(x)$ and its derivative $\frac{\partial U(x)}{\partial x_j}$, $x \in G$ at point $x \in G$.

According to [1], a family of vector-functions $U_{\sigma(\delta)}(x)$ and $\frac{\partial U_{\sigma(\delta)}(x)}{\partial x_j}$ is called a regularized solution of the problem. A regularized solution determines a stable method to find the approximate solution of the problem.

Thus, functionals $U_{\sigma(\delta)}(x)$ and $\frac{\partial U_{\sigma(\delta)}(x)}{\partial x_j}$ determine the regularization of the solution of problems (1) and (2).

Author Contributions: Conceptualization, D.A.J. and S.N.; methodology, D.A.J.; software, D.A.J.; validation, D.A.J. and S.N.; formal analysis, D.A.J.; investigation, D.A.J. and S.N.; resources, D.A.J. and S.N.; data curation, D.A.J.; writing—original draft preparation, D.A.J. and S.N.; writing—review and editing, D.A.J. and S.N.; visualization, D.A.J.; supervision, D.A.J.; project administration, D.A.J.; funding acquisition, D.A.J. and S.N. All authors have read and agreed to the published version of the manuscript.

Funding: This research received no external funding.

Institutional Review Board Statement: Not applicable.

Informed Consent Statement: Not applicable.

Data Availability Statement: Not applicable.

Conflicts of Interest: The authors declare no conflict of interest.

References

1. Tikhonov, A.N. On the solution of ill-posed problems and the method of regularization. *Rep. USSR Acad. Sci.* **1963**, *151*, 501–504.
2. Ivanov, V.K. About incorrectly posed tasks. *Math. Collect.* **1963**, *61*, 211–223.
3. Lavrent'ev, M.M. On the Cauchy problem for second-order linear elliptic equations. *Rep. USSR Acad. Sci.* **1957**, *112*, 195–197.
4. Lavrent'ev, M.M. *On Some Ill-Posed Problems of Mathematical Physics*; Nauka: Novosibirsk, Russia, 1962.
5. Tarkhanov, N.N. *The Cauchy Problem for Solutions of Elliptic Equations*; Akad. Verl.: Berlin, Germany, 1995.
6. Araghi, M.A.F.; Noeiaghdam, S. Fibonacci-regularization method for solving Cauchy integral equations of the first kind. *Ain Shams Eng. J.* **2017**, *8*, 363–369. [CrossRef]
7. Noeiaghdam, S.; Araghi, M.A.F. A novel approach to find optimal parameter in the homotopy-regularization method for solving integral equations. *Appl. Math. Inf. Sci.*, **2020**, *14*, 105–113. [CrossRef]
8. Noeiaghdam, S.; Araghi, M.A.F. Homotopy regularization method to solve the singular Volterra integral equations of the first kind. *Jordan J. Math. Stat.* **2018**, *11*, 1–12.
9. Hadamard, J. *The Cauchy Problem for Linear Partial Differential Equations of Hyperbolic Type*; Nauka: Moscow, Russia, 1978.
10. Bers, A.; John, F.; Shekhter, M. *Partial Differential Equations*; Mir: Moscow, Russia, 1966.
11. Aizenberg, L.A. *Carleman's Formulas in Complex Analysis*; Nauka: Novosibirsk, Russia, 1990.
12. Aleksidze, M.A. *Fundamental Functions in Approximate Solutions of Boundary Value Problems*; Nauka: Moscow, Russia, 1991.
13. Carleman, T. *Les Fonctions Quasi Analytiques*; Gautier-Villars et Cie.: Paris, France, 1926.
14. Tarkhanov, N.N. The solvability criterion for an ill-posed problem for elliptic systems. *Rep. USSR Acad. Sci.* **1989**, *380*, 531–534.
15. Goluzin, G.M.; Krylov, V.M. The generalized Carleman formula and its application to the analytic continuation of functions. *Sb. Math.* **1993**, *40*, 144–149.
16. Yarmukhamedov, S. On the Cauchy problem for the Laplace equation. *Rep. USSR Acad. Sci.* **1977**, *235*, 281–283.
17. Yarmukhamedov, S. On the extension of the solution of the Helmholtz equation. *Rep. Russ. Acad. Sci.* **1997**, *357*, 320–323.
18. Yarmukhamedov, S. The Carleman function and the Cauchy problem for the Laplace equation. *Sib. Math. J.* **2004**, *45*, 702–719. [CrossRef]
19. Yarmukhamedov, S. Representation of Harmonic Functions as Potentials and the Cauchy Problem. *Math. Notes* **2008**, *83*, 763–778. [CrossRef]
20. Ikehata, M. Inverse conductivity problem in the infinite slab. *Inverse Probl.* **2001**, *17*, 437–454. [CrossRef]
21. Ikehata, M. Probe method and a Carleman function. *Inverse Probl.* **2007**, *23*, 1871–1894. [CrossRef]
22. Arbuzov, E.V.; Bukhgeim, A.L. The Carleman formula for the Helmholtz equation. *Sib. Math. J.* **1991**, *47*, 518–526. [CrossRef]
23. Tarkhanov, N.N. Stability of the solutions of elliptic systems. *Funct. Anal. Appl.* **1985**, *19*, 245–247. [CrossRef]
24. Tarkhanov, N.N. On the Carleman matrix for elliptic systems. *Rep. USSR Acad. Sci.* **1985**, *284*, 294–297.
25. Shlapunov, A.A. On the Cauchy problem for the Laplace equation. *Sib. Math. J.* **1992**, *43*, 953–963.
26. Shlapunov, A.A. Boundary problems for Helmholtz equation and the Cauchy problem for Dirac operators. *J. Sib. Fed. Univ. Math. Phys.* **2011**, *4*, 217–228.
27. Niyozov, I.E. On the continuation of the solution of systems of equations of the theory of elasticity. *Uzb. Math. J.* **2015**, *3*, 95–105.
28. Niyozov, I.E. Regularization of a nonstandard Cauchy problem for a dynamic Lame system. *Izv. Vyss. Uchebnykh Zaved.* **2020**, *64*, 54–63. [CrossRef]
29. Juraev, D.A. The construction of the fundamental solution of the Helmholtz equation. *Rep. Acad. Sci. Repub. Uzb.* **2012**, *2*, 14–17.
30. Juraev, D.A. *Regularization of the Cauchy Problem for Systems of Equations of Elliptic Type*; LAP Lambert Academic Publishing: Saarbrucken, Germany, 2014.
31. Juraev, D.A. Regularization of the Cauchy problem for systems of elliptic type equations of first order. *Uzb. Math. J.* **2016**, *2*, 61–71.
32. Juraev, D.A. The Cauchy problem for matrix factorizations of the Helmholtz equation in an unbounded domain. *Sib. Electron. Math. Rep.* **2017**, *14*, 752–764. [CrossRef]
33. Juraev, D.A. Cauchy problem for matrix factorizations of the Helmholtz equation. *Ukr. J.* **2017**, *69*, 1364–1371. [CrossRef]
34. Juraev, D.A. On the Cauchy problem for matrix factorizations of the Helmholtz equation in a bounded domain. *Sib. Electron. Math. Rep.* **2018**, *15*, 11–20.
35. Zhuraev, D.A. Cauchy problem for matrix factorizations of the Helmholtz equation. *Ukr. J.* **2018**, *69*, 1583–1592. [CrossRef]
36. Juraev, D.A. The Cauchy problem for matrix factorizations of the Helmholtz equation in $\mathbb{R}^3$. *J. Univers. Math.* **2018**, *1*, 312–319.
37. Juraev, D.A. On the Cauchy problem for matrix factorizations of the Helmholtz equation in an unbounded domain in $\mathbb{R}^2$. *Sib. Electron. Math. Rep.* **2018**, *15*, 1865–1877.
38. Juraev, D.A. On a regularized solution of the Cauchy problem for matrix factorizations of the Helmholtz equation. *Adv. Math. Model. Appl.* **2019**, *4*, 86–96.
39. Juraev, D.A. On the Cauchy problem for matrix factorizations of the Helmholtz equation. *J. Univers. Math.* **2019**, *2*, 113–126. [CrossRef]

40. Juraev, D.A. The solution of the ill-posed Cauchy problem for matrix factorizations of the Helmholtz equation. *Adv. Math. Model. Appl.* **2020**, *5*, 205–221.
41. Juraev, D.A. Ill-posed problems for first-order elliptic systems with constant coefficients. In *Abstracts of Communications of the Conference: Modern Stochastic Models and Problems of Actuarial Mathematics*; Organizer–Karshi State University: Karshi City, Uzbekistan, 2020; pp. 26–27.
42. Kabanikhin, I.E.; Gasimov, Y.S.; Nurseitov, D.B.; Shishlenin, M.A.; Sholpanbaev, B.B.; Kasemov, S. Regularization of the continuation problem for elliptic equation. *J. Inverse Ill Posed Probl.* **2013**, *21*, 871–874. [CrossRef]

MDPI
St. Alban-Anlage 66
4052 Basel
Switzerland
Tel. +41 61 683 77 34
Fax +41 61 302 89 18
www.mdpi.com

Axioms Editorial Office
E-mail: axioms@mdpi.com
www.mdpi.com/journal/axioms